MIMO-OFDM WIRELESS COMMUNICATIONS WITH MATLAB®

MIMO-OFDM WIRELESS COMMUNICATIONS WITH MATLAB®

Yong Soo Cho
Chung-Ang University, Republic of Korea

Jaekwon Kim
Yonsei University, Republic of Korea

Won Young Yang
Chung-Ang University, Republic of Korea

Chung G. Kang
Korea University, Republic of Korea

IEEE PRESS

John Wiley & Sons (Asia) Pte Ltd

Other Wiley Editorial Offices

John Wiley & Sons, Ltd, The Atrium, Southern Gate, Chichester, West Sussex, PO19 8SQ, UK

John Wiley & Sons Inc., 111 River Street, Hoboken, NJ 07030, USA

Jossey-Bass, 989 Market Street, San Francisco, CA 94103-1741, USA

Wiley-VCH Verlag GmbH, Boschstrasse 12, D-69469 Weinheim, Germany

John Wiley & Sons Australia Ltd, 42 McDougall Street, Milton, Queensland 4064, Australia

John Wiley & Sons Canada Ltd, 5353 Dundas Street West, Suite 400, Toronto, ONT, M9B 6H8, Canada

Wiley also publishes its books in a variety of electronic formats. Some content that appears in print may not be available in electronic books.

Library of Congress Cataloging-in-Publication Data

MIMO-OFDM wireless communications with MATLAB® / Yong Soo Cho ... [et al.].
    p. cm.
  Includes bibliographical references and index.
  ISBN 978-0-470-82561-7 (cloth)
  1. Orthogonal frequency division multiplexing. 2. MIMO systems. 3. MATLAB®. I. Cho, Yong Soo.
  TK5103.484.M56 2010
  621.384–dc22

                              2010013156

Print ISBN: 978-0-470-82561-7
ePDF ISBN: 978-0-470-82562-4
oBook ISBN: 978-0-470-82563-1

Typeset in 10/12pt Times by Thomson Digital, Noida, India.
Printed and bound in Singapore by Markono Print Media Pte Ltd.
This book is printed on acid-free paper responsibly manufactured from sustainable forestry in which at least two trees are planted for each one used for paper production.

To our parents and families
who love and support us
and
to our students
who enriched our knowledge

Contents

Preface

MIMO-OFDM is a key technology for next-generation cellular communications (3GPP-LTE, Mobile WiMAX, IMT-Advanced) as well as wireless LAN (IEEE 802.11a, IEEE 802.11n), wireless PAN (MB-OFDM), and broadcasting (DAB, DVB, DMB). This book provides a comprehensive introduction to the basic theory and practice of wireless channel modeling, OFDM, and MIMO, with MATLAB® programs to simulate the underlying techniques on MIMO-OFDM systems. This book is primarily designed for engineers and researchers who are interested in learning various MIMO-OFDM techniques and applying them to wireless communications. It can also be used as a textbook for graduate courses or senior-level undergraduate courses on advanced digital communications. The readers are assumed to have a basic knowledge on digital communications, digital signal processing, communication theory, signals and systems, as well as probability and random processes.

The first aim of this book is to help readers understand the concepts, techniques, and equations appearing in the field of MIMO-OFDM communication, while simulating various techniques used in MIMO-OFDM systems. Readers are recommended to learn some basic usage of MATLAB® that is available from the MATLAB® help function or the on-line documents at the website www.mathworks.com/matlabcentral. However, they are not required to be an expert on MATLAB® since most programs in this book have been composed carefully and completely, so that they can be understood in connection with related/referred equations. The readers are expected to be familiar with the MATLAB® software while trying to use or modify the MATLAB® codes. The second aim of this book is to make even a novice at both MATLAB® and MIMO-OFDM become acquainted with MIMO-OFDM as well as MATLAB®, while running the MATLAB® program on his/her computer. The authors hope that this book can be used as a reference for practicing engineers and students who want to acquire basic concepts and develop an algorithm on MIMO-OFDM using the MATLAB® program. The features of this book can be summarized as follows:

- Part I presents the fundamental concepts and MATLAB® programs for simulation of wireless channel modeling techniques, including large-scale fading, small-scale fading, indoor and outdoor channel modeling, SISO channel modeling, and MIMO channel modeling.
- Part II presents the fundamental concepts and MATLAB® programs for simulation of OFDM transmission techniques including OFDM basics, synchronization, channel estimation, peak-to-average power ratio reduction, and intercell interference mitigation.
- Part III presents the fundamental concepts and MATLAB® programs for simulation of MIMO techniques including MIMO channel capacity, space diversity and space-time codes,

signal detection for spatially-multiplexed MIMO systems, precoding and antenna selection techniques, and multiuser MIMO systems.

Most MATLAB® programs are presented in a complete form so that the readers with no programming skill can run them instantly and focus on understanding the concepts and characteristics of MIMO-OFDM systems. The contents of this book are derived from the works of many great scholars, engineers, researchers, all of whom are deeply appreciated.

We would like to thank the reviewers for their valuable comments and suggestions, which contribute to enriching this book. We would like to express our heartfelt gratitude to colleagues and former students who developed source programs: Dr. Won Gi Jeon, Dr. Kyung-Won Park, Dr. Mi-Hyun Lee, Dr. Kyu-In Lee, and Dr. Jong-Ho Paik. Special thanks should be given to Ph.D candidates who supported in preparing the typescript of the book: Kyung Soo Woo, Jung-Wook Wee, Chang Hwan Park, Yeong Jun Kim, Yo Han Ko, Hyun Il Yoo, Tae Ho Im, and many MS students in the Digital Communication Lab at Chung-Ang University. We also thank the editorial and production staffs, including Ms. Renee Lee of John Wiley & Sons (Asia) Pte Ltd and Ms. Aparajita Srivastava of Thomson Digital, for their kind, efficient, and encouraging guidance.

Program files can be downloaded from http://comm.cau.ac.kr/MIMO_OFDM/index.html.

Limits of Liability and Disclaimer of Warranty of Software

1

The Wireless Channel: Propagation and Fading

The performance of wireless communication systems is mainly governed by the wireless channel environment. As opposed to the typically static and predictable characteristics of a wired channel, the wireless channel is rather dynamic and unpredictable, which makes an exact analysis of the wireless communication system often difficult. In recent years, optimization of the wireless communication system has become critical with the rapid growth of mobile communication services and emerging broadband mobile Internet access services. In fact, the understanding of wireless channels will lay the foundation for the development of high performance and bandwidth-efficient wireless transmission technology.

In wireless communication, radio propagation refers to the behavior of radio waves when they are propagated from transmitter to receiver. In the course of propagation, radio waves are mainly affected by three different modes of physical phenomena: reflection, diffraction, and scattering [1,2]. *Reflection* is the physical phenomenon that occurs when a propagating electromagnetic wave impinges upon an object with very large dimensions compared to the wavelength, for example, surface of the earth and building. It forces the transmit signal power to be reflected back to its origin rather than being passed all the way along the path to the receiver. *Diffraction* refers to various phenomena that occur when the radio path between the transmitter and receiver is obstructed by a surface with sharp irregularities or small openings. It appears as a bending of waves around the small obstacles and spreading out of waves past small openings. The secondary waves generated by diffraction are useful for establishing a path between the transmitter and receiver, even when a line-of-sight path is not present. Scattering is the physical phenomenon that forces the radiation of an electromagnetic wave to deviate from a straight path by one or more local obstacles, with small dimensions compared to the wavelength. Those obstacles that induce scattering, such as foliage, street signs, and lamp posts, are referred to as the scatters. In other words, the propagation of a radio wave is a complicated and less predictable process that is governed by reflection, diffraction, and scattering, whose intensity varies with different environments at different instances.

A unique characteristic in a wireless channel is a phenomenon called 'fading,' the variation of the signal amplitude over time and frequency. In contrast with the additive noise as the most

MIMO-OFDM Wireless Communications with MATLAB®   Yong Soo Cho, Jaekwon Kim, Won Young Yang and Chung G. Kang
© 2010 John Wiley & Sons (Asia) Pte Ltd

common source of signal degradation, fading is another source of signal degradation that is characterized as a non-additive signal disturbance in the wireless channel. Fading may either be due to multipath propagation, referred to as multi-path (induced) fading, or to shadowing from obstacles that affect the propagation of a radio wave, referred to as shadow fading.

The fading phenomenon in the wireless communication channel was initially modeled for HF (High Frequency, 3~30 MHz), UHF (Ultra HF, 300~3000 GHz), and SHF (Super HF, 3~30 GHz) bands in the 1950s and 1960s. Currently, the most popular wireless channel models have been established for 800MHz to 2.5 GHz by extensive channel measurements in the field. These include the ITU-R standard channel models specialized for a single-antenna communication system, typically referred to as a SISO (Single Input Single Output) communication, over some frequency bands. Meanwhile, spatial channel models for a multi-antenna communication system, referred to as the MIMO (Multiple Input Multiple Output) system, have been recently developed by the various research and standardization activities such as IEEE 802, METRA Project, 3GPP/3GPP2, and WINNER Projects, aiming at high-speed wireless transmission and diversity gain.

The fading phenomenon can be broadly classified into two different types: *large-scale fading* and *small-scale fading*. Large-scale fading occurs as the mobile moves through a large distance, for example, a distance of the order of cell size [1]. It is caused by path loss of signal as a function of distance and shadowing by large objects such as buildings, intervening terrains, and vegetation. Shadowing is a slow fading process characterized by variation of median path loss between the transmitter and receiver in fixed locations. In other words, large-scale fading is characterized by average path loss and shadowing. On the other hand, small-scale fading refers to rapid variation of signal levels due to the constructive and destructive interference of multiple signal paths (multi-paths) when the mobile station moves short distances. Depending on the relative extent of a multipath, frequency selectivity of a channel is characterized (e.g., by frequency-selective or frequency flat) for small-scaling fading. Meanwhile, depending on the time variation in a channel due to mobile speed (characterized by the Doppler spread), short-term fading can be classified as either fast fading or slow fading. Figure 1.1 classifies the types of fading channels.

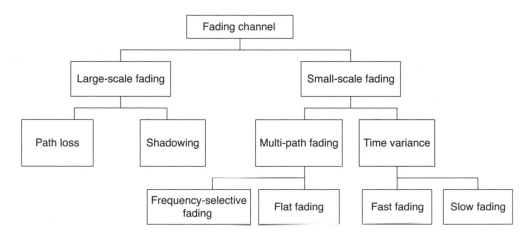

Figure 1.1   Classification of fading channels.

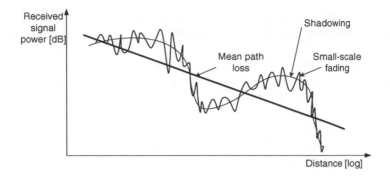

Figure 1.2 Large-scale fading vs. small-scale fading.

The relationship between large-scale fading and small-scale fading is illustrated in Figure 1.2. Large-scale fading is manifested by the mean path loss that decreases with distance and shadowing that varies along the mean path loss. The received signal strength may be different even at the same distance from a transmitter, due to the shadowing caused by obstacles on the path. Furthermore, the scattering components incur small-scale fading, which finally yields a short-term variation of the signal that has already experienced shadowing.

Link budget is an important tool in the design of radio communication systems. Accounting for all the gains and losses through the wireless channel to the receiver, it allows for predicting the received signal strength along with the required power margin. Path loss and fading are the two most important factors to consider in link budget. Figure 1.3 illustrates a link budget that is affected by these factors. The mean path loss is a deterministic factor that can be predicted with the distance between the transmitter and receiver. On the contrary, shadowing and small-scale

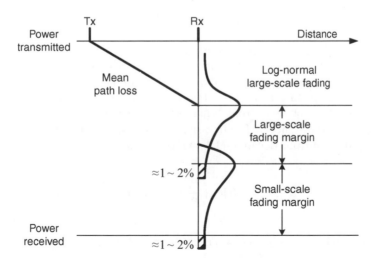

Figure 1.3 Link budget for the fading channel [3]. (© 1994 IEEE. Reproduced from Greenwood, D. and Hanzo, L., "Characterization of mobile radio channels," in *Mobile Radio Communications*, R. Steele (ed.), pp. 91–185, © 1994, with permission from Institute of Electrical and Electronics Engineers (IEEE).)

fading are random phenomena, which means that their effects can only be predicted by their probabilistic distribution. For example, shadowing is typically modeled by a log-normal distribution.

Due to the random nature of fading, some power margin must be added to ensure the desired level of the received signal strength. In other words, we must determine the margin that warrants the received signal power beyond the given threshold within the target rate (e.g., 98–99%) in the design. As illustrated in Figure 1.3, large-scale and small-scale margins must be set so as to maintain the outage rate within 1~2%, which means that the received signal power must be below the target design level with the probability of 0.02 or less [3]. In this analysis, therefore, it is essential to characterize the probabilistic nature of shadowing as well as the path loss.

In this chapter, we present the specific channel models for large-scale and small-scale fading that is required for the link budget analysis.

1.1 Large-Scale Fading

1.1.1 General Path Loss Model

The free-space propagation model is used for predicting the received signal strength in the line-of-sight (LOS) environment where there is no obstacle between the transmitter and receiver. It is often adopted for the satellite communication systems. Let d denote the distance in meters between the transmitter and receiver. When non-isotropic antennas are used with a transmit gain of G_t and a receive gain of G_r, the received power at distance d, $P_r(d)$, is expressed by the well-known Friis equation [4], given as

$$P_r(d) = \frac{P_t G_t G_r \lambda^2}{(4\pi)^2 d^2 L} \tag{1.1}$$

where P_t represents the transmit power (watts), λ is the wavelength of radiation (m), and L is the system loss factor which is independent of propagation environment. The system loss factor represents overall attenuation or loss in the actual system hardware, including transmission line, filter, and antennas. In general, $L > 1$, but $L = 1$ if we assume that there is no loss in the system hardware. It is obvious from Equation (1.1) that the received power attenuates exponentially with the distance d. The free-space path loss, $PL_F(d)$, without any system loss can be directly derived from Equation (1.1) with $L = 1$ as

$$PL_F(d)[dB] = 10 \log\left(\frac{P_t}{P_r}\right) = -10 \log\left(\frac{G_t G_r \lambda^2}{(4\pi)^2 d^2}\right) \tag{1.2}$$

Without antenna gains (i.e., $G_t = G_r = 1$), Equation (1.2) is reduced to

$$PL_F(d)[dB] = 10 \log\left(\frac{P_t}{P_r}\right) = 20 \log\left(\frac{4\pi d}{\lambda}\right) \tag{1.3}$$

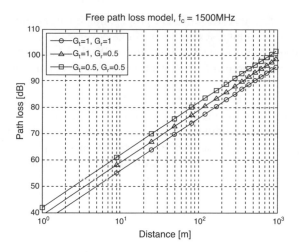

Figure 1.4 Free-space path loss model.

Figure 1.4 shows the free-space path loss at the carrier frequency of $f_c = 1.5\,\text{GHz}$ for different antenna gains as the distance varies. It is obvious that the path loss increases by reducing the antenna gains. As in the aforementioned free-space model, the average received signal in all the other actual environments decreases with the distance between the transmitter and receiver, d, in a logarithmic manner. In fact, a more generalized form of the path loss model can be constructed by modifying the free-space path loss with the path loss exponent n that varies with the environments. This is known as the log-distance path loss model, in which the path loss at distance d is given as

$$PL_{LD}(d)[dB] = PL_F(d_0) + 10n\log\left(\frac{d}{d_0}\right) \tag{1.4}$$

where d_0 is a reference distance at which or closer to the path loss inherits the characteristics of free-space loss in Equation (1.2). As shown in Table 1.1, the path loss exponent can vary from 2 to 6, depending on the propagation environment. Note that $n = 2$ corresponds to the free space. Moreover, n tends to increase as there are more obstructions. Meanwhile, the reference distance

Table 1.1 Path loss exponent [2].

Environment	Path loss exponent (n)
Free space	2
Urban area cellular radio	2.7–3.5
Shadowed urban cellular radio	3–5
In building line-of-sight	1.6–1.8
Obstructed in building	4–6
Obstructed in factories	2–3

(Rappaport, Theodore S., *Wireless Communications: Principles and Practice*, 2nd Edition, © 2002, pg. 76. Reprinted by permission of Pearson Education, Inc., Upper Saddle River, New Jersey.)

d_0 must be properly determined for different propagation environments. For example, d_0 is typically set as 1 km for a cellular system with a large coverage (e.g., a cellular system with a cell radius greater than 10 km). However, it could be 100 m or 1 m, respectively, for a macro-cellular system with a cell radius of 1km or a microcellular system with an extremely small radius [5].

Figure 1.5 shows the log-distance path loss by Equation (1.5) at the carrier frequency of $f_c = 1.5$ GHz. It is clear that the path loss increases with the path loss exponent n. Even if the distance between the transmitter and receiver is equal to each other, every path may have different path loss since the surrounding environments may vary with the location of the receiver in practice. However, all the aforementioned path loss models do not take this particular situation into account. A *log-normal* shadowing model is useful when dealing with a more realistic situation. Let X_σ denote a Gaussian random variable with a zero mean and a standard deviation of σ. Then, the log-normal shadowing model is given as

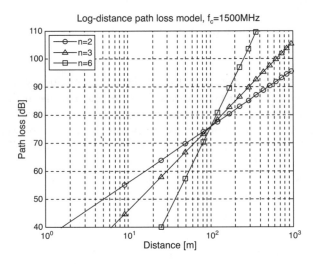

Figure 1.5 Log-distance path loss model.

$$PL(d)[dB] = \overline{PL}(d) + X_\sigma = PL_F(d_0) + 10n\log\left(\frac{d}{d_0}\right) + X_\sigma \qquad (1.5)$$

In other words, this particular model allows the receiver at the same distance d to have a different path loss, which varies with the random shadowing effect X_σ. Figure 1.6 shows the path loss that follows the log-normal shadowing model at $f_c = 1.5$ GHz with $\sigma = 3$ dB and $n = 2$. It clearly illustrates the random effect of shadowing that is imposed on the deterministic nature of the log-distance path loss model.

Note that the path loss graphs in Figures 1.4–1.6 are obtained by running Program 1.3 ("plot_PL_general.m"), which calls Programs 1.1 ("PL_free") and 1.2 ("PL_logdist_or_norm") to compute the path losses by using Equation (1.2), Equation (1.3), Equation (1.4), and Equation (1.5), respectively.

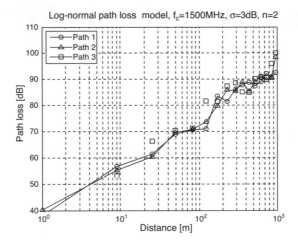

Figure 1.6 Log-normal shadowing path loss model.

Program 1.1 "PL_logdist_or_norm" for log-distance/normal shadowing path loss model

```
function PL=PL_logdist_or_norm(fc,d,d0,n,sigma)
% Log-distance or Log-normal shadowing path loss model
% Inputs:    fc   : Carrier frequency[Hz]
%            d    : Distance between base station and mobile station[m]
%            d0   : Reference distance[m]
%            n    : Path loss exponent
%            sigma : Variance[dB]
lamda=3e8/fc; PL= -20*log10(lamda/(4*pi*d0))+10*n*log10(d/d0); % Eq.(1.4)
if nargin>4, PL = PL + sigma*randn(size(d));   end   % Eq.(1.5)
```

Program 1.2 "PL_free" for free-space path loss model

```
function PL=PL_free(fc,d,Gt,Gr)
% Free Space Path Loss Model
% Inputs: fc    : Carrier frequency[Hz]
%         d     : Distance between base station and mobile station[m]
%         Gt/Gr : Transmitter/Receiver gain
% Output: PL    : Path loss[dB]
lamda = 3e8/fc; tmp = lamda./(4*pi*d);
if nargin>2, tmp = tmp*sqrt(Gt); end
if nargin>3, tmp = tmp*sqrt(Gr); end
PL = -20*log10(tmp); % Eq.(1.2)/(1.3)
```

Program 1.3 "plot_PL_general.m" to plot the various path loss models

```
% plot_PL_general.m
clear, clf
fc=1.5e9; d0=100; sigma=3; distance=[1:2:31].^2;
Gt=[1 1 0.5]; Gr=[1 0.5 0.5]; Exp=[2 3 6];
for k=1:3
    y_Free(k,:)=PL_free(fc,distance,Gt(k),Gr(k));
    y_logdist(k,:)=PL_logdist_or_norm(fc,distance,d0,Exp(k));
    y_lognorm(k,:)=PL_logdist_or_norm(fc,distance,d0,Exp(1),sigma);
end
subplot(131), semilogx(distance,y_Free(1,:),'k-o', distance,y_Free(2,:),
 'k-^', distance,y_Free(3,:),'k-s'), grid on, axis([1 1000 40 110]),
title(['Free Path-loss Model, f_c=',num2str(fc/1e6),'MHz'])
xlabel('Distance[m]'), ylabel('Path loss[dB]')
legend('G_t=1, G_r=1','G_t=1, G_r=0.5','G_t=0.5, G_r=0.5',2)
subplot(132)
semilogx(distance,y_logdist(1,:),'k-o', distance,y_logdist(2,:),'k-^',
 distance,y_logdist(3,:),'k-s'), grid on, axis([1 1000 40 110]),
title(['Log-distance Path-loss Model, f_c=',num2str(fc/1e6),'MHz'])
xlabel('Distance[m]'), ylabel('Path loss[dB]'),
legend('n=2','n=3','n=6',2)
subplot(133), semilogx(distance,y_lognorm(1,:),'k-o', distance,y_lognorm
 (2,:),'k-^', distance,y_lognorm(3,:),'k-s')
grid on, axis([1 1000 40 110]), legend('path 1','path 2','path 3',2)
title(['Log-normal Path-loss Model, f_c=',num2str(fc/1e6),'MHz,
 ', '\sigma=', num2str(sigma), 'dB, n=2'])
xlabel('Distance[m]'), ylabel('Path loss[dB]')
```

1.1.2 Okumura/Hata Model

The Okumura model has been obtained through extensive experiments to compute the antenna height and coverage area for mobile communication systems [6]. It is one of the most frequently adopted path loss models that can predict path loss in an urban area. This particular model mainly covers the typical mobile communication system characteristics with a frequency band of 500–1500 MHz, cell radius of 1–100 km, and an antenna height of 30 m to 1000 m. The path loss at distance d in the Okumura model is given as

$$PL_{Ok}(d)[dB] = PL_F + A_{MU}(f,d) - G_{Rx} - G_{Tx} + G_{AREA} \qquad (1.6)$$

where $A_{MU}(f,d)$ is the medium attenuation factor at frequency f, G_{Rx} and G_{Tx} are the antenna gains of Rx and Tx antennas, respectively, and G_{AREA} is the gain for the propagation environment in the specific area. Note that the antenna gains, G_{Rx} and G_{Tx}, are merely a function of the antenna height, without other factors taken into account like an antenna pattern. Meanwhile, $A_{MU}(f,d)$ and G_{AREA} can be referred to by the graphs that have been obtained empirically from actual measurements by Okumura [6].

The Okumura model has been extended to cover the various propagation environments, including urban, suburban, and open area, which is now known as the Hata model [7]. In fact, the Hata model is currently the most popular path loss model. For the height of transmit antenna, h_{TX}[m], and the carrier frequency of f_c[MHz], the path loss at distance d [m] in an urban area is given by the Hata model as

$$PL_{Hata,U}(d)[dB] = 69.55 + 26.16\log f_c - 13.82\log h_{TX} - C_{RX} + (44.9 - 6.55\log h_{TX})\log d \quad (1.7)$$

where C_{RX} is the correlation coefficient of the receive antenna, which depends on the size of coverage. For small to medium-sized coverage, C_{RX} is given as

$$C_{Rx} = 0.8 + (1.1\log f_c - 0.7)h_{Rx} - 1.56\log f_c \quad (1.8)$$

where h_{RX} [m] is the height of transmit antenna. For large-sized coverage, C_{RX} depends on the range of the carrier frequency, for example,

$$C_{RX} = \begin{cases} 8.29(\log(1.54h_{RX}))^2 - 1.1 & \text{if } 150\,\text{MHz} \le f_c \le 200\,\text{MHz} \\ 3.2(\log(11.75h_{RX}))^2 - 4.97 & \text{if } 200\,\text{MHz} \le f_c \le 1500\,\text{MHz} \end{cases} \quad (1.9)$$

Meanwhile, the path loss at distance d in suburban and open areas are respectively given by the Hata model as

$$PL_{Hata,SU}(d)[dB] = PL_{Hata,U}(d) - 2\left(\log\frac{f_c}{28}\right)^2 - 5.4 \quad (1.10)$$

and

$$PL_{Hata,O}(d)[dB] = PL_{Hata,U}(d) - 4.78(\log f_c)^2 + 18.33\log f_c - 40.97 \quad (1.11)$$

Figure 1.7 presents the path loss graphs for the three different environments – urban, suburban, and open areas – given by the models in Equations (1.7), (1.10), and (1.11), respectively. It is clear that the urban area gives the most significant path loss as compared to the other areas,

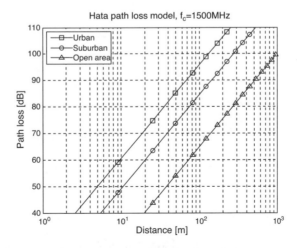

Figure 1.7 Hata path loss model.

simply due to the dense obstructions observed in the urban area. Note that these path loss graphs in Figure 1.7 are obtained by running Program 1.4 ("plot_PL_Hata.m"), which calls Program 1.5 ("PL_Hata") to compute the path losses for various propagation environments by using Equations (1.7)~(1.11).

MATLAB® Programs: Hata Path Loss Model

Program 1.4 "plot_PL_Hata.m" to plot the Hata path loss model

```
% plot_PL_Hata.m
clear, clf
fc=1.5e9; htx=30; hrx=2; distance=[1:2:31].^2;
y_urban=PL_Hata(fc,distance,htx,hrx,'urban');
y_suburban=PL_Hata(fc,distance,htx,hrx,'suburban');
y_open=PL_Hata(fc,distance,htx,hrx,'open');
semilogx(distance,y_urban,'k-s', distance,y_suburban,'k-o', distance,
  y_open,'k-^')
title(['Hata PL model, f_c=',num2str(fc/1e6),'MHz'])
xlabel('Distance[m]'), ylabel('Path loss[dB]')
legend('urban','suburban','open area',2), grid on, axis([1 1000 40 110])
```

Program 1.5 "PL_Hata" for Hata path loss model

```
function PL=PL_Hata(fc,d,htx,hrx,Etype)
% Inputs: fc    : Carrier frequency[Hz]
%         d     : Distance between base station and mobile station[m]
%         htx   : Height of transmitter[m]
%         hrx   : Height of receiver[m]
%         Etype : Environment type('urban','suburban','open')
% Output: PL    : path loss[dB]
if nargin<5, Etype = 'URBAN'; end
fc=fc/(1e6);
if fc>=150&&fc<=200, C_Rx = 8.29*(log10(1.54*hrx))^2 - 1.1;
  elseif fc>200, C_Rx = 3.2*(log10(11.75*hrx))^2 - 4.97; % Eq.(1.9)
  else C_Rx = 0.8+(1.1*log10(fc)-0.7)*hrx-1.56*log10(fc); % Eq.(1.8)
end
PL = 69.55 +26.16*log10(fc) -13.82*log10(htx) -C_Rx ...
    +(44.9-6.55*log10(htx))*log10(d/1000); % Eq.(1.7)
EType = upper(Etype);
if EType(1)=='S', PL = PL -2*(log10(fc/28))^2 -5.4; % Eq.(1.10)
  elseif EType(1)=='O'
    PL=PL+(18.33-4.78*log10(fc))*log10(fc)-40.97; % Eq.(1.11)
end
```

1.1.3 IEEE 802.16d Model

IEEE 802.16d model is based on the log-normal shadowing path loss model. There are three different types of models (Type A, B, and C), depending on the density of obstruction between the transmitter and receiver (in terms of tree densities) in a macro-cell suburban area. Table 1.2 describes these three different types of models in which ART and BRT stand for Above-Roof-Top and Below-Roof-Top. Referring to [8–11], the IEEE 802.16d path loss model is given as

$$PL_{802.16}(d)[dB] = PL_F(d_0) + 10\gamma \log_{10}\left(\frac{d}{d_0}\right) + C_f + C_{RX} \quad \text{for} \quad d > d_0 \tag{1.12}$$

Table 1.2 Types of IEEE 802.16d path loss models.

Type	Description
A	Macro-cell suburban, ART to BRT for hilly terrain with moderate-to-heavy tree densities
B	Macro-cell suburban, ART to BRT for intermediate path loss condition
C	Macro-cell suburban, ART to BRT for flat terrain with light tree densities

In Equation (1.12), $d_0 = 100\,m$ and $\gamma = a - bh_{Tx} + c/h_{TX}$ where a, b, and c are constants that vary with the types of channel models as given in Table 1.3, and h_{TX} is the height of transmit antenna (typically, ranged from 10 m to 80 m). Furthermore, C_f is the correlation coefficient for the carrier frequency f_c [MHz], which is given as

$$C_f = 6 \log_{10}(f_c/2000) \tag{1.13}$$

Table 1.3 Parameters for IEEE 802.16d type A, B, and C models.

Parameter	Type A	Type B	Type C
a	4.6	4	3.6
b	0.0075	0.0065	0.005
c	12.6	17.1	20

Meanwhile, C_{RX} is the correlation coefficient for the receive antenna, given as

$$C_{RX} = \begin{cases} -10.8 \log_{10}(h_{RX}/2) & \text{for Type A and B} \\ -20 \log_{10}(h_{RX}/2) & \text{for Type C} \end{cases} \tag{1.14}$$

or

$$C_{RX} = \begin{cases} -10 \log_{10}(h_{RX}/3) & \text{for } h_{RX} \leq 3m \\ -20 \log_{10}(h_{RX}/3) & \text{for } h_{RX} > 3m \end{cases} \tag{1.15}$$

The correlation coefficient in Equation (1.14) is based on the measurements by AT&T while the one in Equation (1.15) is based on the measurements by Okumura.

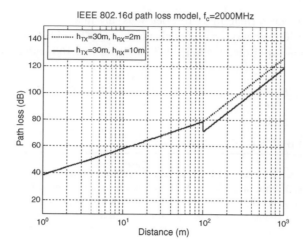

Figure 1.8   IEEE 802.16d path loss model.

Figure 1.8 shows the path loss by the IEEE 802.16d model at the carrier frequency of 2 GHz, as the height of the transmit antenna is varied and the height of the transmit antenna is fixed at 30 m. Note that when the height of the transmit antenna is changed from 2 m to 10 m, there is a discontinuity at the distance of 100 m, causing some inconsistency in the prediction of the path loss. For example, the path loss at the distance of 101 m is larger than that at the distance of 99 m by 8dB, even without a shadowing effect in the model. It implies that a new reference distance d_0' must be defined to modify the existing model [9]. The new reference distance d_0' is determined by equating the path loss in Equation (1.12) to the free-space loss in Equation (1.3), such that

$$20 \log_{10} \left(\frac{4\pi d_0'}{\lambda} \right) = 20 \log_{10} \left(\frac{4\pi d_0'}{\lambda} \right) + 10\gamma \log_{10} \left(\frac{d_0'}{d_0} \right) + C_f + C_{RX} \tag{1.16}$$

Solving Equation (1.16) for d_0', the new reference distance is found as

$$d_0' = d_0 10^{-\frac{C_f + C_{RX}}{10\gamma}} \tag{1.17}$$

Substituting Equation (1.17) into Equation (1.12), a modified IEEE 802.16d model follows as

$$PL_{M802.16}(d)[dB] = \begin{cases} 20 \log_{10} \left(\frac{4\pi d}{\lambda} \right) & \text{for } d \le d_0' \\[3mm] 20 \log_{10} \left(\frac{4\pi d_0'}{\lambda} \right) + 10\gamma \log_{10} \left(\frac{d}{d_0} \right) + C_f + C_{RX} & \text{for } d > d_0' \end{cases} \tag{1.18}$$

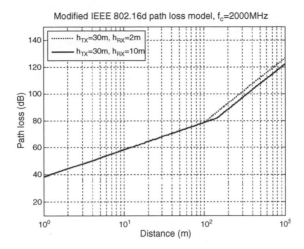

Figure 1.9 Modified IEEE 802.16d path loss model.

Figure 1.9 shows the path loss by the modified IEEE 802.16d model in Equation (1.18), which has been plotted by running the Program 1.7 ("plot_PL_IEEE80216d.m"), which calls Program 1.6 ("PL_IEEE80216d"). Discontinuity is no longer shown in this modified model, unlike the one in Figure 1.8.

MATLAB® Programs: IEEE 802.16d Path Loss Model

Program 1.6 "PL_IEEE80216d" for IEEE 802.16d path loss model

```
function PL=PL_IEEE80216d(fc,d,type,htx,hrx,corr_fact,mod)
% IEEE 802.16d model
% Inputs
%       fc      : Carrier frequency
%       d       : Distance between base and terminal
%       type    : selects 'A', 'B', or 'C'
%       htx     : Height of transmitter
%       hrx     : Height of receiver
%       corr_fact : If shadowing exists, set to 'ATnT' or 'Okumura'.
%                   Otherwise, 'NO'
%       mod     : set to 'mod' to obtain modified IEEE 802.16d model
% Output
%       PL : path loss[dB]
Mod='UNMOD';
if nargin>6, Mod=upper(mod); end
if nargin==6&&corr_fact(1)=='m', Mod='MOD'; corr_fact='NO';
  elseif nargin<6, corr_fact='NO';
    if nargin==5&&hrx(1)=='m', Mod='MOD'; hrx=2;
    elseif nargin<5, hrx=2;
      if nargin==4&&htx(1)=='m', Mod='MOD'; htx=30;
```

```
      elseif nargin<4, htx=30;
        if nargin==3&&type(1)=='m', Mod='MOD'; type='A';
          elseif nargin<3, type='A';
        end
      end
    end
end
d0 = 100;
Type = upper(type);
if Type~='A' && Type~='B' &&Type~='C'
  disp('Error: The selected type is not supported'); return;
end
switch upper(corr_fact)
  case 'ATNT',    PLf=6*log10(fc/2e9);                    % Eq.(1.13)
                  PLh=-10.8*log10(hrx/2);                 % Eq.(1.14)
  case 'OKUMURA', PLf=6*log10(fc/2e9);                    % Eq.(1.13)
                    if hrx<=3, PLh=-10*log10(hrx/3);      % Eq.(1.15)
                      else PLh=-20*log10(hrx/3);
                    end
  case 'NO',      PLf=0; PLh=0;
end
if Type=='A',    a=4.6; b=0.0075; c=12.6; % Eq.(1.3)
  elseif Type=='B', a=4; b=0.0065; c=17.1;
  else     a=3.6; b=0.005; c=20;
end
lamda=3e8/fc; gamma=a-b*htx+c/htx; d0_pr=d0;              % Eq.(1.12)
if Mod(1)=='M'
  d0_pr=d0*10^-((PLf+PLh)/(10*gamma));                   % Eq.(1.17)
end
A = 20*log10(4*pi*d0_pr/lamda) + PLf + PLh;
for k=1:length(d)
  if d(k)>d0_pr, PL(k) = A + 10*gamma*log10(d(k)/d0);    % Eq.(1.18)
    else PL(k) = 20*log10(4*pi*d(k)/lamda);
  end
end
```

Program 1.7 "plot_PL_IEEE80216d.m" to plot the IEEE 802.16d path loss model

```
% plot_PL_IEEE80216d.m
clear, clf, clc
fc=2e9; htx=[30 30]; hrx=[2 10]; distance=[1:1000];
for k=1:2
  y_IEEE16d(k,:)=PL_IEEE80216d(fc,distance,'A',htx(k),hrx(k),'atnt');
  y_MIEEE16d(k,:)=PL_IEEE80216d(fc,distance,'A',htx(k),hrx(k),
    'atnt', 'mod');
end
subplot(121), semilogx(distance,y_IEEE16d(1,:),'k:','linewidth',1.5)
hold on, semilogx(distance,y_IEEE16d(2,:),'k-','linewidth',1.5)
```

```
grid on, axis([1 1000 10 150])
title(['IEEE 802.16d Path-loss Model, f_c=',num2str(fc/1e6),'MHz'])
xlabel('Distance[m]'), ylabel('Pathloss[dB]')
legend('h_{Tx}=30m, h_{Rx}=2m','h_{Tx}=30m, h_{Rx}=10m',2)
subplot(122), semilogx(distance,y_MIEEE16d(1,:),'k:','linewidth',1.5)
hold on, semilogx(distance,y_MIEEE16d(2,:),'k-','linewidth',1.5)
grid on, axis([1 1000 10 150])
title(['Modified IEEE 802.16d Path-loss Model, f_c=', num2str(fc/1e6),'MHz'])
xlabel('Distance[m]'), ylabel('Pathloss[dB]')
legend('h_{Tx}=30m, h_{Rx}=2m','h_{Tx}=30m, h_{Rx}=10m',2)
```

1.2 Small-Scale Fading

Unless confused with large-scale fading, small-scale fading is often referred to as fading in short. Fading is the rapid variation of the received signal level in the short term as the user terminal moves a short distance. It is due to the effect of multiple signal paths, which cause interference when they arrive subsequently in the receive antenna with varying phases (i.e., constructive interference with the same phase and destructive interference with a different phase). In other words, the variation of the received signal level depends on the relationships of the relative phases among the number of signals reflected from the local scatters. Furthermore, each of the multiple signal paths may undergo changes that depend on the speeds of the mobile station and surrounding objects. In summary, small-scale fading is attributed to multi-path propagation, mobile speed, speed of surrounding objects, and transmission bandwidth of signal.

1.2.1 Parameters for Small-Scale Fading

Characteristics of a multipath fading channel are often specified by a power delay profile (PDP). Table 1.4 presents one particular example of PDP specified for the pedestrian channel model by ITU-R, in which four different multiple signal paths are characterized by their relative delay and average power. Here, the relative delay is an excess delay with respect to the reference time while average power for each path is normalized by that of the first path (tap) [12].

Table 1.4 Power delay profile: example (ITU-R Pedestrian A Model).

Tab	Relative delay (*ns*)	Average power (*dB*)
1	0	0.0
2	110	−9.7
3	190	−19.2
4	410	−22.8

Mean excess delay and *RMS delay spread* are useful channel parameters that provide a reference of comparison among the different multipath fading channels, and furthermore, show a general guideline to design a wireless transmission system. Let τ_k denote the channel delay of the kth path while a_k and $P(\tau_k)$ denote the amplitude and power, respectively. Then, the mean excess delay $\bar{\tau}$ is given by the first moment of PDP as

$$\bar{\tau} = \frac{\sum_k a_k^2 \tau_k}{\sum_k a_k^2} = \frac{\sum_k \tau_k P(\tau_k)}{\sum_k P(\tau_k)} \tag{1.19}$$

Meanwhile, RMS delay spread σ_τ is given by the square root of the second central moment of PDP as

$$\sigma_\tau = \sqrt{\overline{\tau^2} - (\bar{\tau})^2} \tag{1.20}$$

where

$$\overline{\tau^2} = \frac{\sum_k a_k^2 \tau_k^2}{\sum_k a_k^2} = \frac{\sum_k \tau_k^2 P(\tau_k)}{\sum_k P(\tau_k)} \tag{1.21}$$

In general, *coherence bandwidth*, denoted as B_c, is inversely-proportional to the RMS delay spread, that is,

$$B_c \approx \frac{1}{\sigma_\tau} \tag{1.22}$$

The relation in Equation (1.22) may vary with the definition of the coherence bandwidth. For example, in the case where the coherence bandwidth is defined as a bandwidth with correlation of 0.9 or above, coherence bandwidth and RMS delay spread are related as

$$B_c \approx \frac{1}{50\sigma_\tau} \tag{1.23}$$

In the case where the coherence bandwidth is defined as a bandwidth with correlation of 0.5 or above, it is given as

$$B_c \approx \frac{1}{5\sigma_\tau} \tag{1.24}$$

1.2.2 Time-Dispersive vs. Frequency-Dispersive Fading

As mobile terminal moves, the specific type of fading for the corresponding receiver depends on both the transmission scheme and channel characteristics. The transmission scheme is specified with signal parameters such as signal bandwidth and symbol period. Meanwhile, wireless channels can be characterized by two different channel parameters, multipath delay spread and Doppler spread, each of which causes time dispersion and frequency dispersion, respectively. Depending on the extent of time dispersion or frequency dispersion, the frequency-selective fading or time-selective fading is induced respectively.

1.2.2.1 Fading Due to Time Dispersion: Frequency-Selective Fading Channel

Due to time dispersion, a transmit signal may undergo fading over a frequency domain either in a selective or non-selective manner, which is referred to as *frequency-selective fading* or

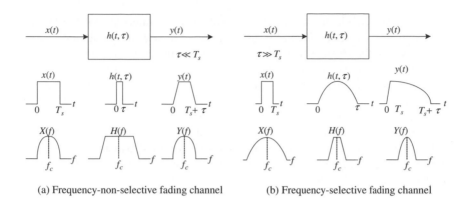

(a) Frequency-non-selective fading channel   (b) Frequency-selective fading channel

Figure 1.10   Characteristics of fading due to time dispersion over multi-path channel [2]. (Rappaport, Theodore S., *Wireless Communications: Principles and Practice*, 2nd Edition, © 2002, pgs. 130–131. Reprinted by permission of Pearson Education, Inc., Upper Saddle River, New Jersey.)

frequency-non-selective fading, respectively. For the given channel frequency response, frequency selectivity is generally governed by signal bandwidth. Figure 1.10 intuitively illustrates how channel characteristics are affected by the signal bandwidth in the frequency domain. Due to time dispersion according to multi-paths, channel response varies with frequency. Here, the transmitted signal is subject to frequency-non-selective fading when signal bandwidth is narrow enough such that it may be transmitted over the flat response. On the other hand, the signal is subject to frequency-selective fading when signal bandwidth is wide enough such that it may be filtered out by the finite channel bandwidth.

As shown in Figure 1.10(a), the received signal undergoes frequency-non-selective fading as long as the bandwidth of the wireless channel is wider than that of the signal bandwidth, while maintaining a constant amplitude and linear phase response within a passband. Constant amplitude undergone by signal bandwidth induces *flat fading*, which is another term to refer to frequency-non-selective fading. Here, a narrower bandwidth implies that symbol period T_s is greater than delay spread τ of the multipath channel $h(t, \tau)$. As long as T_s is greater than τ, the current symbol does not affect the subsequent symbol as much over the next symbol period, implying that inter-symbol interference (ISI) is not significant. Even while amplitude is slowly time-varying in the frequency-non-selective fading channel, it is often referred to as a narrowband channel, since the signal bandwidth is much narrower than the channel bandwidth. To summarize the observation above, a transmit signal is subject to frequency-non-selective fading under the following conditions:

$$B_s \ll B_c \quad \text{and} \quad T_s \gg \sigma_\tau \tag{1.25}$$

where B_s and T_s are the bandwidth and symbol period of the transmit signal, while B_c and σ_τ denote coherence bandwidth and RMS delay spread, respectively.

As mentioned earlier, transmit signal undergoes frequency-selective fading when the wireless channel has a constant amplitude and linear phase response only within a channel bandwidth narrower than the signal bandwidth. In this case, the channel impulse response has a larger delay spread than a symbol period of the transmit signal. Due to the short symbol duration as compared to the multipath delay spread, multiple-delayed copies of the transmit signal is significantly overlapped with the subsequent symbol, incurring inter-symbol interference (ISI). The term *frequency selective channel* is used simply because the amplitude of frequency response varies with the frequency, as opposed to the frequency-flat nature of the frequency-non-selective fading channel. As illustrated in Figure 1.10(b), the occurrence of ISI is obvious in the time domain since channel delay spread τ is much greater than the symbol period. This implies that signal bandwidth B_s is greater than coherence bandwidth B_c and thus, the received signal will have a different amplitude in the frequency response (i.e., undergo frequency-selective fading). Since signal bandwidth is larger than the bandwidth of channel impulse response in frequency-selective fading channel, it is often referred to as a wideband channel. To summarize the observation above, transmit signal is subject to frequency-selective fading under the following conditions:

$$B_s > B_c \quad \text{and} \quad T_s > \sigma_\tau \tag{1.26}$$

Even if it depends on modulation scheme, a channel is typically classified as frequency-selective when $\sigma_\tau > 0.1 T_s$.

1.2.2.2 Fading Due to Frequency Dispersion: Time-Selective Fading Channel

Depending on the extent of the Doppler spread, the received signal undergoes fast or slow fading. In a *fast fading* channel, the coherence time is smaller than the symbol period and thus, a channel impulse response quickly varies within the symbol period. Variation in the time domain is closely related to movement of the transmitter or receiver, which incurs a spread in the frequency domain, known as a Doppler shift. Let f_m be the maximum Doppler shift. The bandwidth of Doppler spectrum, denoted as B_d, is given as $B_d = 2f_m$. In general, the *coherence time*, denoted as T_c, is inversely proportional to Doppler spread, i.e.,

$$T_c \approx \frac{1}{f_m} \tag{1.27}$$

Therefore, $T_s > T_c$ implies $B_s < B_d$. The transmit signal is subject to fast fading under the following conditions:

$$T_s > T_c \quad \text{and} \quad B_s < B_d \tag{1.28}$$

On the other hand, consider the case that channel impulse response varies slowly as compared to variation in the baseband transmit signal. In this case, we can assume that the channel does not change over the duration of one or more symbols and thus, it is referred to as a *static* channel. This implies that the Doppler spread is much smaller than the bandwidth of the baseband transmit signal. In conclusion, transmit signal is subject to slow fading under the following conditions:

$$T_s \ll T_c \quad \text{and} \quad B_s \gg B_d \tag{1.29}$$

In the case where the coherence time is defined as a bandwidth with the correlation of 0.5 or above [1], the relationship in Equation (1.27) must be changed to

$$T_c \approx \frac{9}{16\pi f_m} \tag{1.30}$$

Note that Equation (1.27) is derived under the assumption that a Rayleigh-faded signal varies very slowly, while Equation (1.30) is derived under the assumption that a signal varies very fast. The most common definition of coherence time is to use the geometric mean of Equation (1.27) and Equation (1.30) [1], which is given as

$$T_c = \sqrt{\frac{9}{16\pi f_m^2}} = \frac{0.423}{f_m} \tag{1.31}$$

It is important to note that fast or slow fading does not have anything to do with time dispersion-induced fading. In other words, the frequency selectivity of the wireless channel cannot be judged merely from the channel characteristics of fast or slow fading. This is simply because fast fading is attributed only to the rate of channel variation due to the terminal movement.

1.2.3 Statistical Characterization and Generation of Fading Channel

1.2.3.1 Statistical Characterization of Fading Channel

Statistical model of the fading channel is to Clarke's credit that he statistically characterized the electromagnetic field of the received signal at a moving terminal through a scattering process [12]. In Clarke's proposed model, there are N planewaves with arbitrary carrier phases, each coming from an arbitrary direction under the assumption that each planewave has the same average power [13–16].

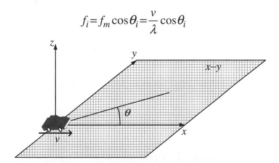

Figure 1.11   Planewave arriving at the receiver that moves in the direction of x with a velocity of v.

Figure 1.11 shows a planewave arriving from angle θ with respect to the direction of a terminal movement with a speed of v, where all waves are arriving from a horizontal direction on x–y plane. As a mobile station moves, all planewaves arriving at the receiver undergo the Doppler shift. Let $x(t)$ be a baseband transmit signal. Then, the corresponding passband transmit signal is given as

$$\tilde{x}(t) = \text{Re}\left[x(t)e^{j2\pi f_c t}\right] \tag{1.32}$$

where $\text{Re}[s(t)]$ denotes a real component of $s(t)$. Passing through a scattered channel of I different propagation paths with different Doppler shifts, the passband received signal can be represented as

$$\tilde{y}(t) = \mathrm{Re}\left[\sum_{i=1}^{I} C_i e^{j2\pi(f_c+f_i)(t-\tau_i)} x(t-\tau_i)\right]$$

$$= \mathrm{Re}\left[y(t)e^{j2\pi f_c t}\right] \tag{1.33}$$

where C_i, τ_i, and f_i denote the channel gain, delay, and Doppler shift for the ith propagation path, respectively. For the mobile speed of v and the wavelength of λ, Doppler shift is given as

$$f_i = f_m \cos \theta_i = \frac{v}{\lambda} \cos \theta_i \tag{1.34}$$

where f_m is the maximum Doppler shift and θ_i is the angle of arrival (AoA) for the ith planewave. Note that the baseband received signal in Equation (1.33) is given as

$$y(t) = \sum_{i=1}^{I} C_i e^{-j\phi_i(t)} x(t-\tau_i) \tag{1.35}$$

where $\phi_i(t) = 2\pi\{(f_c+f_i)\tau_i-f_it_i\}$. According to Equation (1.35), therefore, the corresponding channel can be modeled as a linear time-varying filter with the following complex baseband impulse response:

$$h(t,\tau) = \sum_{i=1}^{I} C_i e^{-j\phi_i(t)} \delta(t-\tau_i) \tag{1.36}$$

where $\delta(\cdot)$ is a Dirac delta function. As long as difference in the path delay is much less than the sampling period T_S, path delay τ_i can be approximated as $\hat{\tau}$. Then, Equation (1.36) can be represented as

$$h(t,\tau) = h(t)\delta(t-\hat{\tau}) \tag{1.37}$$

where $h(t) = \sum_{i=1}^{I} C_i e^{-j\phi_i(t)}$. Assuming that $x(t) = 1$, the received passband signal $\tilde{y}(t)$ can be expressed as

$$\tilde{y}(t) = \mathrm{Re}\left[y(t)e^{j2\pi f_c t}\right]$$

$$= \mathrm{Re}\left[\{h_I(t)+jh_Q(t)\}e^{j2\pi f_c t}\right] \tag{1.38}$$

$$= h_I(t)\cos 2\pi f_c t - h_Q(t)\sin 2\pi f_c t$$

where $h_I(t)$ and $h_Q(t)$ are in-phase and quadrature components of $h(t)$, respectively given as

$$h_I(t) = \sum_{i=1}^{I} C_i \cos \phi_i(t) \tag{1.39}$$

and

$$h_Q(t) = \sum_{i=1}^{I} C_i \sin \phi_i(t) \tag{1.40}$$

Assuming that I is large enough, $h_I(t)$ and $h_Q(t)$ in Equation (1.39) and Equation (1.40) can be approximated as Gaussian random variables by the central limit theorem. Therefore, we conclude that the amplitude of the received signal, $\tilde{y}(t) = \sqrt{h_I^2(t) + h_Q^2(t)}$, over the multipath channel subject to numerous scattering components, follows the Rayleigh distribution. The power spectrum density (PSD) of the fading process is found by the Fourier transform of the autocorrelation function of $\tilde{y}(t)$ and is given by [12]

$$
S_{\tilde{y}\tilde{y}}(f) = \begin{cases} \dfrac{\Omega_p}{4\pi f_m} \dfrac{1}{\sqrt{1 - \left(\dfrac{f-f_c}{f_m}\right)^2}} & |f-f_c| \leq f_m \\[2em] 0 & \text{otherwise} \end{cases} \tag{1.41}
$$

where $\Omega_p = E\{h_I^2(t)\} + E\{h_Q^2(t)\} = \sum_{i=1}^{I} C_i^2$. The power spectrum density in Equation (1.41) is often referred to as the *classical Doppler spectrum*.

Meanwhile, if some of the scattering components are much stronger than most of the components, the fading process no longer follows the Rayleigh distribution. In this case, the amplitude of the received signal, $\tilde{y}(t) = \sqrt{h_I^2(t) + h_Q^2(t)}$, follows the Rician distribution and thus, this fading process is referred to as *Rician fading*.

The strongest scattering component usually corresponds to the *line-of-sight* (LOS) component (also referred to as *specular* components). Other than the LOS component, all the other components are non-line-of-sight (NLOS) components (referred to as *scattering* components). Let $\tilde{p}(\theta)$ denote a probability density function (PDF) of AoA for the scattering components and θ_0 denote AoA for the specular component. Then, the PDF of AoA for all components is given as

$$
p(\theta) = \frac{1}{K+1}\tilde{p}(\theta) + \frac{K}{K+1}\delta(\theta-\theta_0) \tag{1.42}
$$

where K is the Rician factor, defined as a ratio of the specular component power c^2 and scattering component power $2\sigma^2$, shown as

$$
K = \frac{c^2}{2\sigma^2} \tag{1.43}
$$

In the subsequent section, we discuss how to compute the probability density of the above fading processes, which facilitate generating Rayleigh fading and Rician fading.

1.2.3.2 Generation of Fading Channels

In general, the propagation environment for any wireless channel in either indoor or outdoor may be subject to LOS (Line-of-Sight) or NLOS (Non Line-of-Sight). As described in the previous subsection, a probability density function of the signal received in the LOS environment follows the Rician distribution, while that in the NLOS environment follows the

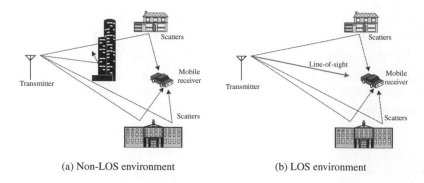

(a) Non-LOS environment   (b) LOS environment

Figure 1.12 Non-LOS and LOS propagation environments.

Rayleigh distribution. Figure 1.12 illustrates these two different environments: one for LOS and the other for NLOS.

Note that any received signal in the propagation environment for a wireless channel can be considered as the sum of the received signals from an infinite number of scatters. By the central limit theorem, the received signal can be represented by a Gaussian random variable. In other words, a wireless channel subject to the fading environments in Figure 1.12 can be represented by a complex Gaussian random variable, $W_1 + jW_2$, where W_1 and W_2 are the independent and identically-distributed (i.i.d.) Gaussian random variables with a zero mean and variance of σ^2. Let X denote the amplitude of the complex Gaussian random variable $W_1 + jW_2$, such that $X = \sqrt{W_1^2 + W_2^2}$. Then, note that X is a Rayleigh random variable with the following probability density function (PDF):

$$f_X(x) = \frac{x}{\sigma^2} e^{-\frac{x^2}{2\sigma^2}} \tag{1.44}$$

where $2\sigma^2 = E\{X^2\}$. Furthermore, X^2 is known as a chi-square (χ^2) random variable.

Below, we will discuss how to generate the Rayleigh random variable X. First of all, we generate two i.i.d. Gaussian random variables with a zero mean and unit variance, Z_1 and Z_2, by using a built-in MATLAB® function, "randn." Note that the Rayleigh random variable X with the PDF in Equation (1.44) can be represented by

$$X = \sigma \cdot \sqrt{Z_1^2 + Z_2^2} \tag{1.45}$$

where $Z_1 \sim \mathcal{N}(0,1)$ and $Z_2 \sim \mathcal{N}(0,1)$[1]. Once Z_1 and Z_2 are generated by the built-in function "randn," the Rayleigh random variable X with the average power of $E\{X^2\} = 2\sigma^2$ can be generated by Equation (1.45).

In the line-of-sight (LOS) environment where there exists a strong path which is not subject to any loss due to reflection, diffraction, and scattering, the amplitude of the received signal can be expressed as $X = c + W_1 + jW_2$ where c represents the LOS component while W_1 and W_2 are the i.i.d. Gaussian random variables with a zero mean and variance of σ^2 as in the non-LOS environment. It has been known that X is the Rician random variable with the

---

[1] $\mathcal{N}(m, \sigma^2)$ represents a Gaussian (normal) distribution with a mean of m and variance of σ^2.

following PDF:

$$f_X(x) = \frac{x}{\sigma^2} e^{-\frac{x^2+c^2}{2\sigma^2}} I_0\left(\frac{xc}{\sigma^2}\right) \qquad (1.46)$$

where $I_0(\,\cdot\,)$ is the modified zeroth-order Bessel function of the first kind. Note that Equation (1.46) can be represented in terms of the Rician K-factor defined in Equation (1.43). In case that there does exist an LOS component (i.e., $K = 0$), Equation (1.46) reduces to the Rayleigh PDF Equation (1.44) as in the non-LOS environment. As K increases, Equation (1.46) tends to be the Gaussian PDF. Generally, it is assumed that $K \sim -40$dB for the Rayleigh fading channel and $K > 15$dB for the Gaussian channel. In the LOS environment, the first path that usually arrives with any reflection can be modeled as a Rician fading channel.

Figure 1.13 has been produced by running Program 1.8 ("plot_Ray_Ric_channel.m"), which calls Program 1.9 ("Ray_model") and Program 1.10 ("Ric_model") to generate the Rayleigh fading and Rician fading channels, respectively. It also demonstrates that the Rician distribution approaches Rayleigh distribution and Gaussian distribution when $K = -40$dB and $K = 15$dB, respectively.

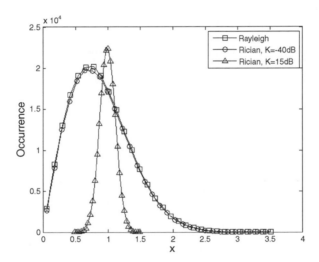

Figure 1.13 Distributions for Rayleigh and Rician fading channels.

Refer to [17–22] for additional information about propagation and fading in wireless channels.

MATLAB® Programs: Rayleigh Fading and Rician Fading Channels

Program 1.8 "plot_Ray_Ric_channel.m" to generate Rayleigh and Rician fading channels

```
% plot_Ray_Ric_channel.m
clear, clf
N=200000; level=30; K_dB=[-40 15];
gss=['k-s'; 'b-o'; 'r-^'];
% Rayleigh model
```

```
Rayleigh_ch=Ray_model(N);
[temp,x]=hist(abs(Rayleigh_ch(1,:)),level);
plot(x,temp,gss(1,:)), hold on
% Rician model
for i=1:length(K_dB);
  Rician_ch(i,:) = Ric_model(K_dB(i),N);
  [temp x] = hist(abs(Rician_ch(i,:)),level);
  plot(x,temp,gss(i+1,:))
end
xlabel('x'), ylabel('Occurrence')
legend('Rayleigh','Rician, K=-40dB','Rician, K=15dB')
```

Program 1.9 "Ray_model" for Rayleigh fading channel model

```
function H=Ray_model(L)
% Rayleigh channel model
% Input : L = Number of channel realizations
% Output: H = Channel vector
H = (randn(1,L)+j*randn(1,L))/sqrt(2);
```

Program 1.10 "Ric_model" for Rician fading channel model

```
function H=Ric_model(K_dB,L)
% Rician channel model
% Input : K_dB = K factor[dB]
% Output: H = Channel vector
K = 10^(K_dB/10);
H = sqrt(K/(K+1)) + sqrt(1/(K+1))*Ray_model(L);
```

2

SISO Channel Models

In Chapter 1, we have considered the general large-scale fading characteristics of wireless channels, including the path loss and shadowing. Furthermore, we have introduced the essential channel characteristics, such as delay spread and coherence time, which are useful for characterizing the short-term fading properties in general. In order to create an accurate channel model in the specific environment, we must have full knowledge on the characteristics of reflectors, including their situation and movement, and the power of the reflected signal, at any specified time. Since such full characterization is not possible in reality, we simply resort to the specific channel model, which can represent a typical or average channel condition in the given environment. The channel model can vary with the antenna configuration in the transmitter and receiver (e.g., depending on single antenna system or multiple antenna system). Especially in the recent development of the multi-input and multi-output (MIMO) systems, a completely different channel model is required to capture their spatio-temporal characteristics (e.g., the correlation between the different paths among the multiple transmit and receive antennas). This chapter deals with the channel models for the single-input and single-output (SISO) system that employs a single transmit antenna and a single receive antenna in the different environments. We consider the short-term fading SISO channel models for two different channel environments: indoor and outdoor channels. Meanwhile, the MIMO channel models will be separately treated in Chapter 3.

2.1 Indoor Channel Models

The indoor channel corresponds to the small coverage areas inside the building, such as office and shopping mall. Since these environments are completely enclosed by a wall, the power azimuth spectrum (PAS) tends to be uniform (i.e., the scattered components will be received from all directions with the same power). Furthermore, the channel tends to be static due to extremely low mobility of the terminals inside the building. Even in the indoor channel environments, however, the channel condition may vary with time and location, which still requires a power delay profile (PDP) to represent the channel delays and their average power. In general, a *static* channel refers to the environment in which a channel condition does not change

MIMO-OFDM Wireless Communications with MATLAB®   Yong Soo Cho, Jaekwon Kim, Won Young Yang and Chung G. Kang
© 2010 John Wiley & Sons (Asia) Pte Ltd

for the duration of data transmission at the given time and location. It is completely opposite to the time-varying environment in which the scattering components (objects or people) surrounding the transmitter or receiver are steadily moving even while a terminal is not in motion. In the wireless digital communication systems, however, the degree of time variation in the signal strength is relative to the symbol duration. In other words, the channel condition can be considered static when the degree of time variation is relatively small with respect to the symbol duration. This particular situation is referred to as a *quasi-static* channel condition. In fact, the indoor channels arc usually modeled under the assumption that they have either static or quasi-static channel conditions. In this subsection, we discuss the useful indoor channel models that deal with the multipath delay subject to the static or quasi-static channel conditions.

2.1.1 General Indoor Channel Models

In this subsection, we consider the two most popular indoor channel models: 2-ray model and exponential model. In the 2-ray model, there are two rays, one for a direct path with zero delay (i.e., $\tau_0 = 0$), and the other for a path which is a reflection with delay of $\tau_1 > 0$, each with the same power (see Figure 2.1(a) for its PDP). In this model, the maximum excess delay is $\tau_m = \tau_1$ and the mean excess delay $\bar{\tau}$ is given as $\bar{\tau} = \tau_1/2$. It is obvious that the RMS delay is the same as the mean excess delay in this case (i.e., $\bar{\tau} = \sigma_\tau = \tau_1/2$). In other words, the delay of the second path is the only parameter that determines the characteristics of this particular model. Due to its simplicity, the 2-ray model is useful in practice. However, it might not be accurate, simply because a magnitude of the second path is usually much less than that of the first path in practice. This model may be acceptable only when there is a significant loss in the first path.

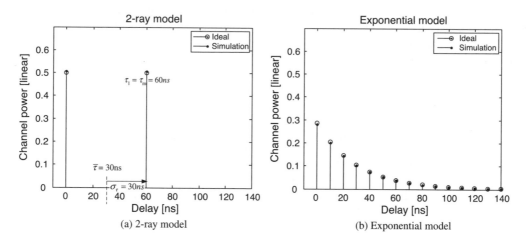

Figure 2.1   2-ray model vs. exponential model: an illustration.

In the exponential model, the average channel power decreases exponentially with the channel delay as follows:

$$P(\tau) = \frac{1}{\tau_d} e^{-\tau/\tau_d} \tag{2.1}$$

where τ_d is the only parameter that determines the power delay profile (PDP). Figure 2.1(b) illustrates a typical PDP of the exponential model. This model is known to be more appropriate for an indoor channel environment. The mean excess delay and RMS delay spread turn out to be equal to each other, that is, $\bar{\tau} = \tau_d$ and $\bar{\tau} = \sigma_\tau = \tau_d$, in the exponential model. Meanwhile, the maximum excess delay is given as

$$\tau_m = -\tau_d \ln A \qquad (2.2)$$

where A is a ratio of non-negligible path power to the first path power, that is, $A = P(\tau_m)/P(0) = \exp(-\tau_m/\tau_d)$. Note that Equation (2.1) can be represented by the following discrete-time model with a sampling period of T_S:

$$P(p) = \frac{1}{\sigma_\tau} e^{-pT_s/\sigma_\tau}, \quad p = 0, 1, \cdots, p_{\max} \qquad (2.3)$$

where p is the discrete time index with $p_{\max}$ as the index of the last path, that is, $p_{\max} = [\tau_m/T_s]$. A total power for the PDP in Equation (2.3) is given as

$$P_{total} = \sum_{p=0}^{p_{\max}} P(p) = \frac{1}{\sigma_\tau} \cdot \frac{1 - e^{-(p_{\max}+1)T_s/\sigma_\tau}}{1 - e^{-T_s/\sigma_\tau}} \qquad (2.4)$$

In order to normalize the total power in Equation (2.4) by one, Equation (2.3) has been modified as

$$P(p) = P(0)e^{-pT_s/\sigma_\tau}, \quad p = 0, 1, \cdots, p_{\max} \qquad (2.5)$$

where $P(0)$ is the first path power, $P(0) = 1/(P_{total} \cdot \sigma_\tau)$ by Equation (2.4) and Equation (2.5).

Figures 2.1(a) and (b) have been obtained by running Program 2.1 ("plot_2ray_exp_model. m"), which calls Program 1.9 ("Ray_model") with 10,000 channel realizations to get the 2-ray model (as depicted in Figure 2.1(a)) and uses Program 2.2 ("exp_PDP") with the RMS delay spread of 30ns and sampling period of 10ns to get the exponential model (as depicted in Figure 2.1(b)).

MATLAB® Programs: General Indoor Channel Model

Program 2.1 "plot_2ray_exp_model.m" to plot a 2-ray channel model and an exponential model

```
% plot_2ray_exp_model.m
clear, clf
scale=1e-9;                              % nano
Ts=10*scale;                             % Sampling time
t_rms=30*scale;                          % RMS delay spread
num_ch=10000; %                          # of channel
% 2-ray model
pow_2=[0.5 0.5];      delay_2=[0 t_rms*2]/scale;
H_2 = [Ray_model(num_ch); Ray_model(num_ch)].'*diag(sqrt(pow_2));
avg_pow_h_2 = mean(H_2.*conj(H_2));
subplot(221)
```

```
stem(delay_2,pow_2,'ko'), hold on, stem(delay_2,avg_pow_h_2,'k.');
xlabel('Delay[ns]'), ylabel('Channel Power[linear]');
title('2-ray Model');
legend('Ideal','Simulation'); axis([-10 140 0 0.7]);
% Exponential model
pow_e=exp_PDP(t_rms,Ts); delay_e=[0:length(pow_e)-1]*Ts/scale;
for i=1:length(pow_e)
  H_e(:,i)=Ray_model(num_ch).'*sqrt(pow_e(i));
end
avg_pow_h_e = mean(H_e.*conj(H_e));
subplot(222)
stem(delay_e,pow_e,'ko'), hold on, stem(delay_e,avg_pow_h_e,'k.');
xlabel('Delay[ns]'), ylabel('Channel Power[linear]');
title('Exponential Model'); axis([-10 140 0 0.7])
legend('Ideal','Simulation')
```

Program 2.2 "exp_PDP" to generate an exponential PDP

```
function PDP=exp_PDP(tau_d,Ts,A_dB,norm_flag)
% Exponential PDP generator
%   Inputs:
%     tau_d         : rms delay spread[sec]
%     Ts            : Sampling time[sec]
%     A_dB          : smallest noticeable power[dB]
%     norm_flag     : normalize total power to unit
%   Output:
%     PDP           : PDP vector
if nargin<4, norm_flag=1; end       % normalization
if nargin<3, A_dB=-20; end          % 20dB below
sigma_tau=tau_d; A=10^(A_dB/10);
lmax=ceil(-tau_d*log(A)/Ts);        % Eq.(2.2)
% compute normalization factor for power normalization
if norm_flag
  p0=(1-exp(-Ts/sigma_tau))/(1-exp((lmax+1)*Ts/sigma_tau)); % Eq.(2.4)
  else p0=1/sigma_tau;
end
% Exponential PDP
l=0:lmax; PDP = p0*exp(-l*Ts/sigma_tau); % Eq.(2.5)
```

2.1.2 IEEE 802.11 Channel Model

IEEE 802.11b Task Group has adopted the exponential model to represent a 2.4 GHz indoor channel [23]. Its PDP follows the exponential model as shown in Section 2.1.1. A channel impulse response can be represented by the output of finite impulse response (FIR) filter. Here, each channel tap is modeled by an independent complex Gaussian random variable with its average power that follows the exponential PDP, while taking the time index of each channel tap by the integer multiples of sampling periods. In other words, the maximum number of paths is determined by the RMS delay spread σ_τ and sampling period T_s as follows:

$$p_{\max} = \lceil 10 \cdot \sigma_\tau / T_s \rceil \tag{2.6}$$

Assuming that the power of the pth channel tap has the mean of 0 and variance of $\sigma_p^2/2$, its impulse response is given as

$$h_p = Z_1 + j \cdot Z_2, \quad p = 0, \cdots, p_{\max} \tag{2.7}$$

where Z_1 and Z_2 are statistically independent and identical Gaussian random variables, each with $\mathcal{N}(0, \sigma_p^2/2)$.

As opposed to the exponential model in which the maximum excess delay is computed by a path of the least non-negligible power level, the maximum excess delay in IEEE 802.11 channel model is fixed to 10 times the RMS delay spread. In this case, the power of each channel tap is given as

$$\sigma_p^2 = \sigma_0^2 e^{-pT_s/\sigma_\tau} \tag{2.8}$$

where σ_0^2 is the power of the first tap, which is determined so as to make the average received power equal to one, yielding

$$\sigma_0^2 = \frac{1 - e^{-T_s/\sigma_\tau}}{1 - e^{-(p_{\max}+1)T_s/\sigma_\tau}} \tag{2.9}$$

In the IEEE 802.11 channel model, a sampling period T_s must be at least as small as 1/4.

Figure 2.2 shows the average channel power and channel frequency response for the IEEE 802.11 channel model. It has been obtained by running Program 2.3 ("plot_IEEE80211_model. m"), which calls Program 1.9 ("Ray_model") with 10,000 channel realizations and Program 2.4 ("IEEE802_11_model") with $\sigma_\tau = 25$ ns and $T_s = 50$ ns. Since the RMS delay spread is relatively small in this example, the power variation in the frequency domain is within at most 15dB, which implies that frequency selectivity is not that significant.

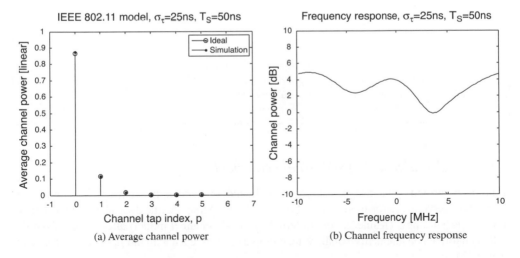

(a) Average channel power   (b) Channel frequency response

Figure 2.2   IEEE 802.11 channel model.

MATLAB® Programs: IEEE 802.11 Channel Model

Program 2.3 "plot_IEEE80211_model.m" to plot an IEEE 802.11 channel model

```
% plot_IEEE80211_model.m
clear, clf
scale=1e-9;                  % nano
Ts=50*scale;                 % Sampling time
t_rms=25*scale;              % RMS delay spread
num_ch=10000;                % Number of channels
N=128;                       % FFT size
PDP=IEEE802_11_model(t_rms,Ts);
for k=1:length(PDP)
   h(:,k) = Ray_model(num_ch).'*sqrt(PDP(k));
   avg_pow_h(k)= mean(h(:,k).*conj(h(:,k)));
end
H=fft(h(1,:),N);
subplot(221)
stem([0:length(PDP)-1],PDP,'ko'), hold on,
stem([0:length(PDP)-1],avg_pow_h,'k.');
xlabel('channel tap index, p'), ylabel('Average Channel Power[linear]');
title('IEEE 802.11 Model, \sigma_\tau=25ns, T_S=50ns');
legend('Ideal','Simulation'); axis([-1 7 0 1]);
subplot(222)
plot([-N/2+1:N/2]/N/Ts/1e6,10*log10(H.*conj(H)),'k-');
xlabel('Frequency[MHz]'), ylabel('Channel power[dB]')
title('Frequency response, \sigma_\tau=25ns, T_S=50ns')
```

Program 2.4 "IEEE802_11_model" for IEEE 802.11 channel model

```
function PDP=IEEE802_11_model(sigma_t,Ts)
% IEEE 802.11 channel model PDP generator
% Input:
%       sigma_t : RMS delay spread
%       Ts      : Sampling time
% Output:
%       PDP     : Power delay profile
lmax = ceil(10*sigma_t/Ts); % Eq.(2.6)
sigma02=(1-exp(-Ts/sigma_t))/(1-exp(-(lmax+1)*Ts/sigma_t)); % Eq.(2.9)
l=0:lmax; PDP = sigma02*exp(-l*Ts/sigma_t); % Eq.(2.8)
```

2.1.3 Saleh-Valenzuela (S-V) Channel Model

It has been verified by intense measurements of the indoor channel that arrivals of the multipath-delayed components can be modeled as a Poisson process. More specifically, Saleh and Valenzuela have proposed a new channel model (referred to as S-V channel model) after finding from the indoor channel measurements that there are multiple clusters, each with multiple rays, in the delay profile [24].

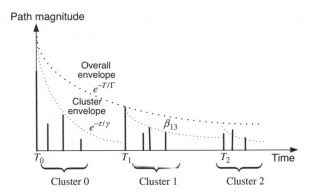

Figure 2.3   Saleh-Valenzuela channel model: an illustration.

Figure 2.3 illustrates the S-V channel model with multiple clusters, each of which is associated with a set of rays. The arrival times of each cluster as well as rays in each cluster follow an individual Poisson process. Therefore, the delay of each path is not spaced in the multiple of sampling periods, but spaced in a rather arbitrary manner. More specifically, the arrival time of the first ray in the mth cluster, denoted by T_m, is modeled by a Poisson process with an average arrival rate of Λ while the arrival times of rays in each cluster is modeled by a Poisson process with an average arrival rate of λ. Then, it can be shown that a distribution of inter-cluster arrival times and a distribution of inter-ray arrival times are given by the following exponential distributions, respectively:

$$f_{T_m}(T_m|T_{m-1}) = \Lambda \exp[-\Lambda(T_m-T_{m-1})], \quad m = 1, 2, \cdots \tag{2.10}$$

and

$$f_{\tau_{r,m}}\big(\tau_{r,m}|\tau_{(r-1),m}\big) = \lambda \exp\big[-\lambda\big(\tau_{r,m}-\tau_{(r-1),m}\big)\big], \quad r = 1, 2, \cdots \tag{2.11}$$

where $\tau_{r,m}$ denotes the arrival time of the rth ray in the mth cluster. In Equation (2.10) and Equation (2.11), the arrival time of the first ray in the mth cluster, $\tau_{0,m}$, is defined as the arrival time of the mth cluster, T_m (i.e., $\tau_{0,m} = T_m$). Let $\beta_{r,m}$ and $\theta_{r,m}$ denote amplitude and phase of the rth ray in the mth cluster, respectively. Then, a channel impulse response is given as

$$h(t) = \sum_{m=0}^{\infty} \sum_{r=0}^{\infty} \beta_{r,m} e^{j\theta_{r,m}} \delta(t-T_m-\tau_{r,m}) \tag{2.12}$$

where $\theta_{r,m}$ is a random variable that is uniformly distributed over $[0, 2\pi)$ and $\beta_{r,m}$ is an independent random variable with the following Rayleigh distribution:

$$f_{\beta_{r,m}}(\beta_{r,m}) = (2\beta_{r,m}/\overline{\beta_{r,m}^2})e^{-\beta_{r,m}^2/\overline{\beta_{r,m}^2}} \tag{2.13}$$

In Equation (2.13), $\overline{\beta_{r,m}^2}$ is the average power of the rth ray in the mth cluster, which is given as

$$\overline{\beta_{r,m}^2} = \overline{\beta_{0,0}^2} e^{-T_m/\Gamma} e^{-\tau_{r,m}/\gamma} \tag{2.14}$$

where Γ and γ denote time constants for exponential power attenuation in the cluster and ray, respectively, while $\overline{\beta_{0,0}^2}$ denotes the average power of the first ray in the first cluster.

As shown in Figure 2.3, the S-V channel model is a double exponential delay model in which average cluster power decays exponentially by following a term $e^{-T_m/\Gamma}$ in Equation (2.14) while average ray power in each cluster also decays exponentially by following a term $e^{-\tau_{r,m}/\gamma}$ in Equation (2.14). Once the average power of the first ray in the first cluster, $\beta_{0,0}^2$, is given, the average power of the rest of rays can be determined by Equation (2.14), which subsequently allows for determining the Rayleigh channel coefficients by Equation (2.13). In case that a path loss is not taken into account, without loss of generality, the average power of the first ray in the first cluster is set to one. Even if there are an infinite number of clusters and rays in the channel impulse response of Equation (2.12), there exist only a finite number of the non-negligible numbers of clusters and rays in practice. Therefore, we limit the number of clusters and rays to M and R, respectively. Meanwhile, a log-normal random variable X, that is, $20\log_{10}(X) \sim \mathcal{N}(0, \sigma_x^2)$, can be introduced to Equation (2.12), so as to reflect the effect of long-term fading as

$$h(t) = X \sum_{m=0}^{M} \sum_{r=0}^{R} \beta_{r,m} e^{j\theta_{r,m}} \delta(t - T_m - \tau_{r,m}) \tag{2.15}$$

Figure 2.4 has been obtained by running Program 2.5 ("plot_SV_model_ct.m"), which calls Program 2.6 ("SV_model_ct") with $\Lambda = 0.023$, $\lambda = 0.023$, $\Gamma = 7.4$, $\gamma = 4.3$, and $\sigma_x = 3$ dB to generate the S-V channel model. Figures 2.4(a) and (b) show the distributions of cluster arrival times and ray arrival times, respectively, including the simulation results to be compared with the analytical ones where the mth cluster arrival time T_m and the rth ray arrival time $\tau_{r,m}$ in the mth cluster are generated in such a way that each of them has an exponential distribution of Equation (2.10) and Equation (2.11), respectively. Figure 2.4(c) shows the channel impulse response of the S-V channel. Figure 2.4(d), showing the channel power distribution, is obtained by simulating 1,000 channels, from which it is clear that the channel power follows a log-normal distribution.

MATLAB® Programs: S-V Channel Model

Program 2.5 "plot_SV_model_ct.m" to plot a Saleh-Valenzuela channel model

```
% plot_SV_model_ct.m
clear, clf
Lam=0.0233; lambda=2.5; Gam=7.4; gamma=4.3;
N=1000; power_nom=1; std_shdw=3; t1=0:300; t2=0:0.01:5;
p_cluster=Lam*exp(-Lam*t1); % ideal exponential pdf
h_cluster=exprnd(1/Lam,1,N); % # of random numbers generated
[n_cluster,x_cluster]=hist(h_cluster,25); % obtain distribution
subplot(221), plot(t1,p_cluster,'k'), hold on,
plot(x_cluster,n_cluster*p_cluster(1)/n_cluster(1),'k:');% plotting
legend('Ideal','Simulation')
title(['Distribution of Cluster Arrival Time, \Lambda=',num2str(Lam)])
xlabel('T_m-T_{m-1}[ns]'), ylabel('p(T_m|T_{m-1})')
p_ray=lambda*exp(-lambda*t2); % ideal exponential pdf
h_ray=exprnd(1/lambda,1,1000); % # of random numbers generated
[n_ray,x_ray]=hist(h_ray,25); % obtain distribution
subplot(222), plot(t2,p_ray,'k'), hold on,
plot(x_ray,n_ray*p_ray(1)/n_ray(1),'k:'); legend('Ideal','Simulation')
title(['Distribution of Ray Arrival Time, \lambda=',num2str(lambda)])
```

```
xlabel('\tau_{r,m}-\tau_{(r-1),m}[ns]'),
  ylabel('p(\tau_{r,m}|\tau_{(r-1),m})')
[h,t,t0,np] = SV_model_ct(Lam,lambda,Gam,gamma,N,power_nom,std_shdw);
subplot(223), stem(t(1:np(1),1),abs(h(1:np(1),1)),'ko');
title('Generated Channel Impulse Response')
xlabel('Delay[ns]'), ylabel('Magnitude')
X=10.^(std_shdw*randn(1,N)./20); [temp,x]=hist(20*log10(X),25);
subplot(224), plot(x,temp,'k-'), axis([-10 10 0 120])
title(['Log-normal Distribution, \sigma_X=',num2str(std_shdw),'dB'])
xlabel('20*log10(X)[dB]'), ylabel('Occasion')
```

Program 2.6 "SV_model_ct" for Saleh-Valenzuela channel [25]

```
function
[h,t,t0,np]=SV_model_ct(Lam,lam,Gam,gam,num_ch,b002,sdi,nlos)
% S-V channel model
% Inputs
%   Lam      : Cluster arrival rate in GHz (avg # of clusters per nsec)
%   lam      : Ray arrival rate in GHz (avg # of rays per nsec)
%   Gam      : Cluster decay factor (time constant, nsec)
%   gam      : Ray decay factor (time constant, nsec)
%   num_ch   : Number of random realizations to generate
%   b002     : Power of first ray of first cluster
%   sdi      : Standard deviation of log-normal shadowing
%            of entire impulse response in dB
%   nlos     : Flag to specify generation of NLOS channels
% Outputs
%   h        : a matrix with num_ch columns, each column having a random
%            realization of channel model (impulse response)
%   t        : Time instances (in nsec) of the paths
%            whose signed amplitudes are stored in h
%   t0       : Arrival time of the first cluster for each realization
%   np       : Bumber of paths for each realization.
if nargin<8, nlos=0;   end      % LOS environment
if nargin<7, sdi=0;   end      % 0dB
if nargin<6, b002=1;   end      %Power of 1st ray of 1st cluster
h_len=1000;
for k=1:num_ch      % Loop over number of channels
  tmp_h = zeros(h_len,1);   tmp_t = zeros(h_len,1);
  if nlos,   Tc = exprnd(1/Lam);
    else    Tc = 0;  % first cluster arrival occurs at time 0
  end
  t0(k) = Tc;   path_ix = 0;
  while (Tc<10*Gam)   % Cluster loop
    Tr=0;
    while (Tr<10*gam)   % Ray loop
      t_val = Tc+Tr;     % time of arrival of this ray
      bkl2 = b002*exp(-Tc/Gam)*exp(-Tr/gam);   % ray power, Eq.(2.14)
      r = sqrt(randn^2+randn^2)*sqrt(bkl2/2);
      h_val=exp(j*2*pi*rand)*r;   % Uniform phase
      path_ix = path_ix+1;         % Row index of this ray
```

```
    tmp_h(path_ix) = h_val; tmp_t(path_ix) = t_val;
    Tr = Tr + exprnd(1/Lam);  % Ray arrival time based on Eq.(2.11)
  end
  Tc = Tc + exprnd(1/lam);  % Cluster arrival time based on Eq.(2.10)
end
np(k)=path_ix;  % Number of rays/paths for this realization
[sort_tmp_t,sort_ix] = sort(tmp_t(1:np(k)));  % in ascending order
t(1:np(k),k) = sort_tmp_t;
h(1:np(k),k) = tmp_h(sort_ix(1:np(k)));
% Log-normal shadowing on this realization
fac = 10^(sdi*randn/20)/sqrt(h(1:np(k),k)'*h(1:np(k),k));
h(1:np(k),k) = h(1:np(k),k)*fac;  % Eq.(2.15)
end
```

(Reproduced with permission from J. Foerster (ed.), © IEEE P802.15-02/490r1-SG3a, "Channel modeling sub-committee report (final)," 2003. © 2003 IEEE.)

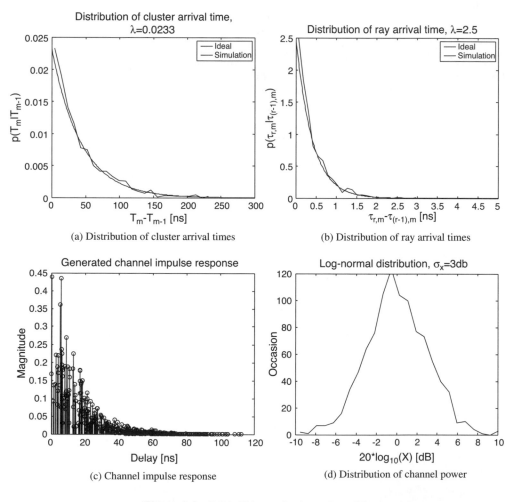

(a) Distribution of cluster arrival times

(b) Distribution of ray arrival times

(c) Channel impulse response

(d) Distribution of channel power

Figure 2.4 Saleh-Valenzuela channel model.

2.1.4 UWB Channel Model

According to measurements of broadband indoor channel, it has been found that amplitudes of multipath fading follow the log-normal or Nakagami distribution rather than the Rayleigh distribution, even if they also show the same phenomenon of clustering as in the Saleh-Valenzuela (S-V) channel model. Based on these results, SG3a UWB multipath model has been proposed by modifying the S-V model in such a way that the multi-cluster signals are subject to independent log-normal fading while the multi-path signals in each cluster are also subject to independent log-normal fading [25].

The ith sample function of a discrete-time impulse response in the UWB multi-path channel model is given as

$$h_i(t) = X_i \sum_{m=0}^{M} \sum_{r=0}^{R} a_{r,m}^{(i)} \delta\left(t - T_m^{(i)} - \tau_{r,m}^{(i)}\right) \qquad (2.16)$$

where X_i, $a_{r,m}^{(i)}$, $T_m^{(i)}$, and $\tau_{r,m}^{(i)}$ are defined as the same as in Equation (2.15), now with the index i to represent the ith generated sample function of the channel. For simplicity of exposition, the index i in Equation (2.16) will be eliminated in the following discussion. As in the S-V channel model, the arrival time distributions of clusters and rays are given by two different Poisson processes of Equation (2.10) and Equation (2.11), respectively. The UWB channel model is different from the S-V channel model in that clusters and rays are subject to independent log-normal fading rather than Rayleigh fading. More specifically, a channel coefficient is given as

$$\alpha_{r,m} = p_{r,m} \xi_m \beta_{r,m} \qquad (2.17)$$

where ξ_m represents log-normal fading of the mth cluster with the variance of σ_1^2 while $\beta_{r,m}$ represents log-normal fading of the rth ray with the variance of σ_2^2 in the mth cluster. Note that independent fading is assumed for clusters and rays. In Equation (2.17), $p_{r,m}$ is a binary discrete random variable to represent an arbitrary inversion of the pulse subject to reflection, that is, taking a value of $+1$ or -1 equally likely. As compared to the channel coefficients of the S-V channel model in Equation (2.12) which has a uniformly-distributed phase over $[0, 2\pi)$, those of UWB channel model have the phase of either $-\pi$ or π, making the channel coefficient always real. Furthermore, we note that amplitude of each ray is given by a product of the independent log-normal random variables, ξ_m and $\beta_{r,m}$. Since a product of the independent log-normal random variables is also a log-normal random variable, a distribution of the channel coefficient $|\xi_m \beta_{r,m}| = 10^{(\mu_{r,m} + z_1 + z_2)/20}$ also follows a log-normal distribution, that is, $20 \log_{10}(\xi_m \beta_{r,m}) \sim N(\mu_{r,m}, \sigma_1^2 + \sigma_2^2)$, with its average power given as

$$E\left[|\xi_m \beta_{r,m}|^2\right] = \Omega_0 e^{-T_m/\Gamma} e^{-\tau_{r,m}/\gamma} \qquad (2.18)$$

where Ω_0 represents the average power of the first ray in the first cluster. Meanwhile, mean of the channel amplitude for the r th ray in the m th cluster can be found as

$$\mu_{r,m} = \frac{10\ln(\Omega_0)-10\,T_m/\Gamma-10\,\tau_{r,m}/\gamma}{\ln(10)} - \frac{(\sigma_1^2+\sigma_2^2)\ln(10)}{20} \qquad (2.19)$$

Besides the same set of channel parameters as in the S-V channel model, including the cluster arrival rate Λ, ray arrival rate λ, cluster attenuation constant Γ, ray attenuation constant γ, standard deviation σ_x of the overall multipath shadowing with a log-normal distribution, additional channel parameters such as the standard deviations of log-normal shadowing for the clusters and rays, denoted as σ_1 and σ_2, respectively, are required for the UWB channel model. Note that a complete model of the multipath channel $h(t)$ in Equation (2.16) is given as a real number. Some proper modifications such as down-conversion and filtering are required for implementing the UWB channel in simulation studies, since its bandwidth cannot be limited due to arbitrary arrival times. All the channel characteristics for UWB channel model, including mean excess delay, RMS delay spread, the number of significant paths within 10dB of peak power (denoted as $\text{NP}_{10\text{dB}}$), and PDP, must be determined so as to be consistent with the measurements in practice.

Table 2.1 summarizes the SG3a model parameters and characteristics that represent the target UWB channel for four different types of channel models, denoted as CM1, CM2, CM3, and CM4. Each of these channel models varies depending on distance and whether LOS exists or not. Here, NP(85%) represents the number of paths that contain 85% of the total energy. CM1 and CM2 are based on measurements for LOS and Non-LOS environments over the distance of 0–4 m, respectively. CM3 is based on measurement for a Non-LOS environment over the distance of 4–10 m. CM4 does not deal with any realistic channel measurement, but it has been set up with the intentionally long RMS delay spread so as to model the worst-case Non-LOS environment. CM1 shows the best channel characteristics with the short RMS delay spread of 5.28 ns since it deals with a short distance under an LOS environment. Due to the Non-LOS environment, CM2 shows the longer RMS delay spread of 8.03 ns, even if it has the short range of distance as in the CM1. Meanwhile, CM3 represents the worse channel with the RMS delay spread of 14.28; it has a longer distance under a Non-LOS environment.

Figure 2.5 has been obtained by running Program 2.7 ("plot_UWB_channel_test.m"), in which Program 2.8 ("UWB_parameters"), Program 2.9 ("convert_UWB_ct"), and Program 2.10 ("UWB_model_ct") are used to set the UWB channel parameters as listed in Table 2.1, to convert the continuous-time UWB channel into the corresponding discrete-time one, and to generate a UWB channel model, respectively. It shows the UWB channel characteristics by simulating 100 CM1 channels. Here, the sampling period has been set to 167 ps. In the current measurement, the RMS delay spread turns out to be around 5 ns, which nearly coincides with the target value of CM1 channel in Table 2.1. The same observation is made for the mean excess delay. In conclusion, it has been verified that the current channel model properly realizes the target channel characteristics.

Table 2.1   UWB channel parameters and model characteristics [25].

Target channel characteristics	CM 1	CM 2	CM 3	CM 4
Mean excess delay (nsec) ($\bar{\tau}$)	5.05	10.38	14.18	
RMS delay (nsec) (σ_τ)	5.28	8.03	14.28	25
NP_{10dB}			35	
NP (85%)	24	36.1	61.54	
Model parameters				
Λ (1/nsec)	0.0233	0.4	0.0667	0.0667
λ (1/nsec)	2.5	0.5	2.1	2.1
Γ	7.1	5.5	14.00	24.00
γ	4.3	6.7	7.9	12
σ_1 (dB)	3.3941	3.3941	3.3941	3.3941
σ_2 (dB)	3.3941	3.3941	3.3941	3.3941
σ_x (dB)	3	3	3	3
Model characteristics				
Mean excess delay (nsec) ($\bar{\tau}$)	5.0	9.9	15.9	30.1
RMS delay (nsec) (σ_τ)	5	8	15	25
NP_{10dB}	12.5	15.3	24.9	41.2
NP (85%)	20.8	33.9	64.7	123.3
Channel energy mean (dB)	−0.4	−0.5	0.0	0.3
Channel energy std (dB)	2.9	3.1	3.1	2.7

(Reproduced with permission from J. Foerster (ed.), © IEEE P802.15-02/490r1-SG3a, "Channel modeling sub-committee report (final)," 2003. © 2003 IEEE.)

MATLAB® Programs: UWB Channel Model

Program 2.7 "plot_UWB_channel.m" to plot a UWB channel model

```
% plot_UWB_channel.m
clear, clf
Ts = 0.167; num_ch=100; randn('state',12); rand('state',12); cm = 1;
[Lam,lam,Gam,gam,nlos,sdi,sdc,sdr] = UWB_parameters(cm);
[h_ct,t_ct,t0,np]= ...
  UWB_model_ct(Lam,lam,Gam,gam,num_ch,nlos,sdi,sdc,sdr);
[hN,N] = convert_UWB_ct(h_ct,t_ct,np,num_ch,Ts);
h = resample(hN,1,N); h = h*N; channel_energy = sum(abs(h).^2);
h_len = size(h,1); t = [0:(h_len-1)]*Ts;
for k=1:num_ch
    sq_h = abs(h(:,k)).^2/channel_energy(k);
    t_norm= t-t0(k); excess_delay(k) = t_norm*sq_h;
    rms_delay(k) = sqrt((t_norm-excess_delay(k)).^2*sq_h);
    temp_h = abs(h(:,k)); threshold_dB = -10;
    temp_thresh = 10^(threshold_dB/20)*max(temp_h);
    num_sig_paths(k) = sum(temp_h>temp_thresh);
    temp_sort = sort(temp_h.^2);
    cum_energy = cumsum(temp_sort(end:-1:1));
    x = 0.85; index_e = min(find(cum_energy >= x*cum_energy(end)));
    num_sig_e_paths(k) = index_e;
```

```
end
energy_mean = mean(10*log10(channel_energy));
energy_stddev = std(10*log10(channel_energy));
mean_excess_delay = mean(excess_delay);
mean_rms_delay = mean(rms_delay);
mean_sig_paths = mean(num_sig_paths);
mean_sig_e_paths = mean(num_sig_e_paths);
temp_average_power - sum(h'.*h')/num_ch;
temp_average_power = temp_average_power/max(temp_average_power);
average_decay_profile_dB = 10*log10(temp_average_power);
fprintf(1,['Model Parameters\n' ' Lam=%.4f, lam=%.4f, Gam=%.4f, gam=%.4f\n
NLOS flag=%d, std_shdw=%.4f, td_ln_1=%.4f,td_ln_2=%.4f\n'],...
   Lam, lam, Gam, gam, nlos, sdi, sdc, sdr);
fprintf(1,'Model Characteristics\n');
fprintf(1,' Mean delays: excess(tau_m)=%.1fns, RMS(tau_rms)=%1.f\n', ...
    mean_excess_delay, mean_rms_delay);
fprintf(1,' # paths: NP_10dB = %.1f, NP_85%% = %.1f\n', ...
    mean_sig_paths, mean_sig_e_paths);
fprintf(1,' Channel energy: mean= %.1fdB, std deviation = %.1fdB\n',
... energy_mean, energy_stddev);
subplot(421), plot(t,h)
title('Impulse response realizations'), xlabel('Time[ns]')
subplot(422), plot([1:num_ch], excess_delay, 'b-', ...
   [1 num_ch],mean_excess_delay*[1 1],'r-' ); title('Excess delay[ns]') sub-
plot(423), title('RMS delay[ns]'), xlabel('Channel number')
plot([1:num_ch],rms_delay,'b-', [1 num_ch],mean_rms_delay*[1 1],'r-');
subplot(424), plot([1:num_ch], num_sig_paths, 'b-', ...
   [1 num_ch], mean_sig_paths*[1 1], 'r-');
grid on, title('Number of significant paths within 10 dB of peak')
subplot(425), plot(t,average_decay_profile_dB); grid on
title('Average Power Decay Profile'), axis([0 t(end) -60 0])
subplot(426), title('Channel Energy');
figh = plot([1:num_ch],10*log10(channel_energy),'b-', ...
   [1 num_ch], energy_mean*[1 1], 'g-', ...
   [1 num_ch], energy_mean+energy_stddev*[1 1], 'r:', ...
   [1 num_ch], energy_mean-energy_stddev*[1 1], 'r:');
legend(figh, 'Per-channel energy', 'Mean', '\pm Std. deviation', 0)
```

Program 2.8 "UWB_parameters" to set the parameters for UWB channel model

```
function [Lam,lam,Gam,gam,nlos,sdi,sdc,sdr]=UWB_parameters(cm)
% Table 2.1:
tmp = 4.8/sqrt(2);
Tb2_1= [0.0233 2.5 7.1 4.3 0 3 tmp tmp; 0.4 0.5 5.5 6.7 1 3 tmp tmp;
    0.0667 2.1 14.0 7.9 1 3 tmp tmp; 0.0667 2.1 24 12 1 3 tmp tmp];
Lam=Tb2_1(cm,1); lam=Tb2_1(cm,2); Gam=Tb2_1(cm,3); gam=Tb2_1(cm,4);
nlos= Tb2_1(cm,5);
sdi= Tb2_1(cm,6); sdc= Tb2_1(cm,7); sdr= Tb2_1(cm,8);
```

Program 2.9 "convert_UWB_ct" to convert a continuous-time channel into a discrete-time one

```
function [hN,N] = convert_UWB_ct(h_ct,t,np,num_ch,Ts)
min_Nfs = 100; N=2^nextpow2(max(1,ceil(min_Nfs*Ts)));
Nfs = N/Ts; t_max = max(t(:));
h_len=1+floor(t_max*Nfs); hN = zeros(h_len,num_ch);
for k = 1:num_ch
  np_k = np(k);
  t_Nfs = 1+floor(t(1:np_k,k)*Nfs);
  for n = 1:np_k
    hN(t_Nfs(n),k) = hN(t_Nfs(n),k) + h_ct(n,k);
  end
end
```

Program 2.10 "UWB_model_ct" for a continuous-time UWB channel model [25]

```
function [h,t,t0,np] = UWB_model_ct(Lam,lam,Gam,gam,num_ch,nlos,
 sdc,sdr,sdi)
std_L=1/sqrt(2*Lam); std_lam=1/sqrt(2*lam);
mu_const=(sdc^2+sdr^2)*log(10)/20; h_len=1000;
for k = 1:num_ch
  tmp_h = zeros(h_len,1);        tmp_t = zeros(h_len,1);
  if nlos
     Tc = (std_L*randn)^2+(std_L*randn)^2;
   else      Tc = 0;
   end
   t0(k) = Tc; path_ix = 0;
   while (Tc<10*Gam)
    Tr = 0; ln_xi = sdc*randn;
    while (Tr<10*gam)
      t_val = Tc + Tr;
      mu = (-10*Tc/Gam-10*Tr/gam)/log(10) - mu_const; % Eq.(2.19)
      ln_beta = mu + sdr*randn;
      pk = 2*round(rand)-1;
      h_val = pk*10^((ln_xi+ln_beta)/20);
      path_ix = path_ix + 1;
      tmp_h(path_ix) = h_val;
      tmp_t(path_ix) = t_val;
      Tr = Tr + (std_lam*randn)^2+(std_lam*randn)^2;
    end
    Tc = Tc + (std_L*randn)^2+(std_L*randn)^2;
  end
  np(k) = path_ix;
  [sort_tmp_t,sort_ix] = sort(tmp_t(1:np(k))); t(1:np(k),k) = sort_tmp_t;
  h(1:np(k),k) = tmp_h(sort_ix(1:np(k)));
  fac = 10^(sdi*randn/20)/sqrt(h(1:np(k),k)'*h(1:np(k),k));
  h(1:np(k),k) = h(1:np(k),k)*fac;
end
```

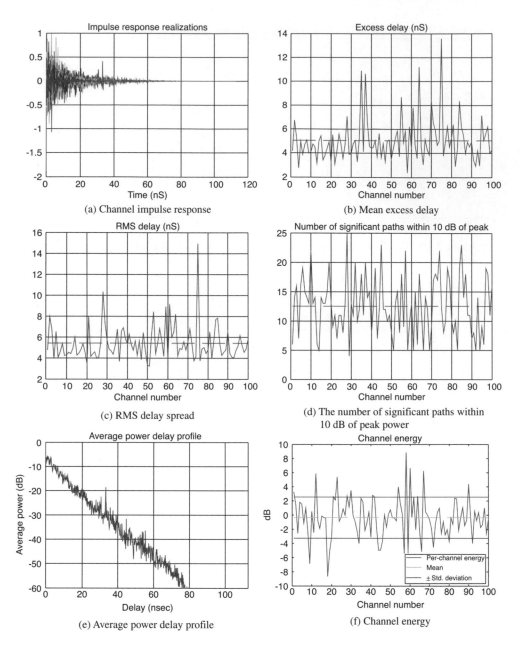

Figure 2.5   Generation of UWB channel: example (CM1).

2.2 Outdoor Channel Models

As opposed to the static or quasi-static nature of the indoor channel, outdoor channels are typically characterized by time variation of the channel gain, which is subject to the mobile speed of terminals. Depending on the mobile speed, time variation of channel gain is governed by Doppler spectrum, which determines the time-domain correlation in the channel gain. In this subsection, we discuss how to model the time-correlated channel variation as the mobile terminal moves. Furthermore, we present some practical methods of implementing the outdoor channel models for both frequency-flat and frequency-selective channels.

2.2.1 FWGN Model

The outdoor channel will be mostly characterized by Doppler spectrum that governs the time variation in the channel gain. Various types of Doppler spectrum can be realized by a filtered white Gaussian noise (FWGN) model. The FWGN model is one of the most popular outdoor channel models. The Clarke/Gans model is a baseline FWGN model that can be modified into various other types, depending on how a Doppler filter is implemented in the time domain or frequency domain. We first discuss the Clarke/Gans model and then, its frequency-domain and time-domain variants.

2.2.1.1 Clarke/Gans Model

The Clarke/Gans model has been devised under the assumption that scattering components around a mobile station are uniformly distributed with an equal power for each component [26]. Figure 2.6 shows a block diagram for the Clarke/Gans model, in which there are two branches, one for a real part and the other for an imaginary part. In each branch, a complex Gaussian noise is first generated in the frequency domain and then, filtered by a Doppler filter such that a frequency component is subject to Doppler shift. Finally, the Doppler-shifted Gaussian noise is transformed into the time-domain signal via an IFFT block. Since output of the IFFT block must be a real signal, its input must be always conjugate symmetric. Constructing a complex channel gain by adding a real part to an imaginary part of the output, a channel with the Rayleigh distributed-magnitude is generated.

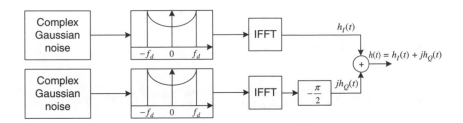

Figure 2.6   Block diagram for Clarke/Gans model.

Figure 2.7 shows the time-domain characteristics of the frequency-non-selective fading channel with a Doppler frequency of $f_m = 100\,\text{Hz}$ and a sampling period of $T_s = 50\,\mu s$. From these results, it is observed that the channel gain is time-varying with the Rayleigh-distributed amplitude and uniformly-distributed phase. Variation of the channel amplitude becomes more significant as the Doppler frequency increases, demonstrating the fast fading characteristics.

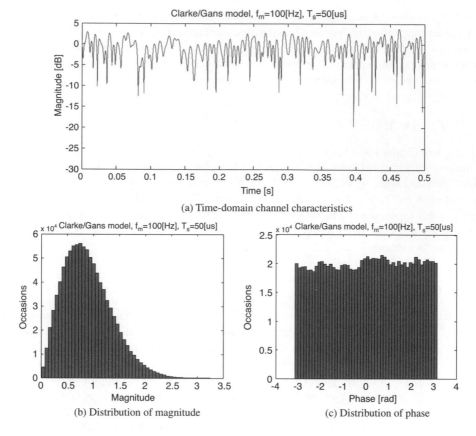

(a) Time-domain channel characteristics

(b) Distribution of magnitude

(c) Distribution of phase

Figure 2.7   Generation of a time-varying channel with Clarke/Gans model.

Figure 2.7 has been obtained by running Program 2.11 ("plot_FWGN.m"), which calls Program 2.12 ("FWGN_model") and Program 2.13 ("Doppler_spectrum") to generate the FWGN model and Doppler spectrum, respectively.

MATLAB® Programs: FWGN Channel Model

Program 2.11 "plot_FWGN.m" to plot an FWGN model

```
% plot_FWGN.m
clear, clf
fm=100; scale=1e-6; % Maximum Doppler frequency and mu
ts_mu=50; ts=ts_mu*scale; fs=1/ts; % Sampling time/frequency
Nd=1e6;   % Number of samples
% obtain the complex fading channel
[h,Nfft,Nifft,doppler_coeff] = FWGN_model(fm,fs,Nd);
```

```
subplot(211),plot([1:Nd]*ts,10*log10(abs(h)))
str=sprintf('Clarke/Gan Model, f_m=%d[Hz], T_s=%d[us]',fm,ts_mu);
title(str), axis([0 0.5 -30 5])
subplot(223), hist(abs(h),50), subplot(224), hist(angle(h),50)
```

Program 2.12 "FWGN_model"

```
function [h,Nfft,Nifft,doppler_coeff]=FWGN_model(fm,fs,N)
% FWGN (Clarke/Gan) Model
% Input:   fm= Maximum Doppler frquency
%             fs= Sampling frequency,    N = Number of samples
% Output:  h = Complex fading channel
Nfft = 2^max(3,nextpow2(2*fm/fs*N));    % Nfft=2^n
Nifft = ceil(Nfft*fs/(2*fm));
% Generate the independent complex Gaussian random process
GI = randn(1,Nfft);         GQ = randn(1,Nfft);
% Take FFT of real signal in order to make hermitian symmetric
CGI = fft(GI);              CGQ = fft(GQ);
% Nfft sample Doppler spectrum generation
doppler_coeff = Doppler_spectrum(fm,Nfft);
% Do the filtering of the Gaussian random variables here
f_CGI = CGI.*sqrt(doppler_coeff); f_CGQ = CGQ.*sqrt(doppler_coeff);
% Adjust sample size to take IFFT by (Nifft-Nfft) sample zero-padding
Filtered_CGI=[f_CGI(1:Nfft/2) zeros(1,Nifft-Nfft) f_CGI(Nfft/2+1:Nfft)];
Filtered_CGQ=[f_CGQ(1:Nfft/2) zeros(1,Nifft-Nfft) f_CGQ(Nfft/2+1:Nfft)];
hI = ifft(Filtered_CGI);   hQ= ifft(Filtered_CGQ);
% Take the magnitude squared of the I and Q components and add them
rayEnvelope = sqrt(abs(hI).^2 + abs(hQ).^2);
% Compute the root mean squared value and normalize the envelope
rayRMS = sqrt(mean(rayEnvelope(1:N).*rayEnvelope(1:N)));
h = complex(real(hI(1:N)),-real(hQ(1:N)))/rayRMS;
```

Program 2.13 "Doppler_spectrum"

```
function y=Doppler_spectrum(fd,Nfft)
% fd = Maximum Doppler frequency
% Nfft= Number of frequency domain points
df = 2*fd/Nfft; % frequency spacing
% DC component first
f(1) = 0; y(1) = 1.5/(pi*fd);
% The other components for one side of the spectrum
for i = 2:Nfft/2,
    f(i)=(i-1)*df; % frequency indices for polynomial fitting
    y([i Nfft-i+2]) = 1.5/(pi*fd*sqrt(1-(f(i)/fd)^2));
end
% Nyquist frequency applied polynomial fitting using last 3 samples
nFitPoints=3 ; kk=[Nfft/2-nFitPoints:Nfft/2];
polyFreq = polyfit(f(kk),y(kk),nFitPoints);
y((Nfft/2)+1) = polyval(polyFreq,f(Nfft/2)+df);
```

2.2.1.2 Modified Frequency-Domain FWGN Model

Since the Clarke/Gans model employs two IFFT blocks, it has disadvantages of computational complexity. Among many other variants of the Clarke/Gans model, we here describe one used in I-METRA model.

Figure 2.8 describes a process of generating the Doppler spectrum. Let f_m denote the Doppler frequency. Since the spectrum repeats with respect to Nyquist frequency, $2f_m$, for normal sampling, it must be folded for an IFFT function to deal with the positive frequency components only (corresponding to "fft" function of MATLAB®) as shown in Figure 2.8(a). When oversampled by a factor of N_{OS}, the bandwidth of Doppler spectrum becomes $B_D = 2N_{OS}f_m$ as shown in Figure 2.8(b). Its inverse $\Delta t = 1/B_D$ is the sample spacing in the time domain, which corresponds to the coherence time of the fading channel. Dividing the Doppler bandwidth into N_{Fading} subbands, each subband of the length $\Delta f_m = B_D/N_{Fading}$ leads to the overall length of the fading channel, given by $T_{Fading} = 1/\Delta f_m = N_{Fading}/B_D$. Meanwhile, Figure 2.8(c) shows a discrete-frequency Doppler spectrum and its equivalent discrete-time fading channel. For the IFFT size of N_{Fading}, the frequency spacing of Doppler spectrum is

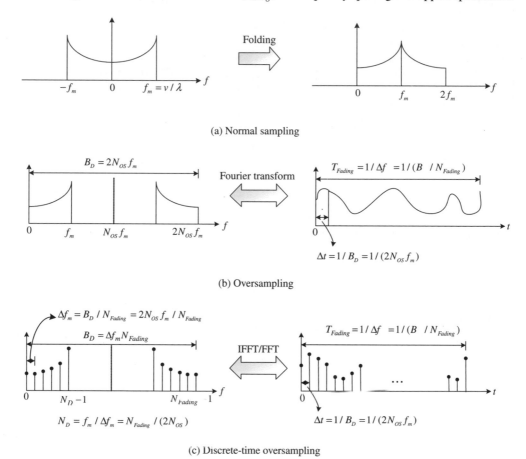

(a) Normal sampling

(b) Oversampling

(c) Discrete-time oversampling

Figure 2.8   Generation of Doppler spectrum.

given by $\Delta f_m = 2 N_{OS} f_m / N_{Fading}$. The number of discrete-time frequency samples in the overall Doppler spectrum is given by $N_D = f_m / \Delta f_m = N_{Fading} / (2 N_{OS})$. This particular method allows for generating the fading signal with a given duration of T_{Fading} without taking the maximum Doppler frequency into account. Furthermore, it is advantageous for simulation since the time-domain signal can be obtained by interpolation with the maximum Doppler frequency f_m. More specifically, the number of discrete-frequency samples within the Doppler bandwidth, N_D, is determined by the number of samples for the fading channel, N_{Fading}, and the oversampling factor N_{OS}. Once the fading channel signal of N_{Fading} samples are generated without taking the maximum Doppler frequency into account, the actual fading channel signal for simulation can be generated by changing the sample spacing Δt according to the maximum Doppler frequency f_m. In fact, $N_D = f_m / \Delta f_m = N_{Fading} / (2 \cdot N_{OS})$ plays a role of the maximum Doppler frequency, while f_m is used only for determining the sample spacing $\Delta t = 1 / (2 \cdot N_{OS} \cdot f_m)$.

Here, a magnitude of channel response can be computed from the Doppler spectrum with an arbitrary phase. A channel response of each path in the time domain is given as

$$h[n] = \sum_{k=-N_{Fading}/2}^{N_{Fading}/2-1} \sqrt{S[k]} e^{j\theta_k} e^{j2\pi nk/N_{Fading}} \tag{2.20}$$

where $S(k)$ is the Doppler spectrum at a discrete frequency $k = f / \Delta f_m$, $n = t / \Delta t$ is the discrete time index, and θ_k is a uniform random variable over $[0, 2\pi)$. By imposing a uniform phase onto Doppler filter, it obtains a more flexible fading process in the time domain than the Clarke/Gans model which is injected with the Gaussian random variables. If a complex Gaussian random variable is imposed in Equation (2.20) rather than the uniform phase, low-pass filtering is required for generating a flexible fading process.

According to the reports on channel measurements, the different channel environments are subject to the different Doppler spectrum and furthermore, the maximum Doppler frequency as well as Doppler spectrum may vary for each path. Figures 2.9(a) and (b) show two different paths, with the maximum Doppler frequency of 100 Hz and 10 Hz, respectively, which are generated by the modified frequency-domain and time-domain FWGN models with the Classical Doppler spectrum, respectively. Figure 2.9(a) has been obtained by running

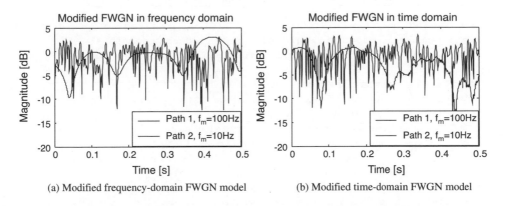

(a) Modified frequency-domain FWGN model   (b) Modified time-domain FWGN model

Figure 2.9   Generation of two different paths with FWGN channel model.

Program 2.14 ("plot_modified_FWGN.m") with $N_{Fading} = 1024$ and $N_{OS} = 8$ where Program 2.15 ("FWGN_ff") and Program 2.17 ("FWGN_tf") are called to generate the modified frequency-domain and time-domain FWGN models. Note that Program 2.15 ("FWGN_ff") deals with the different Doppler frequency in the different path and Program 2.16 ("Doppler_PSD_function") provides the different types of Doppler spectrum to choose, including Classical model and Laplacian model.

MATLAB® Programs: Modified Frequency-Domain FWGN Channel Model

Program 2.14 "plot_modified_FWGN.m" to plot modified FWGN channel models

```
% plot_modified_FWGN.m
clear, clf
Nfading=1024; % IFFT size for Npath x Nfading fading matrix
Nfosf=8;      % Fading oversampling factor
Npath=2;      % Number of paths
N=10000;
FadingType='class';
fm= [100 10]; % Doppler frequency
[FadingMatrix,tf]= FWGN_ff(Npath,fm,Nfading,Nfosf,FadingType);
subplot(211), plot([1:Nfading]*tf,10*log10(abs(FadingMatrix(1,:))),'k:')
hold on, plot([1:Nfading]*tf,10*log10(abs(FadingMatrix(2,:))),'k-')
title(['Modified FWGN in Frequency Domain, Nfading=', num2str(Nfading),',',
Nfosf=',num2str(Nfosf)]);
xlabel('time[s]'), ylabel('Magnitude[dB]'),
legend('Path 1, f_m=100Hz','Path 2, f_m=10Hz'), axis([0 0.5 -20 10])
[FadingMatrix,tf]= FWGN_tf(Npath,fm,N,Nfading,Nfosf,FadingType);
subplot(212), plot([1:N]*tf,10*log10(abs(FadingMatrix(1,:))),'k:')
hold on, plot([1:N]*tf,10*log10(abs(FadingMatrix(2,:))),'k-')
title(['Modified FWGN in Time Domain, Nfading=', num2str(Nfading),',Nfosf=',
 num2str(Nfosf),',T_s=',num2str(tf),'s']);
xlabel('time[s]'), ylabel('Magnitude[dB]')
legend('Path 1, f_m=100Hz','Path 2, f_m=10Hz'), axis([0 0.5 -20 10])
```

Program 2.15 "FWGN_ff" for a modfied frequency-domain FWGN channel

```
function [FadTime,tf]= FWGN_ff(Np,fm_Hz,Nfading,Nfosf,FadingType,varargin)
% Fadng generation based on FWGN method in the frequency domain
% FadTime= FWGN_ff(Np,fm_Hz,Nfading,Nfosf,FadingType,sigma,phi)
% Inputs:
%   Np         : number of multipaths
%   fm_Hz      : a vector of max. Doppler frequency of each path[Hz]
%   Nfading    : Doppler filter size (IFFT size)
%   Nfosf      : oversampling factor of Doppler bandwith
%   FadingType : Doppler type, 'laplacian'/'class'/'flat'
%   sigma      : angle spread of UE in case of 'laplacian' Doppler type
%   phi        : DoM-AoA in case of 'laplacian' Doppler type
% Output:
%   FadTime    : Np x Nfading, fading time matrix
fmax_Hz= max(fm_Hz);
```

```
% Doppler frequency spacing respect to maximal Doppler frequency
dfmax= 2*Nfosf*fmax_Hz/Nfading;
% Obtain a funtion corresponding to Doppler spectrum of "FadingType"
ftn_PSD= Doppler_PSD_function(FadingType);
err_msg= 'The difference between max/min Doppler frequencies is too large.\n
increase the IFFT size';
if isscalar(fm_Hz), fm_Hz= fm_Hz*ones(1,Np); end
if strcmp(lower(FadingType(1:3)),'lap') % Laplacian constrained PAS
  for i=1:Np
    Nd=floor(fm_Hz(i)/dfmax)-1; % Nd=fm_Hz/dfmax=Nfading/(2*Nfosf)
    if Nd<1, error(err_msg); end
    tmp= ftn_PSD([-Nd:Nd]/Nd,varargin{1}(i),varargin{2}(i));
    tmpz= zeros(1,Nfading-2*Nd+1);
    FadFreq(i,:)= [tmp(Nd+1:end-1) tmpz tmp(2:Nd)];
  end
else % symmetric Doppler spectrum
  for i=1:Np
    Nd= floor(fm_Hz(i)/dfmax)-1;
    if Nd<1, error(err_msg); end
    tmp= ftn_PSD([0:Nd]/Nd); tmpz= zeros(1,Nfading-2*Nd+3);
    FadFreq(i,:)= [tmp(1:Nd-1) tmpz fliplr(tmp(2:Nd-1))];
  end
end
% Add a random phase to the Doppler spectrum
FadFreq = sqrt(FadFreq).*exp(2*pi*j*rand(Np,Nfading));
FadTime = ifft(FadFreq,Nfading,2);
FadTime=FadTime./sqrt(mean(abs(FadTime).^2,2)*ones(1,size(FadTime,2)));
 % Normalize to 1
tf=1/(2*fmax_Hz*Nfosf); %fading sample time=1/(Doppler BW*Nfosf)
```

Program 2.16 "Doppler_PSD_function" for Doppler spectrum function

```
function ftn=Doppler_PSD_function(type)
% Doppler spectrum funtion for type =
%  'flat'      : S(f)=1, |f0(=f/fm)|
%  'class'     : S(f)=A/(sqrt(1-f0.^2)), |f0|<1 (A: a real number)
%  'laplacian':
%      S(f)=1./sqrt(1-f0.^2).*(exp(-sqrt(2)/sigma*abs(acos(f0)-phi))
%               +exp(-sqrt(2)/sigma*abs(acos(f0)-phi)))
%            with sigma(angle spread of UE) and phi(=DoM-AoA)
%            in the case of 'laplacian' Doppler type
%  'sui'       : S(f)=0.785*f0.^4-1.72*f0.^2+1, |f0|
%  '3gpprice'  : S(f)=0.41./(2*pi*fm*sqrt(1+1e-9-(f./fm).^2)) +
%                    0.91*delta_ftn(f,0.7*fm), |f|<fm
%  'dr', S(f)=inline('(1./sqrt(2*pi*Dsp/2))*exp(-((f-Dsh).^2)/Dsp)',
%              'f','Dsp','Dsh');
%  f0 : Normalized Doppler frequency defined as f0=f/fm
%  fm : Maximum Doppler frequency
```

```
switch lower(type(1:2))
  case 'fl', ftn=inline('ones(1,length(f0))');
  case 'cl', ftn=inline('1./sqrt(1+1e-9-f0.^2)');
  case 'la', ftn=inline(' (exp(-sqrt(2)/sigma*abs(acos(f0)-phi))+exp(-sqrt(2)/
    sigma*abs(acos(f0)+phi)))./sqrt(1+1e-9-f0.^2)','f0','sigma','phi');
  case 'su', ftn=inline('0.785*f0.^4-1.72*f0.^2+1.');
  case '3g', ftn=inline('0.41./(2*pi*fm*sqrt(1+1e-9-(f./fm).^2))
           +0.91*delta_ftn(f,0.7*fm)','f','fm');
  case 'dr', ftn=inline(' (1./sqrt(2*pi*Dsp/2))*exp(-((f-Dsh).^2)/Dsp)',
             'f','Dsp','Dsh');
  otherwise, error('Unknown Doppler type in Doppler_PSD_function()');
end
```

2.2.1.3 Time-Domain FWGN Model

As shown in Figure 2.10, we can generate the fading channel by filtering the complex Gaussian random process with the time-domain filter whose frequency response corresponds to the Doppler spectrum. Due to the various advantages of the time-domain FWGN model, it is frequently employed in the commercial channel simulators.

Figure 2.10   Time-domain FWGN model: an overview.

As opposed to the frequency-domain FWGN model in which the duration of fading channel is determined by the IFFT size of N_{Fading} and the frequency-domain oversampling factor of N_{OS}, it is determined by the length of the complex Gaussian random signal in the time-domain FWGN model. Since the simulation interval can be extended simply by increasing the number of complex Gaussian random samples subject to the Doppler filter, it is flexible for simulation. In case that the Doppler filter is implemented by an FIR filter, however, its computational complexity may grow exponentially as the number of taps increases. In order to reduce the number of taps, IIR filter can be used as in the TGn channel model of IEEE 802.11n Task Group. Since the IIR filter does not warrant stability, the FIR filter is still frequently used.

Other than the classical model described in Section 1.3, the FWGN channel model allows for employing various other types of Doppler spectrum, including the flat Doppler spectrum and Laplacian Doppler spectrum. The flat Doppler spectrum has the constant power spectral density function, i.e.,

$$S(f) \propto 1, \quad |f| \leq f_m \tag{2.21}$$

Meanwhile, the Laplacian Doppler spectrum is defined by the following power spectrum density function:

$$S(f) \propto \frac{1}{\sqrt{1-(f/f_m)^2}} \cdot \left\{ \exp\left(\frac{\sqrt{2}}{\sigma}|\cos^{-1}(f/f_m)-\phi| \right) + \exp\left(-\frac{\sqrt{2}}{\sigma}|\cos^{-1}(f/f_m)+\phi| \right) \right\}, |f| \leq f_m$$

$$\tag{2.22}$$

where σ is the standard deviation of PAS, and ϕ is the difference between direction of movement (DoM) and direction of arrival (DoA). Unlike the flat or classical Doppler spectrum model, the mobile direction can be accounted into the Laplacian Doppler spectrum.

Note that Program 2.17 ("FWGN_tf"), which calls Program 2.18 ("gen_filter") to generate Doppler filter coefficients, can be used to make the (modified) time-domain FWGN channel model as depicted in Figure 2.9(b).

MATLAB® Programs: Time-Domain FWGN Channel Model

Program 2.17 "FWGN_tf" for time-domain FWGN channel

```
function [FadMtx,tf]=FWGN_tf(Np,fm_Hz,N,M,Nfosf,type,varargin)
% fading generation using FWGN with fitering in the time domain
% Inputs:
%   Np     : Number of multipaths
%   fm_Hz  : A vector of maximum Doppler frequency of each path[Hz]
%   N      : Number of independent random realizations
%   M      : Length of Doppler filter, i.e, size of IFFT
%   Nfosf  : Fading oversampling factor
%   type   : Doppler spectrum type
%       'flat'=flat, 'class'=calssical, 'sui'=spectrum of SUI channel
%       '3gpprice'=rice spectrum of 3GPP
%   Outputs:
%       FadMtx : Np x N fading matrix
if isscalar(fm_Hz), fm_Hz= fm_Hz*ones(1,Np); end
fmax= max(fm_Hz);
path_wgn= sqrt(1/2)*complex(randn(Np,N),randn(Np,N));
for p=1:Np
  filt=gen_filter(fm_Hz(p),fmax,M,Nfosf,type,varargin{:});
  path(p,:)=fftfilt(filt,[path_wgn(p,:) zeros(1,M)]); %filtering WGN
end
FadMtx= path(:,M/2+1:end-M/2);
tf=1/(2*fmax*Nfosf); % fading sample time=1/(Max. Doppler BW*Nfosf)
FadMtx= FadMtx./sqrt(mean(abs(FadMtx).^2,2)*ones(1,size(FadMtx,2)));
```

Program 2.18 "gen_filter" for Doppler filter coefficients

```
function filt=gen_filter(fm_Hz,fmax_Hz,Nfading,Nfosf,type,varargin)
% FIR filter weights generation
%   Inputs:
%     Nfading : Doppler filter size, i.e, IFFT size
%   Outputs:
%     filt    : Filter coefficients
% Doppler BW= 2*fm*Nfosf ==> 2*fmax_Hz*Nfosf
dfmax= 2*Nfosf*fmax_Hz/Nfading;   % Doppler frequency spacing
% respect to maximal Doppler frequency
Nd= floor(fm_Hz/dfmax)-1;
if Nd<1, error('The difference between max/min Doppler frequencies is too
  large.\n increase the IFFT size?');
end
```

```
ftn_PSD=Doppler_PSD_function(type); % Corresponding Doppler function
switch lower(type(1:2))
  case '3g', PSD=ftn_PSD([-Nd:Nd],Nd);
       filt=[PSD(Nd+1:end-1) zeros(1,Nfading-2*Nd+1) PSD(2:Nd)];
  case 'la', PSD=ftn_PSD([-Nd:Nd]/Nd,varargin{:});
       filt=[PSD(Nd+1:end-1) zeros(1,Nfading-2*Nd+1) PSD(2:Nd)];
  otherwise, PSD=ftn_PSD([0:Nd]/Nd);
       filt=[PSD(1:end-1) zeros(1,Nfading-2*Nd+3) PSD(end-1:-1:2)];
       % constructs a symmetric Doppler spectrum
end
filt=real(ifftshift(ifft(sqrt(filt))));
filt=filt/sqrt(sum(filt.^2));
```

2.2.2 Jakes Model

A Rayleigh fading channel subject to a given Doppler spectrum can be generated by synthesizing the complex sinusoids. The number of sinusoids to add must be large enough to approximate the Rayleigh amplitude. Furthermore, each of the sinusoidal generators must be weighted to generate the desired Doppler spectrum. This is the Jakes model that has been originally developed for the hardware simulation, but now frequently used in the software simulation [27].

Figure 2.11 illustrates how the Jakes model is implemented. It has been assumed that all rays of the scattered components arriving in the uniform directions are approximated by N plane waves. Define $N_0 = (N/2-1)/2$ where $N/2$ is limited to an odd number. Let θ_n denote an angle of arrival for the nth plane wave, which is modeled as $\theta_n = 2\pi n/N, n = 1, 2, \cdots, N_0$. As shown in Figure 2.11, a sum of N_0 complex oscillator outputs with the frequencies of $w_n = w_d \cos \theta_n$, $n = 1, 2, \cdots, N_0$, each corresponding to different Doppler shifts, is added to the output of a complex oscillator with a frequency of $w_d = 2\pi f_m$. The real and imaginary parts, $h_I(t)$ and

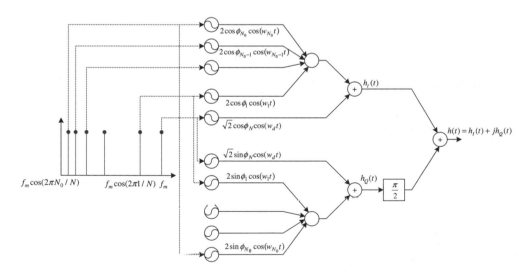

Figure 2.11 Implementation of the Jakes model.

$h_Q(t)$, in the total sum of the complex oscillators can be represented respectively as

$$h_I(t) = 2 \sum_{n=1}^{N_0} \left(\cos \phi_n \cos w_n t \right) + \sqrt{2} \cos \phi_N \cos w_d t \tag{2.23a}$$

and

$$h_Q(t) = 2 \sum_{n=1}^{N_0} \left(\sin \phi_n \cos w_n t \right) + \sqrt{2} \sin \phi_N \cos w_d t \tag{2.23b}$$

where ϕ_n and ϕ_N are the initial phases of the n-th Doppler-shifted sinusoid and the maximum Doppler frequency f_m, respectively. The initial phase must be set to yield a uniform distribution for the phase of fading channel [26]. For example, the initial phase can be set to

$$\begin{aligned} \phi_N &= 0 \\ \phi_n &= \pi n / (N_0 + 1), \qquad n = 1, 2, \cdots, N_0 \end{aligned} \tag{2.24}$$

The complex output of the Jakes model can be represented as

$$h(t) = \frac{E_0}{\sqrt{2N_0 + 1}} \left\{ h_I(t) + j h_Q(t) \right\} \tag{2.25}$$

where E_0 is the average of the fading channel.

Note that the frequency of the Doppler-shifted sinusoid $\{w_n\}_{n=1}^{N_0}$ can be expressed as

$$w_n = w_d \cos \theta_n = 2\pi f_m \cos(2\pi n / N), \qquad n = 1, 2, \cdots, N_0 \tag{2.26}$$

The number of the Doppler-shifted sinusoids, N_0, must be large enough to approximate the amplitude of the fading channel with a Rayleigh distribution. It has been known that $N_0 = 8$ is large enough. Note that the following properties can be shown for Equation (2.23):

$$E\left\{ \left(\frac{E_0 h_I(t)}{\sqrt{2N_0 + 1}} \right)^2 \right\} = E\left\{ \left(\frac{E_0 h_Q(t)}{\sqrt{2N_0 + 1}} \right)^2 \right\} = \frac{E_0^2}{2} \tag{2.27}$$

$$E\left\{ h^2(t) \right\} = E_0^2 \tag{2.28}$$

$$E\left\{ h(t) \right\} = E_0 \tag{2.29}$$

$$E\left\{ h_I(t) h_Q(t) \right\} = 0 \tag{2.30}$$

Equation (2.28) and Equation (2.29) confirm that the Jake model generates the fading signal with the average amplitude of E_0 and average energy of E_0^2. Furthermore, Equation (2.27) and Equation (2.30) show that the real and imaginary parts of the channel are statistically independent with the average power of $E_0^2/2$.

Figures 2.12(a)–(e) show the time-domain characteristics, magnitude/phase distributions, autocorrelation, and Doppler spectrum of the Jakes fading channel. It has been obtained by running Program 2.19 ("plot_Jakes_model.m"), which calls Program 2.20 ("Jakes_Flat") to generate the Jake channel model.

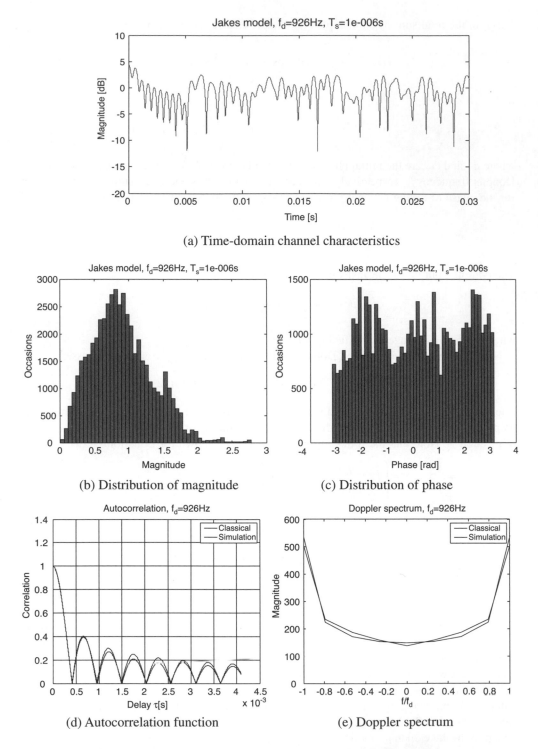

(a) Time-domain channel characteristics

(b) Distribution of magnitude

(c) Distribution of phase

(d) Autocorrelation function

(e) Doppler spectrum

Figure 2.12   Generation of a time-varying channel with Jakes model.

MATLAB® Programs: Jakes Channel Model

Program 2.19 "plot_Jakes_model.m" to plot a Jakes channel model

```
% plot_Jakes_model.m
clear all, close all
% Parameters
fd=926; Ts=1e-6;   % Doppler frequency and Sampling time
M=2^12;   t=[0:M-1]*Ts; f=[-M/2:M/2-1]/(M*Ts*fd);
Ns=50000;   t_state=0;
% Channel generation
[h,t_state]=Jakes_Flat(fd,Ts,Ns,t_state,1,0);
subplot(311), plot([1:Ns]*Ts,10*log10(abs(h))),
title(['Jakes Model, f_d=',num2str(fd),'Hz, T_s=',num2str(Ts),'s']);
axis([0 0.05 -20 10]), xlabel('time[s]'), ylabel('Magnitude[dB]')
subplot(323), hist(abs(h),50);
title(['Jakes Model, f_d=',num2str(fd),'Hz, T_s=',num2str(Ts),'s']);
xlabel('Magnitude'), ylabel('Occasions')
subplot(324), hist(angle(h),50);
title(['Jakes Model, f_d=',num2str(fd),'Hz, T_s=',num2str(Ts),'s']);
xlabel('Phase[rad]'), ylabel('Occasions')
% Autocorrelation of channel
temp=zeros(2,Ns);
for i=1:Ns
  j=i:Ns;
  temp1(1:2,j-i+1)=temp(1:2,j-i+1)+[h(i)'*h(j); ones(1,Ns-i+1)];
end
for k=1:M
    Simulated_corr(k)=real(temp(1,k))/temp(2,k);
end
Classical_corr=besselj(0,2*pi*fd*t);
% Fourier transform of autocorrelation
Classical_Y=fftshift(fft(Classical_corr));
Simulated_Y=fftshift(fft(Simulated_corr));
subplot(325), plot(t,abs(Classical_corr),'b:',t,abs(Simulated_corr),'r:')
title(['Autocorrelation, f_d=',num2str(fd),'Hz'])
grid on, xlabel('delay \tau [s]'), ylabel('Correlation')
legend('Classical','Simulation')
subplot(326), plot(f,abs(Classical_Y),'b:', f,abs(Simulated_Y),'r:')
title(['Doppler Spectrum, f_d=',num2str(fd),'Hz'])
axis([-1 1 0 600]), xlabel('f/f_d'), ylabel('Magnitude')
legend('Classical','Simulation')
```

Program 2.20 "Jakes_Flat" for fading signal with Jakes model

```
function [h,tf]=Jakes_Flat(fd,Ts,Ns,t0,E0,phi_N)
% Inputs:
%   fd,Ts,Ns   : Doppler frequency, sampling time, number of samples
%   t0, E0     : initial time, channel power
%   phi_N      : inital phase of the maximum Doppler frequency sinusoid
```

```
% Outputs:
%   h, tf       : complex fading vector, current time
if nargin<6,    phi_N=0;    end
if nargin<5,    E0=1;       end
if nargin<4,    t0=0;       end
N0 = 8;                     % As suggested by Jakes
N = 4*N0+2;                 % an accurate approximation
wd = 2*pi*fd;               % Maximum Doppler frequency[rad]
t = t0+[0:Ns-1]*Ts; tf = t(end)+Ts; % Time vector and Final time
coswt=[sqrt(2)*cos(wd*t); 2*cos(wd*cos(2*pi/N*[1:N0]')*t)]; % Eq.(2.26)
h = E0/sqrt(2*N0+1)*exp(j*[phi_N pi/(N0+1)*[1:N0]])*coswt; % Eq.(2.23)
```

2.2.3 Ray-Based Channel Model

A ray-based model is frequently used in modeling a MIMO channel, since it can take a spatio-temporal correlation into account [28]. However, it can be also used for a SISO channel. Its fundamental principle is described in this subsection while its extension to the MIMO channel is considered in Chapter 3.

As in the Jake's model, the ray-based model is given by a sum of the arriving plane waves. As shown in Figure 2.13, it can model the plane waves incoming from an arbitrary direction around the mobile terminal, which can deal with the various scattering environments. In general, its power azimuth spectrum (PAS) is not uniform. Unlike the Jakes model, therefore, its Doppler spectrum is not given in the U-shape, but in various forms, depending on the scattering environments.

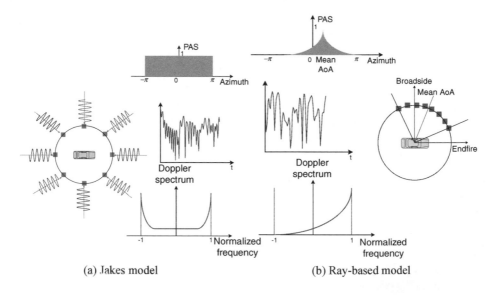

(a) Jakes model   (b) Ray-based model

Figure 2.13   Difference between Jakes model and ray-based model.

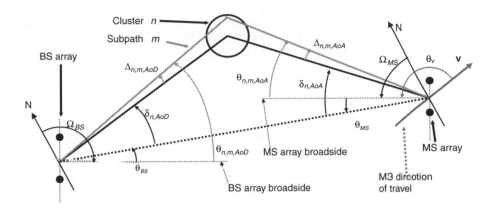

Figure 2.14 Ray-based MIMO channel model. (Reproduced with permission from 3GPP TR 25.996 v7.0.0, "Spatial channel model for multiple input multiple output (MIMO) simulations (release 7)," Technical Specification Group Radio Access Network, 2007. © 2007. 3GPP™ TSs and TRs are the property of ARIB, ATIS, CCSA, ETSI, TTA and TTC who jointly own the copyright in them.)

Figure 2.14 shows one of the Ray-based channel models, known as a spatial channel model (SCM) for a MIMO channel in 3GPP [28]. Let $h_{u,s,n}(t)$ denote a channel impulse response of the nth path (cluster) between the sth transmit antenna and uth receive antenna, which can be represented as [28]

$$h_{u,s,n}(t) = \sqrt{\frac{P_n \sigma_{SF}}{M}} \sum_{m=1}^{M} \left(\begin{array}{l} \sqrt{G_{BS}(\theta_{n,m,AoD})} \exp\left(j\left[kd_s \sin(\theta_{n,m,AoD}) + \Phi_{n,m}\right]\right) \times \\ \sqrt{G_{MS}(\theta_{n,m,AoA})} \exp\left(jkd_u \sin\theta_{n,m,AoA}\right) \exp\left(jk\|\mathbf{v}\| \cos(\theta_{n,m,AoA} - \theta_v) t\right) \end{array} \right)$$

(2.31)

where

P_n: power of the nth path
σ_{SF}: standard deviation of log-normal shadowing
M: the number of subrays per path
$\theta_{n,m,AoD}$: angle of departure of the mth subray for the nth path
$\theta_{n,m,AoA}$: angle of arrival (AoA) of the mth subray for the nth path
$\Phi_{n,m}$: random phase of the mth subray for the nth path
$G_{BS}(\theta_{n,m,AoD})$: BS antenna gain of each array element
$G_{MS}(\theta_{n,m,AoA})$: MS antenna gain of each array element
k: wave number $2\pi/\lambda$ where λ is the carrier wavelength
d_s: distance between antenna element s and reference antenna element ($s = 1$) in BS
d_u: distance between antenna element u and reference antenna element ($u = 1$) in MS
$\|\mathbf{v}\|$: magnitude of MS velocity vector
θ_v: angle of MS velocity vector

Since the SISO channel does not have any spatial correlation, only the Doppler spectrum in Equation (2.31) is meaningful [29]. While the SCM for MIMO channel in Equation (2.31) is detailed in Chapter 3, only the Ray-based SISO channel model is described in this section. Towards the end of this process, we drop all the parameters associated with the spatial correlation and ignore the effect of the log-normal shadowing (i.e., $\sigma_{SF} = 1$), which leads to the following impulse response for the SISO channel:

$$h_n(t) = \sqrt{\frac{P_n}{M}} \sum_{m=1}^{M} \left(\exp(j\Phi_{n,m}) \times \exp\left(j\frac{2\pi}{\lambda} ||\mathbf{v}|| \cos\left(\theta_{n,m,AoA} - \theta_v\right)t \right) \right) \quad (2.32)$$

In the ray-based model, any channel with the given PAS can be modeled by allocating the angle and power to each subray in accordance with the PAS. Two different methods of angle and power allocation to each subray have been considered in SCM Ad-Hoc Group (AHG) of 3GPP: uniform power subray method and discrete Laplacian method. In fact, Equation (2.31) and Equation (2.32) are the channel models that employ the uniform subray method.

2.2.3.1 Uniform Power Subray Method

This method allocates the same power to each subray while arranging the subray angles in a non-uniform manner [30]. Equal power allocation usually simplifies the modeling process. Given M subrays, each of their angles is determined such that the area of each section subject to each subray is equally divided under PAS. Figure 2.15 illustrates the case for two subrays (i.e., $M = 2$), in which three different sections divided by the offset angles have the same area under PAS (i.e., equal power for each section).

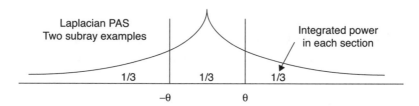

Figure 2.15   Allocation of offset angle in the uniform power subray method: example ($M = 2$).

Consider the Laplacian PAS $P(\theta, \sigma)$ with the average AoA of $0°$ and the RMS angular spread (AS) of σ. Let θ_1 and θ_2 denote the offset angles of adjacent subrays. In order to keep the areas of sections divided by two subsequent offset angles equal to each other, the area of section defined by θ_1 and θ_2 under the given PAS is

$$\int_{\theta_1}^{\theta_2} P(\theta, \sigma)d\theta = \int_{\theta_1}^{\theta_2} \frac{1}{\sqrt{2}\sigma} e^{\frac{-\sqrt{2}|\theta|}{\sigma}} d\theta$$
$$= -\frac{1}{2}\left(e^{\frac{-\sqrt{2}|\theta_2|}{\sigma}} - e^{\frac{-\sqrt{2}|\theta_1|}{\sigma}} \right)$$
$$= \frac{1}{a(M+1)} \quad (2.33)$$

where a is a normalization factor. If M is odd, $a = 1$ and the smallest absolute value of the angle is $0°$ in Equation (2.33). Otherwise, $a = 2$ for two symmetric angles with the smallest absolute value and $a = 1$ for all other subrays in Equation (2.33). In Figure 2.15 where $M = 2$, for example,

$$\int_0^\theta P(\theta, \sigma) d\theta = 1/6 \text{ such that } a = 2.$$

Given σ in Equation (2.33), the offset angle with the same power is found as

$$\theta_{m+1}[\text{deg}] = -\frac{\sigma}{\sqrt{2}}\left[\ln\left(e^{\frac{-\sqrt{2}\theta_m}{\sigma}} - \frac{2}{a(M+1)}\right)\right],$$

$$m = 0, 1, 2, \cdots, \lfloor M/2 \rfloor - 1 \text{ and } \theta_0 = 0° \qquad (2.34)$$

In the case that M is even, θ_1 is found with $\theta_0 = 0°$ as an initial value, while not allocating any subray to $\theta_0 = 0°$. All other subrays, $\theta_2, \theta_3, \cdots, \theta_{\lfloor M/2 \rfloor}$, are subsequently found by Equation (2.34). In case that M is odd, meanwhile, there exists a single subray with the smallest angle at $\theta_0 = 0°$ and then, all other subrays, $\theta_1, \theta_2, \cdots, \theta_{\lfloor M/2 \rfloor}$, are subsequently found by Equation (2.34). Note that Equation (2.34) is defined only for positive angles. The rest of the subrays are defined at the negative symmetric angle, that is, $-\theta_1, -\theta_2, \cdots, -\theta_{\lfloor M/2 \rfloor}$. Figure 2.16 shows the offset angles of subrays that are generated for $M = 20$ with AS of $\sigma = 1°$. Note that the offset angles vary with AS. For example, Table 2.2 shows the offset angles for the different AS in the SCM channel.

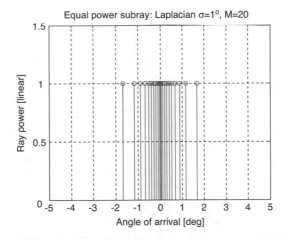

Figure 2.16 Allocation of offset angles in the uniform power subray method: example ($M = 20$ and $\sigma = 1°$).

Table 2.2 Offset angles for SCM [28].

Sub-path # (m)	Offset for a 2 deg AS at BS (Macrocell) (degrees)	Offset for a 5 deg AS at BS (Microcell) (degrees)	Offset for a 35 deg AS at MS (degrees)
1, 2	±0.0894	± 0.2236	±1.5649
3, 4	±0.2826	± 0.7064	±4.9447
5, 6	±0.4984	±1.2461	±8.7224
7, 8	±0.7431	±1.8578	±13.0045
9, 10	±1.0257	± 2.5642	±17.9492
11, 12	±1.3594	± 3.3986	± 23.7899
13, 14	±1.7688	± 4.4220	± 30.9538
15, 16	±2.2961	±5.7403	±40.1824
17, 18	±3.0389	±7.5974	±53.1816
19, 20	±4.3101	±10.7753	±75.4274

(Reproduced with permission from 3GPP TR 25.996 v7.0.0, "Spatial channel model for multiple input multiple output (MIMO) simulations (release 7)," Technical Specification Group Radio Access Network, 2007. © 2007. 3GPP™ TSs and TRs are the property of ARIB, ATIS, CCSA, ETSI, TTA and TTC who jointly own the copyright in them.)

2.2.3.2 Sampled Laplacian Method

As opposed to the uniform power subray method in which each of the subrays has the same power while their offset angles are non-uniformly distributed, the power of the subrays follows the Laplacian PAS with their offset angles asymmetrically centered around the average AoA, as shown in Figure 2.17. First, M reference offset angles are generated by allocating them uniformly with an equal distance of $\delta = 2\alpha/M$ over the range of $[-\alpha, \alpha]$ centered around the average AoA. Here, α varies with the given angular spread. For example, $\alpha = 10°, 15°$, and $179°$ for $\sigma = 2°, 5°$, and $35°$, respectively. Once the reference offset angles are generated, the actual offset angles are determined by adding an arbitrary random number selected over $[-0.5, 0.5]$ to the reference offset angles. A subray is allocated to each of the offset angles. In summary, the offset angle allocation in the sampled Laplacian method is given as

$$\theta_m = -\alpha + m \cdot \delta + \phi \quad \text{for} \quad m = 0, 1, \cdots, M-1 \qquad (2.35)$$

where ϕ is a uniform random variable over $[-0.5, 0.5]$.

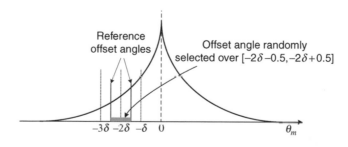

Figure 2.17 Discrete Laplacian method of the offset angle allocation: an illustration (AoA = 0°).

A uniform power subray-based channel model can be generated using Program 2.21 ("plot_ray_fading.m"), which calls Program 2.22 ("equalpower_subray"), Program 2.23 ("assign_offset"), Program 2.24 ("gen_phase"), and Program 2.25 ("ray_fading") to get an angle spacing for equal power Laplacian PAS in SCM (spatial channel model), to assign the AoA/AoD offset to mean AoA/AoD, to generate DoAs at BS/MS and a random phase at BS, and to combine the phases of M subrays to generate complex channel coefficients for each path, respectively.

MATLAB® Programs: Ray-Based Channel with Uniform Power Subray Method

Program 2.21 "plot_ray_fading.m" to plot a ray-based channel model

```
% plot_ray_fading.m
clear, clf
fc=9e8; fs=5e4; speed_kmh=120; Ts=1/fs;
v_ms= speed_kmh/3.6; wl_m= 3e8/fc;
% Channel parameters setting: SCM case 2
PDP_dB=[0. -1. -9. -10. -15. -20]; t_ns=[0 310 710 1090 1730 2510];
BS_theta_LOS_deg=0;    MS_theta_LOS_deg=0;
BS_AS_deg=2;    % Laplacian PAS
BS_AoD_deg=50*ones(size(PDP_dB));
MS_AS_deg=35;    % for Lapalcian PAS
DoT_deg=22.5; MS_AoA_deg=67.5*ones(size(PDP_dB));
% generates the phase of a subray
[BS_theta_deg,MS_theta_deg,BS_PHI_rad]=gen_phase(BS_theta_LOS_deg, ...
  BS_AS_deg,BS_AoD_deg,MS_theta_LOS_deg,MS_AS_deg,MS_AoA_deg);
PDP=dB2w(PDP_dB);
% generates the coefficients
t= [0:1e4-1]*Ts;
h= ray_fading(20,PDP,BS_PHI_rad,MS_theta_deg,v_ms,DoT_deg,wl_m,t);
plot(t,10*log10(abs(h(1,:))))
title(['Ray Channel Model, f_c=',num2str(fc),'Hz, T_s=',num2str(Ts),'s'])
xlabel('time[s]'), ylabel('Magnitude[dB]')
```

Program 2.22 "equalpower_subray": Look-up table for uniform power subray-based offset angles

```
function theta=equalpower_subray(AS_deg)
% Obtain angle spacing for equal power Laplacian PAS in SCM Text
%   Input:
%     AS_deg: angle spread with valid values of 2,5(for BS),35(for MS)
%   Output:
%     theta : offset angle with M=20 as listed in Table 2.2
if AS_deg==2
     theta=[0.0894 0.2826 0.4984 0.7431 1.0257 1.3594 1.7688 2.2961 3.0389
          4.3101];
  elseif AS_deg==5
     theta=[0.2236 0.7064 1.2461 1.8578 2.5642 3.3986 4.4220 5.7403 7.5974
          10.7753];
  elseif AS_deg==35
     theta=[1.5649 4.9447 8.7224 13.0045 17.9492 23.7899 30.9538 40.1824
          53.1816 75.4274];
```

```
  else error('Not support AS');
end
```

Program 2.23 "assign_offset" to allocate the offset angle for each subray

```
function theta_AoA_deg=assign_offset(AoA_deg,AS_deg)
%   Assigns AoA/AoD offset to mean AoA/AoD
%   Inputs:    AoA_deg = mean AoA/AoD, AS is = angle spread
%   Output:    theta_AoA_deg = AoA_deg+offset_deg
offset=equalpower_subray(AS_deg);
theta_AoA_deg=zeros(length(AoA_deg),length(offset));
for n=1:length(AoA_deg)
  for m=1:length(offset),
    theta_AoA_deg(n,[2*m-1:2*m])= AoA_deg(n)+[offset(m) -offset(m)];
  end
end
```

Program 2.24 "gen_phase" to generate the phase for each subray

```
function [BS_theta_deg,MS_theta_deg,BS_PHI_rad]=
        gen_phase(BS_theta_LOS_deg,BS_AS_deg,BS_AoD_deg,
                 MS_theta_LOS_deg,MS_AS_deg,MS_AoA_deg,M)
%   Generates phase at BS and MS
%    Inputs:
%      BS_theta_LOS_deg : AoD of LOS path in degree at BS
%      BS_AS_deg          : AS of BS in degree
%      BS_AoD_deg         : AoD of BS in degree
%      MS_theta_LOS_deg : AoA of LOS path in degree at MS
%      MS_AS_deg          : AS of MS in degree
%      MS_AoA_deg         : AoA of MS in degree
%      M                  : # of subrays
%    Outputs:
%      BS_theta_deg       : (Npath x M) DoA per path in degree at BS
%      MS_theta_deg       : (Npath x M) DoA per path in degree at MS
%      BS_PHI_rad         : (Npath x M) random phase in degree at BS
if nargin==6,   M=20;  end
BS_PHI_rad=2*pi*rand(length(BS_AoD_deg),M);   % uniform phase
BS_theta_deg=assign_offset(BS_theta_LOS_deg+BS_AoD_deg,BS_AS_deg);
MS_theta_deg=assign_offset(MS_theta_LOS_deg+MS_AoA_deg,MS_AS_deg);
% random pairing
index=randperm(M); MS1=size(MS_theta_deg,1);
for n=1:MS1, MS_theta_deg(n,:)= MS_theta_deg(n,index);   end
```

Program 2.25 "ray_fading" to generate the fading for each subray

```
function h=
  ray_fading(M,PDP,BS_PHI_rad,MS_theta_deg,v_ms,theta_v_deg,lam,t)
% Inputs:
%   M          : Number of subrays
```

```
%    PDP              : 1 x Npath Power at delay
%    BS_theta_deg     : (Npath x M) DoA per path in degree at BS
%    BS_PHI_rad       : (Npath x M) random phase in degree at BS
%    MS_theta_deg     : (Npath x M) DoA per path in degree at MS
%    v_ms             : Velocity in m/s
%    theta_v_deg      : DoT of mobile in degree
%    lam              : Wavelength in meter
%    t                : Current time
% Output:
%    h                : length(PDP) x length(t) Channel coefficient matrix
MS_theta_rad=deg2rad(MS_theta_deg); theta_v_rad=deg2rad(theta_v_deg);
% To generate channel coefficients using Eq.(2.32)
for n=1:length(PDP)
  tmph=exp(-j*BS_PHI_rad(n,:)')*ones(size(t)).*
    exp(-j*2*pi/lambda*v_ms*cos(MS_theta_rad(n,:)'-theta_v_rad)*t);
  h(n,:)=sqrt(PDP(n)/M)*sum(tmph);
end
```

Program 2.26 "dB2W" for dB-to-watt conversion

```
function y=dB2w(dB)
y=10.^(0.1*dB);
```

2.2.4 Frequency-Selective Fading Channel Model

As described in Section 1.2.1, a power delay profile (PDP) for the multi-path channel is required for modeling a frequency-selective fading channel. The PDP provides a distribution of the average power for the received signal over individual path, which is represented by the relative power of each path with respect to the power of the earliest path. Tables 2.3–2.6 present the PDP's for ITU-R and COST 207 models, which are most popular among many available PDP models [12, 13].

Table 2.3 Power delay profiles: ITU-R model [12].

Tab	Pedestrian A		Pedestrian B		Vehicular A		Vehicular B		Doppler spectrum
	Relative delay [ns]	Average power [dB]	Relative delay [ns]	Average power [dB]	Relative delay [ns]	Average power [dB]	Relative delay [ns]	Average power [dB]	
1	0	0.0	0.	0.0	0	0.0	0	−2.5	Classic
2	110	−9.7	200	−0.9	310	−1.0	300	0.0	Classic
3	190	−19.2	800	−4.9	710	−9.0	8900	−12.8	Classic
4	410	−22.8	1200	−8.0	1090	−10.0	12 900	−10.0	Classic
5			2300	−7.8	1730	−15.0	17 100	−25.2	Classic
6			3700	−23.9	2510	−20.0	20 000	−16.0	Classic

(Reproduced with permission from Recommendation ITU-R M.1225, "Guidelines for evaluation of radio transmission technologies for IMT-2000," International Telecommunication Union- Radiocommunication, 1997. © 1997 ITU.)

Table 2.4 Power delay profile: COST 207 model (reduced TU, reduced BU) [31].

Tab	Typical urban (TU)			Bad urban (BU)		
	Relative delay [us]	Average power	Doppler spectrum	Relative delay [us]	Average power	Doppler spectrum
1	0.0	0.189	Classic	0.0	0.164	Classic
2	0.2	0.379	Classic	0.3	0.293	Classic
3	0.5	0.239	Classic	1.0	0.147	GAUS1
4	1.6	0.095	GAUS1	1.6	0.094	GAUS1
5	2.3	0.061	GAUS2	5.0	0.185	GAUS2
6	5.0	0.037	GAUS2	6.6	0.117	GAUS2

(Reproduced with permission from M. Failli, *Digital land mobile radio communications - COST 207*, © ECSC-EEC-EAEC, Brussels-Luxembourg, 1989.)

Table 2.5 Power delay profile: COST 207 model (TU, BU) [31].

Tab	Typical urban (TU)			Bad urban (BU)		
	Relative delay [us]	Average power	Doppler spectrum	Relative delay [us]	Average power	Doppler spectrum
1	0.0	0.092	Classic	0.0	0.033	Classic
2	0.1	0.115	Classic	0.1	0.089	Classic
3	0.3	0.231	Classic	0.3	0.141	Classic
4	0.5	0.127	Classic	0.7	0.194	GAUS1
5	0.8	0.115	GAUS1	1.6	0.114	GAUS1
6	1.1	0.074	GAUS1	2.2	0.052	GAUS2
7	1.3	0.046	GAUS1	3.1	0.035	GAUS2
8	1.7	0.074	GAUS1	5.0	0.140	GAUS2
9	2.3	0.051	GAUS2	6.0	0.136	GAUS2
10	3.1	0.032	GAUS2	7.2	0.041	GAUS2
11	3.2	0.018	GAUS2	8.1	0.019	GAUS2
12	5.0	0.025	GAUS2	10.0	0.006	GAUS2

(Reproduced with permission from M. Failli, *Digital land mobile radio communications - COST 207*, © ECSC-EEC-EAEC, Brussels-Luxembourg, 1989.)

Table 2.6 Power delay profile: COST 207 model (RA).

Tab	Typical rural area (RA)		
	Relative delay [us]	Average power	Doppler spectrum
1	0.0	0.602	RICE
2	0.1	0.241	Classic
3	0.2	0.096	Classic
4	0.3	0.036	Classic
5	0.4	0.018	Classic
6	0.5	0.006	Classic

(Reproduced with permission from M. Failli, *Digital land mobile radio communications - COST 207*, © ECSC-EEC-EAEC, Brussels-Luxembourg, 1989.)

In the sequel, we describe how a frequency-selective fading channel can be implemented for the given PDP.

2.2.4.1 Tapped Delay Line (TDL) Model

A TDL model is commonly used for implementing the multi-path channel. It employs a multiple number of frequency-non-selective (flat) fading generators (e.g., using the FWGN model or Jakes model), which are independent of each other, each with the average power of one. As shown in Figure 2.18, the output of independent fading generator is multiplied by the tap power, so as to yield a coefficient of TDL model. In fact, it is implemented as a FIR filter with the following output:

$$y(n) = \sum_{d=0}^{N_D-1} h_d(n)x(n-d) \tag{2.36}$$

where N_D is the number of the taps in the FIR filter. However, implementation of the FIR filter structure is not straightforward if the tapped delay is not the integer multiples of the sampling period t_s. In the sequel, we discuss how to deal with this particular situation.

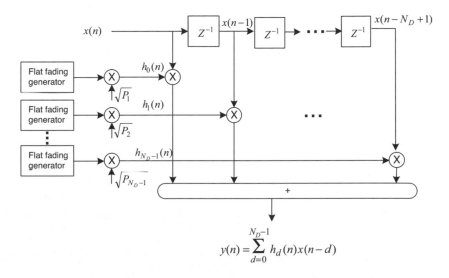

Figure 2.18 TDL-based frequency-selective fading channel model.

2.2.4.2 Tap Adjustment

Since the PDP of the general channel model is based on the actual measurements in the specific environments (e.g., macrocell or indoor), it may not coincide with the integer multiples of the sampling period t_s. In this case, PDP must be adjusted for implementing the discrete-time channel simulator. For example, the tapped delay can be forced into an integer multiple of the sampling period by oversampling, which however may make the number of taps too large for the FIR filter. Instead, tap *interpolation, rounding*, or tap *re-sampling* can be used for simpler implementation. In the sequel, we describe rounding and tap re-sampling methods. We have to make sure that the channel characteristics (e.g., RMS delay spread) are preserved even after adjusting the tap.

A rounding method is to shift the tap into the closest sampling instance. It allows for preserving the number of paths and the power for each path. Figure 2.19 illustrates the tap adjustment by the rounding method. In this method, a new tap delay is expressed as

$$t'_d = floor(t_d/t_s + 0.5) \cdot t_s \tag{2.37}$$

where t_s is the sampling period and t_d is the channel delay. Note that Equation (2.37) can be implemented by using the "round" function in MATLAB®.

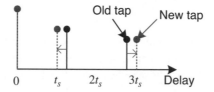

Figure 2.19   Tap adjustment by rounding: an illustration.

Meanwhile, the tap interpolation method is to represent the original channel delay in terms of two neighbor sampling instances, which are weighted by their relative distance with respect to the channel delay. Let t_r denote a relative distance of the channel delay from the discretized delay index t_i, that is,

$$t_r = t_d/t_s - t_i \tag{2.38}$$

where $t_i = floor(t_d/t_s)$, $t_i = 0, 1, 2, 3, \cdots$. Let $\tilde{h}_{t_i}(n)$ denote is the temporary complex channel coefficient for the new tap of delay t_i and assume that $\tilde{h}_0(n) = 0$. Meanwhile, let $h_{t_d}(n)$ denote the complex channel coefficient for the given delay of t_d. Distributing the complex channel coefficient $h_{t_d}(n)$ over two adjacent sampling times as illustrated in Figure 2.20, the new tap of delay t_i is updated as

$$h'_{t_i}(n) = \tilde{h}_{t_i}(n) + \sqrt{1 - t_r} h_{t_d}(n) \tag{2.39}$$

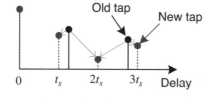

Figure 2.20   Tap adjustment by interpolation: an illustration.

and the temporary complex channel coefficient for the new tap of delay $t_i + 1$ is given as

$$\tilde{h}_{t_i+1}(n) = \sqrt{t_r} h_{t_d}(n) \tag{2.40}$$

where $t_i t_s < h_{t_d}(n) \leq (t_i + 1)t_s$. In case that there exists one or more taps between two consecutive sampling times, they can be superposed at the same sampling instances by the distribution of channel gains in Equation (2.39) and Equation (2.40).

2.2.5 SUI Channel Model

According to the IEEE 802.16d channel model in Section 1.1.3, a suburban path loss environment has been classified into three different terrain types, depending on the tree density and path-loss condition. The SUI (Stanford University Interim) channel model deals with the same environment as in the IEEE 802.16d channel model. Using the different combinations of the channel parameters, it identifies six different channel models that can describe the typical three terrain types in North America [9, 10] (see Table 2.7). The details of the channel parameters for the different SUI models are summarized in Table 2.8. Note that the different K-factors and σ_τ are set for the different antenna types, for example, directional or omni antennas [9, 10].

Table 2.7 SUI channel models for the different terrain types.

Terrain type	SUI channels
A	SUI-5, SUI-6
B	SUI-3, SUI-4
C	SUI-1, SUI-2

In the SUI channel models, the Doppler power spectrum (PSD) is modeled as the following truncated form:

$$S(f) = \begin{cases} 1 - 1.72 f_0^2 + 0.785 f_0^4 & f_0 \leq 1 \\ 0 & f_0 > 1 \end{cases} \tag{2.41}$$

where $f_0 = f/f_m$. Figure 2.21(a) shows the actual Doppler power spectrum that has been measured in 2.5 GHz band. Meanwhile, the truncated Doppler power spectrum in Equation (2.41) is shown in Figure 2.21(b).

Antenna correlation parameter ρ_{ENV} in Table 2.8 is applicable to the MIMO channel model only. Since this chapter intends to deal with the SISO channel only, it will be described later in Chapter 3. As shown in Figure 2.22, a SUI channel modeling process can be summarized into three steps. Once the SUI channel parameters are set, a fading channel is generated by using the FWGN model in Section 2.3.1. Finally, interpolation or re-sampling processes in Section 2.2.4 can be employed so that the fading channel may be suited to the wireless transmission system under consideration.

Figure 2.23 shows the PDP, time-domain characteristic, and power spectra of a SUI channel model. It has been obtained by running Program 2.27 ("plot_SUI_channel.m"), which calls

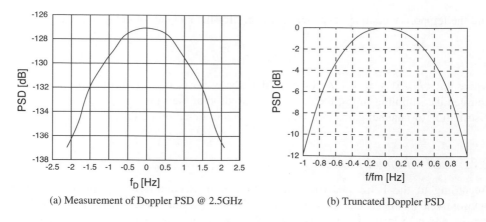

(a) Measurement of Doppler PSD @ 2.5GHz   (b) Truncated Doppler PSD

Figure 2.21   Doppler PSD model.

Table 2.8   SUI channel parameters [10].

	SUI 1/2/3/4/5/6 channel		
	Tap 1	Tap 2	Tap 3
Delay [μs]	0/0/0/0/0/0	0.4/0.4/0.4/1.5/4/14	0.9/1.1/0.9/4/10/20
Power (omni ant.) [dB]	0/0/0/0/0/0	-15/-12/-5/-4/-5/-10	-20/-15/-10/-8/-10/-14
90% K-factor (omni)	4/2/1/0/0/0	0/0/0/0/0/0	0/0/0/0/0/0
75% K-factor (omni)	20/11/7/1/0/0	0/0/0/0/0/0	0/0/0/0/0/0
50% K-factor (omni)	-/-/-/-/2/1	-/-/-/-/0/0	-/-/-/-/0/0
Power (30° ant.) [dB]	0/0/0/0/0/0	-21/-18/-11/-10/-11/-16	-32/-27/-22/-20/-22/-26
90% K-factor (30° ant.)	16/8/3/1/0/0	0/0/0/0/0/0	0/0/0/0/0/0
75% K-factor (30° ant.)	72/36/19/5/2	0/0/0/0/0/0	0/0/0/0/0/0
50% K-factor (30° ant.)	-/-/-/-/7/5	-/-/-/-/0/0	-/-/-/-/0/0
Doppler [Hz]	0.4/0.2/0.4/0.2/2/0.4	0.3/0.15/0.3/0.15/1.5/0.3	0.5/0.25/0.5/0.25/2.5/0.5
Antenna correlation	$\rho_{ENV} = 0.7/0.5/0.4/0.3/0.5/0.3$		
Gain reduction factor	$G_{RF} = 0/2/3/4/4/4\ dB$		
Normalization factor	$F_{omni} = -0.1771/-0.3930/-1.5113/-1.9218/-1.5113/-0.5683dB$		
	$F_{30°} = -0.0371/-0.0768/-0.3573/-0.4532/-0.3573/-0.1184dB$		
Terrain type	C/C/B/B/A/A		
Omni antenna: overall K	$\sigma_\tau = 0.111/0.202/0.264/1.257/2.842/5.240\ \mu s$ K=3.3/16/0.5/0.2/0.1/0.1(90%) K = 10.4/5.1/1.6/0.6/0.3/0.3 (75%), K = -/-/-/-/1.0/1.0 (50%)		
30° antenna: overall K	$\sigma_\tau = 0.042/0.69/0.123/0.563/1.276/2.370\ \mu s$ K = 14.0/6.9/2.2/1.0/0.4/0.4 (90%), K = 44.2/21.8/7.0/3.2/1.3/1.3 (75%), K = -/-/-/-/4.2/4.2 (50%)		

(Reproduced with permission from V. Erceg *et al.*, © IEEE P802.16.3c-01/29r4, "Channel models for fixed wireless applications," 2003.)

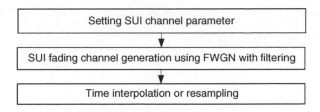

Figure 2.22   SUI channel modeling process.

Program 2.28 ("SUI_parameters") and Program 2.29 ("SUI_fading") to set the SUI channel parameters as listed in Table 2.8 and to generate a SUI fading matrix using FWGN, respectively.

For further information on SISO channel modeling, the reader may consult the references [32–37].

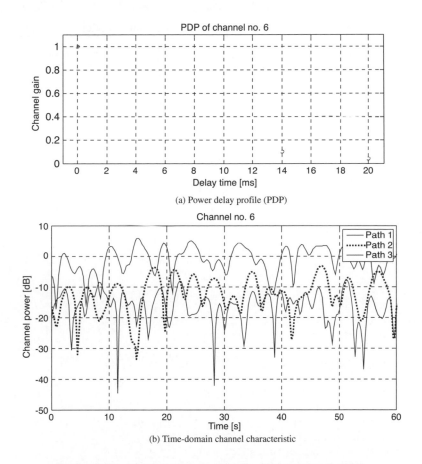

(a) Power delay profile (PDP)

(b) Time-domain channel characteristic

Figure 2.23   Channel characteristics for SUI-6 channel model.

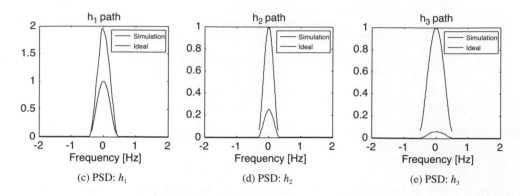

(c) PSD: h_1 (d) PSD: h_2 (e) PSD: h_3

Figure 2.23 (*Continued*)

MATLAB® Programs: SUI Channel Model

Program 2.27 "plot_SUI_channel.m" to plot an SUI channel model

```
% plot_SUI_channel.m
clear, clf
ch_no=6;
fc=2e9;  fs_Hz=1e7;
Nfading=1024;   % Size of Doppler filter
N=10000;
Nfosf=4;
[Delay_us, Power_dB, K_factor, Doppler_shift_Hz, Ant_corr, Fnorm_dB] =SUI_
    parameters(ch_no);
[FadTime tf]=SUI_fading(Power_dB, K_factor, Doppler_shift_Hz, Fnorm_dB, N,
    Nfading, Nfosf);
c_table=['b','r','m','k'];
subplot(311)
stem(Delay_us,10.^(Power_dB/10)), axis([-1 21 0 1.1])
grid on, xlabel('Delay time[ms]'), ylabel('Channel gain');
title(['PDP of Channel No.',num2str(ch_no)]);
subplot(312)
for k=1:length(Power_dB)
  plot((0:length(FadTime(k,:))-1)*tf,20*log10(abs(FadTime(k,:))), c_table
    (k)); hold on
end
grid on, xlabel('Time[s]'), ylabel('Channel Power[dB]');
title(['Channel No.',num2str(ch_no)]), axis([0 60 -50 10])
legend('Path 1','Path 2','Path 3')
idx_nonz=find(Doppler_shift_Hz);
FadFreq =ones(length(Doppler_shift_Hz),Nfading);
for k=1:length(idx_nonz)
  max_dsp=2^Nfosf*max(Doppler_shift_Hz);
  dfmax=max_dsp/Nfading;
```

```
% Doppler frequency spacing respect to maximal Doppler frequency
Nd=floor(Doppler_shift_Hz(k)/dfmax)-1;
f0 = [-Nd+1:Nd]/(Nd); % Frequency vector
f=f0.*Doppler_shift_Hz(k);
tmp=0.785*f0.^4 - 1.72*f0.^2 + 1.0; % Eq.(2.41)
hpsd=psd(spectrum.welch,FadTime(idx_nonz(k),:),'Fs',max_dsp, ...
    'SpectrumType','twosided');
nrom_f=hpsd.Frequencies-mean(hpsd.Frequencies);
PSD_d=fftshift(hpsd.Data);
subplot(3,3,6+k), plot(nrom_f,PSD_d,'b', f,tmp,'r')
xlabel('Frequency[Hz]'), % axis([-1 1 0 1.1*max([P3D_d tmp])])
title(['h_',num2str(idx_nonz(k)),' path']);
end
```

Program 2.28 "SUI_parameters" to set the SUI channel model parameters

```
function [Delay_us,Power_dB,K,Doppler_shift_Hz,Ant_corr,Fnorm_dB]=
          SUI_parameters(ch_no)
% SUI Channel Parameters from Table 2.8
% Inputs:
%   ch_no              : channel scenario number
% Ouptuts:
%   Delay_us           : tap delay[us]
%   Power_dB           : power in each tap[dB]
%   K                  : Ricean K-factor in linear scale
%   Doppler_shift_Hz   : Doppler frequency [Hz]
%   Ant_corr           : antenna (envelope) correlation coefficient
%   Fnorm_dB           : gain normalization factor[dB]
if ch_no<1||ch_no>6, error('No such a channnel number');   end
Delays= [0 0.4 0.9; 0 0.4 1.1; 0 0.4 0.9; 0 1.5 4; 0 4 10; 0 14 20];
Powers= [0 -15 -20;0 -12 -15;0 -5 -10;0 -4 -8;0 -5 -10;0 -10 -14];
Ks = [4 0 0; 2 0 0; 1 0 0; 0 0 0; 0 0 0; 0 0 0];
Dopplers = [0.4 0.3 0.5; 0.2 0.15 0.25; 0.4 0.3 0.5;
          0.2 0.15 0.25; 2 1.5 2.5; 0.4 0.3 0.5];
Ant_corrs = [0.7 0.5 0.4 0.3 0.5 0.3];
Fnorms = [-0.1771 -0.393 -1.5113 -1.9218 -1.5113 -0.5683];
Delay_us= Delays(ch_no,:); Power_dB= Powers(ch_no,:); K=Ks(ch_no,:);
   Doppler_shift_Hz= Dopplers(ch_no,:);
Ant_corr= Ant_corrs(ch_no); Fnorm_dB= Fnorms(ch_no);
```

Program 2.29 "SUI_fading": FWGN (filtered white Gaussian noise) for SUI channel model

```
function [FadMtx,tf]=
  SUI_fading(Power_dB,K_factor,Doppler_shift_Hz, Fnorm_dB, N, M, Nfosf)
%   SUI fading generation using FWGN with fitering in frequency domain
%   FadingMatrixTime=SUI_fading(Power_dB, K_factor,...
%                    Doppler_shift_Hz, Fnorm_dB, N, M, Nfosf)
%   Inputs:
```

```
%     Power_dB          : power in each tap in dB
%     K_factor          : Rician K-factor in linear scale
%     Doppler_shift_Hz  : a vector containing maximum Doppler
%                                 frequency of each path in Hz
%     Fnorm_dB          : gain normalization factor in dB
%     N       : # of independent random realizations
%     M       : length of Doppler filter, i.e, size of IFFT
%     Nfosf   : fading oversampling factor
%  Outputs:
%     FadMtx  : length(Power_dB) x N fading matrix
%     tf      : fading sample time=1/(Max. Doppler BW * Nfosf)
Power = 10.^(Power_dB/10);    % calculate linear power
s2 = Power./(K_factor+1);     % calculate variance
s=sqrt(s2);
m2 = Power.*(K_factor./(K_factor+1));    % calculate constant power
m = sqrt(m2);                     % calculate constant part
L=length(Power);                  % # of tabs
fmax= max(Doppler_shift_Hz);
tf=1/(2*fmax*Nfosf);
if isscalar(Doppler_shift_Hz)
  Doppler_shift_Hz= Doppler_shift_Hz*ones(1,L);
end
path_wgn= sqrt(1/2)*complex(randn(L,N),randn(L,N));
for p=1:L
  filt=gen_filter(Doppler_shift_Hz(p),fmax,M,Nfosf,'sui');
  path(p,:)=fftfilt(filt,[path_wgn(p,:) zeros(1,M)]); % filtering WGN
end
FadMtx= path(:,M/2+1:end-M/2);
for i=1:L , FadMtx(i,:)=FadMtx(i,:)*s(i)+m(i)*ones(1,N); end
FadMtx = FadMtx*10^(Fnorm_dB/20);
```

3

MIMO Channel Models

In Chapter 3, we will first present an overview of a statistical channel model for MIMO system. Then, we will describe a correlation-based I-METRA channel model and a ray-based 3GPP spatial channel model (SCM) as specific methods of implementing the MIMO channel. While the correlation-based channel model can be implemented with a spatial correlation matrix for the spatial channel, temporal correlation must also be generated independently by using the specified Doppler spectrum. On the other hand, the ray-based channel model combines the multiple rays distributed in the angular domain for the given Power Azimuth Spectrum (PAS). This requires neither Doppler spectrum nor spatial correlation matrix, but rather involves a complex computational operation.

3.1 Statistical MIMO Model

Recall that delay spread and Doppler spread are the most important factors to consider in characterizing the SISO system. In the MIMO system which employs multiple antennas in the transmitter and/or receiver, the correlation between transmit and receive antenna is an important aspect of the MIMO channel. It depends on the angle-of-arrival (AoA) of each multi-path component. Consider a SIMO channel with a uniform linear array (ULA) in which M antenna elements are equally spaced apart, with an inter-distance of d as shown in Figure 3.1. Let $y_i(t)$ denote a received signal at the ith antenna element with the channel gain α_i, delay τ_i, and angle of arrival (AoA) ϕ_i. As shown in Figure 3.2(a), the AoA is defined as the azimuth angle of incoming path with respect to the broadside of the antenna element. Note that the received signal of each path consists of the enormous number of unresolvable signals received around the mean of AoA in each antenna element. A vector of the received signals $\mathbf{y}(t) = [y_1(t), y_2(t), \cdots, y_M(t)]^T$ in the uniform linear array (ULA) of M elements can be expressed as

$$\mathbf{y}(t) = \sum_{i=1}^{I} \alpha_i \mathbf{c}(\phi_i) x(t - \tau_i) + \mathbf{N}(t) \tag{3.1}$$

MIMO-OFDM Wireless Communications with MATLAB®   Yong Soo Cho, Jaekwon Kim, Won Young Yang and Chung G. Kang
© 2010 John Wiley & Sons (Asia) Pte Ltd

where I denotes the number of paths in each antenna element and $\mathbf{c}(\phi)$ is an array steering vector. The array steering vector is defined as

$$\mathbf{c}(\phi) = [c_1(\phi), c_2(\phi), \cdots, c_M(\phi)]^T$$

where

$$c_m(\phi) = f_m(\phi)e^{-j2\pi(m-1)(d/\lambda)\sin\phi}, \quad m = 1, 2, \ldots, M \tag{3.2}$$

In Equation (3.2), $f_m(\phi)$ denotes a complex field pattern of the mth array element and λ is the carrier wavelength. The received signal in Equation (3.1) can be expressed in the following integral form:

$$\mathbf{y}(t) = \int\int \mathbf{c}(\phi)h(\phi, \tau)x(t-\tau)d\tau d\phi + \mathbf{N}(t) \tag{3.3}$$

where $h(\phi, \tau)$ represents a channel as a function of ADS (Azimuth-Delay Spread) [38]. The instantaneous power azimuth-delay spectrum (PADS) is given as

$$P_{\text{inst}}(\phi, \tau) = \sum_{i=1}^{I} |\alpha_i|^2 \delta(\phi-\phi_i, \tau-\tau_i) \tag{3.4}$$

The average PADS is defined as an expected value of Equation (3.4), such that

$$P(\phi, \tau) = E\{P_{\text{inst}}(\phi, \tau)\} \tag{3.5}$$

By taking an integral of PADS over delay, PAS (Power Azimuth Spectrum or Power Angular Spectrum) is obtained as

$$P_A(\phi) = \int P(\phi, \tau)\,d\tau \tag{3.6}$$

Meanwhile, AS (Azimuth Spread or Angular Spread) is defined by the central moment of PAS, that is

$$\sigma_A = \sqrt{\int (\phi-\phi_0)^2 P_A(\phi)\,d\phi} \tag{3.7}$$

where ϕ_0 is the mean AoA (i.e., $\phi_0 = \int \phi P_A(\phi)\,d\phi$) [38]. Similarly, by taking an integral of PADS over AoA, PDS (Power Delay Spectrum) is obtained as

$$P_D(\tau) = \int P(\phi, \tau)\,d\phi \tag{3.8}$$

Furthermore, DS (Delay Spread) is defined as the central moment of PDS, that is,

$$\sigma_D = \sqrt{\int (\tau-\tau_0)^2 P_D(\tau)\,d\tau} \tag{3.9}$$

where τ_0 is an average delay spread (i.e., $\tau_0 = \int \tau P_D(\tau)\,d\tau$).

Once a joint PDF of AoA and delay is given by $f(\phi, \tau)$, the marginal PDF of AoA and delay spread is respectively given as

$$f_A(\phi) = \int f(\phi, \tau) \, d\tau \tag{3.10}$$

and

$$f_D(\tau) = \int f(\phi, \tau) \, d\phi \tag{3.11}$$

While Clarke's channel model assumes that AoA is uniformly distributed in the mobile station (MS), its distribution is significantly different in the base station (BS). In general, spatial correlation in the MS turns out to be almost zero for the antenna elements that are equally spaced by $\lambda/2$. In order to warrant a low spatial correlation in the BS, however, the antenna elements must be spaced by roughly $10\lambda \sim 40\lambda$, even if it depends on AS. Meanwhile, the PDF of delay spread is typically approximated by an exponential function.

Figure 3.2(a) illustrates a MIMO channel model by magnifying a channel environment in Figure 3.1. It shows three resolvable paths. Each path has M_r unresolvable paths, each of which arrives centered around the mean AoA. The AoA for these unresolvable paths follow a Gaussian distribution in the microcell or macrocell environment (See Figure 3.2(d)). Furthermore, a power distribution in AoA (i.e., PAS) follows a Laplacian distribution, even though it varies with the cell environment. Note that the natures of AoA and PAS distributions are different from each other. In other words, a distribution of AoA does not take the power of each path into account, while PAS deals with a distribution of the power with respect to AoA. Finally, a distribution of the power for the resolvable paths, that is, PDS or PDP (Power Delay Profile), usually follows an exponential distribution (see Figure 3.2(b)).

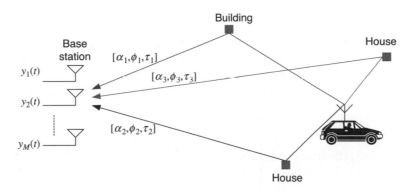

Figure 3.1   SIMO channel environment: an illustration.

3.1.1 Spatial Correlation

In general the received signals for each path of the different antenna elements may be spatially correlated, especially depending on the difference in their distances traveled. In this subsection, let us investigate the spatial correlation between the received signals in the different antennas.

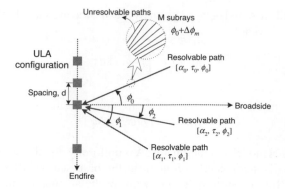

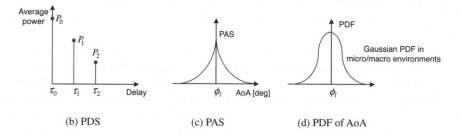

(a) Multi-path signals in the multiple antenna elements: an illustration for ULA

(b) PDS   (c) PAS   (d) PDF of AoA

Figure 3.2 MIMO channel model: an illustration.

Consider two omni-directional antennas, a and b, that are spaced apart by d as shown in Figure 3.3. For the baseband received signals with the mean AoA of ϕ_0, the difference in their distance traveled is given by $d \sin\phi_0$ and the corresponding delay becomes $\tau_0 = (d/c) \sin\phi_0$. Let α and β denote the amplitude and phase of each path, which follow the Rayleigh distribution and uniform distribution over $[0, 2\pi)$, respectively. Assuming a narrowband

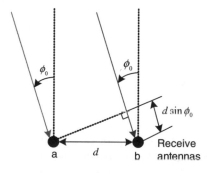

Figure 3.3 Signal models for two omni-directional antennas.

channel, their impulse responses can be respectively represented as

$$h_a(\phi) = \alpha e^{j\beta} \sqrt{P(\phi)} \qquad (3.12a)$$

and

$$h_b(\phi) = \alpha e^{j(\beta + 2\pi d \, \sin(\phi)/\lambda)} \sqrt{P(\phi)} \qquad (3.12b)$$

where $P(\phi)$ denotes the PAS defined in Equation (3.6), such that $P(\phi) = P_A(\phi)$.

Let us define a spatial correlation function of the received signals with the mean AoA of ϕ_0 in two antenna elements spaced apart by d as

$$\rho_c(d, \phi_0) = E_\phi\{h_a(\phi)h_b^*(\phi)\}$$
$$= \int_{-\pi}^{\pi} h_a(\phi)h_b^*(\phi)P(\phi-\phi_0)d\phi. \qquad (3.13)$$

Consider an extreme case that has the mean AoA of $\phi_0 = 0°$ and AS of $\sigma_A = 0°$, that is, $P(\phi-\phi_0) = \delta(\phi)$, implying that there exists only one sub-ray in a perpendicular direction for each antenna element. In this particular case, AoA does not incur any time difference between h_a and h_b. Therefore, spatial correlation is always equal to 1, that is, $\rho_c(d) = E_\phi\{h_a h_b^*\} = E\{|\alpha|^2\} = 1$. However, in the case that both AoA and AS are not equal to $0°$, there is a time difference between $h_a(\phi)$ and $h_b(\phi)$ as shown in Equation (3.12). This yields the following spatial correlation function:

$$\rho_c(d, \phi_0) = E_\phi\{h_a(\phi)h_b^*(\phi)\}$$
$$= \int_{-\pi}^{\pi} e^{-\frac{j2\pi d \, \sin(\phi-\phi_0)}{\lambda}} P(\phi-\phi_0)d\phi$$
$$= R_{xx}(d, \phi_0) + jR_{xy}(d, \phi_0) \qquad (3.14)$$

where we assume that PAS $P(\phi)$ has been normalized as $\int_{-\pi}^{\pi} P(\phi) \, d\phi = 1$ while $R_{xx}(d, \phi_0)$ and $R_{xy}(d, \phi_0)$ represent correlations between real parts of two received signals and between their real and imaginary parts, respectively [39]. Defining a normalized antenna distance by $D = 2\pi d/\lambda$, the spatial correlation functions can be written as

$$R_{xx}(D, \phi_0) = E\{\mathrm{Re}(h_a) \cdot \mathrm{Re}(h_b)\}$$
$$= \int_{-\pi}^{\pi} \cos(D \sin \phi)P(\phi-\phi_0)d\phi \qquad (3.15)$$

and

$$R_{xy}(D, \phi_0) = E\{\mathrm{Re}(h_a) \cdot \mathrm{Im}(h_b)\}$$
$$= \int_{-\pi}^{\pi} \sin(D \sin \phi)P(\phi-\phi_0)d\phi \qquad (3.16)$$

As seen in Equation (3.15) and Equation (3.16), spatial correlation between antenna elements depends mainly on the mean AoA and PAS as well as antenna spacing d. In particular, when AS of PAS is small, most sub-rays that compose each path arrive at each antenna from the same angle. It implies that they are correlated with each other since while the magnitudes of two signals become nearly equal, their phases are different by their AoAs.

Since channel capacity and diversity gain decrease as the correlation between the antenna elements increases (as will be discussed in Chapter 9), antenna spacing must be set large enough to reduce the correlation.

3.1.2 PAS Model

As discussed in the previous subsection, PAS is an important factor in determining the spatial correlation between antenna elements. It is clear from Equation (3.14)–Equation (3.16). In fact, a mathematical analysis for spatial correlation requires a distribution of PAS for the real environments. We find that there are various types of PAS models available from the actual measurements of the different channel environments (e.g., indoor or outdoor, macrocell or microcell), including those summarized in Table 3.1 [40]. A pattern of PAS depends mainly on the distribution of the locally-scattered components. In general, enormous amounts of locally-scattered components are observed by the MS in all different environments. Therefore, its PAS usually follows a uniform distribution. For the BS, however, the different PAS distributions are observed depending on the characteristics of terrain in a cell, which is usually shown to have a small AS. Note that they still show a uniform PAS distribution for the BS in picocells or indoor environments. In the sequel, we will explain in detail the other type of PAS models in Table 3.1.

Table 3.1 PAS model for the different environments [40].

		BS	MS
Outdoor	Macrocell	• Truncated Laplacian • n-th power of a cosine function	Uniform
	Microcell	• Truncated Gaussian • Uniform	
	Picocell	Almost Uniform	
Indoor		Uniform	

(Reproduced with permission from L. Schumacher *et al.*, "MIMO channel characterisation," Technical Report IST-1999-11729 METRA, D2, 2001. © 2001 IST-METRA.)

Meanwhile, Table 3.2 shows the average AS measured from indoor and outdoor environments at different carrier frequencies. Even if multiple clusters may be observed in indoor environments, it presents the result for only one of them. It is clear from Table 3.2 that AS is rather large in the indoor environment while it is usually less than 10° in the outdoor environments.

3.1.2.1 PAS Models

n-th Power of a Cosine Function PAS Model   The PAS can be represented by the n-th power of a cosine function as follows:

$$P(\phi) = \frac{Q}{\pi} \cos^n(\phi), \qquad -\frac{\pi}{2} + \phi_0 \leq \phi \leq \frac{\pi}{2} + \phi_0 \tag{3.17}$$

where n is an even integer related to beamwidth, and Q is a factor used to normalize PAS into 1 [41]. Figure 3.4 illustrates PAS with the mean AoA of 0° (i.e., $\phi_0 = 0°$) for $n = 2, 4,$ and 8. Note

Table 3.2 Average AS for the different environments [40].

| Reference | Carrier frequency (MHz) | Outdoor | | | | | Indoor |
| | | Macrocell | | | Microcell | | |
		Urban	Suburban	Rural	LOS	NLOS	
[41]	1000						20–60°
[42]	1800	5–10°					
[43]	1800	8°	5°				
[44]	1845			<10°			
[45]	1873	3–15°					
[46]	2100	7–12°	13–18°				
[47]	2154		10.3°				
[48]	2200			3°	<10°	<20°	
[49]	7000						22–26°

(Reproduced with permission from L. Schumacher *et al.*, "MIMO channel characterisation," Technical Report IST-1999-11729 METRA, D2, 2001. © 2001 IST-METRA.)

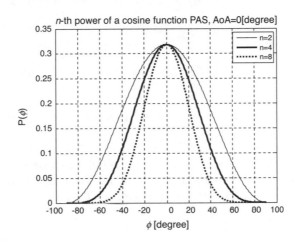

Figure 3.4 The *n*-th power of a cosine function PAS model.

that the width of PAS becomes narrow as n increases. Substituting Equation (3.17) into Equation (3.15) and Equation (3.16), we find that the spatial correlation functions are given as

$$R_{xx}(D, \phi_0) = \int_{-\pi/2}^{\pi/2} \cos(D \sin \phi) \cdot \frac{Q}{\pi} \cos^n(\phi - \phi_0) d\phi \qquad (3.18)$$

and

$$R_{xy}(D, \phi_0) = \int_{-\pi/2}^{\pi/2} \sin(D \sin \phi) \cdot \frac{Q}{\pi} \cos^n(\phi - \phi_0) d\phi \qquad (3.19)$$

Due to the n-th power of cosine function, there are no explicit expressions for Equation (3.18) and Equation (3.19) and thus, they can be solved only by numerical analysis.

Uniform PAS Model   The uniform PAS model is suited for modeling a rich-scattering environment, such as an indoor environment. It represents a situation with a uniform power distribution over the specified range of angle as

$$P(\phi) = Q \cdot 1, \qquad -\Delta\phi + \phi_0 \le \phi \le \Delta\phi + \phi_0 \qquad (3.20)$$

where $\Delta\phi = \sqrt{3}\sigma_A$ [39] and Q is the normalization factor to PAS, which is found as

$$Q = 1/(2\Delta\phi) \qquad (3.21)$$

Substituting Equation (3.20) into Equation (3.15) and Equation (3.16), we find the spatial correlation functions given as

$$R_{xx}(D, \phi_0) = J_0(D) + 4Q \sum_{m=1}^{\infty} J_{2m}(D)\cos(2m\phi_0)\sin(2m \cdot \Delta\phi)/2m \qquad (3.22)$$

and

$$R_{xy}(D) = 4Q \sum_{m=1}^{\infty} J_{2m+1}(D, \phi_0)\sin((2m+1)\phi_0)\sin((2m+1) \cdot \Delta\phi)/(2m+1) \qquad (3.23)$$

where $J_m(\cdot)$ is the first-kind mth order Bessel function. Note that $R_{xx}(D) \to J_0(D)$ and $R_{xy}(D) \to 0$ as $\sigma_A \to \infty$.

Figure 3.5 illustrates the PAS distribution and spatial correlation coefficients for the Uniform PAS model. As shown in Figure 3.5(b), values of the spatial correlation coefficients decrease as AS increases for the same antenna spacing. The spatial correlation coefficients become nearly zero in a certain interval. In Figure 3.5(b), for example, we find that they

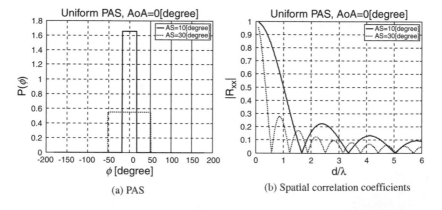

(a) PAS   (b) Spatial correlation coefficients

Figure 3.5   PAS and spatial correlation coefficients for Uniform PAS model.

become nearly zero at the integer multiples of 1.7λ and 0.6λ when AS is $10°$ and $30°$, respectively.

Truncated Gaussian PAS Model   For the Truncated Gaussian PAS model, a power distribution is given as

$$P(\phi) = \frac{Q}{\sqrt{2\pi}\sigma}e^{-\frac{(\phi-\phi_0)^2}{2\sigma^2}}, \qquad -\Delta\phi + \phi_0 \leq \phi \leq \Delta\phi + \phi_0 \qquad (3.24)$$

where σ is the standard deviation of PAS and $\Delta\phi$ is uniformly distributed over $[-\pi, \pi]$ (typically, it is set to $\Delta\phi = \pi$) [39]. The normalization factor Q can be found by the following constraint:

$$\int_{-\pi}^{\pi} P(\phi)d\phi = \frac{Q}{\sqrt{2\pi}\sigma}\int_{-\Delta\phi}^{\Delta\phi} e^{-\frac{\phi^2}{2\sigma^2}}\,d\phi = 1 \qquad (3.25)$$

By solving Equation (3.25) for the normalization factor Q, it is found as

$$Q = 1/erf\left(\frac{\Delta\phi}{\sqrt{2}\sigma}\right) \qquad (3.26)$$

where $erf\,(\,\cdot\,)$ denotes an error function[1].

   Substituting Equation (3.24) and Equation (3.26) into Equation (3.15) and Equation (3.16), the spatial correlation coefficients are found as

$$R_{xx}(D) = J_0(D) + Q\sum_{m=1}^{\infty} J_{2m}(D)e^{-2\sigma^2 m^2}\cos(2m\phi_0)$$

$$\cdot \,\text{Re}\left[erf\left(\frac{\Delta\phi}{\sigma\sqrt{2}} - jm\sigma\sqrt{2}\right) - erf\left(-\frac{\Delta\phi}{\sigma\sqrt{2}} - jm\sigma\sqrt{2}\right)\right] \qquad (3.27)$$

and

$$R_{xy}(D) = Q\sum_{m=1}^{\infty} J_{2m+1}(D)e^{-2\sigma^2(m+1/2)^2}\sin((2m+1)\phi_0)$$

$$\cdot \,\text{Re}\left[erf\left(\frac{\Delta\phi}{\sigma\sqrt{2}} - j\sigma\sqrt{2}\left(m+\frac{1}{2}\right)\right) - erf\left(-\frac{\Delta\phi}{\sigma\sqrt{2}} - j\sigma\sqrt{2}\left(m+\frac{1}{2}\right)\right)\right] \qquad (3.28)$$

---

[1]The error function is defined as $erf(x) = \frac{2}{\sqrt{\pi}}\int_0^x e^{-t^2}\,dt$.

Figure 3.6 illustrates the PAS and spatial correlation coefficients for the Truncated Gaussian PAS model. As shown here, a power distribution becomes broad over the large angle as AS increases, which decreases the spatial correlation coefficient at the same antenna spacing. Furthermore, the spatial correlation coefficient becomes nearly zero at the integer multiples of 0.6λ when AS is 30°.

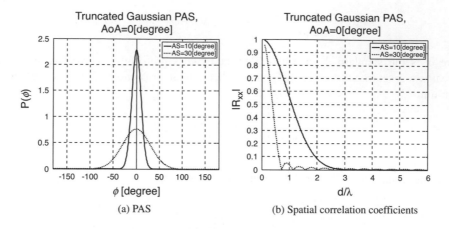

(a) PAS   (b) Spatial correlation coefficients

Figure 3.6   PAS and spatial correlation coefficients for Truncated Gaussian PAS model.

Truncated Laplacian PAS Model   The Truncated Laplacian PAS model is commonly employed for macrocell or microcell environments. Its power distribution is given as

$$P(\phi) = \frac{Q}{\sqrt{2}\sigma} e^{-\frac{\sqrt{2}|\phi-\phi_0|}{\sigma}}, \qquad -\Delta\phi + \phi_0 \leq \phi \leq \Delta\phi + \phi_0 \tag{3.29}$$

where σ is the standard deviation of PAS [39]. The normalization factor Q can be found by

$$\int_{-\pi}^{\pi} P(\phi)d\phi = \frac{Q}{\sqrt{2}\sigma} \int_{-\pi}^{\pi} e^{-\frac{\sqrt{2}|\phi-\phi_0|}{\sigma}} d\phi$$

$$= \frac{2Q}{\sqrt{2}\sigma} \int_{0}^{\Delta\phi} e^{-\frac{\sqrt{2}\phi}{\sigma}} d\phi \tag{3.30}$$

$$= Q\left(1 - e^{-\sqrt{2}\Delta\phi/\sigma}\right) = 1$$

which gives

$$Q = \frac{1}{1 - e^{-\sqrt{2}\Delta\phi/\sigma}} \tag{3.31}$$

Substituting Equation (3.29) and Equation (3.31) into Equation (3.15) and Equation (3.16), the spatial correlation coefficients are found as

$$R_{xx}(D,\phi_0) = J_0(D) + 4Q\sum_{m=1}^{\infty} J_{2m}(D)\cos(2m\phi_0)$$

$$\cdot \frac{\dfrac{\sqrt{2}}{\sigma} + e^{\frac{-\sqrt{2}\Delta\phi}{\sigma}}\left\{2m\cdot\sin(2m\cdot\Delta\phi) - \sqrt{2}\cos(2m\cdot\Delta\phi)/\sigma\right\}}{\sqrt{2}\upsilon\left[\left(\dfrac{\sqrt{2}}{\sigma}\right)^2 + (2m)^2\right]} \qquad (3.32)$$

and

$$R_{xy}(D,\phi_0) = 4Q\sum_{m=1}^{\infty} J_{2m+1}(D)\sin((2m+1)\phi_0)$$

$$\cdot \frac{\dfrac{\sqrt{2}}{\sigma} + e^{\frac{-\sqrt{2}\Delta\phi}{\sigma}}\left\{(2m+1)\cdot\sin((2m+1)\Delta\phi) - \sqrt{2}\cos((2m+1)\Delta\phi)/\sigma\right\}}{\sqrt{2}\sigma\left[\left(\dfrac{\sqrt{2}}{\sigma}\right)^2 + (2m+1)^2\right]} \qquad (3.33)$$

Figure 3.7 illustrates the PAS and spatial correlation coefficient for the Truncated Laplacian PAS model. It is obvious from Figure 3.7 that antenna spacing must be large enough to reduce the spatial correlation. In order to ensure the spatial correlation coefficient below 0.1 when AS is 30°, for example, antenna spacing must be set to be roughly greater than 1.3λ. When AS is reduced to 10°, the antenna spacing must be further increased beyond 4λ so as to maintain the spatial correlation coefficient below 0.1.

3.1.2.2 Relationship between Standard Deviation and AS in PAS Model

Note that the Truncated Gaussian or Laplacian PAS model limits its range to a finite interval as opposed to the general Gaussian or Laplacian distribution function which takes an interval of $[-\infty,\infty]$. The standard deviation in the Gaussian or Laplacian PAS model corresponds to that in the general Gaussian or Laplacian distribution, which is given as

$$\sigma = \sqrt{\int_{-\infty}^{\infty} (\phi-\phi_{00})^2 P(\phi)d\phi} \qquad (3.34)$$

where $\phi_{00} = \int_{-\infty}^{\infty} \phi P(\phi)d\phi$. Recall that AS is defined by the central moment as in Equation (3.7). AS in the Truncated Gaussian or Laplacian PAS model is truncated to $[-\pi,\pi]$ and thus,

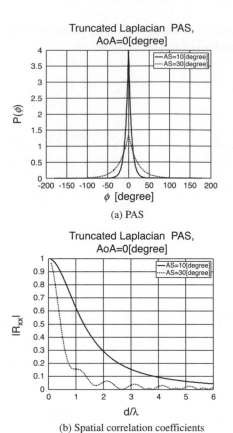

(a) PAS

(b) Spatial correlation coefficients

Figure 3.7 PAS and spatial correlation coefficients for Truncated Laplacian PAS model.

$$\sigma_A = \sqrt{\int_{-\pi}^{\pi} (\phi - \phi_0)^2 P(\phi) d\phi} \qquad (3.35)$$

where $\phi_0 = \int_{-\pi}^{\pi} \phi P_A(\phi) d\phi$. From Equation (3.34) and Equation (3.35), it is clear that the standard deviation of PAS and AS are different only in their intervals of integral. Let us assume that the mean AoA is zero. Since the Truncated Gaussian PAS and Truncated Laplacian PAS are symmetric, the following relationship holds between PAS and AS:

$$\begin{aligned}
\sigma^2 &= \int_{-\infty}^{\infty} \phi^2 P(\phi) d\phi \\
&= \int_{-\pi}^{\pi} \phi^2 P(\phi) d\phi + 2\int_{\pi}^{\infty} \phi^2 P(\phi) d\phi \qquad (3.36) \\
&= \sigma_A^2 + 2\int_{\pi}^{\infty} \phi^2 P(\phi) d\phi
\end{aligned}$$

The relationship in Equation (3.36) is shown in Figure 3.8. It shows that the standard deviation of PAS is almost equal to AS when AS is below 30°. However, their difference becomes significant when AS increases beyond 30°. Note that the Laplacian PAS decreases more dramatically further away from its mean and thus, the second term in Equation (3.36) becomes less significant than the Gaussian PAS. It implies that the broader range of linearity can be provided by the Laplacian PAS model.

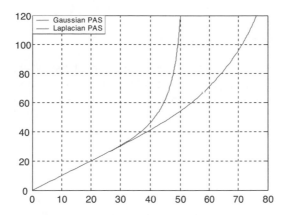

Figure 3.8 Standard deviations of PAS model as AS varies.

3.1.2.3 Multiple Clusters

In cases that there exist the near-by major scattering components in addition to the local scattering components in the MS, the multiple clusters of the signals with the different angles of arrival and PAS are received. In particular, this phenomenon is typically observed in the indoor environments. Their characteristics of spatial correlation can be expressed in terms of a sum of the spatial correlation coefficients associated with those clusters.

Figure 3.9 illustrates the Laplacian PAS model with two clusters. In Figure 3.9(a), the first cluster has the AoA of −60°, AS of 30°, and a unity amplitude, while the second cluster has the

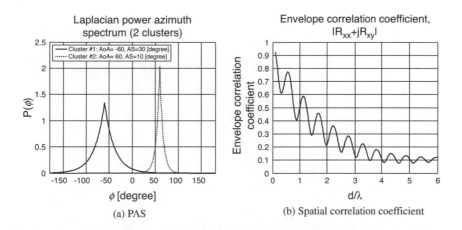

(a) PAS

(b) Spatial correlation coefficient

Figure 3.9 PAS and spatial correlation coefficient for Truncated Laplacian PAS: two clusters.

AoA of 60°, AS of 10°, and the amplitude of 0.5. Meanwhile, Figure 3.9(b) shows the correlation coefficient of the received signal for these two clusters.

The MATLAB® programs for generating the PAS and computing the spatial correlation coefficients can be downloaded from the web site [42].

3.2 I-METRA MIMO Channel Model

A MIMO fading channel can be implemented by using the statistical characteristics, including spatial correlation for PAS, that are explained in Section 3.1. We first discuss the statistical model of the correlated fading channel, which explains a general concept of implementing a MIMO fading channel. Then, we present the I-METRA model which is often adopted for MIMO channel modeling [42–45].

3.2.1 Statistical Model of Correlated MIMO Fading Channel

Consider a MIMO system with a base station of M antennas and the mobile stations of N antennas as illustrated in Figure 3.10. A narrowband MIMO channel $\mathbf{H}$ can be statistically expressed with an $M \times N$ matrix (i.e., $\mathbf{H} \in \mathbf{C}^{M \times N}$) as

$$\mathbf{H} = \Theta_R^{1/2} \mathbf{A}_{iid} \Theta_T^{1/2} \tag{3.37}$$

where Θ_R and Θ_T are the correlation matrices for the receive antennas and transmit antennas, respectively, while $\mathbf{A}_{iid}$ represents an i.i.d. (independent and identically distributed) Rayleigh fading channel. The basic assumption behind the correlation matrix-based MIMO channel model in Equation (3.37) is that the correlation matrices for the transmitter and receiver can be separated. That particular assumption holds when antenna spacing in the

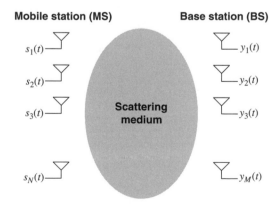

Figure 3.10   Antenna arrays for MIMO communication.

transmitter and receiver is sufficiently smaller than a distance between the transmitter and receiver, which is usually true for most of wireless communication environments. The various types of MIMO channels can be generated by adjusting the correlation matrices Θ_R and Θ_T. As an extreme case, a complete i.i.d. channel can be generated when Θ_R and Θ_T are the identity matrices. Furthermore, other extreme types of MIMO channel, including a rank-1 Rician channel and a Rician channel with an arbitrary phase, can also be easily generated with this model.

Now, a broadband MIMO channel can be modeled by a tapped delay line (TDL), which is an extension of the narrowband MIMO channel in Equation (3.37), as follows:

$$\mathbf{H}(\tau) = \sum_{l=1}^{L} \mathbf{A}_l \delta(\tau - \tau_l) \qquad (3.38)$$

where $\mathbf{A}_l$ is the complex channel gain matrix for the lth path with delay τ_l [46]. Let $\alpha_{mn}^{(l)}$ be the channel coefficient between the mth BS antenna and the nth MS antenna for the lth path. Assume that $\alpha_{mn}^{(l)}$ is zero-mean complex Gaussian-distributed, and thus, $|\alpha_{mn}^{(l)}|$ is Rayleigh-distributed. The complex channel gain matrix $\mathbf{A}_l$ in Equation (3.38) is given as

$$\mathbf{A}_l = \begin{bmatrix} \alpha_{11}^{(l)} & \alpha_{12}^{(l)} & \cdots & \alpha_{1N}^{(l)} \\ \alpha_{21}^{(l)} & \alpha_{22}^{(l)} & \cdots & \alpha_{2N}^{(l)} \\ \vdots & \vdots & \ddots & \vdots \\ \alpha_{M1}^{(l)} & \alpha_{M2}^{(l)} & \cdots & \alpha_{MN}^{(l)} \end{bmatrix} \qquad (3.39)$$

Let $y_m(t)$ denote the received signal at the mth antenna element in BS. Then, the received signals at the BS antenna are denoted as $\mathbf{y}(t) = [y_1(t), y_2(t), \cdots, y_M(t)]^T$. Similarly, the transmitted signals at the MS are denoted as $\mathbf{x}(t) = [x_1(t), x_2(t), \cdots, x_N(t)]^T$ where $x_n(t)$ is the signal transmitted at the nth antenna element. The relation between the MS and BS signals can be expressed as

$$\mathbf{y}(t) = \int \mathbf{H}(\tau)\mathbf{s}(t - \tau)d\tau \qquad (3.40)$$

Consider a downlink of MIMO system in Figure 3.11. As the antenna spacing at BS is relatively small when Tx and Rx are sufficiently apart, the spatial correlation at MS does not depend on the Tx antenna. In other words, spatial correlation at MS is independent of BS antennas. Then, the correlation coefficient of a channel gain for two different MS antennas, n_1 and n_2, can be expressed as

$$\rho_{n_1 n_2}^{MS} = \left\langle |\alpha_{mn_1}^{(l)}|^2, |\alpha_{mn_2}^{(l)}|^2 \right\rangle, \quad m = 1, 2, \ldots, M \qquad (3.41)$$

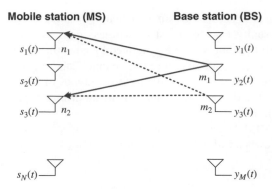

Figure 3.11   Downlink MIMO system.

where

$$\langle x,y \rangle = (E\{xy\}-E\{x\}E\{y\})/\sqrt{(E\{x^2\}-E\{x\}^2)(E\{y^2\}-E\{y\}^2)} \qquad (3.42)$$

For the MS in an environment surrounded by local scatters, the spatial correlation becomes negligible when the MS antennas are separated by more than $\lambda/2$, which means $\rho_{n_1 n_2}^{MS} = \langle |\alpha_{mn_1}^{(l)}|^2, |\alpha_{mn_2}^{(l)}|^2 \rangle \approx 0$ for $n_1 \neq n_2, m = 1, 2, \ldots, M$. In spite of such a theory, however, the experimental results often show that the channel coefficients with the antennas separated by $\lambda/2$ can be highly correlated in some situations, especially in an indoor environment. Let us define a symmetric spatial correlation matrix for the MS as

$$\mathbf{R}_{MS} = \begin{bmatrix} \rho_{11}^{MS} & \rho_{12}^{MS} & \cdots & \rho_{1N}^{MS} \\ \rho_{21}^{MS} & \rho_{22}^{MS} & \cdots & \rho_{2N}^{MS} \\ \vdots & \vdots & \ddots & \vdots \\ \rho_{N1}^{MS} & \rho_{N2}^{MS} & \cdots & \rho_{NN}^{MS} \end{bmatrix} \qquad (3.43)$$

where $\rho_{ij}^{MS} = \rho_{ij}^{MS}$, $i,j = 1,2,\ldots,N$. Note that a diagonal component of $\mathbf{R}_{MS}$ corresponds to the auto-correlation, which is always given by a correlation coefficient of one (i.e., $\rho_{ii}^{MS} = 1$, $i = 1, 2, \cdots, N$).

Meanwhile, consider a typical urban environment, where there are no local scatters in the vicinity of the BS antennas, as opposed to MS antennas surrounded by the local scatters. It is attributed to the fact that the BS antennas are usually elevated above the local scatters so as to reduce the path loss. In this case, the PAS at the BS is subject to a relatively narrow beamwidth. Consider an uplink MIMO system in Figure 3.12. As long as Tx and Rx are sufficiently apart, all MS antennas that are closely co-located tend to have the same radiation pattern, illuminating the same surrounding scatters [46]. Then, it also makes the spatial correlation of BS antennas independent of MS antennas and thus, the

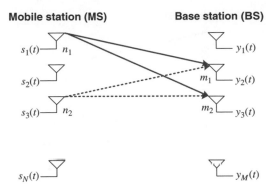

Figure 3.12   Uplink MIMO system.

correlation coefficient of a channel gain for two different BS antennas, m_1 and m_2, can be expressed as

$$\rho_{m_1 m_2}^{BS} = \left\langle \left| \alpha_{m_1 n}^{(l)} \right|^2, \left| \alpha_{m_2 n}^{(l)} \right|^2 \right\rangle, \quad n = 1, 2, \ldots, N \tag{3.44}$$

Using Equation (3.44), let us define the spatial correlation matrix for the BS as

$$\mathbf{R}_{BS} = \begin{bmatrix} \rho_{11}^{BS} & \rho_{12}^{BS} & \cdots & \rho_{1M}^{BS} \\ \rho_{21}^{BS} & \rho_{22}^{BS} & \cdots & \rho_{2M}^{BS} \\ \vdots & \vdots & \ddots & \vdots \\ \rho_{M1}^{BS} & \rho_{M2}^{BS} & \cdots & \rho_{MM}^{BS} \end{bmatrix} \tag{3.45}$$

which is again a symmetric matrix with the unit diagonal components as in Equation (3.43). Note that the correlation coefficients, $\{\rho_{n_1 n_2}^{MS}\}$ and $\{\rho_{m_1 m_2}^{BS}\}$, in Equation (3.42) and Equation (3.45), can be analytically determined by the spatial correlation function for the given PAS model, for example, Equation (3.22) and Equation (3.23) for the Uniform PAS model.

In order to generate the channel gain matrix $\mathbf{A}_l$ in Equation (3.39), information on the channel correlation between the Tx and Rx antennas is required. In fact, the spatial correlation matrices at BS and MS, $\mathbf{R}_{BS}$ and $\mathbf{R}_{MS}$, do not provide sufficient information to generate $\mathbf{A}_l$. In fact, as illustrated in Figure 3.13, the correlation of coefficients between the pairs of Tx and Rx, $\alpha_{m_1 n_1}^{(l)}$ and $\alpha_{m_2 n_2}^{(l)}$, is additionally required, i.e.,

$$\rho_{n_2 m_2}^{n_1 m_1} = \left\langle \left| \alpha_{m_1 n_1}^{(l)} \right|^2, \left| \alpha_{m_2 n_2}^{(l)} \right|^2 \right\rangle \tag{3.46}$$

where $n_1 \neq n_2$ and $m_1 \neq m_2$. In general, there is no known theoretical solution to Equation (3.46). However, it can be approximated as

$$\rho_{n_2 m_2}^{n_1 m_1} \cong \rho_{n_1 n_2}^{MS} \rho_{m_1 m_2}^{BS} \tag{3.47}$$

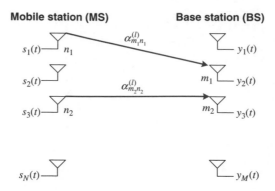

Figure 3.13   Correlation between Tx and Rx antennas.

under the assumption that the average power of the channel coefficient $\alpha_{mn}^{(l)}$ is the same for all paths.

3.2.2 Generation of Correlated MIMO Channel Coefficients

Let the MIMO fading channel for the lth path be represented by an $MN \times 1$ vector $\mathbf{a}_l = \left[a_1^{(l)}, a_2^{(l)}, \cdots, a_{MN}^{(l)} \right]^T$, which is a vector-form representation of the uncorrelated MIMO channel gain matrix $\mathbf{A}_l$ in Equation (3.39). Here, $a_x^{(l)}$ is a complex Gaussian random variable with a zero mean such that $E\left\{ \left| a_x^{(l)} \right|^2 \right\} = 1$ and $\langle |a_{x_1}^{(l_1)}|^2, |a_{x_2}^{(l_2)}|^2 \rangle = 0$ for $x_1 \neq x_2$ or $l_1 \neq l_2$ (i.e., $\{a_x^{(l)}\}$ are the uncorrelated channel coefficients). The correlated MIMO channel coefficients can now be generated by multiplying the uncorrelated MIMO fading channel vector by an $MN \times MN$ matrix $\mathbf{C}$, which is referred to as the following correlation-shaping matrix or symmetric mapping matrix, that is,

$$\tilde{\mathbf{A}}_l = \sqrt{P_l} \mathbf{C} \mathbf{a}_l \tag{3.48}$$

where P_l is the average power of the lth path [40,46]. Note that $\tilde{\mathbf{A}}_l$ is the correlated $MN \times 1$ MIMO channel vector with the correlated MIMO fading channel coefficients, given as

$$\tilde{\mathbf{A}} = \left[\underbrace{\alpha_{11}^{(l)}, \alpha_{21}^{(l)}, \cdots, \alpha_{M1}^{(l)}}, \ \underbrace{\alpha_{12}^{(l)}, \alpha_{22}^{(l)}, \cdots, \alpha_{M2}^{(l)}}, \ \underbrace{\alpha_{13}^{(l)}, \cdots, \alpha_{MN}^{(l)}} \right]^T \tag{3.49}$$

In fact, the correlation-shaping matrix $\mathbf{C}$ in Equation (3.48) defines the spatial correlation coefficients. In the sequel, we describe how the correlation-shaping matrix $\mathbf{C}$ is generated. First by using Equation (3.47), a spatial correlation matrix is given as

$$\mathbf{R} = \begin{cases} \mathbf{R}_{BS} \otimes \mathbf{R}_{MS} : & \text{downlink} \\ \mathbf{R}_{MS} \otimes \mathbf{R}_{BS} : & \text{uplink} \end{cases} \tag{3.50}$$

where $\otimes$ denotes the Kronocker product. Using $\mathbf{R}$ in Equation (3.50), a root-power correlation matrix Γ is given as

$$\Gamma = \begin{cases} \sqrt{\mathbf{R}}, & \text{for field type} \\ \mathbf{R}, & \text{for complex type} \end{cases}$$

where Γ is a non-singular matrix which can be decomposed into a symmetric mapping matrix (or correlation-shaping matrix) with the Cholesky or square-root decomposition as follows:

$$\Gamma = \mathbf{C}\mathbf{C}^T \tag{3.51}$$

Note that $\mathbf{C}$ in Equation (3.51) can be obtained by Cholesky decomposition or square-root decomposition, depending on whether $\mathbf{R}_{BS}$ and $\mathbf{R}_{MS}$ are given as the complex matrices or real matrices, respectively [43].

MATLAB® Program: Generation of Correlated MIMO Fading Channel

The SUI channel model in Section 2.2.5 can be modified for generating a MIMO channel. Note that the spatial correlation coefficients ρ_{ENV} are given by the real numbers in Section 2.2.5. Therefore, $\mathbf{C}$ can be obtained by the square-root decomposition. Program 3.1 ("channel_coeff") is a MATLAB® code to generate the correlated MIMO fading channel.

Program 3.1 "channel_coeff" to generate a correlated MIMO fading channel

```
function hh=channel_coeff(NT,NR,N,Rtx,Rrx,type)
% correlated Rayleigh MIMO channel coefficient
% Inputs:
%   NT   : number of transmitters
%   NR      : number of receivers
%   N   : length of channel matrix
%   Rtx  : correlation vector/matrix of Tx
%                    e.g.) [1 0.5], [1 0.5;0.5 1]
%   Rrx  : correlation vector/matrix of Rx
%   type : correlation type: 'complex' or 'field'
% Outputs:
%   hh      : NR x NT x N correlated channel
% uncorrelated Rayleigh fading channel, CN(1,0)
h=sqrt(1/2)*(randn(NT*NR,N)+j*randn(NT*NR,N));
if nargin, hh=h; return; end % Uncorrelated channel
if isvector(Rtx), Rtx=toeplitz(Rtx); end
if isvector(Rrx), Rrx=toeplitz(Rrx); end
% Narrow band correlation coefficient
if strcmp(type,'complex')
  C =chol(kron(Rtx,Rrx))'; % Complex correlation
 else
  C =sqrtm(sqrt(kron(Rtx,Rrx))); % Power (field) correlation
end
% Apply correlation to channel matrix
hh=zeros(NR,NT,N);
for i=1:N, tmp=C*h(:,i); hh(:,:,i)=reshape(tmp,NR,NT); end
```

3.2.3 I-METRA MIMO Channel Model

I-METRA (Intelligent Multi-element Transmit and Receive Antennas) MIMO channel model has been proposed by a consortium of the industries and universities, including Nokia [42–45]. It is based on a stochastic MIMO channel model discussed in Section 3.2.2, which generates a correlated MIMO fading channel using the spatial correlation derived for ULA (Uniform Linear Array) subject to a single or multiple cluster with the Uniform Truncated Gaussian, or Truncated Laplacian PAS. In the current discussion, the Truncated Gaussian and Truncated Laplacian PAS models are simply referred to as Gaussian and Laplacian PAS, respectively.

The overall procedure for the I-METRA MIMO channel modeling consists of two main steps as shown in Figure 3.14. In the first step, the BS and MS spatial correlation matrices ($\mathbf{R}_{BS}$ and $\mathbf{R}_{MS}$) and normalization factor are determined for the specified channel configuration, including the number of BS and MS antennas, antenna spacing, the number of clusters, PAS, AS, and AoA. The spatial correlation matrix $\mathbf{R}$ for uplink or downlink is determined by Equation (3.50). In the second step, a symmetric mapping matrix $\mathbf{C}$ is found by Equation (3.51) and then, the correlated fading MIMO channel is generated by multiplying it by the power per path and the uncorrelated fading signal as in Equation (3.48).

Once the correlated MIMO channel coefficients are generated for each path by following the procedure in Figure 3.14, the overall MIMO channel is simulated by using a tapped delay line.

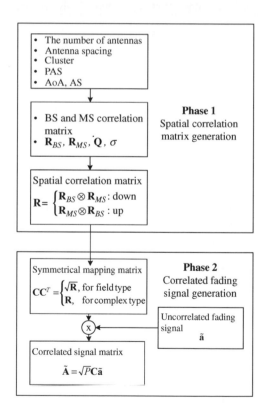

Figure 3.14 I-METRA MIMO channel modeling procedure: an overview.

Figure 3.15 shows a functional block diagram that implements the overall MIMO channel characteristics, taking the delay and power profiles into account [40]. Here, an uncorrelated fading channel is generated with the pre-stored Doppler spectrum. It is multiplied by a spatial correlation mapping matrix to generate a correlated fading channel as detailed in Figure 3.14. The given PDP characteristics are implemented by passing the correlated fading signal through a FIR filter that is designed to satisfy the given average power and delay specification for each path. Furthermore, antenna radio pattern can be adjusted by generating a steering matrix. Some of these attributes are detailed in the sequel.

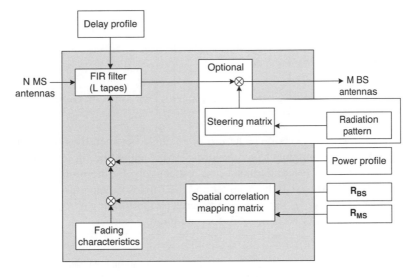

Figure 3.15 Functional block diagram for I-METRA MIMO channel model. (Reproduced with permission from L. Schumacher *et al.*, "MIMO channel characterisation," Technical Report IST-1999-11729 METRA, D2, 2001. © 2001 IST-METRA.)

3.2.3.1 Doppler Spectrum

A fading process is independently generated by any fading channel model for the SISO channel. Among many different methods of generating the independent fading process, the FWGN channel model will be useful, simply because it easily accommodates various types of Doppler spectrum, for example, flat, classical, Laplacian Doppler spectrum that is provided with the I-METRA MIMO channel. Furthermore, the MIMO channels with different Doppler spectrum models can also be generated by modifying the I-METRA MIMO channel as desired.

3.2.3.2 Rician Fading for a MIMO Channel

In contrast with Rician fading in the SISO channel, a phase change in each LOS path between the different antennas must be taken into account for modeling a Rician fading process in the

MIMO channel. As shown in Figure 3.16, the Rician fading process can be modeled by the sum of two matrices, each of which are weighed by the power ratio between the LOS signal and the scattered signal, respectively. A matrix for the Rician fading channel of the first path, $\mathbf{H}_1$, can be represented as

$$\mathbf{H}_1 = \sqrt{K}\sqrt{P_1}\mathbf{H}_{\mathrm{LOS}} + \sqrt{P_1}\mathbf{H}_{\mathrm{Rayleigh}} \tag{3.52}$$

where P_1 denotes the average power of the first path and K is a power ratio of the LOS to Rayleigh components. Furthermore, $\mathbf{H}_{\mathrm{LOS}}$ represents the LOS component [45,54], which is defined as

$$\mathbf{H}_{LOS}(t) = e^{j2\pi f_d t}\begin{bmatrix} 1 \\ e^{j2\pi\frac{d_{Rx}}{\lambda}\sin(AoA_{Rx})} \\ \vdots \\ e^{j2\pi\frac{d_{Rx}}{\lambda}(M-1)\sin(AoA_{Rx})} \end{bmatrix} \cdot \begin{bmatrix} 1 \\ e^{j2\pi\frac{d_{Tx}}{\lambda}\sin(AoD_{Tx})} \\ \vdots \\ e^{j2\pi\frac{d_{Tx}}{\lambda}(N-1)\sin(AoD_{Tx})} \end{bmatrix}^T \tag{3.53}$$

In Equation (3.53), $f_d = (v/\lambda)\cos\alpha$, where α is an angle between the Direction-of-Movement (DoM) and LOS component (see Figure 3.16). Meanwhile, d_{rx} and d_{tx} denote the antenna spacing in the receiver and transmitter, respectively. Furthermore, AoA_{rx} and AoD_{tx} represent the Angle-of-Arrival at the receiver and the Angle-of-Departure at the transmitter, respectively. Note that $\mathbf{H}_{\mathrm{LOS}}(t)$ has been constructed by multiplying $e^{j2\pi f_d t}$ by a matrix of the unit-magnitude components with an AoA-specific phase.

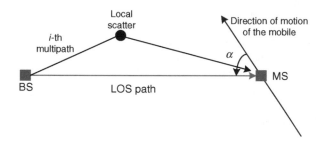

Figure 3.16   Signal model for a Rician fading MIMO channel.

3.2.3.3 Steering Matrix

Let φ denote a Direction-of-Arrival (DoA). Referring to Figure 3.17, AoA is the angle of arrival for each individual multipath component, while mean DoA refers to the mean of these AoAs. When $\varphi \neq 0°$, and thus, mean DoA of the incidental field is not on a broadside, an antenna radio pattern incurs a phase difference of $d\sin\varphi$ between two adjacent antenna elements. Up to this point, we have just described how a correlation matrix and fading signal are generated without taking this particular aspect into account. In the beamforming system which deals with

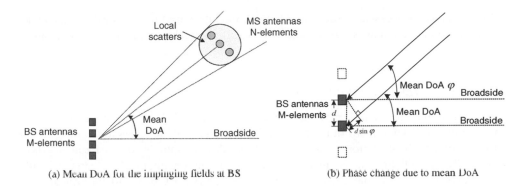

(a) Mean DoA for the impinging fields at BS   (b) Phase change due to mean DoA

Figure 3.17   Effect of DoA. (Reproduced with permission from L. Schumacher *et al.*, "MIMO channel characterisation," Technical Report IST-1999-11729 METRA, D2, 2001. © 2001 IST-METRA.)

the phase difference between antenna elements, however, it must be reflected into the channel model.

Let us consider the effects of the mean DoA when all scatters are located near the MS as illustrated in Figure 3.17(a). It shows that the impinging field at the BS tends to be confined to a narrow azimuth region with well-defined mean DoA [46]. When $\varphi \neq 0°$, the received signals in two adjacent antenna elements of BS are subject to the delay of $\tau = (d/c)\sin \varphi$. By modifying Equation (3.40) for this situation, the received signal can be represented as

$$\mathbf{y}(t) = \mathbf{W}(\varphi_{BS}) \int \mathbf{H}(\tau)\mathbf{s}(t-\tau)d\tau \tag{3.54}$$

where $\mathbf{W}(\varphi)$ is the steering diagonal matrix for the given mean DoA of φ. The steering diagonal matrix is defined as

$$\mathbf{W}(\varphi) = \begin{bmatrix} w_1(\varphi) & 0 & \cdots & 0 \\ 0 & w_2(\varphi) & \cdots & 0 \\ \vdots & \vdots & \ddots & \vdots \\ 0 & 0 & \cdots & w_M(\varphi) \end{bmatrix}_{M \times M} \tag{3.55}$$

where $w_m(\varphi)$ represents the average phase shift relative to the first antenna element for the mean azimuth DoA of the impinging field equal to φ. For a uniform linear antenna (ULA) array with element spacing of d, $w_m(\varphi)$ is given as

$$w_m(\varphi) = f_m(\varphi)e^{-j2\pi(m-1)(d/\lambda)\sin\varphi} \tag{3.56}$$

where $f_m(\varphi)$ is the complex radiation pattern of the mth antenna element. Note that Equation (3.56) represents the combined effect of the phase differences due to DoA and the antenna radiation pattern. When the antenna signals at the array are statistically independent (uncorrelated), we expect random variation in phase between two antennas.

In this case, it is not necessary to define mean DoA and introduce the steering diagonal matrix as in Equation (3.55), that is, Equation (3.40) is still applicable without any modification.

3.2.4 3GPP MIMO Channel Model

Table 3.3 presents the channel parameters for the I-METRA MIMO channel that had been proposed in 3GPP. UE (User Equipment) and Node B in this table refers to the MS (mobile station) and BS (base station), respectively. Case A corresponds to the frequency-non-selective Rayleigh fading environment without any correlation among all antenna elements, which can be used as a simple reference model. Case B and Case C deal with the typical urban macrocell environment with the different delay spread, in which each delay component is coming from the same AoA.

Case D models the microcell and bad urban environments with each delay component from the different AoA. We assume that all channel taps (paths) are subject to Rayleigh fading. In order to model the LOS situation, however, the first tap of the channel models for Case B and Case C may be subject to Rician fading. Meanwhile, the different PDPs of ITU mobility model are employed for the different cases.

For 4×4 MIMO channel model with antenna spacing of 0.5λ in Table 3.3, the complex spatial correlation matrices in the BS and MS are found by Equation (3.15) and Equation (3.16). For the different cases in Table 3.3 where BS and MS have the different spatial characteristics in terms of PAS and AoA, the specific spatial correlation matrices are given as follows:

- **Case B:** BS with Laplacian PAS for AS $= 5°$ and AoA $= 20°$, MS with Uniform PAS for AoA $= 22.5°$

$$
\mathbf{R}_{BS} = \begin{bmatrix}
1 & 0.4640+j0.8499 & -0.4802+j0.7421 & -0.7688-j0.0625 \\
0.4640-j0.8499 & 1 & 0.4640+j0.8499 & -0.4802+j0.7421 \\
-0.4802-j0.7421 & 0.4640-j0.8499 & 1 & 0.4640+j0.8499 \\
-0.7688+j0.0625 & -0.4802-j0.7421 & 0.4640-j0.8499 & 1
\end{bmatrix}
$$

(3.57)

$$
\mathbf{R}_{MS} = \begin{bmatrix}
1 & -0.3043 & 0.2203 & -0.1812 \\
-0.3043 & 1 & -0.3043 & 0.2203 \\
0.2203 & -0.3043 & 1 & -0.3043 \\
-0.1812 & 0.2203 & -0.3043 & 1
\end{bmatrix}
$$

(3.58)

- **Case C:** BS with Laplacian PAS for AS $= 10°$ and AoA $= 20°$, MS with Laplacian PAS for AS $= 35°$ and AoA $= 67.5°$

Table 3.3 I-METRA channel parameters in 3GPP [47].

		Case A Rayleigh uncorrelated	Case B macrocell	Case C macrocell	Case D microcell/bad-urban
	Number of paths	1	4	6	6
	PDP	N/A	ITU Pedestrian A	ITU Vehicular A	ITU Pedestrian B
	Doppler spectrum	Classical	Classical	Laplacian	Laplacian
	Speed (km/h)	3/40/120	3/40/120	3/40/120	3/40/120
UE (MS)	Topology	N/A	0.5λ spacing	0.5λ spacing	0.5λ spacing
	PAS	N/A	Path #1, Rician, $K = 6\,dB$ (uniform over 360°)	Laplacian, $AS = 35°$ (uniform over 360°)	Laplacian, $AS = 35°$ (uniform over 360°)
	DoM (deg)	N/A	0	22.5	−22.5
	AoA (deg)	N/A	22.5 (all paths)	67.5 (all paths)	22.5 (odd paths) −67.5 (even paths)
Node B (BS)	Topology	N/A	ULA: (1) 0.5λ spacing (2) 4.0λ spacing	ULA: (1) 0.5λ spacing (2) 4.0λ spacing	ULA: (1) 0.5λ spacing (2) 4.0λ spacing
	PAS	N/A	Laplacian, $AS = 5°$	Laplacian, $AS = 10°$	Laplacian, $AS = 15°$
	AoA (deg)	N/A	20,50[a]	20,50[a]	2,−20,10,−8,−3,31[b]

[a] AoA identical to all paths for Case B and Case C: 20° for Case B and 50° for Case C.

[b] AoA varying with each path for Case D, for example, 2° for the first path and 31° for the last path.

(Reproduced with permission from 3GPP TR25.876 v1.1.0, "Multiple-input multiple-output (MIMO) antenna processing for HSDPA," Technical Specification Group Radio Access Network, 2002. © 2002. 3GPP™ TSs and TRs are the property of ARIB, ATIS, CCSA, ETSI, TTA and TTC who jointly own the copyright in them. They are subject to further modifications and are therefore provided to you "as is" for information purposes only. Further use is strictly prohibited.)

$$\mathbf{R}_{BS} = \begin{bmatrix} 1 & 0.4290+j0.7766 & -0.3642+j0.5472 & -0.4527-j0.0521 \\ 0.4290-j0.7766 & 1 & 0.4290+j0.7766 & -0.3642+j0.5472 \\ -0.3642-j0.5472 & 0.4290-j0.7766 & 1 & 0.464+j0.8499 \\ -0.4527+j0.0521 & -0.3642-j0.5472 & 0.4290-j0.7766 & 1 \end{bmatrix}$$

$$(3.59)$$

$$\mathbf{R}_{MS} = \begin{bmatrix} 1 & -0.6906+j0.3419 & 0.4903-j0.3626 & -0.3733+j0.3450 \\ -0.6906-j0.3419 & 1 & -0.6906+j0.3419 & 0.4903-j0.3626 \\ 0.4903+j0.3626 & -0.6906-j0.3419 & 1 & -0.6906+j0.3419 \\ -0.3733-j0.3450 & 0.4903+j0.3626 & -0.6906-j0.3419 & 1 \end{bmatrix}$$

$$(3.60)$$

- **Case D:** BS with Laplacian PAS for AS $= 15°$ and AoA $= 2°$, MS with Laplacian PAS for AS $= 35°$ and AoA $= 22.5°/-67.5°$ (even path/odd path)

$$\mathbf{R}_{BS} = \begin{bmatrix} 1 & 0.7544+j0.0829 & 0.4109+j0.0938 & 0.2313+j0.0803 \\ 0.7544-j0.0829 & 1 & 0.7544+j0.0829 & 0.4109+j0.0938 \\ 0.4109-j0.0938 & 0.7544-j0.0829 & 1 & 0.7544+j0.0829 \\ 0.2313-j0.0803 & 0.4109-j0.0938 & 0.7544-j0.0829 & 1 \end{bmatrix}$$

$$(3.61)$$

$$\mathbf{R}_{MS} = \begin{bmatrix} 1 & 0.0819+j0.4267 & -0.0719+j0.0124 & -0.0863+j0.0124 \\ 0.0819-j0.4267 & 1 & 0.0819+j0.4267 & -0.0719+j0.0124 \\ -0.0719-j0.0124 & 0.0819-j0.4267 & 1 & 0.0819+j0.4267 \\ -0.0863-j0.0124 & -0.0719-j0.0124 & 0.0819-j0.4267 & 1 \end{bmatrix}$$

$$(3.62)$$

The spatial correlation functions for Equations (3.57)–(3.62) follow from Equation (3.22) and Equation (3.23) for the Uniform PAS model, and from Equation (3.32) and Equation (3.33) for the Truncated Laplacian PAS model. It is clear that the smaller the AS, the larger the correlation coefficient at the BS (e.g., Case B with the smallest AS has the largest value of the correlation coefficient at the BS).

Figures 3.18–3.21 present the downlink simulation results for Case B with two transmit and two receive antennas at mobile speed of 120km/h, which have been generated by the MATLAB® program provided by I-METRA [45]. Figure 3.18 shows the time-varying channel coefficients between each pair of transmit and receive antennas. Figure 3.19 shows the average

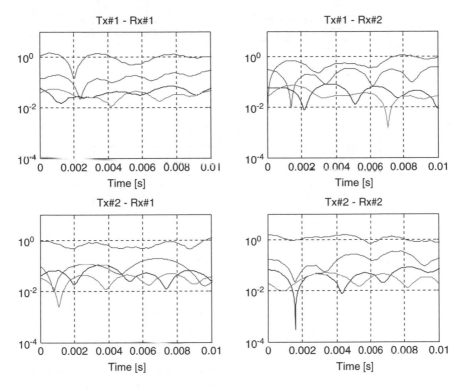

Figure 3.18 Time-varying channel coefficients: 2Tx-2Rx Case B channel. (Reproduced with permission from L. Schumacher *et al.*, "Channel characterisation," Technical Report IST-2000-30148 I-METRA, D2, v1.2, 2002. © 2002 IST-METRA.)

PDP (denoted by circles) and ideal PDP subject to ITU-R Pedestrian A model (denoted by squares) for each pair of transmit and receive antennas. Figure 3.20 shows the correlation coefficients for each pair of multipaths between the transmit and receive antennas. Figure 3.21 shows the Doppler spectrum over the transmit and receive antennas. We observe the Laplacian Doppler spectrum for the first path which is subject to Rician fading with $K = 6$ dB while all other paths show the classical Doppler spectrum.

MATLAB® program for the I-METRA MIMO channel model is available from a web site [42].

3.3 SCM MIMO Channel Model

SCM has been proposed by a joint work of Ad Hoc Group (AHG) in 3GPP and 3GPP2, which aimed at specifying the parameters for spatial channel model and developing a procedure for channel modeling [28,48]. It is a ray-based channel model, which superposes sub-ray components on the basis of PDP, PAS, and antenna array structure.

Even if it is applicable to both SISO and MIMO channel modeling as discussed in Section 2.2.3, the ray-based model, which generates the sub-ray components based up on the PDP, PAS, and a structure of antenna array, is more useful with the following advantages:

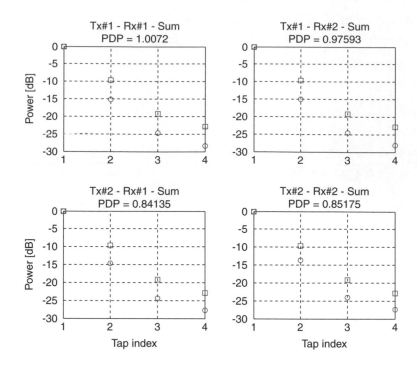

Figure 3.19 Normalized PDP: 2Tx-2Rx Case B channel. (Reproduced with permission from L. Schumacher *et al.*, "Channel characterisation," Technical Report IST-2000-30148 I-METRA, D2, v1.2, 2002. © 2002 IST-METRA.)

- Directly models the statistical characteristics of MIMO channel.
- Maintains the statistical characteristics in the time, space, and frequency domains.
- Simple to implement.
- Flexible in changing the various types of PDP and PAS.
- Supports both LOS and NLOS channels.
- Its effective rank of channel matrix **H** depending on the number of sub-rays in each path, M.

3.3.1 SCM Link-Level Channel Parameters

Since only one snapshot of the channel characteristics can be captured by the link-level channel model, link-level simulations are not enough for understanding the typical behavior of the system and evaluating the system-level performance, such as average system throughput and outage rate. For example, the functional processes of system-level attributes, such as scheduling and HARQ, cannot be represented by the link-level simulation. In fact, the link-level simulation is used for the purpose of calibration, which compares the performance results from different implementations of the given algorithm [28].

Note that both SCM link-level channel model and I-METRA channel model employ a similar set of parameters. Table 3.4 presents the parameters for a link-level channel model in

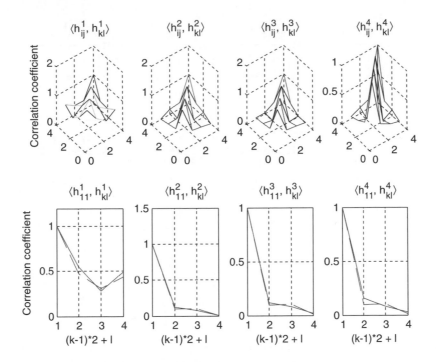

Figure 3.20 Correlation coefficients for individual paths: 2Tx-2Rx Case B channel. (Reproduced with permission from L. Schumacher *et al.*, "Channel characterisation," Technical Report IST-2000-30148 I-METRA, D2, v1.2, 2002. © 2002 IST-METRA.)

SCM. It is obvious that the SCM model follows the existing 3GPP and 3GPP2 MIMO models while assuming the uniform linear array (ULA).

The Power Azimuth Spectrum (PAS) of a path arriving at the BS follows a Laplacian distribution [28]. For an AoD (Angle of Departure) $\bar{\theta}$ and RMS angular spread σ, the per-path BS PAS at an angle θ is given by

$$P(\theta, \sigma, \bar{\theta}) = N_o e^{\frac{-\sqrt{2}|\theta - \bar{\theta}|}{\sigma}} G(\theta) \tag{3.63}$$

where $G(\theta)$ is the BS antenna gain and N_0 is the normalization constant given as

$$\frac{1}{N_0} = \int_{-\pi + \bar{\theta}}^{\pi + \bar{\theta}} e^{\frac{-\sqrt{2}|\theta - \bar{\theta}|}{\sigma}} G(\theta) d\theta, \quad -\pi + \bar{\theta} \leq \theta \leq \pi + \bar{\theta} \tag{3.64}$$

In case that a sector antenna (e.g., 3-sector or 6-sector antenna) is employed in the BS, the antenna gain must be taken into account for computing the PAS. A typical antenna pattern for the BS is given as

$$A(\theta) = -\min \left[12 \left(\frac{\theta}{\theta_{3\,\mathrm{dB}}} \right)^2, A_m \right] \text{ [dB]}, \quad \text{for} \quad -180° \leq \theta \leq -180° \tag{3.65}$$

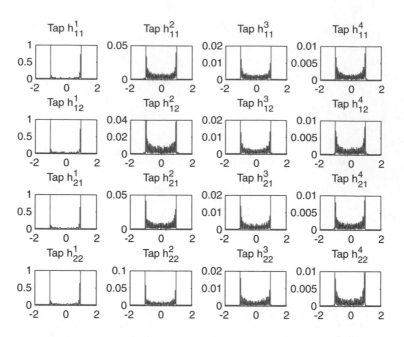

Figure 3.21 Doppler spectrum for each path: 2Tx-2Rx Case B channel. (Reproduced with permission from L. Schumacher *et al.*, "Channel characterisation," Technical Report IST-2000-30148 I-METRA, D2, v1.2, 2002. © 2002 IST-METRA.)

where $\theta_{3\,dB}$ and A_m denote the 3dB beamwidth and the maximum attenuation, respectively [28]. Note that the antenna gain is defined as $G(\theta) = 10^{A(\theta)/10}$. For the 3-sector antenna with $\theta_{3\,dB} = 70°, A_m = 20$ dB. It has the antenna pattern as shown in Figure 3.22 and antenna gain of 14dBi. For 6-sector antenna with $\theta_{3\,dB} = 35°$, meanwhile, $A_m = 23$ dB and its antenna gain is 17dBi. For AoA = 22.5 and AS = 35, Figure 3.22 shows the overall PAS, which is given by a product of PAS and antenna gain as varying the direction of sub-ray, θ.

On the other hand, the PAS of a path arriving at the MS is assumed to follow a Laplacian distribution or uniform distribution over $(0, 2\pi]$. As an omnidirectional antenna is used by the MS, no antenna gain is taken into PAS. Consider an incoming path that arrives at an AoA (Angle of Arrival) $\bar{\theta}$ with RMS angular spread σ. Then, the Laplacian PAS of the sub-ray arriving at an angle θ for MS is given by

$$P(\theta, \sigma, \bar{\theta}) = N_o e^{\frac{-\sqrt{2}|\theta - \bar{\theta}|}{\sigma}}, \qquad -\pi + \bar{\theta} \le \theta \le \pi + \bar{\theta} \tag{3.66}$$

where N_0 is the normalization constant given by

$$\frac{1}{N_0} = \int_{-\pi + \bar{\theta}}^{\pi + \bar{\theta}} e^{\frac{-\sqrt{2}|\theta - \bar{\theta}|}{\sigma}} d\theta = \sqrt{2}\sigma \left(1 - e^{-\sqrt{2}\pi/\sigma}\right) \tag{3.67}$$

Table 3.4 SCM link-level parameters for calibration purpose [28].

Model		Case I		Case II		Case III		Case IV	
Corresponding 3GPP Designator*		Case B		Case C		Case D		Case A	
Corresponding 3GPP2 Designator*		Model A, D, E		Model C		Model B		Model F	
PDP		Modified Pedestrian A		Vehicular A		Pedestrian B		Single Path	
# of Paths		1) 4+1 (LOS on, K = 6dB) 2) 4 (LOS off)		6		6		1	
Relative Path Power (dB) / Delay (ns)		1) 0.0 2) −Inf	0	0,0	0	0.0	0	0	0
		1) −6.51 2) 0.0	0	−1.0	310	−0.9	200		
		1) −16.21 2) −9.7	110	−9.0	710	−4.9	800		
		1) −25.71 2) −19.2	190	−10.0	1090	−8.0	1200		
		1) −29.31 2) −22.8	410	−15.0	1730	−7.8	2300		
				−20.0	2510	−23.9	3700		
Speed (km/h)		1) 3 2) 30, 120		3, 30, 120		3, 30, 120		3	
UE/Mobile Station	Topology	Reference 0.5λ		Reference 0.5λ		Reference 0.5λ		N/A	
	PAS	1) LOS on: Fixed AoA for LOS component, remaining power has 360 degree uniform PAS. 2) LOS off: PAS with a Laplacian distribution, RMS angle spread of 35 degrees per path		RMS angle spread of 35 degrees per path with a Laplacian distribution or 360 degree uniform PAS.		RMS angle spread of 35 degrees per path with a Laplacian distribution		N/A	
	DoT (degrees)	0		22.5		-22.5		N/A	
	AoA (degrees)	22.5 (LOS component) 67.5 (all other paths)		67.5 (all paths)		22.5 (odd numbered paths), -67.5 (even numbered paths)		N/A	
Node B/ Base Station	Topology	Reference: ULA with 0.5λ-spacing　or　4λ-spacing　or　10λ-spacing						N/A	
	PAS	Laplacian distribution with RMS angle spread of 2 degrees　or　5 degrees, per path depending on AoA/AoD						N/A	
	AoD/AoA (degrees)	50° for 2° RMS angle spread per path 20° for 5° RMS angle spread per path						N/A	
Note: Designers correspond to channel models previously proposed in 3GPP and 3GPP2 ad-hoc groups.									

(Reproduced with permission from 3GPP TR 25.996 v7.0.0, "Spatial channel model for multiple input multiple output (MIMO) simulations (release 7)," Technical Specification Group Radio Access Network, 2007. © 2007. 3GPP™ TSs and TRs are the property of ARIB, ATIS, CCSA, ETSI, TTA and TTC who jointly own the copyright in them.)

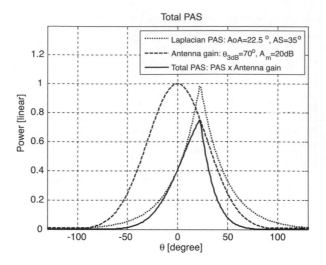

Figure 3.22 Overall PAS with the antenna gain.

Meanwhile, the uniform PAS of the sub-ray arriving at an angle θ for MS is given by

$$P(\theta, \sigma, \bar{\theta}) = N_0 \cdot 1, \qquad -\sqrt{3}\sigma + \bar{\theta} \le \theta \le \sqrt{3}\sigma + \bar{\theta} \tag{3.68}$$

where $N_0 = 1/2\sqrt{3}\sigma$. The PAS in Equations (3.66) and (3.68) do not take the antenna gain into account.

3.3.2 SCM Link-Level Channel Modeling

Figure 3.23 shows the overall procedure of implementing the ray-based SCM link-level channel model [28]. Generation of the channel with a specific PAS has been described in Section 3.3.3. Figure 3.23 also includes the channel parameters that are specific to the ray-based SCM model. Ray-based SCM model parameters are defined in Figure 3.24.

Consider the transmit antenna with S array elements and receive antenna with U array elements. Then, the channel coefficient of the nth path between the sth element and the uth element can be conceptually represented as

$$h_{s,u,n}(t) = \sqrt{\frac{n\text{th Path}}{\text{Power}}} \sum_{m=1}^{M} \left\{ \begin{pmatrix} \text{BS} \\ \text{PAS} \end{pmatrix} \cdot \begin{pmatrix} \text{Phase due to} \\ \text{BS Array} \end{pmatrix} \cdot \begin{pmatrix} \text{MS} \\ \text{PAS} \end{pmatrix} \cdot \begin{pmatrix} \text{Phase due to} \\ \text{MS Array} \end{pmatrix} \right\}$$

$$\tag{3.69}$$

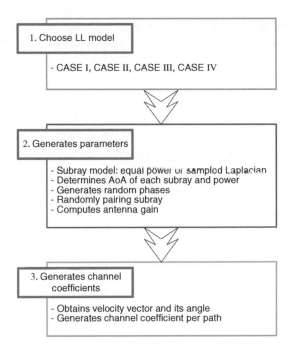

Figure 3.23 SCM link-level channel model: an overview.

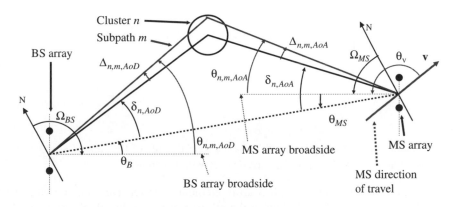

Figure 3.24 Ray-based SCM angle parameters [28]. (Reproduced with permission from 3GPP TR 25.996 v7.0.0, "Spatial channel model for multiple input multiple output (MIMO) simulations (release 7)," Technical Specification Group Radio Access Network, 2007. © 2007. 3GPP™ TSs and TRs are the property of ARIB, ATIS, CCSA, ETSI, TTA and TTC who jointly own the copyright in them.)

where M denotes the number of sub-rays in each path and the quantities in the bracket correspond to the attributes for each sub-ray. More specifically for the uniform-power sub-ray case, Equation (3.69) is given as

$$h_{u,s,n}(t) = \sqrt{\frac{P_n}{M}} \sum_{m=1}^{M} \begin{pmatrix} \sqrt{G_{BS}(\theta_{n,m,AoD})}\exp\left(j\left[kd_s\sin(\theta_{n,m,AoD}) + \Phi_{n,m}\right]\right) \times \\ \sqrt{G_{MS}(\theta_{n,m,AoA})}\exp\left(jkd_u\sin(\theta_{n,m,AoA})\right) \times \\ \exp\left(jk\|\mathbf{v}\|\cos(\theta_{n,m,AoA}-\theta_v)t\right) \end{pmatrix} \qquad (3.70)$$

where $k = 2\pi/\lambda$, d_s is the distance (in meters) from BS antenna element s with respect to the reference element $(s=1)$, d_u is the distance (in meters) from MS antenna element u with respect to the reference element $(u=1)$, $\theta_{n,m,AoD}$ and $\theta_{n,m,AoA}$ denote the AoD and AoA of the mth sub-ray in the nth path, respectively, $\Phi_{n,m}$ is a random phase of the mth sub-ray in the nth path, and θ_v is the Direction of Travel (DoT) with the magnitude of the MS velocity $\|\mathbf{v}\|$ [28].

For the case of the discretized (sampled) Laplacian model, the channel coefficient Equation (3.69) is given as

$$h_{u,s,n}(t) = \sqrt{P_n} \sum_{m=1}^{M} \begin{pmatrix} \sqrt{P_{BS}(\theta_{n,m,AoD})G_{BS}(\theta_{n,m,AoD})}\exp\left(j\left[kd_s\sin\theta_{n,m,AoD} + \Phi_{n,m}\right]\right) \times \\ \sqrt{P_{MS}(\theta_{n,m,AoD})G_{MS}(\theta_{n,m,AoA})}\exp\left(jkd_u\sin\theta_{n,m,AoA}\right) \\ \times\exp\left(jk\|\mathbf{v}\|\cos(\theta_{n,m,AoA}-\theta_v)t\right) \end{pmatrix}$$

$$(3.71)$$

For the Rician channel model [28], the channel coefficients of the line-of-sight (LOS) and non-line-of-sight (NLOS) components are respectively given as

$$h_{s,u,n=1}^{LOS}(t) = \sqrt{\frac{1}{K+1}}h_{s,u,1}(t) + \sqrt{\frac{K}{K+1}} \begin{pmatrix} \sqrt{G_{BS}(\theta_{BS})}\exp(jkd_s\sin\theta_{BS}) \times \\ \sqrt{G_{MS}(\theta_{MS})}\exp(jkd_u\sin\theta_{MS} + \Phi_{LOS}) \times \\ \exp(jk\|\mathbf{v}\|\cos(\theta_{MS}-\theta_v)t) \end{pmatrix}$$

$$(3.72)$$

and

$$h_{s,u,n}^{LOS}(t) = \sqrt{\frac{1}{K+1}} h_{s,u,n}(t), \quad \text{for} \quad n \neq 1 \tag{3.73}$$

where K denotes the Rician factor. As indicated in Equation (3.72), only one sub-ray in the first path is subject to LOS while Equation (3.70) or (3.71) is still applicable to all paths except the first one.

3.3.3 Spatial Correlation of Ray-Based Channel Model

In general, the channel coefficients of two antenna elements can be correlated both in the temporal and spatial domains. Spatial correlation is the cross-correlation of the signals that originate from the same source and are received at two spatially separated antenna elements [49]. Consider a spatial channel model for the uniform linear array antenna in which all antenna elements are equally spaced apart with a distance of d, as illustrated in Figure 3.25. Here, spatial correlation between two adjacent antenna elements for the nth path is given by

$$\rho(d) = E\left\{h_{1,u,n}(t) \cdot h_{2,u,n}^{*}(t)\right\} = \int_{-\pi}^{\pi} e^{\frac{j2\pi d \sin \theta}{\lambda}} P(\theta) d\theta \tag{3.74}$$

In the case where all sub-rays have equal power, Equation (3.74) can be simplified into

$$\rho_{SCM}^{sc}(d) = \frac{1}{M} \sum_{m=1}^{M} e^{\frac{j2\pi d \sin \theta_{n,m,AoA}}{\lambda}} \tag{3.75}$$

where $\theta_{n,m,AoA}$ is the AoA of the mth sub-ray [58]. For the mean AoA of $\bar{\theta}$, the AoA of the mth sub-ray can be represented as $\theta_{n,m,AoA} = \bar{\theta} + \Delta_{n,m,AoA}$ where $\Delta_{n,m,AoA}$ is the offset angle associated with the desired Angular Spread (AS).

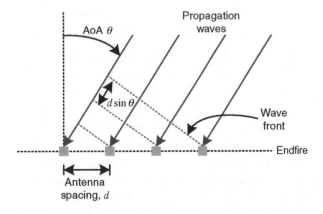

Figure 3.25   Signal model for uniform linear array (ULA) antenna.

Figure 3.26 shows the spatial correlation for the uniform power sub-ray method as mean AoA and antenna spacing are varied. It is clear that the correlation decreases as AS or antenna spacing increases. Furthermore, the correlation value is close to 1 for the mean AoA of $\bar{\theta} = \pm 90°$ since each antenna experiences almost the same wave-front while it decreases as the mean AoA approaches zero. Figure 3.27 illustrates these two extreme cases of AoA: $\bar{\theta} = 0°$ and $\bar{\theta} = 90°$. Note that the spatial correlation is equal to 1 when AoA is $\pm 90°$ and antenna space d is $\lambda/2$. In case where the MS moves, the channel correlation is time-varying due to the Doppler spread. Figure 3.28 shows the phase difference between antenna elements as the direction of travel (DoT) varies. Meanwhile, the temporal correlation cannot be derived directly from the PAS, but can be derived for the uniform-power sub-ray as

$$\rho_{SCM}^{tc}(\tau) = E\left\{ h_{s,u,n}(t+\tau) \cdot h_{s,u,n}^*(t) \right\}$$

$$= \frac{1}{M} \sum_{m=1}^{M} e^{\frac{j2\pi\tau\|\mathbf{v}\|\cos(\theta_{n,m,AoA}-\theta_v)}{\lambda}} \tag{3.76}$$

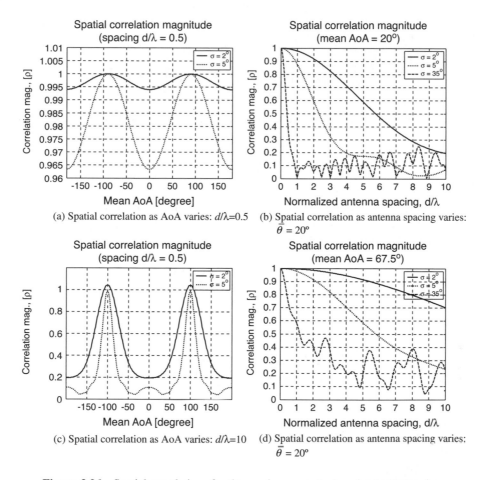

(a) Spatial correlation as AoA varies: d/λ=0.5

(b) Spatial correlation as antenna spacing varies: $\bar{\theta} = 20°$

(c) Spatial correlation as AoA varies: d/λ=10

(d) Spatial correlation as antenna spacing varies: $\bar{\theta} = 20°$

Figure 3.26   Spatial correlations for the varying mean AoA and antenna spacing.

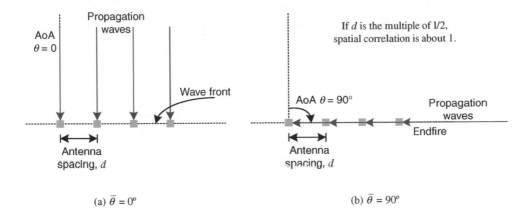

(a) $\bar{\theta} = 0°$

(b) $\bar{\theta} = 90°$

Figure 3.27 Extreme cases of AoA: an illustration.

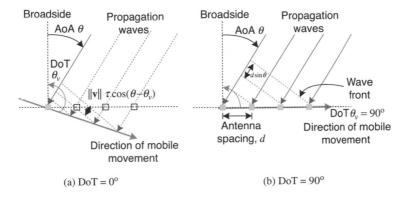

(a) DoT = 0°

(b) DoT = 90°

Figure 3.28 Phase difference between antenna elements for the varying direction of travel (DoT).

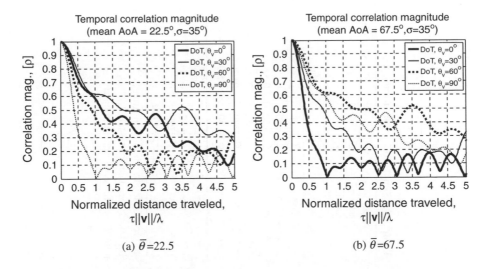

(a) $\bar{\theta}$=22.5

(b) $\bar{\theta}$=67.5

Figure 3.29 Temporal correlation for the varying distance traveled (AS $\sigma = 35°$).

Table 3.5 Simulation results for the link-level SCM MIMO channel model [28].

	Normalized antenna spacing	AS (deg)	AoA (deg)	(1) Reference in SCM text		(2) Numerical integral of Laplacian PAS		(3) Discrete integral using equal power subray in SCM text		(4) Simulation result using equal power subray in SCM text	
				Mag.	Complex	Mag.	Complex	Mag.	Complex	Mag.	Complex
BS	0.5	5	20	0.9688	0.4743 + 0.8448i	0.9683	0.4640 + 0.8499i	0.9676	0.4640 + 0.8491i	0.9678	0.4652 + 0.8487i
	0.5	2	50	0.9975	-0.7367 + 0.6725i	0.9975	-0.7390 + 0.6700i	0.9975	-0.7390 + 0.6700i	0.9976	-0.7388 + 0.6703i
	4	5	20	0.3224	-0.2144 + 0.2408i	0.3198	-0.2203 + 0.2318i	0.1968	-0.1433 + 0.1350i	0.2008	-0.1434 + 0.1406i
	4	2	50	0.8624	0.8025 + 0.3158i	0.8631	0.7954 + 0.3350i	0.8530	0.7872 + 0.3285i	0.8541	0.7891 + 0.3268i
	10	5	20	0.0704	-0.0617 + 0.034i	0.5020	-0.0619 + 0.0327i	0.0708	-0.0590 + 0.0391i	0.0655	-0.0529 + 0.0386i
	10	2	50	0.5018	-0.2762 - 0.4190i	0.0700	-0.2615 - 0.4285i	0.3858	-0.1933 - 0.3339i	0.3885	-0.1972 - 0.3347i
MS	0.5	104	0	0.3042	-0.3042						
	0.5	35	-67.5	0.7744	-0.6948 - 0.342i	0.7771	0.6966 - 0.3445i	0.7326	-0.6324 - 0.3697i	0.7313	-0.6310 - 0.3696i
	0.5	35	22.5	0.4399	0.0861 + 0.431i	0.4397	0.0858 + 0.4312i	0.3426	0.0238 + 0.3418i	0.3394	0.0227 + 0.3386i
	0.5	35	67.5	0.7744	-0.6948 + 0.342i	0.7771	-0.6966 + 0.3445i	0.7326	-0.6324 + 0.3697i	0.7313	-0.6310 + 0.3696i

(Reproduced with permission from 3GPP TR 25.996 v7.0.0, "Spatial channel model for multiple input multiple output (MIMO) simulations (release 7)," Technical Specification Group Radio Access Network, 2007. © 2007. 3GPP™ TSs and TRs are the property of ARIB, ATIS, CCSA, ETSI, TTA and TTC who jointly own the copyright in them.)

where θ_v and $||\mathbf{v}||$ denote the DoT and mobile speed, respectively [49]. Note that if DoT $\theta_v = 90°$, the temporal correlation is equal to the spatial correlation. It can then be shown as

$$
\begin{aligned}
\rho_{SCM}^{tc}(\tau) &= \frac{1}{M}\sum_{m=1}^{M} e^{\frac{j2\pi\tau||\mathbf{v}||\cos(\theta_{n,m,AoA}-\theta_v)}{\lambda}} \\
&= \frac{1}{M}\sum_{m=1}^{M} e^{\frac{j2\pi\tau||\mathbf{v}||\sin\theta_{n,m,AoA}}{\lambda}} \\
&\stackrel{d=\tau||\mathbf{v}||}{=} \frac{1}{M}\sum_{m=1}^{M} e^{\frac{j2\pi d\sin\theta_{n,m,AoA}}{\lambda}} \\
&= \rho_{SCM}^{sc}(d)
\end{aligned}
\tag{3.77}
$$

Figure 3.29 shows that the temporal correlation for the different DoT and mean AoA as the distance traveled varies with respect to the wavelength. It is clear that the temporal correlation increases as $\bar{\theta}-\theta_v$ is reduced.

Table 3.5 presents the simulation results for the link-level channel model, which are compared with the reference values of correlation provided by SCM text in the column (1). Column (2) in Table 3.5 corresponds to the result of the numerical integral of Equation (3.74) for Laplacian PAS while the column (3) is the result of the discrete integral of Equation (3.75).

Finally, column (4) in Table 3.5 is generated by simulating the Ray-based channel model with the given offset angles. As the SCM approximates the Laplacian PAS with 20 sub-rays with equal power, the correlation coefficients almost coincide with the simulation results as long as they are large enough (e.g., greater than 0.3).

The MATLAB® program for the SCM channel model can be downloaded from the web site [50]. See [51–67] for additional information about MIMO channel modeling.

4

Introduction to OFDM

4.1 Single-Carrier vs. Multi-Carrier Transmission

4.1.1 Single-Carrier Transmission

4.1.1.1 Single-Carrier Baseband Transmission: System Model

Figure 4.1 shows a typical end-to-end configuration for a single-carrier communication system. Consider a band-limited channel $h(t)$ with a bandwidth of W. The transmit symbols $\{a_n\}$, each with a symbol period of T seconds, that is, a data rate of $R = 1/T$, are pulse-shaped by a transmit filter $g_T(t)$ in the transmitter. After receiving them through the channel, they are processed with the receive filter, equalizer, and detector in the receiver. Let $g_T(t)$, $g_R(t)$, and $h^{-1}(t)$ denote the impulse response of the transmit filter, receive filter, and equalizer, respectively. The output of the equalizer can be expressed as

$$y(t) = \sum_{m=-\infty}^{\infty} a_m g(t - mT) + z(t) \tag{4.1}$$

where $z(t)$ is an additive noise and $g(t)$ is the impulse response of overall end-to-end system given as

$$g(t) = g_T(t) * h(t) * g_R(t) * h^{-1}(t) \tag{4.2}$$

The equalizer is designed to compensate the effect of channel. In this section, we just assume that the effect of the channel is perfectly compensated by the equalizer as given in Equation (4.2). Therefore, the overall impulse response is subject to transmit and receive filters only. When the noise term is ignored, the sampled output signal of the equalizer can be expressed as

$$y(t_n) = \sum_{m=-\infty}^{\infty} a_m g((n - m)T) \quad \text{with} \quad t_n = nT \tag{4.3}$$

MIMO-OFDM Wireless Communications with MATLAB®   Yong Soo Cho, Jaekwon Kim, Won Young Yang and Chung G. Kang
© 2010 John Wiley & Sons (Asia) Pte Ltd

<figure>

Figure 4.1   Single-carrier baseband communication system model.
</figure>

Isolating the nth sample to detect a_n, Equation (4.3) can be written as

$$y(t_n) = a_n g(0) + \sum_{m=-\infty, m \neq n}^{\infty} a_m g((n-m)T) \tag{4.4}$$

Note that $g(t)$ cannot be time-limited due to the finite channel bandwidth. In case that $g((n-m)T) \neq 0$ for $\forall m \neq n$, the second term in Equation (4.4) remains as an inter-symbol interference (ISI) to a_n. In fact, the ISI is caused by a trail of the overall impulse response, which could degrade the performance of a digital communication system. Therefore, the transmit filter and receive filter must be designed deliberately so as to minimize or completely eliminate the ISI in a practical system. Figure 4.2 illustrates how the ISI is incurred by the trail of the overall impulse response in the receiver. As illustrated here, the extent of ISI depends on the duration of a symbol period T: the shorter the symbol period is, the larger the influence of the ISI may become. This implies that unless $g((n-m)T) \neq 0$ for $\forall m \neq n$, the ISI becomes significant as the data rate increases (i.e., decreasing T in Figure 4.2) in a single-carrier system.

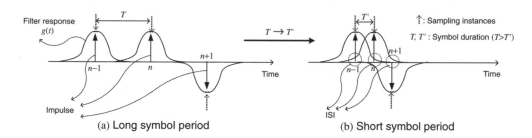

(a) Long symbol period   (b) Short symbol period

Figure 4.2   Illustration: inter-symbol interference (ISI) and symbol period.

4.1.1.2 ISI and Nyquist Criterion

In Equation (4.4), ISI can be completely eliminated by fulfilling the following time-domain condition on the overall impulse response:

$$g(nT) = \delta[n] = \begin{cases} 1, & n = 0 \\ 0, & n \neq 0 \end{cases} \tag{4.5}$$

Note that the condition in Equation (4.5) is equivalent to the following frequency-domain condition:

$$\sum_{i=-\infty}^{\infty} G\left(f - \frac{i}{T}\right) = T \tag{4.6}$$

where $G(f)$ is the Fourier transform of $g(t)$, which represents the overall frequency response. The condition Equation (4.5) or Equation (4.6) is known as the Nyquist criterion [68,69], which guarantees an ISI-free communication even with a short symbol period T for high-rate transmission in a single-carrier transmission system. Note that filters satisfying the Nyquist criterion are referred to as Nyquist filters. One obvious Nyquist filter is an ideal LPF (Low Pass Filter), which has a sinc function-type of impulse response or equivalently, rectangular pulse (or brick-wall) type of frequency response as described by

$$G_I(f) = \frac{1}{2W}\text{rect}\left(\frac{f}{2W}\right) = \begin{cases} T, & |f| \le \dfrac{1}{2T} \\[2mm] 0, & |f| > \dfrac{1}{2T} \end{cases} \tag{4.7}$$

where $W = R/2 = 1/(2T)$. In Equation (4.7), R and W correspond to the Nyquist rate and Nyquist bandwidth, respectively. Note that the Nyquist bandwidth W is the minimum possible bandwidth that is required to realize the date rate R without ISI. However, the ideal filter in Equation (4.7) is not *physically realizable* because its impulse response is not causal (i.e., $g(t) \ne 0$ for some $t < 0$ and its duration is infinite). Another well-known, yet physically realizable Nyquist filter is the raised-cosine filter, which is specified by the following frequency response:

$$G_{RC}(f) = \begin{cases} T, & |f| \le \dfrac{1-r}{2T} \\[3mm] \dfrac{T}{2}\left\{1 + \cos\dfrac{\pi T}{r}\left(|f| - \dfrac{1-r}{2T}\right)\right\}, & \dfrac{1-r}{2T} < |f| \le \dfrac{1+r}{2T} \\[3mm] 0, & |f| > \dfrac{1+r}{2T} \end{cases} \tag{4.8}$$

where r is the roll-off factor that tailors the total bandwidth and $0 \le r \le 1$. It is clear that Equation (4.8) satisfies the ISI-free condition Equation (4.6), but is not as sharp as the frequency response of an ideal LPF. Note that the raised cosine frequency response in Equation (4.8) occupies a frequency range wider than the Nyquist bandwidth. The actual bandwidth is governed by the roll-off factor r. Figures 4.3(a1) and (a2) show the impulse and frequency responses of raised cosine filters with the roll-off factors of $r = 0, 0.5$, and 1, respectively. Note that the raised cosine filter with $r = 0$ happens to be identical to the ideal LPF, and the raised cosine filter with $r = 1$ occupies twice the Nyquist bandwidth. In the special case where the channel is ideal, we require $G_R(f) = G_T^*(f)$ where $G_T(f)$ and $G_R(f)$ are the frequency response of the transmit filter $g_T(t)$ and receive filter $g_R(t)$, respectively. Since $G_R(f) = G_T^*(f)$, $G_{RC}(f) = |G_T(f)|^2$ or $G_T(f) = \sqrt{G_{RC}(f)}$ and thus, the transmit filter $g_T(t)$ must have the following frequency response, which is known as the square-root raised cosine filters:

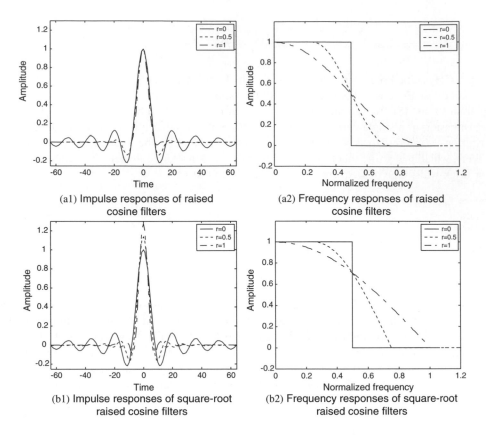

(a1) Impulse responses of raised
cosine filters

(a2) Frequency responses of raised
cosine filters

(b1) Impulse responses of square-root
raised cosine filters

(b2) Frequency responses of square-root
raised cosine filters

Figure 4.3   Raised cosine and square-root raised cosine filters.

$$
G_{SRRC}(f) = \begin{cases} \sqrt{T}, & |f| \le \dfrac{1-r}{2T} \\[2mm] \sqrt{\dfrac{T}{2}\left\{1+\cos\dfrac{\pi T}{r}\left(|f|-\dfrac{1-r}{2T}\right)\right\}}, & \dfrac{1-r}{2T} < |f| \le \dfrac{1+r}{2T} \\[2mm] 0, & |f| > \dfrac{1+r}{2T} \end{cases} \tag{4.9}
$$

Figures 4.3(b1) and (b2) show the impulse and frequency responses of square-root raised cosine filters with the roll-off factors of $r = 0, 0.5$, and 1, respectively. Note that if two identical square-root raised cosine filters are used as transmit and receive filters, respectively, the latter plays the role of the matched filter for the former [70] and thus, the combined frequency response satisfies the Nyquist criterion even though each of them does not.

4.1.1.3 Limitation of Single-Carrier Transmission for High Data Rate

In order to support the symbol rate of R_s symbols per second, the minimum required bandwidth is the Nyquist bandwidth, which is given by $R_s/2$[Hz]. It implies that wider bandwidth is required to support a higher data rate in a single-carrier transmission. So far, it has been assumed that the channel is perfectly compensated by the equalizer. However, as the symbol rate increases, the signal bandwidth becomes larger. When the signal bandwidth becomes larger than the coherence bandwidth in the wireless channel, the link suffers from multi-path fading, incurring the inter-symbol interference (ISI). In general, adaptive equalizers are employed to deal with the ISI incurred by the time-varying multi-path fading channel. Furthermore, the complexity of an equalizer increases with the data rate. More specifically, adaptive equalizers are implemented by finite impulse response (FIR) filters with the adaptive tap coefficients that are adjusted so as to minimize the effect of ISI. In fact, more equalizer taps are required as the ISI becomes significant, for example, when the data rate increases.

The optimum detector for the multi-path fading channel is a maximum-likelihood sequence detector (MLSD), which bases its decisions on the observation of a sequence of received symbols over successive symbol intervals, in favor of maximizing the posteriori probability. Note that its complexity depends on the modulation order and the number of multi-paths. Let M and L denote the number of possible signal points for each modulation symbol and the span of ISI incurred over the multi-path fading channel, respectively. Due to a memory of length L for the span of ISI, M^L corresponding Euclidean distance path metrics must be evaluated to select the best sequence in the MLSD. When a more efficient transmission is sought by increasing M and a high data rate is implemented, the complexity of the optimum equalizer becomes prohibitive, for example, $M^L = 64^{16}$ for $L \approx 16$ with 64-QAM at the data rate of 10Mbps over the multi-path fading channel with a delay spread of 10 μs. When M and L are too large, other more practical yet suboptimum equalizers, such as MMSE or LS equalizer, can be used. However, the complexity of these suboptimum equalizers is still too enormous to be implemented as the ISI increases with the data rate. This particular situation can be explained by the fact that the inverse function (a frequency-domain response of equalizer) becomes sharper as the frequency-selectivity of the channel increases [1,17]. In conclusion, a high data rate single-carrier transmission may not be feasible due to too much complexity of the equalizer in the receiver.

4.1.2 Multi-Carrier Transmission

4.1.2.1 Basic Structure of a Multi-Carrier Transmission Scheme

To overcome the frequency selectivity of the wideband channel experienced by single-carrier transmission, multiple carriers can be used for high rate data transmission. Figure 4.4(a) shows the basic structure and concept of a multi-carrier transmission system [72–74]. Here, a wideband signal is analyzed (through multiple narrowband filter $H_k(f)$'s) into several narrowband signals at the transmitter and is synthesized (through multiple narrowband filter $G_k(f)$'s, each being matched to $H_k(f)$) at the receiver so that the frequency-selective wideband channel can be approximated by multiple frequency-flat narrowband channels as depicted in Figure 4.4(b). Note that the frequency-nonselectivity of narrowband channels reduces the complexity of the equalizer for each subchannel. As long as the orthogonality among the subchannels is maintained, the ICI (inter-carrier interference) can be suppressed, leading to distortionless transmission [72,83,84].

In the multichannel system, let the wideband be divided into N narrowband subchannels, which have the subcarrier frequency of $f_k, k = 0, 1, 2, \cdots, N-1$. Figure 4.5(a) shows the basic structure of a multi-carrier communication scheme, which is one specific form of the multichannel

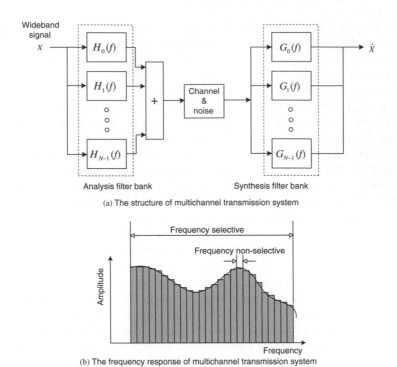

(a) The structure of multichannel transmission system

(b) The frequency response of multichannel transmission system

Figure 4.4 Structure and frequency characteristic of multichannel transmission system.

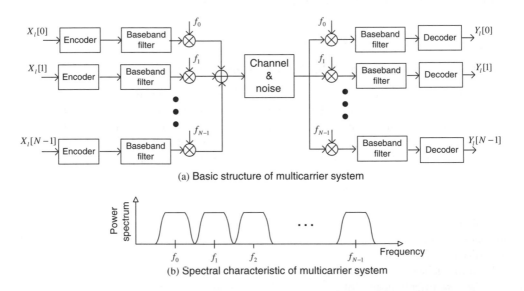

(a) Basic structure of multicarrier system

(b) Spectral characteristic of multicarrier system

Figure 4.5 Structure and spectral characteristic of multicarrier transmission system.

system [75], where the different symbols are transmitted with orthogonal subchannels in parallel form. Let $X_l[k]$ and $Y_l[k]$ denote the transmitted and received signals carried at the carrier frequency f_k in the lth symbol interval, respectively. It implies that multi-carrier transmission can be regarded as a kind of FDMA (frequency division multiple access) method. Figure 4.5(b) illustrates a transmitted signal spectrum in the multi-carrier transmission system, which occupies multiple subbands of equal bandwidth, each centered at the different carrier frequency. If each subchannel is bandlimited as depicted in Figure 4.5(b), it becomes an FMT (Filtered Multi-Tone) transmission, which will be discussed in Section 4.1.2.3.

While a FMT type of multicarrier transmission system can cope with the frequency-selectivity of a wideband channel, its implementation becomes complex since it involves more encoders/decoders and oscillators, and higher quality filters as the number of subcarriers increases.

4.1.2.2 OFDM Transmission Scheme

Orthogonal frequency division multiplexing (OFDM) transmission scheme is another type of a multichannel system, which is similar to the FMT transmission scheme in the sense that it employs multiple subcarriers. As shown in Figure 4.6(a), it does not use individual bandlimited filters and oscillators for each subchannel and furthermore, the spectra of subcarriers are overlapped for bandwidth efficiency, unlike the FMT scheme where the wideband is fully divided into N orthogonal narrowband subchannels. The multiple orthogonal subcarrier signals, which are overlapped in spectrum, can be produced by generalizing the single-carrier Nyquist criterion in Equation (4.6) into the multi-carrier criterion. In practice, discrete Fourier transform (DFT) and inverse DFT (IDFT) processes are useful for implementing these orthogonal signals. Note that DFT and IDFT can be implemented efficiently by using fast Fourier transform (FFT) and inverse fast Fourier transform (IFFT), respectively. In the OFDM transmission system, N-point IFFT is taken for the transmitted symbols $\{X_l[k]\}_{k=0}^{N-1}$, so as to generate $\{x[n]\}_{n=0}^{N-1}$, the samples for the sum of N orthogonal subcarrier signals. Let $y[n]$ denote the received sample that corresponds to $x[n]$ with the additive noise $w[n]$ (i.e., $y[n] = x[n] + w[n]$). Taking the N-point FFT of the received samples, $\{y[n]\}_{n=0}^{N-1}$, the noisy version of transmitted symbols $\{Y_l[k]\}_{k=0}^{N-1}$ can be obtained in the receiver. Figure 4.6(c) shows the OFDM transmission structure implemented by IDFT/DFT. The inherent advantages of the OFDM transmission will be detailed later in this chapter. As all subcarriers are of the finite duration T, the spectrum of the OFDM signal can be considered as the sum of the frequency-shifted sinc functions in the frequency domain as illustrated in Figure 4.6(c), where the overlapped neighboring sinc functions are spaced by $1/T$. The discrete multi-tone (DMT) scheme used in ADSL (Asymmetric Digital Subscriber Line) and Zipper-based VDSL (Very high-rate Data digital Subscriber Line) also has the same structure as OFDM [77].

Since each subcarrier signal is time-limited for each symbol (i.e., not band-limited), an OFDM signal may incur out-of-band radiation, which causes non-negligible adjacent channel interference (ACI). It is clearly seen from Figure 4.6(d) that the first sidelobe is not so small as compared to the main lobe in the spectra. Therefore, OFDM scheme places a guard band at outer subcarriers, called virtual carriers (VCs), around the frequency band to reduce the out-of-band radiation. The OFDM scheme also inserts a guard interval in the time domain, called cyclic prefix (CP), which mitigates the inter-symbol interference (ISI) between OFDM symbols [78]. Details of these issues will be discussed in Section 4.2.

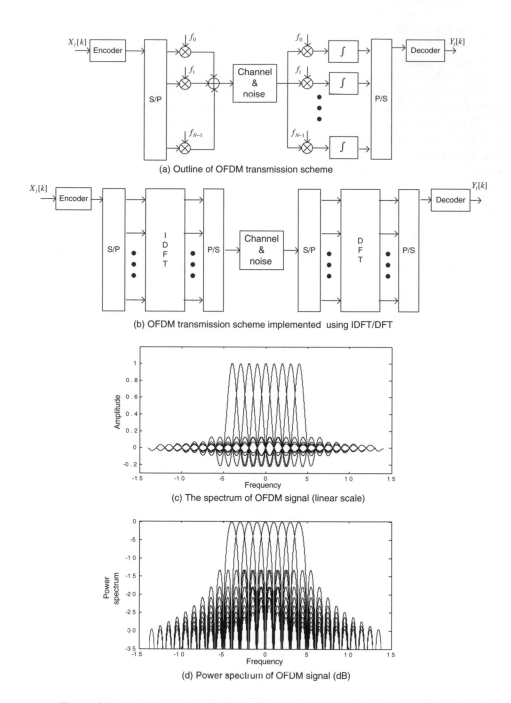

(a) Outline of OFDM transmission scheme

(b) OFDM transmission scheme implemented using IDFT/DFT

(c) The spectrum of OFDM signal (linear scale)

(d) Power spectrum of OFDM signal (dB)

Figure 4.6   Structure and spectral characteristic of OFDM transmission scheme.

4.1.2.3 FMT Transmission Scheme

The filtered multi-tone (FMT) scheme is another type of multichannel system. As shown in Figure 4.7(a), each of the transmitted symbols $\{X_l[k]\}_{k=0}^{N-1}$ is $(N+v)$-times oversampled and then filtered by a band pass filter (BPF) where v is a non-negative integer. Let $h(t)$ denote the impulse response of baseband filter. Then, the band pass filter of kth subchannel at the transmitter is given by $h_k(t) = h(t)e^{j2\pi f_k t}$ where f_k is the center frequency of the kth subchannel. Due to the oversampling at the transmitter, the sampling frequency is $(N+v)/T$ and thus, $f_k = (N+v)k/(NT)$. Note that the frequency spacing between subcarriers is $\Delta f = (N+v)/(NT)$. The received signal passes through each matched filter corresponding to $h_k(t)$ and $(N+v)$-times downsampled to yield a received symbol $\{Y_l[k]\}$.

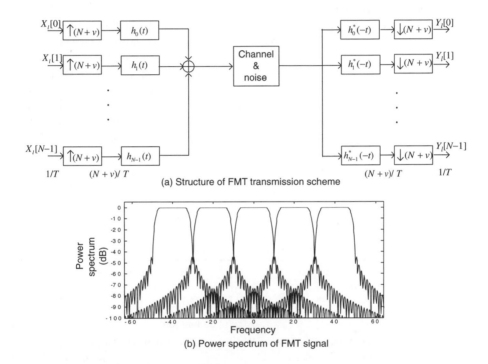

(a) Structure of FMT transmission scheme

(b) Power spectrum of FMT signal

Figure 4.7   Structure and power spectrum of FMT transmission scheme.

There are two different types of realizations for FMT scheme. One is a direct realization, which uses a bank of upsampler-filters for analysis at the transmitter and another bank of downsampler-filters for synthesis at the receiver. The other is a more efficient realization, which uses IDFT/DFT, together with polyphase filter banks for analysis/synthesis to reduce hardware complexity [79,80].

Figure 4.7(b) shows that contiguous subchannels have been further separated out by the bandlimited filters, reducing the spectral overlap between them. As opposed to the OFDM schemes, therefore, it does not require employing the virtual carriers as a guard interval.

4.1.3 Single-Carrier vs. Multi-Carrier Transmission

In the previous subsections, we have discussed the single-carrier and multi-carrier transmission schemes. It is clear that each of these schemes has its own advantages and disadvantages. The single-carrier scheme may not be useful for a high rate wireless transmission, simply because it requires a high-complexity equalizer to deal with the inter-symbol interference problem in the multi-path fading channel or equivalently, frequency-selective fading channel. Meanwhile, the multi-carrier scheme is useful for a high rate wireless transmission, which does not involve the complexity of channel equalization.

We have also discussed OFDM and FMT as two different types of multi-carrier schemes. These schemes differ from each other in the way of dividing the frequency band into subbands. OFDM does not need filters to separate the subbands since the orthogonality is kept among subcarriers, but it requires a guard band such as VCs (Virtual Carriers) to combat the ACI. In contrast, FMT uses filters to separate the subbands to reduce the ACI at the sacrifice of spectral efficiency, but it does not need the guard band. From the spectral efficiency viewpoint, FMT is known to be advantageous over OFDM only in a case where the number of subcarriers is less than 64. FMT has been classified as a scalable adaptive/advanced modulation (SAM) series, which can change the number of subcarriers and accordingly, the data rate. It has been adopted as the transmission scheme in the TETRA (TErrestrial Trunked RAdio) II standard in the European Telecommunications Standards Institute (ETSI). Other than OFDM and FMT, there are different types of multi-carrier transmission schemes, including DWMT (Discrete Wavelet Multi-Tone), OFDM/OQAM-IOTA, and so on [81,82]. Table 4.1 summarizes the differences between the single-carrier and multi-carrier transmission schemes, including their advantages and disadvantages.

Table 4.1 Comparison between the single-carrier and multi-carrier transmission schemes.

	Single-carrier transmission	Multi-carrier transmission	
		OFDM/DMT	FMT
Subcarrier spacing	—	1/(symbol duration)	$\geq$1/(symbol duration)
Pulse shaping	Nyquist filter (e.g., raised-cosine filter)	Window (e.g., rectangular)	Nyquist filter (e.g., raised-cosine filter)
Subchannel separation	—	Orthogonality	Bandpass filter
Guard interval	Not required	Required (CP)	Not required
Guard band	Not required	Required (VC)	Not required
Advantages	Simple in flat fading channels	High bandwidth efficiency for a large number of subcarriers ($\geq$64)	Small ACI
Disadvantages	High-complexity equalizer required for frequency-selective channels	Low bandwidth efficiency and large ACI for a small number of subcarriers	High bandwidth efficiency for a small number of subcarriers ($<$64)

4.2 Basic Principle of OFDM

4.2.1 OFDM Modulation and Demodulation

4.2.1.1 Orthogonality

Consider the time-limited complex exponential signals $\{e^{j2\pi f_k t}\}_{k=0}^{N-1}$ which represent the different subcarriers at $f_k = k/T_{sym}$ in the OFDM signal, where $0 \leq t \leq T_{sym}$. These signals are defined to be orthogonal if the integral of the products for their common (fundamental) period is zero, that is,

$$\frac{1}{T_{sym}} \int_0^{T_{sym}} e^{j2\pi f_k t} e^{-j2\pi f_i t} dt = \frac{1}{T_{sym}} \int_0^{T_{sym}} e^{j2\pi \frac{k}{T_{sym}} t} e^{-j2\pi \frac{i}{T_{sym}} t} dt$$

$$= \frac{1}{T_{sym}} \int_0^{T_{sym}} e^{j2\pi \frac{(k-i)}{T_{sym}} t} dt$$

$$= \begin{cases} 1, & \forall \text{ integer } k = i \\ 0, & \text{otherwise} \end{cases} \qquad (4.10)$$

Taking the discrete samples with the sampling instances at $t = nT_s = nT_{sym}/N$, $n = 0, 1, 2, \cdots, N-1$, Equation (4.10) can be written in the discrete time domain as

$$\frac{1}{N} \sum_{n=0}^{N-1} e^{j2\pi \frac{k}{T_{sym}} \cdot nT_s} e^{-j2\pi \frac{i}{T} \cdot nT_s} = \frac{1}{N} \sum_{n=0}^{N-1} e^{j2\pi \frac{k}{T_{sym}} \cdot \frac{nT}{N}} e^{-j2\pi \frac{i}{T_{sym}} \cdot \frac{nT_{sym}}{N}}$$

$$= \frac{1}{N} \sum_{n=0}^{N-1} e^{j2\pi \frac{(k-i)}{N} n}$$

$$= \begin{cases} 1, & \forall \text{ integer } k = i \\ 0, & \text{otherwise} \end{cases} \qquad (4.11)$$

The above orthogonality is an essential condition for the OFDM signal to be ICI-free.

MATLAB® Program: Checking the Orthogonality

To get familiar with the concept of orthogonality, let us check the orthogonality among the following six time-limited sinusoidal signals, along with their discrete Fourier transforms (DFTs):

$$x_1(t) = \exp(j2\pi t), \quad \text{Re}\{x_1[n] = \exp(j2\pi nT_s), T_s = 0.1 \, s\} \qquad (4.12.\text{a1})$$

$$\to X_1[k] = \text{DFT}_{16}\{x_1(nT_s), n = 0 : 15\} \qquad (4.12.\text{b1})$$

$$x_2(t) = \exp(j2\pi 2t), \text{Re}\{x_2[n] = \exp(j2\pi 2nT_s), T_s = 0.1 \, s\} \qquad (4.12.\text{a2})$$

$$\to X_2[k] = \text{DFT}_{16}\{x_2(nT_s), n = 0 : 15\} \qquad (4.12.\text{b2})$$

$$x_3(t) = \exp(j2\pi 3(t-0.1)), \text{Re}\{x_3[n] = \exp(j2\pi 3(n-1)T_s), T_s = 0.1 \, s\} \qquad (4.12.\text{a3})$$

$$\to X_3[k] = \text{DFT}_{16}\{x_3(nT_s), n = 0 : 15\} \qquad (4.12.\text{b3})$$

$$x_4(t) = \exp(j2\pi 4(t-0.1)), \text{Re}\{x_4[n] = \exp(j2\pi 4(n-1)T_s), T_s = 0.1 \, s\} \qquad (4.12.\text{a4})$$

$$\to X_4[k] = \text{DFT}_{16}\{x_4(nT_s), n = 0 : 15\} \qquad (4.12.\text{b4})$$

$$x_5(t) = \exp(j2\pi 3.9t), \mathrm{Re}\{x_5[n] = \exp(j2\pi 3.9nT_s), T_s = 0.1\,s\} \qquad (4.12.a5)$$

$$\rightarrow X_5[k] = \mathrm{DFT}_{16}\{x_5(nT_s), n = 0:15\} \qquad (4.12.b5)$$

$$x_6(t) = \exp(j2\pi 4(t-\delta)), \mathrm{Re}\{x_6[n] = \exp(j(2\pi 4nT_s - 2\pi 4\delta)), T_s = 0.1\,s\} \qquad (4.12.a6)$$

$$\text{with } \delta = \begin{cases} 0.1, & \text{if } t \ge 1.4 \\ 0.15, & \text{elsewhere} \end{cases}$$

$$\rightarrow X_6[k] = \mathrm{DFT}_{16}\{x_6(nT_s), n = 0:15\} \qquad (4.12.b6)$$

The above signals and their DFT spectra in Equation (4.12) are depicted in Figures 4.8(a1)–(a6) and (b1)–(b6). Program 4.1 ("test_orthogonality.m") intends to check the orthogonality among these different signals. It generates a matrix of the sample signal vectors $\mathbf{x}_i = \{x_i(nT_s), n = 0:15\}$ in each row, then computes the product of itself and its transpose, which checks the orthogonality among the signal vectors. Running this program yields

```
>> test_orthogonality
  ans =
  1.00          -0.00+0.00i  0.00-0.00i  -0.00-0.00i  -0.02- 0.03i  0.06-0.08i
 -0.00-0.00i     1.00        -0.00-0.00i  -0.00+0.00i  -0.04- 0.03i -0.01-0.11i
  0.00+0.00i    -0.00+0.00i   1.00         0.00+0.00i  -0.08+0.07i  -0.11+0.03i
 -0.00+0.00i    -0.00-0.00i   0.00-0.00i   1.00         0.29-0.94i  -0.76-0.58i
  0.02+0.03i    -0.04+0.03i  -0.08-0.07i   0.29+0.94i   1.00         0.33-0.85i
  0.06+0.08i    -0.01+0.11i  -0.11-0.03i  -0.76+0.58i   0.33+0.85i   1.00
```

The upper-left 4×4 submatrix is a diagonal (identity) matrix, which implies that the first four signals $\mathbf{x}_1, \mathbf{x}_2, \mathbf{x}_3$, and $\mathbf{x}_4$ (with radian frequency of an integer times the fundamental frequency $2\pi/T_{sym}$ (rad/sec) and for duration $T = 1.6[s]$) are orthogonal to each other regardless of some delay. In contrast, all the entries in the fifth/sixth rows and columns are not zero, which implies that the four signals $\mathbf{x}_1, \mathbf{x}_2, \mathbf{x}_3, \mathbf{x}_4$ and the last two signals $\mathbf{x}_5, \mathbf{x}_6$ are not mutually orthogonal since the frequency of $\mathbf{x}_5$ is not a multiple of the fundamental frequency and $\mathbf{x}_6$ has a discontinuity as can be seen from Figures 4.8(a5) and (a6), respectively. Such orthogonality can be also revealed from the DFT spectra of $\mathbf{x}_1, \mathbf{x}_2, \mathbf{x}_3, \mathbf{x}_4, \mathbf{x}_5$, and $\mathbf{x}_6$ in Figures 4.8(b1)–(b6). Note that the DFT spectra of $\mathbf{x}_1, \mathbf{x}_2, \mathbf{x}_3$, and $\mathbf{x}_4$ in Figures 4.8 (b1)–(b4) are so clear as to show the frequency $2\pi k/T_{sym}(k = 1,2,3,4)$ of each signal.[1] However, those of $\mathbf{x}_5$ and $\mathbf{x}_6$ in Figures 4.8(b5)–(b6) are not clear due to spectral leakage.

[1] In Figures 4.8(b1)–(b6), the discrete frequencies are normalized by $2\pi/T_{sym}$.

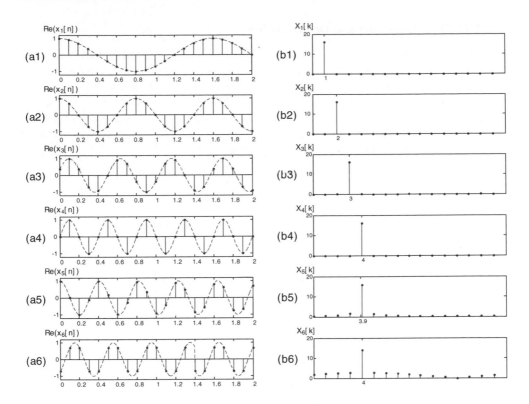

Figure 4.8 Sinusoidal signals with different frequencies/phases and their DFTs.

Program 4.1 "test_orthogonality.m" to test the orthogonality among sinusoidal signals

```
% test_orthogonality.m
% To check the orthogonality among some sinusoidal signals
% with different frequencies/phases
clear, clf
T=1.6; ND=1000; nn=0:ND; ts=0.002; tt=nn*ts; % Time interval
Ts = 0.1; M = round(Ts/ts); % Sampling period in continuous/discrete-time
nns = [1:M:ND+1]; tts = (nns-1)*ts; % Sampling indices and times
ks = [1:4 3.9 4]; tds = [0 0 0.1 0.1 0 0.15]; % Frequencies and delays
K = length(ks);
for i=1:K
   k=ks(i); td=tds(i); x(i,:) = exp(j*2*pi*k*(tt-td)/T);
   if i==K, x(K,:) = [x(K,[302:end]) x(K-3,[1:301])]; end
   subplot(K,2,2*i-1), plot(tt,real(x(i,:))),
   hold on, plot(tt([1 end]),[0 0],'k'), stem(tts,real(x(i,nns)),'.')
end
N = round(T/Ts); xn = x(:,nns(1:N));
xn*xn'/N % check orthogonality
Xk = fft(xn.').'; kk = 0:N-1;
for i=1:K,
   k=ks(i); td=tds(i); subplot(K,2,2*i), stem(kk,abs(Xk(i,:)),'.');
end
```

4.2.1.2 OFDM Modulation and Demodulation

OFDM transmitter maps the message bits into a sequence of PSK or QAM symbols which will be subsequently converted into N parallel streams. Each of N symbols from serial-to-parallel (S/P) conversion is carried out by the different subcarrier. Let $X_l[k]$ denote the lth transmit symbol at the kth subcarrier, $l = 0, 1, 2, \cdots, \infty$, $k = 0, 1, 2, \cdots, N-1$. Due to the S/P conversion, the duration of transmission time for N symbols is extended to NT_s, which forms a single OFDM symbol with a length of T_{sym} (i.e., $T_{sym} = NT_s$). Let $\Psi_{l,k}(t)$ denote the lth OFDM signal at the kth subcarrier, which is given as

$$\Psi_{l,k}(t) = \begin{cases} e^{j2\pi f_k(t - lT_{sym})}, & 0 < t \le T_{sym} \\ 0, & \text{elsewhere} \end{cases} \tag{4.13}$$

Then the passband and baseband OFDM signals in the continuous-time domain can be expressed respectively as

$$x_l(t) = \text{Re}\left\{ \frac{1}{T_{sym}} \sum_{l=0}^{\infty} \left\{ \sum_{k=0}^{N-1} X_l[k]\Psi_{l,k}(t) \right\} \right\} \tag{4.14}$$

and

$$x_l(t) = \sum_{l=0}^{\infty} \sum_{k=0}^{N-1} X_l[k] e^{j2\pi f_k(t - lT_{sym})}$$

The continuous-time baseband OFDM signal in Equation (4.14) can be sampled at $t = lT_{sym} + nT_s$ with $T_s = T_{sym}/N$ and $f_k = k/T_{sym}$ to yield the corresponding discrete-time OFDM symbol as

$$x_l[n] = \sum_{k=0}^{N-1} X_l[k] e^{j2\pi kn/N} \quad \text{for} \quad n = 0, 1, \ldots, N-1 \tag{4.15}$$

Note that Equation (4.15) turns out to be the N-point IDFT of PSK or QAM data symbols $\{X_l[k]\}_{k=0}^{N-1}$ and can be computed efficiently by using the IFFT (Inverse Fast Fourier Transform) algorithm.

Consider the received baseband OFDM symbol $y_l(t) = \sum_{k=0}^{N-1} X_l[k] e^{j2\pi f_k(t-lT_{sym})}$, $lT_{sym} < t \le lT_{sym} + nT_s$, from which the transmitted symbol $X_l[k]$ can be reconstructed by the orthogonality among the subcarriers in Equation (4.10) as follows:

$$\begin{aligned} Y_l[k] &= \frac{1}{T_{sym}} \int_{-\infty}^{\infty} y_l(t)\, e^{-j2\pi k f_k(t - lT_{sym})} dt \\ &= \frac{1}{T_{sym}} \int_{-\infty}^{\infty} \left\{ \sum_{i=0}^{N-1} X_l[i] e^{j2\pi f_i(t - lT_{sym})} \right\} e^{-j2\pi f_k(t - lT_{sym})} dt \\ &= \sum_{i=0}^{N-1} X_l[i] \left\{ \frac{1}{T_{sym}} \int_{0}^{T_{sym}} e^{j2\pi(f_i - f_k)(t - lT_{sym})} dt \right\} = X_l[k] \end{aligned} \tag{4.16}$$

where the effects of channel and noise are not taken into account. Let $\{y_l[n]\}_{n=0}^{N-1}$ be the sample values of the received OFDM symbol $y_l(t)$ at $t = lT_{sym} + nT_s$. Then, the integration in the modulation process of Equation (4.16) can be represented in the discrete time as follows:

$$
\begin{aligned}
Y_l[k] &= \sum_{n=0}^{N-1} y_l[n] e^{-j2\pi kn/N} \\
&= \sum_{n=0}^{N-1} \left\{ \frac{1}{N} \sum_{i=0}^{N-1} X_l[i] e^{j2\pi in/N} \right\} e^{-j2\pi kn/N} \\
&= \frac{1}{N} \sum_{n=0}^{N-1} \sum_{i=0}^{N-1} X_l[i] e^{j2\pi(i-k)n/N} = X_l[k]
\end{aligned}
\tag{4.17}
$$

In fact, Equation (4.17) is the N-point DFT of $\{y_l[n]\}_{n=0}^{N-1}$ and can be computed efficiently by using the FFT (Fast Fourier Transform) algorithm.

According to the above discussion, OFDM modulation and demodulation can be illustrated by the block diagram in Figure 4.9, which shows that the frequency-domain symbol $X[k]$ modulates the subcarrier with a frequency of $f_k = k/T_{sym}$, for $N = 6$ (i.e., $k = 0, 1, 2, \cdots, 5$), while it can be demodulated by using the orthogonality among the subcarriers in the receiver. Note that the original symbol $X[k]$ has a duration of T_s, but its length has been extended to $T_{sym} = NT_s$ by transmitting N symbols in a parallel form. The OFDM symbol corresponds to a composite signal of N symbols in a parallel form, which now has a duration of T_{sym}. Meanwhile, Figure 4.9(b) illustrates a typical realization of orthogonality among all subcarriers. Furthermore, it has been shown that this multi-carrier modulation can be implemented by IFFT and FFT in the transmitter and receiver, respectively. Figure 4.10 shows a complete block diagram

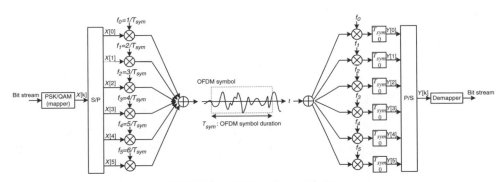

(a) OFDM modulation/demodulation

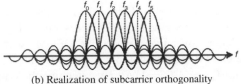

(b) Realization of subcarrier orthogonality

Figure 4.9 Illustrative block diagram of OFDM modulation and demodulation: $N = 6$.

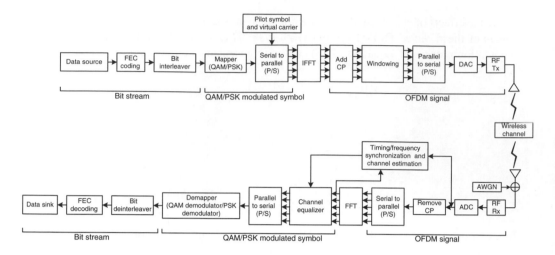

Figure 4.10 Block diagram of transmitter and receiver in an OFDM system.

of the typical OFDM transmitter and receiver, including the IFFT and FFT operations. Each detail of blocks in this diagram will be discussed in the following subsections.

4.2.2 OFDM Guard Interval

4.2.2.1 Effect of Multipath Channel on OFDM Symbols

Consider the lth OFDM signal, $x_l(t) = \sum_{k=0}^{N-1} X_l[k]e^{j2\pi f_k(t-lT_{sym})}$, $lT_{sym} < t \le lT_{sym} + nT_s$. For the channel with an impulse response of $h_l(t)$, the received signal is given as

$$y_l(t) = x_l(t)*h_l(t) + z_l(t) = \int_0^\infty h_l(\tau)x_l(t-\tau)dt + z_l(t), \quad lT_{sym} < t \le lT_{sym} + nT_s \quad (4.18)$$

where $z_l(t)$ is the additive white Gaussian noise (AWGN) process. Taking the samples of Equation (4.18) at $nT_s = nT_{sym}/N$, Equation (4.18) can be represented in a discrete time as follows:

$$y_l[n] = x_l[n]*h_l[n] + z_l[n] = \sum_{m=0}^{\infty} h_l[m]x_l[n-m] + z_l[n] \quad (4.19)$$

where $x_l[n] = x_l(nT_s)$, $y_l[n] = y_l(nT_s)$, $h_l[n] = h_l(nT_s)$, and $z_l[n] = z_l(nT_s)$.

In order to understand an ISI effect of the multipath channel, we consider the illustrative examples for the discrete-time channel in Figure 4.11, where two impulse responses with different lengths are shown along with their frequency responses. Figure 4.12 illustrates an ISI effect of the multipath channel over two consecutive OFDM symbols. Let T_{sub} denote the duration of the effective OFDM symbol without guard interval. Since $W = 1/T_s$ and thus, $\Delta f - W/N = 1/(NT_s)$ and $T_{sub} = NT_s = 1/\Delta f$. By extending the symbol duration by N times (i.e., $T_{sub} = NT_s$), the effect of the multipath fading channel is greatly reduced on the OFDM symbol. However, its effect still remains as a harmful factor that may break the orthogonality

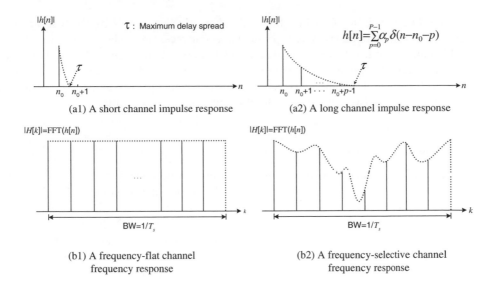

(a1) A short channel impulse response   (a2) A long channel impulse response

(b1) A frequency-flat channel   (b2) A frequency-selective channel
     frequency response   frequency response

Figure 4.11   Impulse/frequency responses of a discrete-time channel: examples.

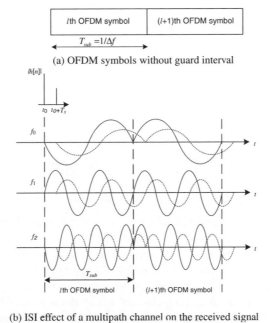

(a) OFDM symbols without guard interval

(b) ISI effect of a multipath channel on the received signal

Figure 4.12   Effect of a multipath channel on the received signal without guard interval.

among the subcarriers in the OFDM scheme. As shown in Figure 4.12(b), the first received symbol (plotted in a solid line) is mixed up with the second received symbol (plotted in a dotted line), which incurs the ISI. It is obvious that all subcarriers are no longer orthogonal over the duration of each OFDM symbol. To warrant a performance of OFDM, there must be some means of dealing with the ISI effect over the multipath channel. As discussed in the sequel, a guard interval between two consecutive OFDM symbols will be essential.

4.2.2.2 Cyclic Prefix (CP)

The OFDM guard interval can be inserted in two different ways. One is the zero padding (ZP) that pads the guard interval with zeros. The other is the cyclic extension of the OFDM symbol (for some continuity) with CP (cyclic prefix) or CS (cyclic suffix). CP is to extend the OFDM symbol by copying the last samples of the OFDM symbol into its front. Let T_G denote the length of CP in terms of samples. Then, the extended OFDM symbols now have the duration of $T_{sym} = T_{sub} + T_G$. Figure 4.13(a) shows two consecutive OFDM symbols, each of which has the CP of length T_G, while illustrating the OFDM symbol of length $T_{sym} = T_{sub} + T_G$. Meanwhile, Figure 4.13(b) illustrates them jointly in the time and frequency domains. Figure 4.13(c) shows the ISI effects of a multipath channel on some subcarriers of the OFDM symbol. It can be seen from this figure that if the length of the guard interval (CP) is set longer than or equal to the maximum delay of a multipath channel, the ISI effect of an OFDM symbol (plotted in a dotted line) on the next symbol is confined within the guard interval so that it may not affect the FFT of the next OFDM symbol, taken for the duration of T_{sub}. This implies that the guard interval longer than the maximum delay of the multipath channel allows for maintaining the orthogonality among the subcarriers. As the continuity of each delayed subcarrier has been warranted by the CP, its orthogonality with all other subcarriers is maintained over T_{sub}, such that

$$\frac{1}{T_{sub}} \int_0^{T_{sub}} e^{j2\pi f_k(t-t_0)} e^{-j2\pi f_i(t-t_0)} dt = 0, \quad k \neq i$$

for the first OFDM signal that arrives with a delay of t_0, and

$$\frac{1}{T_{sub}} \int_0^{T_{sub}} e^{j2\pi f_k(t-t_0)} e^{-j2\pi f_i(t-t_0-T_s)} dt = 0, \quad k \neq i$$

for the second OFDM signal that arrives with a delay of $t_0 + T_s$.

Figure 4.14 shows that if the length of the guard interval (CP) is set shorter than the maximum delay of a multipath channel, the tail part of an OFDM symbol (denoted by a quarter circle) affects the head part of the next symbol, resulting in the ISI. In practice, symbol timing offset (STO) may occur, which keeps the head of an OFDM symbol from coinciding with the FFT window start point. In this context, Figure 4.15 shows that even if the length of CP is set longer than the maximum delay of the multipath channel, ISI and/or ICI may occur depending on the timing of the FFT window start point. More specifically, if the FFT window start point is earlier than the lagged end of the previous symbol, ISI occurs; if it is later than the beginning of a symbol, not only ISI (caused by the next symbol), but ICI also occurs [78].

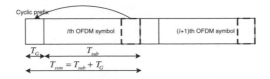

(a) OFDM symbols with CP

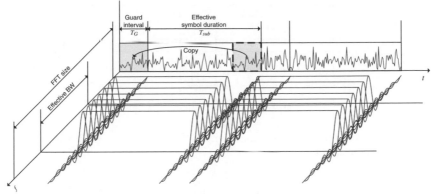

(b) Time/frequency-domain description of OFDM symbols with CP

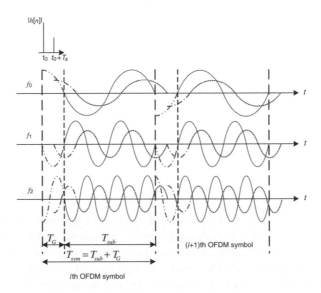

(c) ISI effect of a multipath channel for each subcarrier

Figure 4.13 Effect of a multipath channel on OFDM symbols with CP.

Now we suppose that the CP length is set not shorter than the maximum delay of the channel and the FFT window start point of an OFDM symbol is determined within its CP interval (i.e., unaffected by the previous symbol). Then the OFDM receiver takes the FFT of the received samples $\{y_l[n]\}_{n=0}^{N-1}$ to yield

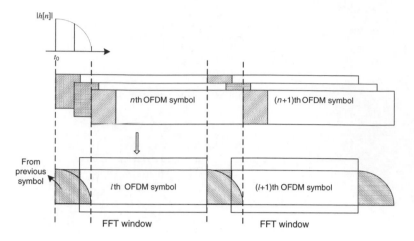

Figure 4.14 ISI effect of a multipath channel on OFDM symbols with CP length shorter than the maximum delay of the channel.

$$Y_l[k] = \sum_{n=0}^{N-1} y_l[n] e^{-j2\pi kn/N}$$

$$= \sum_{n=0}^{N-1} \left\{ \sum_{m=0}^{\infty} h_l[m] x_l[n-m] + z_l[n] \right\} e^{-j2\pi kn/N}$$

$$= \sum_{n=0}^{N-1} \left\{ \sum_{m=0}^{\infty} h_l[m] \left\{ \frac{1}{N} \sum_{i=0}^{N-1} X_l[i] e^{j2\pi i(n-m)/N} \right\} \right\} e^{-j2\pi kn/N} + Z_l[k] \qquad (4.20)$$

$$= \frac{1}{N} \sum_{i=0}^{N-1} \left\{ \left\{ \sum_{m=0}^{\infty} h_l[m] e^{-j2\pi im/N} \right\} X_l[i] \sum_{n=0}^{\infty} e^{-j2\pi(k-i)n/N} \right\} e^{-j2\pi kn/N} + Z_l[k]$$

$$= H_l[k] X_l[k] + Z_l[k]$$

where $X_l[k]$, $Y_l[k]$, $H_l[k]$, and $Z_l[k]$ denote the kth subcarrier frequency components of the lth transmitted symbol, received symbol, channel frequency response, and noise in the frequency domain, respectively. The last identity in Equation (4.20) implies that the OFDM system can be simply thought of as multiplying the input (message) symbol by the channel frequency response in the frequency domain. In other words, it can be equivalently represented as in Figure 4.16. Since $Y_l[k] = H_l[k] X_l[k]$ under no noise condition, the transmitted symbol can be detected by one-tap equalization, which simply divides the received symbol by the channel (i.e., $X_l[k] = Y_l[k]/H_l[k]$). Note that $Y_l[k] \neq H_l[k] X_l[k]$ without CP, since $\mathrm{DFT}\{y_l[n]\} \neq \mathrm{DFT}\{x_l[n]\} \cdot \mathrm{DFT}\{h_l[n]\}$ when $\{y_l[n]\} = \{x_l[n]\} * \{h_l[n]\}$ for the convolution operation $*$. In fact, $Y_l[k] = H_l[k] X_l[k]$ when $\{y_l[n]\} = \{x_l[n]\} \otimes \{h_l[n]\}$ where $\otimes$ denotes the operation for circular convolution. In other words, insertion of CP in the transmitter makes the transmit samples circularly-convolved with the channel samples, which yields $Y_l[k] = H_l[k] X_l[k]$ as desired in the receiver.

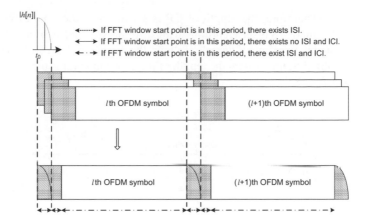

Figure 4.15   ISI/ICI effect depending on the FFT window start point.

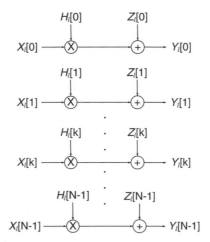

Figure 4.16   Frequency-domain equivalent model of OFDM system.

4.2.2.3 Cyclic Suffix (CS)

Cyclic suffix (CS) is also a cyclic extension of the OFDM system. It is different from CP only in that CS is the copy of the head part of an effective OFDM symbol, and it is inserted at the end of the symbol. CS is used to prevent the interference between upstream and downstream, and is also used as the guard interval for frequency hopping or RF convergence, and so on. Both CP and CS are used in Zipper-based VDSL systems in which the Zipper duplexing technique is a form of FDD (Frequency-Division Duplexing) that allocates different frequency bands (subcarriers) to downstream or upstream transmission in an OFDM symbol, allowing for bidirectional signal flow at the same time. Here, the purpose of CP and CS is to suppress the ISI effect of the multipath channel, while ensuring the orthogonality between the upstream and downstream signals. Therefore, the length of CP is set to cover the time dispersion of the channel, while the length of CS is set according to the difference between the upstream transmit time and downstream receive time. Figure 4.17 shows the structure of the OFDM symbol used in Zipper-based VDSL systems, where the length of the guard interval is the sum of CP length T_{CP} and CS length T_{CS} [85].

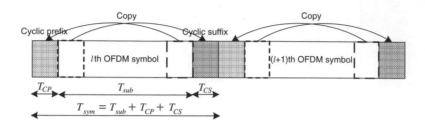

Figure 4.17   OFDM symbol with both CP and CS.

4.2.2.4 Zero Padding (ZP)

We may insert zero into the guard interval. This particular approach is adopted by multiband-OFDM (MB-OFDM) in an Ultra Wide-band (UWB) system [86]. Figures 4.18(a) and (b) show OFDM symbols with ZP and the ISI effect of a multipath channel on OFDM symbols for each subcarrier, respectively. Even with the length of ZP longer than the maximum delay of the multipath channel, a small STO causes the OFDM symbol of an effective duration to have a discontinuity within the FFT window and therefore, the guard interval part of the next OFDM symbol is copied and added into the head part of the current symbol to prevent ICI as described in Figure 4.19.

Since the ZP is filled with zeros, the actual length of an OFDM symbol containing ZP is shorter than that of an OFDM symbol containing CP or CS and accordingly, the length of a rectangular window for transmission is also shorter, so that the corresponding sinc-type spectrum may be wider. This implies that compared with an OFDM symbol containing CP or CS, an OFDM symbol containing ZP has PSD (Power Spectral Density) with the smaller in-band ripple and the larger out-of-band power as depicted in Figure 4.20, allowing more power to be used for transmission with the peak transmission power fixed.

Note that the data rate of the OFDM symbol is reduced by $T_{sub}/T_{sym} = T_{sub}/(T_{sub} + T_G)$ times due to the guard interval.

4.2.3 OFDM Guard Band

Each subcarrier component of an OFDM symbol with the effective duration T_{sub} can be regarded as a single-tone signal multiplied by a rectangular window of length T_{sub}, whose spectrum is a sinc function with zero-crossing bandwidth of $2/T_{sub}$. Therefore, the power spectrum of an OFDM signal is the sum of many frequency-shifted sinc functions, which has large out-of-band power such that ACI (adjacent channel interference) is incurred. As a result, a guard band is required to reduce the effect of ACI in the OFDM system.

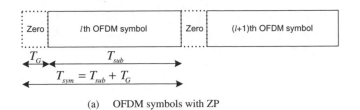

(a)   OFDM symbols with ZP

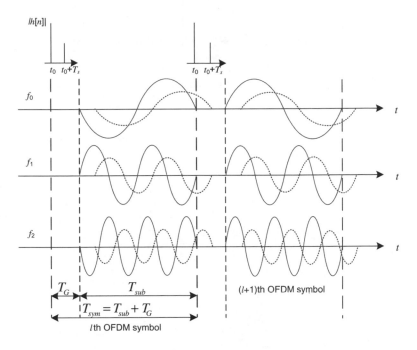

(b) The ISI effect of a multipath channel for each subcarrier

Figure 4.18   Effect of a multipath channel on OFDM symbols with ZP.

To reduce the out-of-band power of OFDM symbols, a BPF can be used, but it may require enormous computation and high complexity to make the filtering performance acceptable. As an alternative, a time-domain shaping function like raised cosine (RC) windowing can be used. The passband and baseband signals for the lth OFDM symbol, shaped by an RC window $\Psi_{l,k}(t)$ with a roll-off factor β, can be written respectively as follows:

$$x_l(t) = \mathrm{Re}\left\{ h_{RC}(t - lT_{sym}) \sum_{k=0}^{N-1} X_{l,k} \Psi_{l,k}(t) \right\} \tag{4.21}$$

and

$$x_l^{rc}(t) = h_{RC}(t - lT_{sym}) \sum_{k=0}^{N-1} X_{l,k} e^{j2\pi k \Delta f(t - lT_{sym})} \tag{4.22}$$

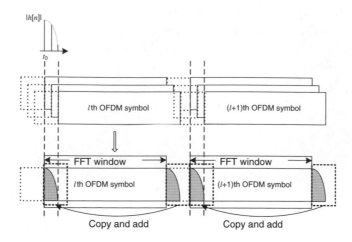

Figure 4.19 Copying-and-adding the guard interval of the next symbol into the head part of the current symbol to prevent ICI.

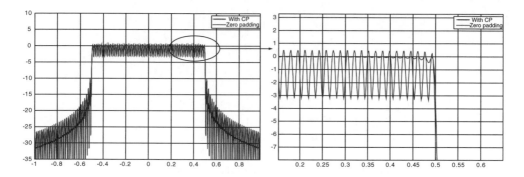

Figure 4.20 Power spectra of two OFDM symbols, one with ZP and one with CP.

where

$$\Psi_{l,k}(t) = \begin{cases} e^{j2\pi f_k(t-lT_{sym})} & \text{for } -(T_G+T_W/2) \le t \le (T_{sub}+T_W/2) \\ 0 & \text{otherwise} \end{cases} \tag{4.23}$$

and

$$h_{RC}(t) = \begin{cases} 0.5+0.5\cos(\pi(t+\beta T_{sym}+T_G)/\beta T_{sym}) & \text{for } -(T_G+\beta T_{sym}/2) \le t < -(T_G-\beta T_{sym}/2) \\ 1.0 & \text{for } -(T_G+\beta T_{sym}/2) \le t < (T_{sub}-\beta T_{sym}/2) \\ 0.5+0.5\cos(\pi(t-T_{sub}+\beta T_{sym})/\beta T_{sym}) & \text{for } (T_{sub}-\beta T_{sym}/2) \le t \le (T_{sub}+\beta T_{sym}/2) \end{cases} \tag{4.24}$$

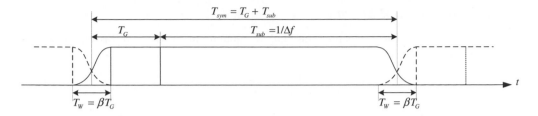

Figure 4.21 Raised cosine window for OFDM symbol.

Figure 4.21 shows a raised cosine window that is used to shape OFDM symbols for reducing their out-of-band powers. As the roll-off factor β increases, the transition part of the RC window becomes smoother so that ACI can be reduced in return for a longer effective guard interval.

Another measure against the ACI is to employ the virtual carriers (VCs), which are the unused subcarriers at both ends of the transmission band. No additional processing is required when the virtual carriers are employed. However, the spectral (bandwidth) efficiency is reduced by N_{used}/N times due to the unused subcarriers, where N_{used} is the number of subcarriers used for data transmission. The virtual carriers can be used in combination with the (RC) windowing to reduce the out-of-band power and eventually to combat the ACI. Figure 4.22 shows the power spectrum of RC windows with different roll-off factors, where 54 out of the total 64 subcarriers (excluding 10 virtual subcarriers) are used for data transmission. It can be seen from this figure that the out-of-band power decreases as the roll-off factor becomes larger.

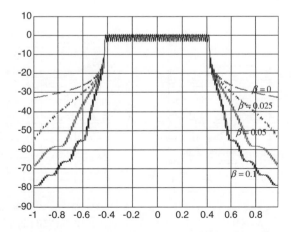

Figure 4.22 Power spectrum of raised cosine (RC).

4.2.4 BER of OFDM Scheme

The analytical BER expressions for M-ary QAM signaling in AWGN and Rayleigh channels are respectively given as

$$P_e = \frac{2(M-1)}{M \log_2 M} Q\left(\sqrt{\frac{6 E_b}{N_0} \cdot \frac{\log_2 M}{M^2-1}}\right) : \text{AWGN channel} \qquad (4.25)$$

$$P_e = \frac{M-1}{M \log_2 M}\left(1 - \sqrt{\frac{3\gamma \log_2 M/(M^2-1)}{3\gamma \log_2 M/(M^2-1)+1}}\right) : \text{Rayleigh fading channel} \qquad (4.26)$$

where γ and M denote E_b/N_0 and the modulation order, respectively [87], while $Q(\cdot)$ is the standard Q-function defined as

$$Q(x) = \frac{1}{\sqrt{2\pi}} \int_x^\infty e^{-t^2/2} dt. \qquad (4.27)$$

Note that if N_{used} subcarriers out of total N (FFT size) subcarriers (except $N_{vc} = N-N_{used}$ virtual subcarriers) are used for carrying data, the time-domain SNR, SNR_t, differs from the frequency-domain SNR, SNR_f, as follows:

$$SNR_t = SNR_f + 10 \log \frac{N_{used}}{N} [dB] \qquad (4.28)$$

MATLAB® Programs: BER of OFDM-QAM System on Varying the Length of GI

Program 4.2 ("OFDM_basic.m") can be used to simulate the effect of ISI as the length of a guard interval (CP, CS, or ZP) varies. It considers the BER performance of an OFDM system with 64-point FFT ($N = 64$) and 16 virtual carriers ($N_{vc} = N-N_{sub} = 16$), for 16-QAM signaling in the AWGN or a multipath Rayleigh fading channel (with the maximum delay of 15 samples). In Figure 4.23(a), it is clear that the BER performance with CP or ZP of length 16 samples is consistent with that of the analytic result in the Rayleigh fading channel. This implies that the OFDM system is just subject to a flat fading channel as long as CP or ZP is large

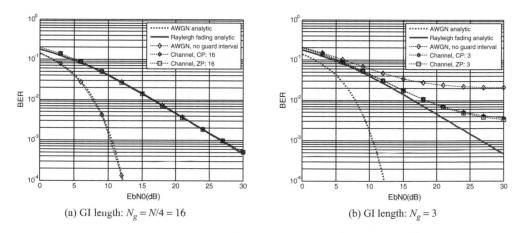

(a) GI length: $N_g = N/4 = 16$ (b) GI length: $N_g = 3$

Figure 4.23 BER performance for OFDM system with 16-QAM.

enough. It is also clear that the BER performance in an AWGN channel is consistent with the analytical results. This is true regardless of how long GI is, because there is no multipath delay in the AWGN channel. As illustrated in Figure 4.23(b), however, the effect of ISI on the BER performance becomes significant in the multipath Rayleigh fading channel as the length of GI decreases, which eventually leads to an error floor.

Program 4.2 "OFDM_basic.m" to simulate an OFDM transmission system

```
% OFDM_basic.m
clear all
NgType=1; % NgType=1/2 for cyclic prefix/zero padding
if NgType==1, nt='CP'; elseif NgType==2, nt='ZP'; end
Ch=0; % Ch=0/1 for AWGN/multipath channel
if Ch==0, chType='AWGN'; Target_neb=100; else chType='CH'; Target_neb=500; end
figure(Ch+1), clf
PowerdB=[0 -8 -17 -21 -25]; % Channel tap power profile 'dB'
Delay=[0 3 5 6 8];          % Channel delay 'sample'
Power=10.^(PowerdB/10);     % Channel tap power profile 'linear scale'
Ntap=length(PowerdB);       % Chanel tap number
Lch=Delay(end)+1;           % Channel length
Nbps=4; M=2^Nbps;     % Modulation order=2/4/6 for QPSK/16QAM/64QAM
Nfft=64;              % FFT size
Ng=Nfft/4;            % GI (Guard Interval) length (Ng=0 for no GI)
Nsym=Nfft+Ng;         % Symbol duration
Nvc=Nfft/4;           % Nvc=0: no VC (virtual carrier)
Nused=Nfft-Nvc;
EbN0=[0:5:30];        % EbN0
N_iter=1e5;           % Number of iterations for each EbN0
Nframe=3;             % Number of symbols per frame
sigPow=0;             % Signal power initialization
file_name=['OFDM_BER_' chType '_' nt '_' 'GL' num2str(Ng) '.dat'];
fid=fopen(file_name, 'w+');
norms=[1 sqrt(2)   0 sqrt(10)   0 sqrt(42)];    % BPSK 4-QAM 16-QAM
for i=0:length(EbN0)
   randn('state',0); rand('state',0); Ber2=ber(); % BER initialization
   Neb=0; Ntb=0; % Initialize the number of error/total bits
   for m=1:N_iter
       % Tx_____
       X= randint(1,Nused*Nframe,M); % bit: integer vector
       Xmod= qammod(X,M,0,'gray')/norms(Nbps);
       if NgType~=2, x_GI=zeros(1,Nframe*Nsym);
         elseif NgType==2, x_GI= zeros(1,Nframe*Nsym+Ng);
           % Extend an OFDM symbol by Ng zeros
       end
       kk1=[1:Nused/2]; kk2=[Nused/2+1:Nused]; kk3=1:Nfft; kk4=1:Nsym;
       for k=1:Nframe
           if Nvc~=0, X_shift= [0 Xmod(kk2) zeros(1,Nvc-1) Xmod(kk1)];
           else X_shift= [Xmod(kk2) Xmod(kk1)];
           end
```

```
      x= ifft(X_shift);
      x_GI(kk4)= guard_interval(Ng,Nfft,NgType,x);
      kk1=kk1+Nused; kk2= kk2+Nused; kk3=kk3+Nfft; kk4=kk4+Nsym;
   end
   if Ch==0, y= x_GI; % No channel
      else % Multipath fading channel
      channel=(randn(1,Ntap)+j*randn(1,Ntap)).*sqrt(Power/2);
      h=zeros(1,Lch); h(Delay+1)=channel; % cir: channel impulse response
      y = conv(x_GI,h);
   end
   if i==0 % Only to measure the signal power for adding AWGN noise
      y1=y(1:Nframe*Nsym); sigPow = sigPow + y1*y1'; continue;
   end
   % Add AWGN noise_____
   snr = EbN0(i)+10*log10(Nbps*(Nused/Nfft)); % SNR vs. Eb/N0 by Eq.(4.28)
   noise_mag = sqrt((10.^(-snr/10))*sigPow/2);
   y_GI = y + noise_mag*(randn(size(y))+j*randn(size(y)));
   % Rx_____
   kk1=(NgType==2)*Ng+[1:Nsym]; kk2=1:Nfft;
   kk3=1:Nused; kk4=Nused/2+Nvc+1:Nfft; kk5=(Nvc~=0)+[1:Nused/2];
   if Ch==1
      H= fft([h zeros(1,Nfft-Lch)]); % Channel frequency response
      H_shift(kk3)= [H(kk4) H(kk5)];
   end
   for k=1:Nframe
      Y(kk2)= fft(remove_GI(Ng,Nsym,NgType,y_GI(kk1)));
      Y_shift=[Y(kk4) Y(kk5)];
      if Ch==0, Xmod_r(kk3) = Y_shift;
         else Xmod_r(kk3)=Y_shift./H_shift; % Equalizer - channel compensation
      end
      kk1=kk1+Nsym; kk2=kk2+Nfft; kk3=kk3+Nused; kk4=kk4+Nfft;
      kk5=kk5+Nfft;
   end
   X_r=qamdemod(Xmod_r*norms(Nbps),M,0,'gray');
   Neb=Neb+sum(sum(de2bi(X_r,Nbps)~=de2bi(X,Nbps)));
   Ntb=Ntb+Nused*Nframe*Nbps; %[Ber,Neb,Ntb]=ber(bit_Rx,bit,Nbps);
   if Neb>Target_neb, break; end
 end
 if i==0, sigPow= sigPow/Nsym/Nframe/N_iter;
 else
    Ber = Neb/Ntb;
    fprintf('EbN0=%3d[dB], BER=%4d/%8d =%11.3e\n', EbN0(i), Neb,Ntb,Ber)
    fprintf(fid, '%d\t%11.3e\n', EbN0(i), Ber);
    if Ber<1e-6, break; end
 end
end
if (fid~=0), fclose(fid); end
plot_ber(file_name,Nbps);
```

Program 4.3 Routines for GI (guard interval) insertion, GI removal, and BER plotting

```
function y = guard_interval(Ng,Nfft,NgType,ofdmSym)
if NgType==1, y=[ofdmSym(Nfft-Ng+1:Nfft) ofdmSym(1:Nfft)];
   elseif NgType==2, y=[zeros(1,Ng) ofdmSym(1:Nfft)];
end
```

```
function y=remove_GI(Ng,Lsym,NgType,ofdmSym)
if Ng~=0
   if NgType==1, y=ofdmSym(Ng+1:Lsym); % cyclic prefix
     elseif NgType==2 % cyclic suffix
        y=ofdmSym(1:Lsym-Ng)+[ofdmSym(Lsym-Ng+1:Lsym) zeros (1,Lsym-2*Ng)];
   end
  else y=ofdmSym;
end
```

```
function plot_ber(file_name,Nbps)
EbN0dB=[0:1:30]; M=2^Nbps;
ber_AWGN = ber_QAM(EbN0dB,M,'AWGN');
ber_Rayleigh = ber_QAM(EbN0dB,M,'Rayleigh');
semilogy(EbN0dB,ber_AWGN,'r:'), hold on,
semilogy(EbN0dB,ber_Rayleigh,'r-')
a= load(file_name); semilogy(a(:,1),a(:,2),'b-s'); grid on
legend('AWGN analytic','Rayleigh fading analytic', 'Simulation');
xlabel('EbN0[dB]'), ylabel('BER'); axis([a(1,1) a(end,1) 1e-5 1])
```

```
function ber=ber_QAM(EbN0dB,M,AWGN_or_Rayleigh)
% Find analytical BER of M-ary QAM in AWGN or Rayleigh channel
% EbN0dB=EbN0dB: Energy per bit-to-noise power[dB] for AWGN channel
% =rdB : Average SNR(2*sigma Eb/N0) [dB] for Rayleigh channel
% M = Modulation order (Alphabet or Constellation size)
N= length(EbN0dB); sqM= sqrt(M);
a= 2*(1-power(sqM,-1))/log2(sqM); b= 6*log2(sqM)/(M-1);
if nargin<3, AWGN_or_Rayleigh='AWGN'; end
if lower(AWGN_or_Rayleigh(1))=='a'
  ber = a*Q(sqrt(b*10.^(EbN0dB/10))); % ber=berawgn(EbN0dB,'QAM',M) Eq.(4.25)
  else % diversity_order=1; ber=berfading(EbN0dB,'QAM',M,diversity_order)
    rn=b*10.^(EbN0dB/10)/2; ber = 0.5*a*(1-sqrt(rn./(rn+1))); % Eq.(4.26)
end
```

```
function y=Q(x)
% co-error function: 1/sqrt(2*pi) * int_x^inf exp(-t^2/2) dt. % Eq.(4.27)
y=erfc(x/sqrt(2))/2;
```

4.2.5 Water-Filling Algorithm for Frequency-Domain Link Adaptation

In general, a data rate can be adaptively varied with the channel variation. Such link adaptation is a useful means of maximizing the system bandwidth efficiency. One particular example of link adaptation is time-domain AMC (Adaptive Modulation and Coding) technique. It has been widely adopted for the packet data systems (e.g., cdma2000 1x EV-DO), in which a multiple number of time slots with different channel gains are dynamically shared among the different users in an opportunistic manner which allows for increasing the average system throughput. In addition to time-domain link adaptation, the AMC technique can also be applied to the frequency domain in the OFDM system. In fact, it allows for fully taking advantage of the time-varying characteristics of the subcarriers in the frequency-selective channel for link adaptation. The underlying key principle in the frequency-domain AMC technique is the water-filling (or water pouring) algorithm that allocates more (or less) bits and power to some subcarriers with larger (or smaller) SNR for maximizing the channel capacity.

Consider the OFDM system with N_{used} subcarriers, each with a subcarrier spacing of Δf. The capacity of a subchannel corresponding to the kth subcarrier, f_k, is given by the Hartley-Shannon channel capacity, such that

$$C(f_k) = \Delta f \log_2 \left(1 + |H[k]|^2 \frac{P[k]}{N_0} \right) \tag{4.29}$$

where $H[k]$, $P[k]$, and N_0 denote the frequency response, transmission power, and noise variance of the kth subchannel, respectively. Then, the total channel capacity is given by the sum of the capacity for individual subcarriers, that is,

$$C = \sum_{k=0}^{N_{used}-1} C(f_k) \tag{4.30}$$

If N_{used} is large enough, which means Δf is small enough to be narrower than the coherence bandwidth, $H[k]$ can be considered constant for each subcarrier k. In other words, SNR can be approximated as a constant for each subcarrier. Given the SNR for each subcarrier, we may allocate different powers to different subcarriers so as to maximize the total system capacity. In other words, it can be formulated as the following optimization problem:

$$\max_{P_0,\ldots,P_{N_{used}-1}} \sum_{k=0}^{N_{used}-1} C(f_k) = \sum_{k=0}^{N_{used}-1} \log \left(1 + \frac{|H[k]|^2 P[k]}{N_0} \right) \tag{4.31}$$

subject to

$$\sum_{k=0}^{N_{used}-1} P[k] = N_{used} \cdot P \tag{4.32}$$

where P is the average power per subcarrier available in the transmitter. Employing the Lagrange multiplier method for optimization with equality constraint in Equation (4.32), the following solution is obtained:

$$P^*[k] = \left(\frac{1}{\lambda} - \frac{N_0}{|H[k]|^2} \right)^+ = \begin{cases} \dfrac{1}{\lambda} - \dfrac{N_0}{|H[k]|^2}, & \text{if } \dfrac{1}{\lambda} - \dfrac{N_0}{|H[k]|^2} \geq 0 \\ 0, & \text{otherwise} \end{cases} \tag{4.33}$$

where λ is the Lagrange multiplier that is chosen to meet the power constraint in Equation (4.32). This solution implies that the sums of power and NSR (noise-to-signal ratio) for each subcarrier must be the same for all subcarriers (i.e., $P^*[k] + N_0/|H[k]|^2 = 1/\lambda = $ constant), except for the subcarriers that have been assigned no power just because $1/\lambda - N_0/|H[k]|^2 < 0$ [87]. According to this algorithm, a subcarrier with larger SNR is allocated more transmission power. Figure 4.24 presents a graphical description of the optimal power allocation solution in Equation (4.33). The NSR $N_0/|H[k]|^2$, given in a function of the subcarrier index k, can be considered as the bottom of a water tank with an irregular shape. If each subcarrier is poured with P units of water in the tank, the depth of the water at subcarrier n corresponds to the power allocated to that subcarrier, while $1/\lambda$ is the height of the water level. Since the algorithm has been described by filling a tank with water, it is called a water-filling algorithm. In this water-filling analogy, it is interesting to note that no power must be allocated to subcarriers with the bottom of the tank above the given water level. In fact, this corresponds to a situation in which a poor channel must not be used for transmitting data.

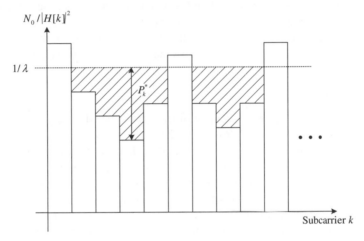

Figure 4.24   Optimal power allocation with the water-filling algorithm.

As discussed by the water-filling algorithm, the total data rate and BER in the OFDM system can be varied by the power allocation to the subcarriers. In other words, the data rate, average transmission power, and BER are design parameters that are tightly associated with each other. Therefore, one or two parameters among data rate, average transmission power, and BER can be optimized, subject to the constraint associated with the rest of the parameters [88–91]. For example, if the transmission power and BER should be the same over all subcarriers, the number of bits that can be allocated to each subcarrier is given by

$$b[k] = \log_2\left(1 + \frac{SNR_k}{\Gamma}\right) \qquad (4.34)$$

where SNR_k is the received SNR of the kth subcarrier and Γ is the SNR gap[2]. Figure 4.25 illustrates an adaptive bit-loading situation where different numbers of bits are allocated to different subcarriers subject to the frequency-selective characteristics. According to Equation (4.34), more bits can be allocated to the subcarriers with better SNR.

---

[2] The SNR gap is defined as a ratio of ideal SNR at which the system can transmit at C (channel capacity) bits/transmission to a practical SNR at which the system can transmit R bits/transmission.

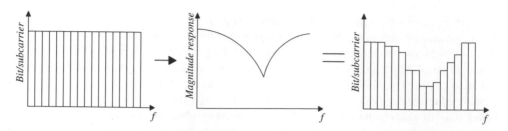

Figure 4.25   Adaptive bit loading to subcarriers in an OFDM system.

As explained above, the transmission power or bit allocation to each subcarrier can be optimized by taking its SNR into account [72]. This implies that the water-filling algorithm requires a full knowledge of channel state information for each subcarrier on the transmitter side. In other words, the channel quality information must be exchanged between the transmitter and receiver through a feedback loop. In general, however, this is not straightforward, especially in mobile environments where the channel may be steadily time-varying.

4.3 Coded OFDM

As depicted in Figure 4.26, some consecutive subcarriers in the OFDM system may suffer from deep fading, in which the received SNR is below the required SNR level. This is still true even if the required SNR is set much below the average SNR. In order to deal with the burst symbol errors due to deep fading in this multi-carrier situation, it may be essential to employ FEC (Forward Error Correction) codes. In other words, unless the OFDM system is protected by FEC coding, the required SNR must be set too low, unnecessarily reducing the overall data rate. Therefore, most of the practical OFDM systems are the coded OFDM systems. The popular FEC codes associated with the coded OFDM systems include RS (Reed-Solomon) code, convolutional code, TCM (Trellis-Coded Modulation), concatenated code, turbo code, and LDPC code. The FEC codes can make error corrections only as far as the errors are within the error-correcting capability (that is defined as the maximum number of guaranteed correctable errors per codeword), but they may fail with burst symbol errors. In practice, *interleaving* is often employed to convert the burst errors into random errors. There are two types of interleaving: block interleaving and convolutional interleaving. Bit-wise, data symbol-wise, or OFDM symbol-wise interleavings can be used for block interleaving. Interleaving type and

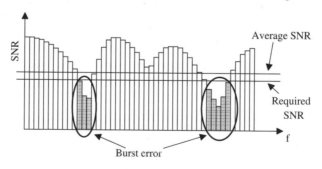

Figure 4.26   Burst errors subject to frequency selectivity.

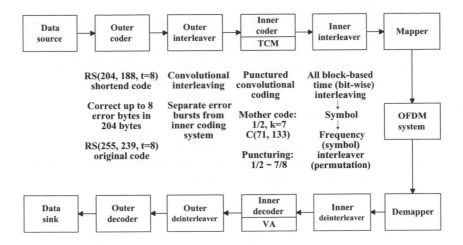

Figure 4.27 Example of coded OFDM: Eureka-147-based DAB.

size (depth) must be determined by the type of FEC code, degree of frequency and time fading, and delay due to interleaving.

Figure 4.27 shows FEC codes and interleaving techniques used in Eureka-147-based DAB (digital audio broadcasting), which is a typical example of the coded OFDM system. Here, concatenated coding is used for FEC. More specifically, an RS code is used for the outer code while TCM is used for the inner code. The outer interleaver and inner interleaver are used to separate error bursts from inner coding and to interleave bits/symbols in the time/frequency domain, respectively.

4.4 OFDMA: Multiple Access Extensions of OFDM

In general, OFDM is a transmission technique in which all subcarriers are used for transmitting the symbols of a single user. In other words, OFDM is not a multiple access technique by itself, but it can be combined with existing multiple access techniques such as TDMA (Time Division Multiple Access), FDMA (Frequency Division Multiple Access), and CDMA (Code Division Multiple Access) for a multi-user system. As depicted in Figure 4.28 [97], all subcarriers can be

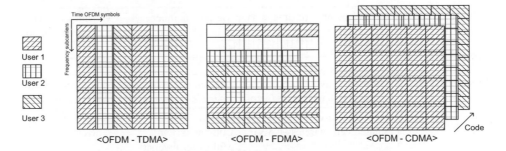

Figure 4.28 Multiple access techniques used in OFDM systems.

shared by multiple users in the forms of OFDM-TDMA, OFDMA (OFDM-FDMA), or MC-CDMA (OFDM-CDMA).

The OFDM-TDMA system allocates all subcarriers only to a single user for the duration of several OFDM symbols, where the number of OFDM symbols per user can be adaptively changed in each frame (Figure 4.28(a)). In this case, the resource allocation among the users is orthogonal in time. The OFDMA system assigns a subset of subcarriers (not all subcarriers in each OFDM symbol) to each user, where the number of subcarriers for a specific user can be adaptively varied in each frame (Figure 4.28(b)). In other words, the subcarriers in each OFDM symbol are orthogonally divided among the multiple users. Meanwhile, an OFDM-CDMA system allows for sharing both time and subcarriers among all users (not in an orthogonal manner) where a subset of orthogonal codes is assigned to each user and the information symbols are spread in the frequency domain (Figure 4.28(c)).

Among these multiple access techniques associated with OFDM, OFDMA is one of the most useful approaches in the mobile cellular system. As users in the same cell may have different signal-to-noise and interference ratios (SINRs), it would be more efficient to allow multiple users to select their own subset of subcarriers with better channel conditions, rather than selecting a single user that uses all the subcarriers at the same time. In other words, there may be one or more users with significantly better channel conditions, especially when the number of users increases. Improvement in the bandwidth efficiency, achieved by selecting multiple users with better channel conditions, is referred to as multi-user diversity gain. OFDMA is a technique that can fully leverage the multi-user diversity gain inherent to the multi-carrier system. The amount of physical resources (i.e., time slots, subcarrier, and spreading codes, assigned to each user in these techniques) depends not only on the required data rate of each user, but also on the multi-user diversity gain among the users. A concept of multi-user diversity will be further detailed in Section 4.5. Note that the aforementioned multiple access techniques associated with OFDM systems differ from each other in many aspects (e.g., flexibility and multiple access interference (MAI)), as compared in Table 4.2.

Table 4.2 Multiple access techniques associated with OFDM: comparison.

Attributes		TDMA	FDMA	CDMA
Method		One user/subset of time slots/ all subcarriers	Multiple users/same time/subset of subcarriers	All users/same time/all subcarriers
Flexibility		Variable number of time slots	Variable number of subcarriers	Variable number of spreading code
MAI	Intra-cell	None	None	Present
	Inter-cell	Present	Present	Present
MAI suppression		Interference avoidance (low frequency reuse factor)	Interference avoidance Interference averaging	Multi-user detection Interference averaging
Others		Small FFT size Isolated cell (wireless LAN)	Large FFT size Cellular system Multi-user diversity Power concentration	Inherent frequency diversity

4.4.1 Resource Allocation – Subchannel Allocation Types

As mentioned before, a subset of subcarriers is allocated to each user in OFDMA and thus, the number of subcarriers to be allocated to each user must be scheduled by the system. To facilitate a basic unit of resource allocation in OFDMA, a group of subcarriers is defined as a *subchannel*. Depending on how the subcarriers are allocated to construct each subchannel, the resource allocation methods are classified into a block type, a comb type, and a random type. As depicted in Figure 4.29, the different types of resource allocation differ in their distribution of the subcarriers.

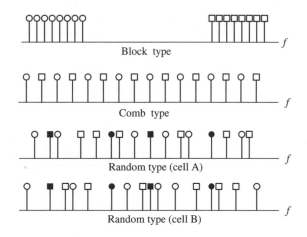

Figure 4.29   Types of resource allocation in OFDMA system.

In a block type of resource allocation, each subchannel is constructed by a set of adjacent subcarriers. It is also called a cluster type, a localized type, or a band type. This type is often used in an environment of low mobility and stable channel condition. It allows for link adaptation by an adaptive modulation and coding (AMC) scheme for the different subchannels, subject to their own instantaneous channel conditions. Furthermore, the average system throughput can be improved by allowing users to select their own preferred subchannels, which fully leverages the multi-user diversity gain of OFDMA. Another advantage of using a block type of resource allocation is the simplicity of channel estimation, since each block is constructed within the coherence bandwidth. In spite of the contiguous allocation of subcarriers in block type subchannels, diversity gain can still be introduced by frequency-hopping the subchannels throughout the whole frequency band.

In a comb type of resource allocation, each subchannel is composed of a set of equi-spaced subcarriers. Since the subcarriers are regularly interleaved throughout the whole frequency band in this case, it is sometimes referred to as an interleaved type. In fact, a diversity gain is sought by distributing the subcarriers over the whole band.

In a random type of resource allocation, each subchannel is composed of a set of subcarriers distributed randomly over the whole frequency band. It is also referred to as a distributed type. If a random type of subchannels is used, an interference averaging effect can be achieved in addition to a diversity gain. In this case, all pilots located over the whole bandwidth shall be used for channel estimation. This type of subchannels tends to average out the channel quality over the whole band. Therefore, it can accommodate high mobility, even when the quality of each subcarrier steadily varies from one symbol (or frame) to the other. In the cellular system, furthermore, it is useful for reducing the co-channel interference by randomly distributing subcarriers in such a way that the probability of subcarrier collision among adjacent cells is minimal. Figure 4.29 illustrates the case of a collision (filled in black) between the subcarriers of two adjacent cells, cell A and cell B, using a random type of subchannel.

Three types of resource allocation are compared to each other in Table 4.3. We note that each type has its own advantages and disadvantages in many different aspects.

Table 4.3 Resource allocation types in OFDMA: comparison.

Type	Block (cluster)	Comb (interleaved)	Random
Method	Adjacent subcarriers	Equidistant subcarriers	Random subcarriers
Diversity	Frequency hopping required	Frequency diversity gain	Frequency diversity gain
Mobility	Slow	Fast	Fast
Channel estimation	Simple (coherence BW)	Whole band	Whole band
Inter-cell interference	Coordination required	Coordination required	Interference averaging

4.4.2 Resource Allocation – Subchannelization

In the practical system, a specific subchannelization scheme is specified in terms of the resource allocation types discussed in the above. For example, let us take a look at subchannels defined in the Mobile WiMAX system [298–302]. Depending on whether the subcarriers are scattered (distributed) or clustered (localized in a unit of band), the subchannelization is broadly classified into two different classes, a *diversity* subchannel and a *band AMC* subchannel. As shown in Table 4.4, there are three different types of diversity subchannels in the downlink, PUSC (Partial Usage SubChannel), FUSC (Full Usage SubChannel), and OFUSC (Optional Full Usage SubChannel), depending on whether each subchannel is constructed by the

Table 4.4 Diversity subchannels and band AMC subchannels in mobile WiMAX system.

Subchannels	Name	Configuration	Usage
Diversity subchannel: downlink	PUSC	Scattered 48 tones	FCH/DL-MAP (reuse 3), data transmission
	FUSC	Scattered 48 tones	Data transmission, broadcasting
	OFUSC	Scattered 48 tones	Data transmission, broadcasting
Diversity subchannel: uplink	PUSC	6 distributed tiles 8 tones/tiles	Data transmission, broadcasting
	OPUSC	6 distributed tiles 8 tones/tiles	Data transmission, broadcasting
Band AMC: downlink/uplink	AMC	6 adjacent bins 8 tones/bins	Data transmission, AMC, AAS

subcarriers that are scattered throughout the whole band or not. Similarly, there are two different types of diversity subchannels in the uplink, PUSC and OPUSC. Band AMC channels are used in both uplink and downlink. In all types of subchannels, one subchannel is composed of 48 subcarriers.

Table 4.5 shows major parameters for downlink subchannels used in the Mobile WiMAX system. As an example, the procedure for FUSC subchannel generation is described. One FUSC symbol uses 851 subcarriers (including DC subcarriers), 768 data subcarriers and 82 pilot subcarriers, out of 1024 subcarriers. In one FUSC symbol, there are 16 subchannels, each consisting of 48 subcarriers. In order to generate FUSC subchannels, 768 data subcarriers are divided into 48 groups as shown in Figure 4.30. Then, one subcarrier is selected from each group to construct a subchannel. The specific rule for allocating subcarriers to subchannels is determined by the permutation formula with the parameter "DL_PermBase." The subchannels constructed in this way have all the subcarriers distributed over the entire band. They are designed such that collision (hit) between the subcarriers in the adjacent cells does not occur when different subchannels generated by the permutation formula with the same DL_Perm-Base are used. However, collision (hit) between two subcarriers may occur when each cell uses different DL_PermBase to produce FUSC symbols. Figure 4.31 shows the hit distribution at the cell boundary when FUSC symbols are used at both cells. Here, DL_PermBases used for cell A and cell B are 13 and 7, respectively. Hit distribution varies depending on the DL_PermBase, cell loading factor (the number of subchannels allocated), and the indices of subchannels allocated.

Table 4.5 Parameters for downlink subchannels in mobile WiMAX system.

Parameters	PUSC	FUSC	OFUSC	Band AMC
No. of left guard subcarriers	92	87	80	80
No. of right guard subcarriers	91	86	79	79
No. of used subcarriers (no DC)	840	850	864	864
No. of DC subcarrier (#512)	1	1	1	1
No. of pilot subcarriers	120	82	96	96
No. of data subcarriers	720	768	768	468
No. of subcarriers per subchannel	48	48	48	48
No. of symbols per subchannel	2	1	1	2, 3, 6
No. of subchannels/symbols	30	16	16	48
Pilot configuration	Common	Common	Common	Dedicated
Data configuration	Distributed	Distributed	Distributed	Adjacent

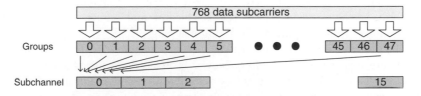

Figure 4.30 Structure for FUSC subchannel generation.

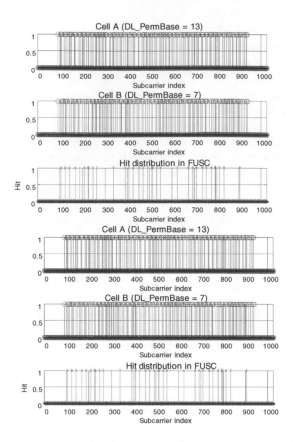

Figure 4.31 Hit distribution at the cell boundary when FUSC symbols are used.

The band-AMC subchannel corresponds to the block-type resource allocation type, which allows each user to select its own preferred blocks. Depending on the channel SNR, different AMC will be implemented for a different user (see Figure 4.32). When the channel varies rapidly with time, however, channel information may not be matched with the actual channel condition due to the feedback delay. Such degradation in reliability of channel feedback information may degrade the system throughput. Therefore, the band-AMC subchannel may be useful only in the environment of low mobility and a stable channel condition.

From a viewpoint of frequency- and time-domain resources available in OFDMA systems, subchannel mapping can be classified into one-dimensional mapping or two-dimensional mapping. Also, depending on the priority between the time and frequency domain mappings, it is classified into vertical mapping or horizontal mapping [99].

Figure 4.33(a) illustrates one-dimensional vertical mapping. In this mapping scheme, subchannels are mapped along the frequency axis first, where the offset and length of a subchannel are the mapping parameters to specify the subchannel allocation for data region. Also, all subcarriers in a slot are used first so that transmission time and decoding delay can be as short as possible at the transmitter and receiver, respectively. However, it requires a

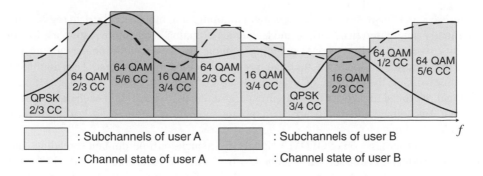

Figure 4.32   User diversity effect in a band-AMC subchannel: an illustration.

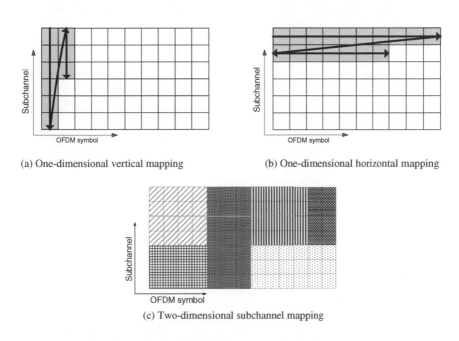

(a) One-dimensional vertical mapping   (b) One-dimensional horizontal mapping

(c) Two-dimensional subchannel mapping

Figure 4.33   Subchannel mapping schemes in OFDMA system.

high power concentrated instantly. Therefore, this mapping scheme is often used for transmitting the control information that must be decoded as fast as possible in the downlink transmission.

Figure 4.33(b) illustrates one-dimensional horizontal mapping. In this mapping scheme, subchannels are mapped along the time axis first, where the offset and length of a subchannel are the mapping parameters as in one-dimensional vertical mapping. Also, a minimum number of subcarriers are used for transmission during a long time slot and consequently, it takes a long decoding delay corresponding to the number of slots used in subchannel mapping to decode each data burst at the receiver. Accordingly, it has a disadvantage of long transmission time and decoding delay. Since it spends a low transmit power, however, it is appropriate for uplink transmission at the mobile station where power consumption is a critical issue.

Figure 4.33(c) illustrates a two-dimensional mapping scheme. Since the data region can be dynamically configured in both time and frequency domains, it has more resource allocation flexibility over a one-dimensional mapping scheme. However, additional mapping parameters are required to specify actual data region allocated to each burst. They include the subchannel offset, the number of used symbols, and symbol offset. Therefore, it incurs additional overhead to provide the mapping parameters associated with each data burst. We note that two-dimensional subcarrier mapping scheme is often used for transmitting downlink data burst.

For a robust transmission of OFDMA signals, frequency- or time-domain spreading can be applied. Consider a QPSK data burst encoded with a forward error correction code (FEC) of 1/2 coding rate in Figure 4.34(a). It can be repeated three times by frequency-domain spreading as shown in Figure 4.34(b), which yields a QPSK data burst with a FEC of 1/6 coding rate. From this frequency-domain spreading, frequency diversity can be obtained at the expense of a reduced coding rate. In a similar manner, time diversity can be obtained, now by time-domain spreading as shown in Figure 4.34(c).

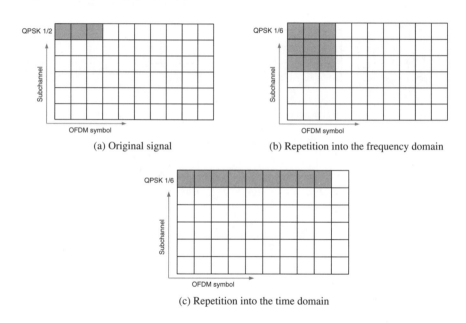

(a) Original signal   (b) Repetition into the frequency domain

(c) Repetition into the time domain

Figure 4.34 Repetition in OFDMA system.

4.5 Duplexing

Duplexing refers to the mechanism of dividing a communication link for downlink and uplink. Two different duplexing schemes are mainly used for cellular systems: FDD (Frequency Division Duplexing) and TDD (Time Division Duplexing). Besides FDD and TDD, there exist some variants, for example, Zipper and HDD (Hybrid Division Duplexing). Zipper is a digital duplexing technique that performs downlink and uplink transmissions simultaneously. HDD (Hybrid Division Duplexing) is a technique that combines TDD and FDD in a single unified frame structure, so as to inherit their own individual advantage in different environments.

Table 4.6 Comparison between FDD and TDD.

	FDD	TDD
Spectrum	2 separate band for each link (guard band)	Single channel (guard time)
Duplexing	Full duplex	Half duplex
Flexibility (DL/UL)	Low	High
Complexity/cost	High (RX filter, etc.)	Low
Coverage	Wide coverage	Small coverage
Mobility	High	Low
Etc.	Low latency	Symmetric channel

FDD uses different bands (separated by a guard band) for simultaneous downlink and uplink signal transmission, respectively. Although it requires downlink and uplink RF parts separately and has a disadvantage of low flexibility, FDD has been widely used in cellular systems since it enables a mobile station to move with high speed due to its short delay time and wide cell coverage.

TDD is a duplexing technique that allocates different time slots (separated by a guard time) and uses a common frequency band for downlink and uplink. In TDD, downlink and uplink transmissions cannot be performed simultaneously, but it has an advantage of high flexibility when downlink and uplink transmissions have different traffic loads because assignment of downlink and uplink time slots can be controlled dynamically. Besides, many techniques such as MIMO, smart antenna, link adaptation, pre-compensation, and so on, can easily be applied to FDD, owing to the reciprocity between downlink and uplink channels. However, TDD requires accurate time synchronization because its downlink and uplink are separated in time. In cellular systems, TDD requires synchronization in Tx/Rx timing between cells so as to minimize interferences from adjacent cells. Table 4.6 summarizes a comparison between FDD and TDD.

5

Synchronization for OFDM

As discussed in Chapter 4, the OFDM system carries the message data on orthogonal subcarriers for parallel transmission, combating the distortion caused by the frequency-selective channel or equivalently, the inter-symbol-interference in the multi-path fading channel. However, the advantage of the OFDM can be useful only when the orthogonality is maintained. In case the orthogonality is not sufficiently warranted by any means, its performance may be degraded due to inter-symbol interference (ISI) and inter-channel interference (ICI) [1]. In this chapter, we will analyze the effects of symbol time offset (STO) and carrier frequency offset (CFO), and then discuss the synchronization techniques to handle the potential STO and CFO problems in OFDM systems. Let ε and δ denote the normalized CFO and STO, respectively. Referring to Equation (4.20), the received baseband signal under the presence of CFO ε and STO δ can be expressed as

$$
\begin{aligned}
y_l[n] &= \mathrm{IDFT}\{Y_l[k]\} = \mathrm{IDFT}\{H_l[k]\,X_l[k] + Z_l[k]\} \\
&= \frac{1}{N}\sum_{k=0}^{N-1} H_l[k]\,X_l[k]\,e^{j2\pi(k+\varepsilon)(n+\delta)/N} + z_l\,[n]
\end{aligned}
\tag{5.1}
$$

where $z_l[n] = \mathrm{IDFT}\{Z_l[k]\}$.

5.1 Effect of STO

IFFT and FFT are the fundamental functions required for the modulation and demodulation at the transmitter and receiver of OFDM systems, respectively. In order to take the N-point FFT in the receiver, we need the exact samples of the transmitted signal for the OFDM symbol duration. In other words, a symbol-timing synchronization must be performed to detect the starting point of each OFDM symbol (with the CP removed), which facilitates obtaining the exact samples. Table 5.1 shows how the STO of δ samples affects the received symbols in the time and frequency domain where the effects of channel and noise are neglected for simplicity of exposition. Note that the STO of δ in the time domain incurs the phase offset of $2\pi k\delta/N$ in the frequency domain, which is proportional to the subcarrier index k as well as the STO δ.

MIMO-OFDM Wireless Communications with MATLAB®   Yong Soo Cho, Jaekwon Kim, Won Young Yang and Chung G. Kang
© 2010 John Wiley & Sons (Asia) Pte Ltd

Table 5.1   The effect of symbol time offset (STO).

	Received signal	STO (δ)
Time domain	$y[n]$	$x[n+\delta]$
Frequency domain	$Y[k]$	$e^{j2\pi k\delta/N}X[k]$

Depending on the location of the estimated starting point of OFDM symbol, the effect of STO might be different. Figure 5.1 shows four different cases of timing offset, in which the estimated starting point is exact, a little earlier, too early, or a little later than the exact timing instance. Here, we assume that the multi-path delay spread incurs the lagged channel response of $\tau_{\max}$. In the current analysis, the effects of the noise and channel are ignored. Referring to Figure 5.1, let us discuss the effects of STO for these four different cases below.

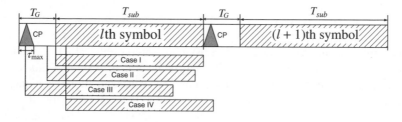

Figure 5.1   Four different cases of OFDM symbol starting point subject to STO.

- **Case I:** This is the case when the estimated starting point of OFDM symbol coincides with the exact timing, preserving the orthogonality among subcarrier frequency components. In this case, the OFDM symbol can be perfectly recovered without any type of interference.
- **Case II:** This is the case when the estimated starting point of OFDM symbol is before the exact point, yet after the end of the (lagged) channel response to the previous OFDM symbol. In this case, the lth symbol is not overlapped with the previous $(l\text{-}1)$th OFDM symbol, that is, without incurring any ISI by the previous symbol in this case. In order to see the effects of the STO, consider the received signal in the frequency domain by taking the FFT of the time-domain received samples $\{x_l[n+\delta]\}_{n=0}^{N-1}$, given as

$$
\begin{aligned}
Y_l[k] &= \frac{1}{N}\sum_{n=0}^{N-1} x_l[n+\delta]\, e^{-j2\pi nk/N} \\
&= \frac{1}{N}\sum_{n=0}^{N-1}\left\{\sum_{p=0}^{N-1} X_l[p]\, e^{j2\pi(n+\delta)p/N}\right\} e^{-j2\pi nk/N} \\
&= \frac{1}{N}\sum_{p=0}^{N-1} X_l[p]\, e^{j2\pi p\delta/N}\sum_{n=0}^{N-1} e^{j2\pi\frac{(p-k)}{N}n} \\
&= X_l[k]\, e^{j2\pi k\delta/N}
\end{aligned}
\tag{5.2}
$$

where the last line follows from the following identity:

$$\sum_{n=0}^{N-1} e^{j2\pi\frac{(p-k)}{N}n} = e^{j\pi(p-k)\frac{N-1}{N}} \cdot \frac{\sin[\pi(k-p)]}{\sin[\pi(k-p)/N]}$$

$$= \begin{cases} N & \text{for} \quad k=p \\ 0 & \text{for} \quad k \neq p \end{cases}$$

The expression in Equation (5.2) implies that the orthogonality among subcarrier frequency components can be completely preserved. However, there exists a phase offset that is proportional to the STO δ and subcarrier index k, forcing the signal constellation to be rotated around the origin. Figure 5.2(a) and (b) show the received symbols in the signal constellation for Case I and Case II, respectively. As expected, the phase offset due to STO is observed in Case II. Note that it is straightforward to compensate for the phase offset simply by a single-tap frequency-domain equalizer.

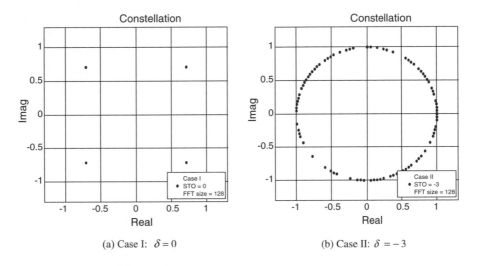

(a) Case I: $\delta = 0$ (b) Case II: $\delta = -3$

Figure 5.2   Signal constellation subject to STO.

- **Case III:** This is the case when the starting point of the OFDM symbol is estimated to exist prior to the end of the (lagged) channel response to the previous OFDM symbol, and thus, the symbol timing is too early to avoid the ISI. In this case, the orthogonality among subcarrier components is destroyed by the ISI (from the previous symbol) and furthermore, ICI (Inter-Channel Interference) occurs.
- **Case IV:** This is the case when the estimated starting point of the OFDM symbol is after the exact point, which means the symbol timing is a little later than the exact one. In this case, the signal within the FFT interval consists of a part of the current OFDM symbol $x_l[n]$ and a part of next one $x_{l+1}[n]$; more specifically,

$$y_l[n] = \begin{cases} x_l[n+\delta] & \text{for} \quad 0 \leq n \leq N-1-\delta \\ x_{l+1}[n+2\delta-N_g] & \text{for} \quad N-\delta \leq n \leq N-1 \end{cases} \tag{5.3}$$

where N_g is the GI length. Taking the FFT of this composite signal $\{y_l[n]\}_{n=0}^{N-1}$ for demodulation,

$$Y_l[k] = \text{FFT}\{y_l[n]\}$$

$$= \sum_{n=0}^{N-1-\delta} x_l[n+\delta]\, e^{-j2\pi nk/N} + \sum_{n=N-\delta}^{N-1} x_{l+1}[n+2\delta-N_g]\, e^{-j2\pi nk/N}$$

$$= \sum_{n=0}^{N-1-\delta} \left(\frac{1}{N} \sum_{p=0}^{N-1} X_l[p]\, e^{j2\pi(n+\delta)p/N} \right) e^{-j2\pi nk/N}$$

$$+ \sum_{n=N-\delta}^{N-1} \left(\frac{1}{N} \sum_{p=0}^{N-1} X_{l+1}[p]\, e^{j2\pi(n+2\delta-N_g)p/N} \right) e^{-j2\pi nk/N} \tag{5.4}$$

$$= \frac{1}{N} \sum_{p=0}^{N-1} X_l[p]\, e^{j2\pi p\delta/N} \sum_{n=0}^{N-1-\delta} e^{j2\pi\frac{(p-k)}{N}n} + \frac{1}{N} \sum_{p=0}^{N-1} X_{l+1}[p]\, e^{j2\pi p(2\delta-N_g)/N} \sum_{n=N-\delta}^{N-1} e^{j2\pi\frac{(p-k)}{N}n}$$

$$= \frac{N-\delta}{N} X_l[p] e^{j2\pi p\delta/N} + \sum_{p=0,p\neq k}^{N-1} X_l[p]\, e^{j2\pi p\delta/N} \sum_{n=0}^{N-1-\delta} e^{j2\pi\frac{(p-k)}{N}n}$$

$$+ \frac{1}{N} \sum_{p=0}^{N-1} X_{l+1}[p]\, e^{j2\pi p(2\delta-N_g)/N} \sum_{n=N-\delta}^{N-1} e^{j2\pi\frac{(p-k)}{N}n}$$

Considering the following identity:

$$\sum_{n=0}^{N-1-\delta} e^{j2\pi\frac{(p-k)}{N}n} = e^{j\pi(p-k)\frac{N-1-\delta}{N}} \cdot \frac{\sin[(N-\delta)\pi(k-p)/N]}{\sin[\pi(k-p)/N]} = \begin{cases} N-\delta & \text{for} \quad p=k \\ \text{Nonzero} & \text{for} \quad p\neq k \end{cases}$$

the second term in the last line of Equation (5.4) corresponds to ICI, which implies that the orthogonality has been destroyed. Furthermore, it is also clear from the third term in the last line of Equation (5.4) that the received signal involves the ISI (from the next OFDM symbol $X_{l+1}[p]$).

Figures 5.3(a) and (b) show the signal constellation for Case III and Case IV, respectively. Note that the distortion (including the phase offset) in Case IV is too severe to be compensated. It implies that a symbol timing scheme is essential for preventing STOs in this case.

5.2 Effect of CFO

The baseband transmit signal is converted up to the passband by a carrier modulation and then, converted down to the baseband by using a local carrier signal of (hopefully) the same carrier frequency at the receiver. In general, there are two types of distortion associated with the carrier signal [94]. One is the phase noise due to the instability of carrier signal generators used at the transmitter and receiver, which can be modeled as a zero-mean Wiener random process [100–102]. The other is the carrier frequency offset (CFO) caused by Doppler frequency

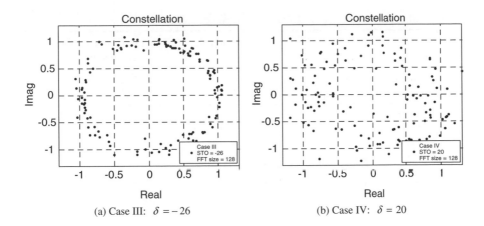

Figure 5.3 Signal constellation subject to STO: Case III and Case IV.

shift f_d. Furthermore, even if we intend to generate exactly the same carrier frequencies in the transmitter and receiver, there may be an unavoidable difference between them due to the physically inherent nature of the oscillators. Let f_c and f_c' denote the carrier frequencies in the transmitter and receiver, respectively. Let f_{offset} denote their difference (i.e., $f_{\text{offset}} = f_c - f_c'$). Meanwhile, Doppler frequency f_d is determined by the carrier frequency f_c and the velocity v of the terminal (receiver) as

$$f_d = \frac{v \cdot f_c}{c} \tag{5.5}$$

where c is the speed of light. Let us define the normalized CFO, ε, as a ratio of the CFO to subcarrier spacing Δf, shown as

$$\varepsilon = \frac{f_{\text{offset}}}{\Delta f} \tag{5.6}$$

Let ε_i and ε_f denote the integer part and fractional part of ε, respectively, and therefore, $\varepsilon = \varepsilon_i + \varepsilon_f$, where $\varepsilon_i = \lfloor \varepsilon \rfloor$. Table 5.2 presents examples of the Doppler frequency and normalized CFO at the mobile speed of 120 km/h for different commercial systems, each of which employs different carrier frequency. For the time-domain signal $x[n]$, a CFO of ε causes a phase offset of $2\pi n\varepsilon$, that is, proportional to the CFO ε and time index n. Note that it is equivalent to a frequency shift of $-\varepsilon$ on the frequency-domain signal $X[k]$. For the transmitted signal $x[n]$, the effect of CFO ε on the received signal $y[n]$ is summarized in Table 5.3.

Table 5.2 Doppler frequency and normalized CFO: an example.

System	Carrier frequency (f_c)	Subcarrier spacing (Δf)	Velocity (v)	Maximum Doppler frequency (f_d)	Normalized CFO (ε)
DMB	375 MHz	1 kHz	120 km/h	41.67 Hz	0.042
3GPP	2 GHz	15 kHz	120 km/h	222.22 Hz	0.0148
Mobile WiMAX	2.3 GHz	9.765 kHz	120 km/h	255.55 Hz	0.0263

Table 5.3 The effect of CFO on the received signal.

	Received signal	Effect of CFO ε on the received signal
Time-domain signal	$y[n]$	$e^{j2\pi n\varepsilon/N}x[n]$
Frequency-domain signal	$Y[k]$	$X[k-\varepsilon]$

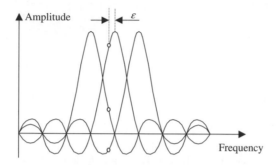

Figure 5.4 Inter-carrier interference (ICI) subject to CFO.

Figure 5.4 shows that the frequency shift of $-\varepsilon$ in the frequency-domain signal $X[k]$ subjects to the CFO of ε and leads to an inter-carrier interference (ICI), which means a subcarrier frequency component is affected by other subcarrier frequency components. To look into the effect of CFO, we assume that only a CFO of ε exists between transmitter and receiver, without any phase noise [100,103]. From Equation (5.1), the time-domain received signal can be written as

$$y_l[n] = \frac{1}{N}\sum_{k=0}^{N-1} H[k]X_l[k]\, e^{j2\pi(k+\varepsilon)n/N} + z_l[n] \tag{5.7}$$

Figures 5.5(a), (b), and (c) show that the phase of a time-domain signal is affected by the CFO as can be anticipated from Table 5.3 or Equation (5.7). Here, we assume the FFT size of $N = 32$ with QPSK modulation subject to no noise. The solid and dotted lines in the graphs on the left-hand side of Figure 5.5 represent the ideal case without CFO (i.e., $\varepsilon = 0$) and the case with CFO (i.e., $\varepsilon \neq 0$), respectively. The graphs on the right-hand side show the phase differences between them. From these figures, we can see that the received signal rotates faster in the time domain as CFO increases. Meanwhile, the phase differences increase linearly with time, with their slopes increasing with the CFO. As illustrated in Figure 5.5(c), if $\varepsilon > 0.5$, the phase difference exceeds π within an OFDM symbol, which results in a phase ambiguity. This is related to the range of CFO estimation, which will be discussed in Section 5.4.1.

Recall that the normalized CFO can be divided into two parts: integer CFO (IFO) ε_i and fractional CFO (FFO) ε_f (i.e., $\varepsilon = \varepsilon_i + \varepsilon_f$). In the following subsections, let us take a look at how each of these affects the frequency-domain received signal.

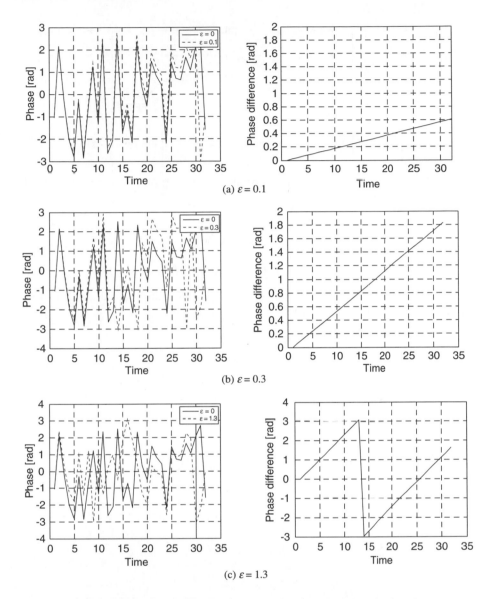

(a) $\varepsilon = 0.1$

(b) $\varepsilon = 0.3$

(c) $\varepsilon = 1.3$

Figure 5.5   Effects of CFO ε on the phase of the time-domain signal.

5.2.1 *Effect of Integer Carrier Frequency Offset (IFO)*

Figure 5.6 illustrates how the transmit samples $\{x_l[n]\}_{n=0}^{N-1}$ experience the IFO of ε_i. This leads to the signal of $e^{j2\pi\varepsilon_i n/N}x[n]$ in the receiver. Due to the IFO, the transmit signal $X[k]$ is cyclic-shifted by ε_i in the receiver, and thus producing $X[k-\varepsilon_i]$ in the kth subcarrier. Unless the cyclic-shift is compensated, it will incur a significant degradation in the BER performance. However, we note that the orthogonality among the subcarrier frequency components is not destroyed and thus, ICI does not occur.

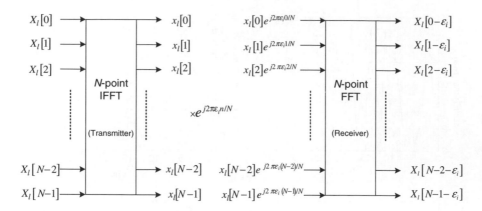

Figure 5.6   Effect of integer CFO on the received signal.

5.2.2 *Effect of Fractional Carrier Frequency Offset (FFO)*

Taking the FFT of $\{y_l[n]\}$ in Equation (5.7), the frequency-domain received signal with an FFO of ε_f can be written as follows [100]:

$$
Y_l[k] = \mathrm{FFT}\{y_l[n]\} = \sum_{n=0}^{N-1} y_l[n] e^{-j2\pi kn/N}
$$

$$
= \sum_{n=0}^{N-1} \frac{1}{N}\sum_{m=0}^{N-1} H[m]X_l[m]\, e^{j2\pi(m+\varepsilon_f)n/N} e^{-j2\pi kn/N} + \sum_{n=0}^{N-1} z_l[n] e^{-j2\pi kn/N}
$$

$$
= \frac{1}{N}\sum_{m=0}^{N-1} H[m]X_l[m] \sum_{n=0}^{N-1} e^{j2\pi(m-k+\varepsilon_f)n/N} + Z_l[k]
$$

$$
= \frac{1}{N} H[k]X_l[k]\sum_{n=0}^{N-1} e^{j2\pi\varepsilon_f n/N} + \frac{1}{N}\sum_{m=0,m\neq k}^{N-1} H[m]X_l[m] \sum_{n=0}^{N-1} e^{j2\pi(m-k+\varepsilon_f)n/N} + Z_l[k]
$$

$$
= \frac{1}{N}\frac{1-e^{j2\pi\varepsilon_f}}{1-e^{j2\pi\varepsilon_f/N}} H[k]X_l[k] + \frac{1}{N}\sum_{m=0,m\neq k}^{N-1} H[m]X_l[m] \frac{1-e^{j2\pi(m-k+\varepsilon_f)}}{1-e^{j2\pi(m-k+\varepsilon_f)/N}} + Z_l[k]
$$

$$
= \frac{1}{N}\frac{e^{j\pi\varepsilon_f}\left(e^{-j\pi\varepsilon_f}-e^{j\pi\varepsilon_f}\right)}{e^{j\pi\varepsilon_f/N}\left(e^{-j\pi\varepsilon_f/N}-e^{j\pi\varepsilon_f/N}\right)} H[k]X_l[k]
$$

$$
\quad + \frac{1}{N}\sum_{m=0,\ m\neq k}^{N-1} H[m]X_l[m] \frac{e^{j\pi(m-k+\varepsilon_f)}\left(e^{-j\pi(m-k+\varepsilon_f)}-e^{j\pi(m-k+\varepsilon_f)}\right)}{e^{j\pi(m-k+\varepsilon_f)/N}\left(e^{-j\pi(m-k+\varepsilon_f)/N}-e^{j\pi(m-k+\varepsilon_f)/N}\right)} + Z_l[k]
$$

$$
= e^{j\pi\varepsilon_f(N-1)/N}\left\{\frac{\sin(\pi\varepsilon_f)}{N\sin(\pi\,\varepsilon_f/N)}\right\} H_l[k]X_l[k]
$$

$$
\quad + e^{j\pi\varepsilon_f(N-1)/N}\sum_{m=0,\ m\neq k}^{N-1}\frac{\sin(\pi(m-k+\varepsilon_f))}{N\sin(\pi(m-k+\varepsilon_f)/N)} H[m]X_l[m]\, e^{j\pi(m-k)(N-1)/N} + Z_l[k]
$$

$$
= \frac{\sin\pi\varepsilon_f}{N\sin(\pi\varepsilon_f/N)} \cdot e^{j\pi\varepsilon_f(N-1)/N} H_l[k]X_l[k] + I_l[k] + Z_l[k] \tag{5.8}
$$

where

$$I_l[k] = e^{j\pi\varepsilon_f(N-1)/N} \sum_{m=0,\ m\neq k}^{N-1} \frac{\sin(\pi(m-k+\varepsilon_f))}{N\sin(\pi(m-k+\varepsilon_f)/N)} H[m]X_l[m]\, e^{j\pi(m-k)(N-1)/N} \qquad (5.9)$$

The first term of the last line in Equation (5.8) represents the amplitude and phase distortion of the kth subcarrier frequency component due to FFO. Meanwhile, $I_l[k]$ in Equation (5.8) represents the ICI from other subcarriers into kth subcarrier frequency component, which implies that the orthogonality among subcarrier frequency components is not maintained any longer due to the FFO. Figure 5.7 shows three consecutively received OFDM symbols with different FFO values where the effects of channel, STO, and noise are ignored. It is clear from this figure that amplitude and phase distortion becomes severe as FFO increases, which is attributed to the ICI term in Equation (5.9).

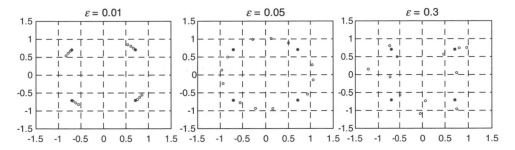

Figure 5.7   Constellation of received symbols with CFO ε.

The routine "add_STO()" in Program 5.1 and "add_CFO()" in Program 5.2 can be used to add the effects of STO and CFO to the received signal, respectively.

MATLAB® Programs: Adding STO and CFO

Program 5.1 "add_STO" for adding STO to the received signal

```
function y_STO=add_STO(y, nSTO)
% add STO (symbol time offset)
% y : Received signal
% nSTO : Number of samples corresponding to STO
if nSTO>=0, y_STO=[y(nSTO+1:end) zeros(1,nSTO)]; % advance
  else y_STO=[zeros(1,-nSTO) y(1:end+nSTO)]; % delay
end
```

Program 5.2 "add_CFO" for adding CFO to the received signal

```
function y_CFO=add_CFO(y,CFO,Nfft)
% add CFO (carrier frequency offset)
% y : Received signal
% CFO = IFO (integral CFO) + FFO (fractional CFO)
% Nfft = FFT size
nn=0:length(y)-1; y_CFO = y.*exp(j*2*pi*CFO*nn/Nfft); % Eq.(5.7)
```

5.3 Estimation Techniques for STO

As shown in Section 5.1, an STO may cause not only phase distortion (that can be compensated by using an equalizer) but also ISI (that cannot be corrected once occurred) in OFDM systems. In order to warrant its performance, therefore, the starting point of OFDM symbols must be accurately determined by estimating the STO with a synchronization technique at the receiver. In this section, we discuss how to estimate the STO. In general, STO estimation can be implemented either in the time or frequency domain.

5.3.1 Time-Domain Estimation Techniques for STO

Consider an OFDM symbol with a cyclic prefix (CP) of N_G samples over T_G seconds and effective data of N_{sub} samples over T_{sub} seconds. In the time domain, STO can be estimated by using CP or training symbols. In the sequel, we discuss the STO estimation techniques with CP or training symbols.

5.3.1.1 STO Estimation Techniques Using Cyclic Prefix (CP)

Recall that CP is a replica of the data part in the OFDM symbol. It implies that CP and the corresponding data part will share their similarities that can be used for STO estimation. Figure 5.8 denotes N_G samples of CP and another N_G samples of the data part by B and B', respectively. Note that the two identical blocks of samples in B and B' are spaced N_{sub} samples apart. As shown in Figure 5.8, consider two sliding windows, W1 and W2, which are spaced N_{sub} samples apart. These windows can slide to find the similarity between the samples within W1 and W2. The similarity between two blocks of N_G samples in W1 and W2 is maximized when CP of an OFDM symbol falls into the first sliding window. In fact, this maximum point can be used to identify the STO.

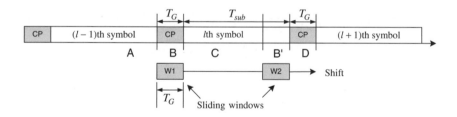

Figure 5.8 STO estimation technique using double sliding windows.

Since the similarity between two blocks in W1 and W2 is maximized when the difference between them is minimized, the STO can be found by searching the point where the difference between two blocks of N_G samples within these two sliding windows is minimized [104], that is,

$$\hat{\delta} = \arg\min_{\delta} \left\{ \sum_{i=\delta}^{N_G-1+\delta} |y_l[n+i] - y_l[n+N+i]| \right\} \tag{5.10}$$

In spite of the simplicity of this technique, its performance can be degraded when CFO exists in the received signal. Another STO estimation technique, which can also deal with CFO, is to minimize the squared difference between a N_G-sample block (seized in window W1) and the conjugate of another N_G-sample block (seized in window W2) [105], shown as

$$\hat{\delta} = \arg\min_{\delta} \left\{ \sum_{i=\delta}^{N_G-1+\delta} \left(|y_l[n+i]| - |y_l^*[n+N+i]| \right)^2 \right\} \tag{5.11}$$

Another approach is to consider the correlation between those two blocks in W1 and W2. Toward this end, a maximum-likelihood estimation scheme can be applied to yield

$$\hat{\delta} = \arg\max_{\delta} \left\{ \sum_{i=\delta}^{N_G-1+\delta} |y_l[n+i]y_l^*[n+N+i]| \right\} \tag{5.12}$$

which corresponds to maximizing the correlation between a block of N_G samples (seized in window W1) and another block of N_G samples (seized in window W2). However, the performance of Equation (5.12) is degraded when CFO exists in the received signal. To deal with the CFO in the received signal, we utilize another ML technique that maximizes the log-likelihood function, given as

$$\hat{\delta}_{ML} = \arg\max_{\delta} \sum_{i=\delta}^{N_G-1+\delta} \left[2(1-\rho)\text{Re}\{y_l[n+i]y_l^*[n+N+i]\} - \rho \sum_{i=\delta}^{N_G-1+\delta} |y_l[n+i]-y_l[n+N+i]| \right]$$
$$\tag{5.13}$$

where $\rho = \text{SNR}/(\text{SNR}+1)$[106]. We can also think of another ML technique that estimates both STO and CFO at the same time as derived in [107]. In this technique, the STO is estimated as

$$\hat{\delta}_{ML} = \arg\max_{\delta}\{|\gamma[\delta]| - \rho\, \Phi[\delta]\} \tag{5.14}$$

where

$$\gamma[m] = \sum_{n=m}^{m+L-1} y_l[n]y_l^*[n+N],$$

$$\text{and}\quad \Phi[m] = \frac{1}{2}\sum_{n=m}^{m+L-1}\left\{ |y_l[n]|^2 + |y_l[n+N]|^2 \right\} \tag{5.15}$$

using L to denote the actual number of samples used for averaging in windows. Taking the absolute value of the correlation $\gamma[m]$, STO estimation in Equation (5.14) can be robust even under the presence of CFO.

5.3.1.2 STO Estimation Techniques Using Training Symbol

Training symbols can be transmitted to be used for symbol synchronization in the receiver. In contrast with CP, it involves overhead for transmitting training symbols, but it does not suffer from the effect of the multi-path channel. Two identical OFDM training symbols, or a single OFDM symbol with a repetitive structure can be used. Figures 5.9 and 5.10 illustrate the example of a single OFDM symbol with a repetitive structure of different repetition periods, which are periods of $T_{sub}/2$ and $T_{sub}/4$, respectively. The repetitive pattern in the time domain can be generated by inserting 0s between subcarriers. Once the transmitter sends the repeated training signals over two blocks within the OFDM symbol, the receiver attempts to find the CFO by maximizing the similarity between these two blocks of samples received within two sliding windows. The similarity between two sample blocks can be computed by an auto-correlation property of the repeated training signal.

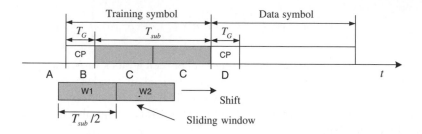

Figure 5.9   STO estimation using the repetitive training symbol (period $= T_{sub}/2$).

For the structure with a period of $T_{sub}/2$ in Figure 5.9, using the fact that the parts of the received signal, denoted by A, B, and D, are different from that in the other part, denoted by C, two sliding windows, W1 and W2, can be formed to estimate STO. As in the STO estimation technique using CP, for example, STO can be estimated by minimizing the squared difference between two blocks of samples received in W1 and W2 [108,109], such that

$$\hat{\delta} = \arg\min_{\delta} \left\{ \sum_{i=\delta}^{\frac{N}{2}-1+\delta} \left| y_l[n+i] - y_l^*\left[n+\frac{N}{2}+i\right] \right|^2 \right\} \tag{5.16}$$

or by maximizing the likelihood function [109], that is,

$$\hat{\delta} = \arg\max_{\delta} \left\{ \frac{\left| \sum_{i=\delta}^{\frac{N}{2}-1+\delta} y_l[n+i]y_l^*\left[n+\frac{N}{2}+i\right] \right|^2}{\left| \sum_{i=\delta}^{\frac{N}{2}-1+\delta} y_l\left[n+\frac{N}{2}+i\right] \right|^2} \right\} \tag{5.17}$$

Due to the effect of multi-path channel, the minimum-squared difference or the maximum correlation can occur in the interval, denoted by C. In order to avoid such erroneous STO estimation, we can set a threshold. For example, STO can be located by finding the first dip (although it is not the minimum) where the squared difference is below the given threshold, or by finding the first peak (although it is not the maximum) where the correlation is above the given threshold. The techniques in Equation (5.16) and Equation (5.17) have advantage of estimating STO without being affected by CFO. Since the sliding windows have a length of $N/2$, however, the estimated difference or correlation has a flat interval (plateau) over the length of CP corresponding to interval B, which does not lend itself to locating the STO. The difficulty in locating STO due to the flat interval can be handled by taking the average of correlation values over the length of CP [103], shown as

$$\hat{\delta} = \arg\max_{\delta} \left(\frac{1}{N_G + 1} \sum_{m=-N_G+i}^{i} s_l[n+m] \right) \tag{5.18}$$

where

$$s_l[n] = \frac{\left| \sum_{i=\delta}^{N/2-1+\delta} y_l[n+i] y_l^* \left[n + \frac{N}{2} + i \right] \right|^2}{\left(\frac{1}{2} \left| \sum_{i=\delta}^{N-1+\delta} y_l[n+i] \right|^2 \right)^2} \tag{5.19}$$

Meanwhile, the accuracy of STO estimation can be improved by changing the period of the repetitive pattern of the training symbol as in Figure 5.10. In this example, the training signal is repeated four times, yet inverting the signs of training signals in the third and fourth period [103]. Now, the accuracy of STO estimation in Equation (5.17) can be further improved with

$$\hat{\delta} = \arg\max_{\delta} \frac{\left| \sum_{i=\delta}^{N/4-1+\delta} y_l \left[n+i+\frac{N}{2}m \right] y_l^* \left[n+i+\frac{N}{4}+\frac{N}{2}m \right] \right|^2}{\left(\frac{1}{2} \sum_{m=0}^{1} \sum_{i=\delta}^{N/4-1+\delta} \left| y_l \left[n+i+\frac{N}{4}+\frac{N}{2}m \right] \right|^2 \right)^2} \tag{5.20}$$

Another type of STO estimation technique is to use the cross-correlation between the training symbol and received signal, since the training symbol is known to the receiver. In this case, we do not need to use two sliding windows, W1 and W2, in Figure 5.9. In fact, only one sliding window which corresponds to the locally generated training symbol with a period of $T_{sub}/2$ is enough. Its performance can be degraded when CFO exists. In general, however, it provides better accuracy than the one using the auto-correlation property when the effect of CFO is not significant.

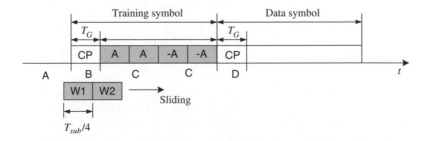

Figure 5.10 STO estimation using the repetitive training symbol (period $= T_{sub}/4$).

'Program 5.3 ("STO_estimation.m") performs STO estimation using CP. It employs the
maximum correlation-based technique by Equation (5.12) in Program 5.4 and the minimum
difference-based technique by Equation (5.11) in Program 5.5. Figure 5.11 illustrates the
results obtained by these programs, in which CFO is located at the point of minimizing the
difference between the sample blocks of CP and that of data part or maximizing their
correlation.

Program 5.3 "STO_estimation.m": CP-based symbol timing offset (STO) estimation

```
% STO_estimation.m
clear, clf
nSTOs = [-3 -3 2 2]; % Number of samples corresponding to STO
CFOs = [0 0.5 0 0.5]; SNRdB=40; MaxIter=10; % CFOs, SNR, # of iteration
Nfft=128; Ng=Nfft/4; % FFT size and GI (CP) length
Nofdm=Nfft+Ng; % OFDM symbol length
Nbps=2; M=2^Nbps; % Number of bits per (modulated) symbol
mod_object = modem.qammod('M',M,'SymbolOrder','gray');
Es=1; A=sqrt(3/2/(M-1)*Es); % Signal energy and QAM normalization factor
N=Nfft; com_delay=Nofdm/2; Nsym=100;
rand('seed',1); randn('seed',1);
for i=1:length(nSTOs)
  nSTO=nSTOs(i); CFO=CFOs(i);
  x = []; % Initialize a block of OFDM signals
  for m=1:Nsym % Transmit OFDM signals
     msgint=randint(1,N,M);
     Xf = A*modulate(mod_object,msgint);
     xt = ifft(Xf,Nfft); x_sym = [xt(end-Ng+1:end) xt]; % IFFT & Add CP
     x = [x x_sym];
  end
  y = x; % No channel effect
  y_CFO = add_CFO(y,CFO,Nfft); y_CFO_STO= add_STO(y_CFO,-nSTO);
  Mag_cor= 0; Mag_dif= 0;
  for iter=1:MaxIter
     y_aw = awgn(y_CFO_STO,SNRdB,'measured'); % AWGN added
     % Symbol Timing Acqusition
     [STO_cor,mag_cor]=STO_by_correlation(y_aw,Nfft,Ng,com_delay);
     [STO_dif,mag_dif]=STO_by_difference(y_aw,Nfft,Ng,com_delay);
     Mag_cor= Mag_cor+mag_cor; Mag_dif= Mag_dif+mag_dif;
  end % End of for loop of iter
  [Mag_cor_max,ind_max] = max(Mag_cor); nc= ind_max-1-com_delay;
  [Mag_dif_min,ind_min] = min(Mag_dif); nd= ind_min-1-com_delay;
  nn=-Nofdm/2+[0:length(Mag_cor)-1];
  subplot(220+i), plot(nn,Mag_cor,'b:', nn,Mag_dif,'r'), hold on
  str1=sprintf('Cor(b-)/Dif(r:) for nSTO=%d, CFO=%1.2f',nSTO,CFO);
  title(str1); xlabel('Sample'), ylabel('Magnitude');
  stem(nc,Mag_cor(nc+com_delay+1),'b') % Estimated STO from correlation
```

```
    stem(nd,Mag_dif(nd+com_delay+1),'r') % Estimated STO from difference
    stem(nSTO,Mag_dif(nSTO+com_delay+1),'k.') % True STO
end % End of for loop of i
```

Program 5.4 "STO_by_correlation": CP-based symbol synchronization using the correlation

```
function [STO_est, Mag]=STO_by_correlation(y,Nfft,Ng,com_delay)
% estimates STO by maximizing the correlation between CP (cyclic prefix)
%       and rear part of OFDM symbol
% Input: y          = Received OFDM signal including CP
%          Ng        = Number of samples in Guard Interval (CP)
%       com_delay    = Common delay
% Output: STO_est= STO estimate
%           Mag       = Correlation function trajectory varying with time
Nofdm=Nfft+Ng; % OFDM symbol length
if nargin<4, com_delay = Nofdm/2; end
nn=0:Ng-1;
yy = y(nn+com_delay)*y(nn+com_delay+Nfft)'; % Correlation
maximum=abs(yy);
for n=1:Nofdm
 n1 = n-1;
 yy1 = y(n1+com_delay)*y(n1+com_delay+Nfft)';
 yy2 = y(n1+com_delay+Ng)*y(n1+com_delay+Nfft+Ng)';
 yy = yy-yy1+yy2; Mag(n)=abs(yy); % Eq.(5.12)
 if Mag(n)>maximum, maximum=Mag(n); STO_est=Nofdm-com_delay-n1; end
end
```

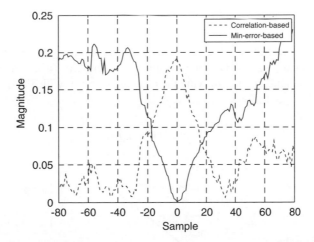

Figure 5.11 Performance of CP-based STO estimation: maximum correlation-based vs. minimum difference-based estimation.

Program 5.5 "STO_by_difference": CP-based symbol synchronization using the difference

```
function [STO_est,Mag]=STO_by_difference(y,Nfft,Ng,com_delay)
% estimates STO by minimizing the difference between CP (cyclic prefix)
%      and rear part of OFDM symbol
% Input: y        = Received OFDM signal including CP
%        Ng        = Number of samples in CP (Guard Interval)
%     com_delay    = Common delay
% Output: STO_est = STO estimate
%         Mag      = Correlation function trajectory varying with time
Nofdm=Nfft+Ng; minimum=100; STO_est=0;
if nargin<4, com_delay = Nofdm/2; end
for n=1:Nofdm
    nn = n+com_delay+[0:Ng-1]; tmp0 = abs(y(nn))-abs(y(nn+Nfft));
    Mag(n) = tmp0*tmp0'; % Squared difference by Eq.(5.11)
    if Mag(n), minimum=Mag(n); STO_est=Nofdm-com_delay-(n-1); end
end
```

5.3.2 Frequency-Domain Estimation Techniques for STO

As implied by Equation (5.2), the received signal subject to STO suffers from a phase rotation. Since the phase rotation is proportional to subcarrier frequency, the STO can be estimated by the phase difference between adjacent subcarrier components of the received signal in the frequency domain. For example, if $X_l[k] = X_l[k-1]$ and $H_l[k] \approx H_l[k-1]$ for all k, $Y_l[k]Y_l^*[k-1] \approx |X_l[k]|^2 e^{j2\pi\delta/N}$ and thus, the STO can be estimated as

$$\hat{\delta} = \frac{N}{2\pi}\arg\left(\sum_{k=1}^{N-1} Y_l[k]Y_l^*[k-1]\right) \tag{5.21}$$

Figure 5.12 shows another technique for STO estimation by using the effect of phase rotation [110]. More specifically, an STO can be estimated from the (delayed) channel impulse response, which is obtained by multiplying the received symbol (with STO) by the conjugated training symbol $X_l^*[k]$ as follows:

$$\hat{\delta} = \arg\max_n \left(y_l^X[n]\right)$$

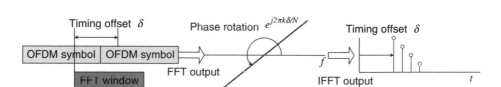

Figure 5.12   STO estimation using the channel impulse response.

where

$$y_l^X[n] = \text{IFFT}\{Y_l[k]e^{j2\pi\delta k/N}X_l^*[k]\}$$

$$= \frac{1}{N}\sum_{k=0}^{N-1}Y_l[k]e^{j2\pi\,\delta k/N}X_l^*[k]e^{j2\pi nk/N}$$

$$= \frac{1}{N}\sum_{k=0}^{N-1}H_l[k]X_l[k]X_l^*[k]e^{j2\pi(\delta+n)k/N} \tag{5.22}$$

$$= \frac{1}{N}\sum_{k=0}^{N-1}H_l[k]e^{j2\pi(\delta+n)k/N}$$

$$= h_l[n+\delta]$$

In Equation (5.22), it is assumed that the power of training symbol $X_l[k]$ is equal to one (i.e., $X_l[k]X_l^*[k] = |X_l[k]|^2 = 1$). Figure 5.13 shows two examples of STO estimation using the channel impulse response, one with $\delta = 0$ and the other with $\delta = 10$ [samples]. In this figure, the first channel impulse response, indicated by a solid line, starts at 0th sampling point and the second channel impulse response, indicated by a dotted line, starts at 10th sampling point. The STO can be correctly estimated from this. The frequency-domain STO estimation techniques as discussed above can be used for fine symbol synchronization, since they usually produce fairly accurate STO estimates.

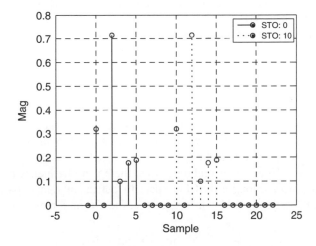

Figure 5.13   Examples of STO estimation using channel impulse response.

The STO can be compensated differently, depending on the detection scheme (coherent or non-coherent) when the value of STO is small, that is, less than a fraction of sampling interval. In the case of coherent detection using pilot symbols, the channel estimate includes the phase rotation caused by STO as well as the channel itself. Since the frequency-domain equalizer, designed to compensate the effect of channel, can absorb the effect of STO as long as the STO is small, a separate symbol synchronizer may not be needed in the case of coherent detection. In the case of non-coherent detection, however, the phase difference among the subcarrier components in the received signal can be used for detecting the transmitted symbols. Since the STO causes a phase rotation proportional to subcarrier frequency, the STO effect should be removed by a symbol synchronizer before symbol detection is made in this case.

5.4 Estimation Techniques for CFO

Like STO estimation, CFO estimation can also be performed either in the time or the frequency domain.

5.4.1 Time-Domain Estimation Techniques for CFO

For CFO estimation in the time domain, cyclic prefix (CP) or training symbol is used. Each of these techniques is described as below.

5.4.1.1 CFO Estimation Techniques Using Cyclic Prefix (CP)

With perfect symbol synchronization, a CFO of ε results in a phase rotation of $2\pi n\varepsilon/N$ in the received signal (Table 5.3). Under the assumption of negligible channel effect, the phase difference between CP and the corresponding rear part of an OFDM symbol (spaced N samples apart) caused by CFO ε is $2\pi N\varepsilon/N = 2\pi\varepsilon$. Then, the CFO can be found from the phase angle of the product of CP and the corresponding rear part of an OFDM symbol, for example, $\hat{\varepsilon} = (1/2\pi)\arg\{y_l^*[n]y_l[n+N]\}$, $n = -1, -2, \ldots, -N_g$. In order to reduce the noise effect, its average can be taken over the samples in a CP interval as

$$\hat{\varepsilon} = \frac{1}{2\pi}\arg\left\{\sum_{n=-N_G}^{-1} y_l^*[n]y_l[n+N]\right\} \tag{5.23}$$

Since the argument operation arg( ) is performed by using $\tan^{-1}( )$, the range of CFO estimation in Equation (5.23) is $[-\pi, +\pi)/2\pi = [-0.5, +0.5)$ so that $|\hat{\varepsilon}| < 0.5$ and consequently, integral CFO cannot be estimated by this technique.

Note that $y_l^*[n]y_l[n+N]$ becomes real only when there is no frequency offset. This implies that it becomes imaginary as long as the CFO exists. In fact, the imaginary part of $y_l^*[n]y_l[n+N]$ can be used for CFO estimation [111]. In this case, the estimation error is defined as

$$e_\varepsilon = \frac{1}{L}\sum_{n=1}^{L} \text{Im}\{y_l^*[n]y_l[n+N]\} \tag{5.24}$$

where L denotes the number of samples used for averaging. Note that the expectation of the error function in Equation (5.24) can be approximated as

$$E\{e_\varepsilon\} = \frac{\sigma_d^2}{N}\sin\left(\frac{2\pi\varepsilon}{N}\right) \sum_{k \text{ corresponding to useful carriers}}^{L} |H_k|^2 \approx K\varepsilon \tag{5.25}$$

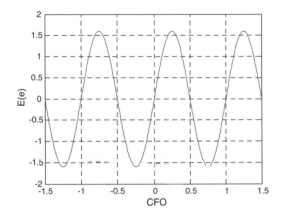

Figure 5.14 Characteristic curve of the error function Equation (5.25).

where σ_d^2 is the transmitted signal power, H_k is the channel frequency response of the kth subcarrier, and K is a term that comprises transmit and channel power. Figure 5.14 shows that the error function in Equation (5.25) has an S-curve around the origin, which is required for synchronization. Note that frequency synchronization can be maintained by controlling VCO in accordance with the sine function in Equation (5.25). This particular approach also provides $|\hat{\varepsilon}| < 0.5$ as with Equation (5.23).

5.4.1.2 CFO Estimation Techniques Using Training Symbol

We have seen that the CFO estimation technique using CP can estimate the CFO only within the range $\{|\varepsilon| \leq 0.5\}$. Since CFO can be large at the initial synchronization stage, we may need estimation techniques that can cover a wider CFO range. The range of CFO estimation can be increased by reducing the distance between two blocks of samples for correlation. This is made possible by using training symbols that are repetitive with some shorter period. Let D be an integer that represents the ratio of the OFDM symbol length to the length of a repetitive pattern. Let a transmitter send the training symbols with D repetitive patterns in the time domain, which can be generated by taking the IFFT of a comb-type signal in the frequency domain given as

$$X_l[k] = \begin{cases} A_m, & \text{if} \quad k = D \cdot i, i = 0, 1, \ldots, (N/D-1) \\ 0, & \text{otherwise} \end{cases} \tag{5.26}$$

where A_m represents an M-ary symbol and N/D is an integer. As $x_l[n]$ and $x_l[n+N/D]$ are identical (i.e., $y_l^*[n]y_l[n+N/D] = |y_l[n]|^2 e^{j\pi\varepsilon}$), a receiver can make CFO estimation as follows [108, 109]:

$$\hat{\varepsilon} = \frac{D}{2\pi} \arg \left\{ \sum_{n=0}^{N/D-1} y_l^*[n] y_l[n+N/D] \right\} \tag{5.27}$$

The CFO estimation range covered by this technique is $\{|\varepsilon| \leq D/2\}$, which becomes wider as D increases. Note that the number of samples for the computation of correlation is reduced by $1/D$, which may degrade the MSE performance. In other words, the increase in estimation range is obtained at the sacrifice of MSE (mean square error) performance. Figure 5.15 shows the estimation range of CFO vs. MSE performance for $D = 1$ and 4. Here, a trade-off relationship between the MSE performance and estimation range of CFO is clearly shown. As the estimation range of CFO increases, the MSE performance becomes worse. By taking the average of the estimates with the repetitive patterns of the shorter period as

$$\hat{\varepsilon} = \frac{D}{2\pi} \arg \left\{ \sum_{m=0}^{D-2} \sum_{n=0}^{N/D-1} y_l^*[n+mN/D] y_l[n+(m+1)N/D] \right\} \tag{5.28}$$

the MSE performance can be improved without reducing the estimation range of CFO.

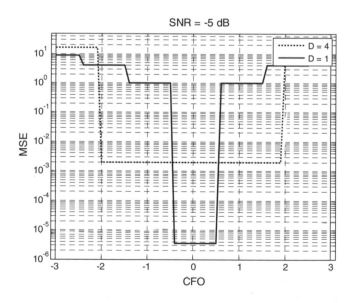

Figure 5.15   Estimation range of CFO vs. MSE performance.

5.4.2 Frequency-Domain Estimation Techniques for CFO

If two identical training symbols are transmitted consecutively, the corresponding signals with CFO of ε are related with each other as follows:

$$y_2[n] = y_1[n]e^{j2\pi N\varepsilon/N} \leftrightarrow Y_2[k] = Y_1[k]e^{j2\pi\varepsilon} \tag{5.29}$$

Using the relationship in Equation (5.29), the CFO can be estimated as

$$\hat{\varepsilon} = \frac{1}{2\pi}\tan^{-1}\left\{\sum_{k=0}^{N-1}\text{Im}\left[Y_1^*[k]Y_2[k]\right] \bigg/ \sum_{k=0}^{N-1}\text{Re}\left[Y_1^*[k]Y_2[k]\right]\right\} \tag{5.30}$$

which is a well-known approach by Moose [112]. Although the range of CFO estimated by Equation (5.30) is $|\varepsilon| \leq \pi/2\pi = 1/2$, it can be increased D times by using a training symbol with D repetitive patterns. The repetitive patterns in the time-domain signal can be generated by Equation (5.26). In this case, Equation (5.30) is applied to the subcarriers with non-zero value and then, averaged over the subcarriers. As discussed in the previous subsection, the MSE performance may deteriorate due to the reduced number of non-zero samples taken for averaging in the frequency domain. Note that this particular CFO estimation technique requires a special period, usually known as a preamble period, in which the consecutive training symbols are provided for facilitating the computation in Equation (5.30). In other words, it is only applicable during the preamble period, for which data symbols cannot be transmitted.

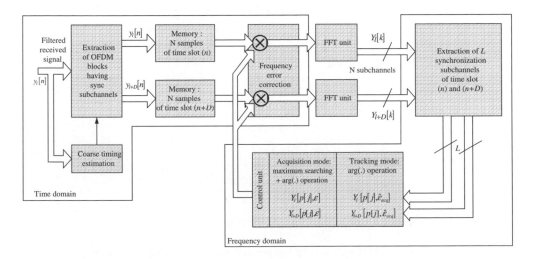

Figure 5.16   CFO synchronization scheme using pilot tones.

We can think about another technique that allows for transmitting the data symbols while estimating the CFO. As proposed by Classen [113], pilot tones can be inserted in the frequency domain and transmitted in every OFDM symbol for CFO tracking. Figure 5.16 shows a structure of CFO estimation using pilot tones. First, two OFDM symbols, $y_l[n]$ and $y_{l+D}[n]$, are saved in the memory after synchronization. Then, the signals are transformed into $\{Y_l[k]\}_{k=0}^{N-1}$ and $\{Y_{l+D}[k]\}_{k=0}^{N-1}$ via FFT, from which pilot tones are extracted. After estimating CFO from pilot tones in the frequency domain, the signal is compensated with the estimated CFO in the time domain. In this process, two different estimation modes for CFO estimation are implemented: acquisition and tracking modes. In the acquisition mode, a large range of CFO including an integer CFO is estimated. In the tracking mode, only fine CFO is estimated. The integer CFO is estimated by

$$\hat{\varepsilon}_{acq} = \frac{1}{2\pi \cdot T_{sub}} \max_{\varepsilon} \left\{ \left| \sum_{j=0}^{L-1} Y_{l+D}[p[j], \varepsilon] Y_l^*[p[j], \varepsilon] \, X_{l+D}^*[p[j]] X_l[p[j]] \right| \right\} \quad (5.31)$$

where $L, p[j]$, and $X_l[p[j]]$ denote the number of pilot tones, the location of the jth pilot tone, and the pilot tone located at $p[j]$ in the frequency domain at the lth symbol period, respectively. Meanwhile, the fine CFO is estimated by

$$\hat{\varepsilon}_f = \frac{1}{2\pi \cdot T_{sub} \cdot D} \arg \left\{ \sum_{j=1}^{L-1} Y_{l+D}[p[j], \hat{\varepsilon}_{acq}] Y_l^*[p[j], \hat{\varepsilon}_{acq}] \, X_{l+D}^*[p[j]] X_l[p[j]] \right\} \quad (5.32)$$

In the acquisition mode, $\hat{\varepsilon}_{acq}$ and $\hat{\varepsilon}_f$ are estimated and then, the CFO is compensated by their sum. In the tracking mode, only $\hat{\varepsilon}_f$ is estimated and then compensated.

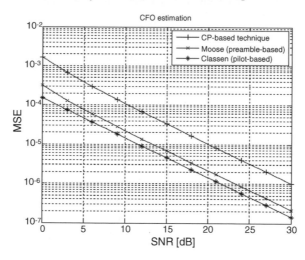

Figure 5.17 MSE of CFO estimation techniques (obtained by running "CFO_estimation.m").

MATLAB® Programs: CFO Estimation

Program 5.6 ("CFO_estimation.m") performs CFO estimation by using three different techniques, once by using Equation (5.28) or the corresponding routine "CFO_CP" in Program

5.7 using the phase difference between CP and the corresponding rear part of an OFDM symbol, once by using Equation (5.30) or the corresponding routine "CFO_Moose" in Program 5.8 using the phase difference between two repetitive preambles, and once by using Equation (5.31) or the corresponding routine "CFO_Classen" in Program 5.9 using the phase difference between pilot tones in two consecutive OFDM symbols. Interested readers are recommended to run the program and see the mean squared CFO estimation errors decrease as the SNR of received signal increases (see Figure 5.17). Performances of estimation techniques vary depending on the number of samples in CP, the number of samples in preamble, and the number of pilot tones, used for CFO estimation.

Program 5.6 "CFO_estimation.m" using time/frequency-domain techniques

```
%CFO_estimation.m
%Time-domain CP based method and Frequency-domain (Moose/Classen) method
clear, clf
CFO = 0.15;
Nfft=128; % FFT size
Nbps=2; M=2^Nbps; % Number of bits per (modulated) symbol
h = modem.qammod('M',M,'SymbolOrder','gray');
Es=1; A=sqrt(3/2/(M-1)*Es); % Signal energy and QAM normalization factor
N=Nfft; Ng=Nfft/4; Nofdm=Nfft+Ng; Nsym=3;
x=[]; % Transmit signal
for m=1:Nsym
    msgint=randint(1,N,M);
    if i<=2, Xp= add_pilot(zeros(1,Nfft),Nfft,4); Xf=Xp; % add_pilot
    else    Xf = A*modulate(h,msgint);
    end
    xt = ifft(Xf,Nfft); % IFFT
    x_sym = add_CP(xt,Ng); % add CP
    x = [x x_sym];
end
y=x; % No channel effect
sig_pow= y*y'/length(y); % Signal power calculation
SNRdBs= 0:3:30; MaxIter = 100;
for i=1:length(SNRdBs)
  SNRdB = SNRdBs(i);
  MSE_CFO_CP = 0; MSE_CFO_Moose = 0; MSE_CFO_Classen = 0;
  rand('seed',1); randn('seed',1); % Initialize seed for random number
  y_CFO= add_CFO(y,CFO,Nfft); % Add CFO
  for iter=1:MaxIter
    y_aw = awgn(y_CFO,SNRdB,'measured'); % AWGN added
    CFO_est_CP = CFO_CP(y_aw,Nfft,Ng); % CP-based
      MSE_CFO_CP = MSE_CFO_CP + (CFO_est_CP-CFO)^2;
    CFO_est_Moose = CFO_Moose(y_aw,Nfft); % Moose
      MSE_CFO_Moose = MSE_CFO_Moose + (CFO_est_Moose-CFO)^2;
    CFO_est_Classen = CFO_Classen(y_aw,Nfft,Ng,Xp); % Classen
      MSE_CFO_Classen = MSE_CFO_Classen+(CFO_est_Classen-CFO)^2;
  end % End of (iter) loop
```

```
  MSE_CP(i)=MSE_CFO_CP/MaxIter; MSE_Moose(i)=MSE_CFO_Moose/MaxIter;
  MSE_Classen(i)=MSE_CFO_Classen/MaxIter;
end % End of SNR loop
semilogy(SNRdBs,MSE_CP,'-+'), grid on, hold on
semilogy(SNRdBs,MSE_Moose,'-x'), semilogy(SNRdBs,MSE_Classen,'-*')
xlabel('SNR[dB]'), ylabel('MSE'); title('CFO Estimation');
legend('CP-based technique','Moose','Classen')
```

Program 5.7 "CFO_CP": Time-domain technique using CP

```
function CFO_est=CFO_CP(y,Nfft,Ng)
% Time-domain CFO estimation based on CP (Cyclic Prefix)
nn=1:Ng; CFO_est = angle(y(nn+Nfft)*y(nn)')/(2*pi); % Eq.(5.27)
```

Program 5.8 "CFO_Moose": Frequency-domain technique using preamble

```
function CFO_est=CFO_Moose(y,Nfft)
% Frequency-domain CFO estimation using Moose method
%    based on two consecutive identical preambles (OFDM symbols)
for i=0:1, Y(i+1,:)=fft(y(Nfft*i+1:Nfft*(i+1)),Nfft); end
CFO_est = angle(Y(2,:)*Y(1,:)')/(2*pi); % Eq.(5.30)
```

Program 5.9 "CFO_Classen": Frequency-domain technique using pilot tones

```
function CFO_est=CFO_Classen(yp,Nfft,Ng,Nps)
% Frequency-domain CFO estimation using Classen method
%    based on pilot tones in two consecutive OFDM symbols
if length(Nps)==1, Xp=add_pilot(zeros(1,Nfft),Nfft,Nps); % Pilot signal
  else Xp=Nps; % If Nps is an array, it must be a pilot sequence Xp
end
Nofdm=Nfft+Ng; kk=find(Xp~=0); Xp=Xp(kk); % Extract pilot tones
for i=1:2
  yp_without_CP = remove_CP(yp(1+Nofdm*(i-1):Nofdm*i),Ng);
  Yp(i,:) = fft(yp_without_CP,Nfft);
end
CFO_est = angle(Yp(2,kk).*Xp*(Yp(1,kk).*Xp)')/(2*pi); % Eq.(5.31)
CFO_est = CFO_est*Nfft/Nofdm; % Eq.(5.31)
```

Program 5.10 "add_pilot" to generate and insert a pilot sequence

```
function xp=add_pilot(x,Nfft,Nps)
% CAZAC (Constant Amplitude Zero AutoCorrelation) sequence -> pilot
% Nps : Pilot spacing
if nargin <3, Nps=4; end
Np=Nfft/Nps; % Number of pilots
xp=x; % Prepare an OFDM signal including pilot signal for initialization
for k=1:Np
```

```
    xp((k-1)*Nps+1)=exp(j*pi*(k-1)^2/Np); % Eq.(7.17) for Pilot boosting
end
```

Program 5.11 "add_CP" to add CP

```
function y=add_CP(x,Ng)
% Add CP (Cyclic Prefix) of length Ng
y = [x(:,end-Ng+1:end) x];
```

Program 5.12 "remove_CP" to remove CP

```
function y=remove_CP(x,Ng,Noff)
% Remove CP (Cyclic Prefix) of length Ng
if nargin<3, Noff=0; end
y=x(:,Ng+1-Noff:end-Noff);
```

5.5 Effect of Sampling Clock Offset

In this section, we will investigate the effect of the sampling clock offset, which includes the phase offset in the sampling clocks and frequency offset in the sampling clocks.

5.5.1 Effect of Phase Offset in Sampling Clocks

The phase offset in the sampling clocks can be viewed as the symbol timing error, which occurs when the sampling rates or frequencies at the transmitter and receiver are identical. In the presence of phase offset, their sampling times do not coincide with each other. In this situation, the sampling instants at the receiver differ from the optimal ones by some constant time [114]. Figure 5.18 illustrates a phase offset in the sampling clocks.

Just like the STO discussed in Section 5.1, a phase rotation is incurred in the frequency domain due to the phase offset in the sampling clocks. It is proportional to the timing offset and subcarrier index. Since the effect of phase offset in the sampling clocks is usually small, it is often considered just as a part of STO, without requiring any additional compensator.

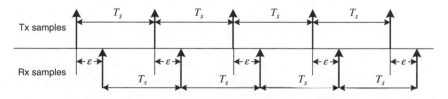

Figure 5.18   Phase offset in sampling clocks: an illustration.

5.5.2 Effect of Frequency Offset in Sampling Clocks

The SFO (frequency offset in the sampling clocks) between the transmitter and receiver occurs due to mismatch between the transmitter and receiver oscillators, or due to Doppler frequency shift [115]. Figure 5.19 illustrates the SFO that causes the phase offset in clocks to vary with time, resulting in an ICI (Inter-Carrier Interference).

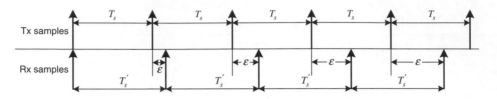

Figure 5.19   Frequency offset in the sampling clocks: an illustration.

The frequency-domain received signal with SFO can be written as

$$\tilde{Y}[k] = Y[k] \cdot \frac{\sin(\pi \Delta k)}{\sin(\pi \Delta k / N)} \cdot e^{\frac{-j\pi \Delta k (N-1)}{N}} + z_{ICI}[k] \quad \text{with} \quad \Delta = \frac{T_s - T'_s}{T_s} \quad (5.33)$$

where $z_{ICI}[k]$ denotes the ICI caused by SFO. The first term in Equation (5.33) corresponds to the amplitude and phase distortion of the signal received at each subcarrier. In [114], the variance of $z_{ICI}[k]$ is given by

$$\text{var}\{z_{ICI}[k]\} = \frac{\pi^2}{3}(\Delta k)^2 \quad (5.34)$$

When SFO is present, the sampling timing offset may vary in every OFDM symbol and a periodic insertion or loss of one sample may occur in one symbol period. Figure 5.19 shows an example of one sample loss occurring in the case where the sampling clock frequency of the receiver is lower than that of the transmitter.

5.6 Compensation for Sampling Clock Offset

A digital OFDM receiver samples the received continuous-time signal at instants determined by the clock in the receiver [116]. Depending on the presence of control mechanism of the sampling clock, the compensation scheme of the sampling clock offset can be divided into two types: synchronous sampling and non-synchronous sampling schemes. The synchronous sampling scheme controls the sampling timing instants in the analog domain via VCXO (Voltage Controlled Crystal Oscillator) and DPLL (Digital Phase Locked Loop). Meanwhile, the non-synchronous sampling scheme compensates the sampling clock offset in the digital domain after free-running sampling. Figure 5.20(a) shows a synchronous sampling system

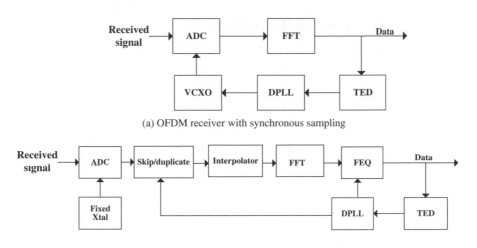

(a) OFDM receiver with synchronous sampling

(b) OFDM receiver with non-synchronous sampling

Figure 5.20   Block diagrams for OFDM receivers with synchronous/non-synchronous sampling.

where the phase rotation at pilot tones, measured by TED (Timing Error Detector), is used to control a VCXO via DPLL for alignment of the receiver clock with the transmitter clock. Figure 5.20(b) shows a non-synchronous sampling system where the sampling rate is fixed and the sampling time offset is compensated by using digital devices such as an FIR interpolating filter. Since it does not require a feedback signal for adjusting the sampling frequency (at ADC), it is simpler to implement than the synchronous sampling systems. However, the non-synchronous sampling scheme is more vulnerable to SFO if it is not compensated properly. Since a sample can be inserted or lost in one OFDM symbol when SFO is present, the non-synchronous sampling scheme performs the operations of skip/duplication/interpolation before the FFT operation and compensates for the effect of phase rotation by using FEQ (Frequency-domain EQualizer). Table 5.4 shows a comparison between synchronous and non-synchronous sampling schemes in OFDM systems.

Table 5.4   Synchronous sampling vs. non-synchronous sampling in OFDM systems.

	Synchronous sampling	Non-synchronous sampling
Required components	VCXO (NCO) PLL (DPLL)	XO FIR interpolation filter
Disadvantages	Usage of VCXO is undesirable • Higher cost and noise jitter than XO • Discrete component	Performance degradation • Slowly time varying ICI • Usually requires oversampling or high-order interpolation filter
Synchronization scheme	▪ Pilot tone-based synchronization	▪ Discrete time-domain correction ▪ Hybrid time-domain/frequency-domain correction

5.7 Synchronization in Cellular Systems

This section presents the synchronization techniques between the BS (Base Station) and MS (Mobile Station) in cellular systems. In TDD (Time Division Duplexing) systems, the information on synchronization and channel estimation, obtained by downlink, can be applied to uplink since downlink and uplink channels tend to be symmetric. In FDD (Frequency Division Duplexing) cellular systems, however, synchronization and channel estimation are performed separately for downlink and uplink since they operate in different frequency bands.

Figure 5.21 shows a basic flow of information between the BS and MS for cell searching and synchronization. Referring to Figure 5.21, the detailed procedure is as follows:

① BS broadcasts a downlink preamble periodically, which can be used for maintaining a connection to any MS with alignment in time and frequency.

② MS acquires frame timing, symbol timing, carrier frequency, and Cell ID by using the preamble transmitted from BS. MS also acquires information on timing and resource for random access.

③ MS transmits a random access preamble using the resource information (time, frequency, code) broadcasted by BS.

④ Upon reception of the random access preamble, BS estimates the uplink symbol timing offset, carrier frequency offset (only for FDD case), and power level for MS.

⑤ BS sends a ranging response (as a response to random access) indicating the value of timing advance (TA), carrier frequency offset, and power level for MS.

⑥ MS compensates the uplink timing, carrier frequency offset, and power level by using the information in the ranging response.

⑦ MS transmits a compensated uplink signal.

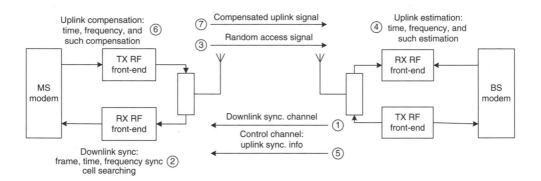

Figure 5.21   Cell search and synchronization process in a cellular system.

5.7.1 Downlink Synchronization

Downlink synchronization for OFDM-based communication and broadcasting systems can be either preamble-based or pilot-based. For example, DAB (Digital Audio Broadcasting; Eureka-147) systems use a null symbol and phase reference symbol for synchronization.

Since DAB employs non-coherent transmission scheme (DQPSK), pilots or channel estimation is not needed. In DVB-T (Digital Video Broadcasting – Terrestrial) and DVB-H (Digital Video Broadcasting – Handheld) systems, pilot tones are used for synchronization and channel estimation. Meanwhile, IEEE 802.11a wireless LAN (Local Area Network) employs the OFDM burst modem, which uses the short/long preambles for synchronization and channel estimation. Once synchronization and channel estimation are performed at the beginning of transmission in wireless LAN, the estimated parameters are used during transmission of the burst. Unlike the wireless LAN, the cellular system requires a process of cell searching in addition to synchronization and channel estimation. After initial synchronization, MS searches for the target BS with the best link connection among adjacent BSs, and continues to search for the possibility of handover. Depending on the standards in the cellular systems, slightly different terminologies are used for cell searching and downlink synchronization: preamble in Mobile WiMAX systems, and PSS (Primary Synchronization Signal) and SSS (Secondary Synchronization Signal) in 3GPP-LTE systems.

In the design of the preamble, the following must be taken into consideration. The preamble sequence for cellular systems should have a low PAPR as well as good (low) cross-correlation and (impulse-like) auto-correlation characteristics. Well-known preamble sequences for cellular systems include PN sequence, Gold sequence, and CAZAC sequence [117]. Also, the period of preamble depends on the range of CFO to be estimated. As discussed in Section 5.4, there exists a trade-off relationship between the MSE performance and estimation range of CFO. Other system requirements such as the number of cells to be distinguished should be also considered along with the required performance and complexity when the preamble sequence needs to be designed for cell searching and synchronization. In the sequel, we consider specific synchronization procedures for the commercial cellular systems, including the mobile WiMAX and 3GPP LTE systems.

Consider the Mobile WiMAX system in which each cell has three sectors [118,119]. The orthogonal subsets of subcarriers are allocated at different sectors, that is, providing the frequency reuse factor of 3, so that co-channel interference can be avoided. Figure 5.22(a) shows the cell structure of the Mobile WiMAX system while Figure 5.22(b) shows the corresponding preamble structure in the frequency domain where preamble sequences are allocated at a different set of subcarriers for each segment in a cell. Here, the term *segment* has a similar meaning to sector. Out of 114 different preamble sequences, 96 sequences are used to identify 32 cells, each with three unique segments. The preamble repeats approximately three times within the OFDM symbol period in the time domain, because only one out of three segments is used as shown in Figure 5.22(b). The MS can detect the starting point of a frame by using the auto-correlation property of the preamble. After the frame detection, cyclic prefix is used to estimate FFO (Fractional carrier-Frequency Offset). After the time-domain operation is completed, the next preamble is converted into a frequency-domain one, where IFO (Integer carrier Frequency Offset), Cell ID, segment ID, and CINR (Carrier-to-Interference and Noise Ratio) are estimated. After the initial synchronization process is completed, fine symbol timing and carrier-frequency tracking are performed. In order to improve the performance of synchronization at the cell boundary, two OFDM symbols can be used for a preamble, as discussed in the upcoming standard for the next generation Mobile WiMAX (e.g., IEEE 802.16 m standard).

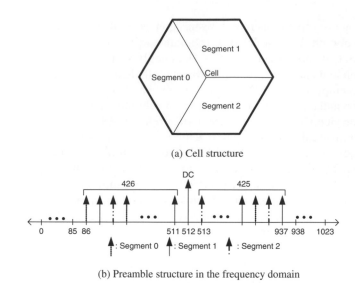

(a) Cell structure

(b) Preamble structure in the frequency domain

Figure 5.22   Cell and preamble structures in Mobile WiMAX system.

As another example, let us consider a downlink synchronization procedure for 3GPP LTE (Long-Term Evolution) system [120,121]. Figure 5.23 shows a downlink frame structure in the LTE system for the normal cyclic prefix case. One frame of duration 10 ms consists of a group of 10 subframes (corresponding to 20 slots). Depending on whether the normal cyclic prefix or extended cyclic prefix is used, a slot of duration 0.5 ms consists of 6 or 7 OFDM symbols. A subset of subcarriers for symbol #6 in the slot #0 and slot #10 are designated as a primary synchronization channel (P-SCH). Furthermore, a subset of the subcarriers for symbol #5 in the slot #0 and slot #10 are designated as secondary synchronization channel 1 (S-SCH1) and secondary synchronization channel 2 (S-SCH2), respectively. The P-SCH and S-SCH are used to transmit the PSS (Primary Synchronization Signal) and SSS (Secondary Synchronization Signal), respectively. The PSS and SSS are transmitted on the central 1.08 MHz band, consisting of 72 subcarriers (subcarrier spacing of 15 kHz) including guard band, no matter which bandwidth is used among the scalable transmission bandwidths of 1.25, 2.5, 5, 10, 15, and 20 MHz. In the 3GPP LTE system, 504

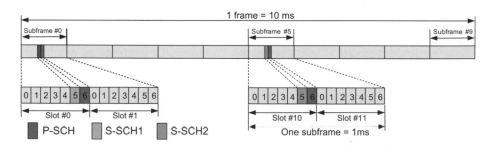

Figure 5.23   Downlink frame structure in 3GPP-LTE system (normal CP).

unique cells can be distinguished by combining three physical layer cell-identities, provided by PSS, with 168 physical layer cell identity groups, provided by SSS.

As the first step of downlink synchronization in the 3GPP LTE system, the MS uses PSS with the period of 5ms, transmitted twice in a frame, to estimate symbol timing and CFO. PSS is also used to detect the physical layer cell identity. As the second step of downlink synchronization, SSS is used to detect the physical layer cell ID group and frame timing. As shown in Figure 5.23, the physical layer cell ID group is identified by combining SSS1 in the S-SCH1 with SSS2 in the S-SCH2.

5.7.2 Uplink Synchronization

In downlink transmission, the MS is synchronized with BS in an open loop control mechanism as discussed in Section 5.7.1. Once downlink synchronization is complete, the MS receives the signal transmitted from BS after propagation delay between BS and MS. Since the propagation speed is 3.33 µs/km, the MS located 3km away from BS will receive the transmitted signal 16.7 µs later. For broadcasting systems where no interaction takes place between downlink and uplink, such a small propagation delay does not cause any performance degradation. Unlike downlink synchronization, however, uplink synchronization needs to be carried out in a closed loop control mechanism for cellular systems. Since the MS cannot estimate the propagation delay from the downlink preamble, the MS sends a random access preamble to BS. Upon reception of the random access preamble, BS estimates the round-trip propagation delay by using the techniques discussed in Section 5.3.2. Then, the BS sends back to the MS a ranging response indicating the value of timing advance (TA).

In the OFDMA system, multiple MSs may transmit their uplink bursts. Each burst uses a different frequency band (resource unit) allocated by the scheduler in BS. Although the resources are allocated in an orthogonal manner, multiple access interference (MAI) may occur due to the loss of orthogonality if the bursts transmitted from MSs do not arrive at the BS simultaneously. Figure 5.24 illustrates timing misalignment among MS bursts due to the different distances between the different users and BS. Once MAI occurs, it is difficult to

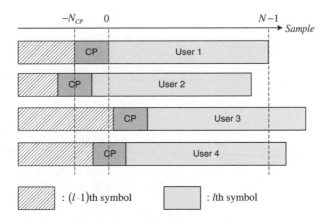

Figure 5.24   Timing misalignments among MS bursts received at BS in OFDMA system.

recover the transmitted bursts, resulting in significant performance degradation. In the OFDMA system, therefore, the BS needs to send back to each MS the corresponding value of TA. At the initial stage of uplink synchronization, the value of TA is not available to MS. In order for the random access preamble to be detected at the BS, the length of the initial ranging preamble should usually be longer than duration of one OFDM symbol. In order to increase the cell coverage, we need to increase the length of the initial ranging preamble. As shown in Figure 5.25, two repetitive OFDM symbols are used for initial ranging preamble in the Mobile WiMAX system. In the 3GPP-LTE system, several different formats for random access preamble are used to support the different sizes of cell coverage.

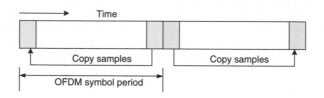

Figure 5.25   Initial ranging symbol in mobile WiMAX system.

Figure 5.26 shows a procedure of uplink timing acquisition that uses the initial ranging symbol in OFDM cellular systems. When MS wakes up from a sleep mode or performs handoff from one cell to another, it acquires downlink synchronization by using downlink preamble. Once downlink synchronization is completed, the MS receives the signal transmitted from BS after propagation delay t_p. The MS also receives the broadcast control channel from BS, and acquires a set of cell and system-specific information. This information includes transmission bandwidth, the number of transmit antenna ports, cyclic prefix length, and random access-specific parameters. MS is advised regarding the availability of resources and which resource

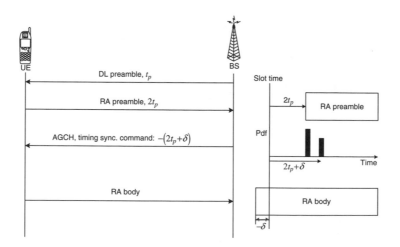

Figure 5.26   Uplink timing acquisition process.

(time, frequency, code) to transmit the random access preamble. Then, since MS is not synchronized with BS in uplink, a random access mechanism is used before actual data transmission. The MS transmits the random access preamble so that the BS can estimate the information on TA, power, and CFO, associated with the MS. Here, δ denotes the estimated symbol timing offset. Also, the BS determines whether the random access attempt was successful or not by matching the random access preamble number (the matching process can be performed at MS). If the BS successfully receives the random access preamble, it sends an access grant message along with the TA information to the MS. Finally, the MS transmits a data burst after compensating uplink timing with the TA information received from BS.

A flow diagram in Figure 5.27 summarizes an uplink synchronization procedure that has been discussed above. After a successful connection with BS through the initial synchronization process, the MS is required to maintain the quality of connection with the BS. After the initial synchronization, periodic ranging is performed to allow the MS to adjust transmission parameters such as symbol timing, CFO, and power, so that the MS can steadily maintain uplink communication with the BS.

See [122–125] for additional information about synchronization for ODFM systems.

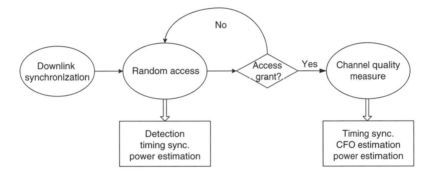

Figure 5.27   Uplink synchronization procedure.

6

Channel Estimation

In an OFDM system, the transmitter modulates the message bit sequence into PSK/QAM symbols, performs IFFT on the symbols to convert them into time-domain signals, and sends them out through a (wireless) channel. The received signal is usually distorted by the channel characteristics. In order to recover the transmitted bits, the channel effect must be estimated and compensated in the receiver [126–128]. As discussed in Chapters 4 and 5, each subcarrier can be regarded as an independent channel, as long as no ICI (Inter-Carrier Interference) occurs, and thus preserving the orthogonality among subcarriers. The orthogonality allows each subcarrier component of the received signal to be expressed as the product of the transmitted signal and channel frequency response at the subcarrier. Thus, the transmitted signal can be recovered by estimating the channel response just at each subcarrier. In general, the channel can be estimated by using a preamble or pilot symbols known to both transmitter and receiver, which employ various interpolation techniques to estimate the channel response of the subcarriers between pilot tones. In general, data signal as well as training signal, or both, can be used for channel estimation. In order to choose the channel estimation technique for the OFDM system under consideration, many different aspects of implementations, including the required performance, computational complexity and time-variation of the channel must be taken into account.

6.1 Pilot Structure

Depending on the arrangement of pilots, three different types of pilot structures are considered: block type, comb type, and lattice type [129–132].

6.1.1 Block Type

A block type of pilot arrangement is depicted in Figure 6.1. In this type, OFDM symbols with pilots at all subcarriers (referred to as pilot symbols herein) are transmitted periodically for channel estimation. Using these pilots, a time-domain interpolation is performed to estimate

MIMO-OFDM Wireless Communications with MATLAB®   Yong Soo Cho, Jaekwon Kim, Won Young Yang and Chung G. Kang
© 2010 John Wiley & Sons (Asia) Pte Ltd

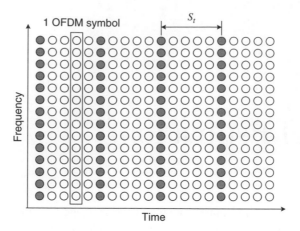

Figure 6.1   Block-type pilot arrangement.

the channel along the time axis. Let S_t denote the period of pilot symbols in time. In order to keep track of the time-varying channel characteristics, the pilot symbols must be placed as frequently as the coherence time is. As the coherence time is given in an inverse form of the Doppler frequency $f_{Doppler}$ in the channel, the pilot symbol period must satisfy the following inequality:

$$S_t \leq \frac{1}{f_{Doppler}} \tag{6.1}$$

Since pilot tones are inserted into all subcarriers of pilot symbols with a period in time, the block-type pilot arrangement is suitable for frequency-selective channels. For the fast-fading channels, however, it might incur too much overhead to track the channel variation by reducing the pilot symbol period.

6.1.2 Comb Type

Comb-type pilot arrangement is depicted in Figure 6.2. In this type, every OFDM symbol has pilot tones at the periodically-located subcarriers, which are used for a frequency-domain interpolation to estimate the channel along the frequency axis. Let S_f be the period of pilot tones in frequency. In order to keep track of the frequency-selective channel characteristics, the pilot symbols must be placed as frequently as coherent bandwidth is. As the coherence bandwidth is determined by an inverse of the maximum delay spread σ_{max}, the pilot symbol period must satisfy the following inequality:

$$S_f \leq \frac{1}{\sigma_{max}} \tag{6.2}$$

As opposed to the block-type pilot arrangement, the comb-type pilot arrangement is suitable for fast-fading channels, but not for frequency-selective channels.

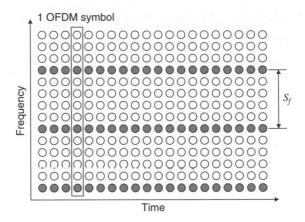

Figure 6.2 Comb-type pilot arrangement.

6.1.3 Lattice Type

Lattice-type pilot arrangement is depicted in Figure 6.3. In this type, pilot tones are inserted along both the time and frequency axes with given periods. The pilot tones scattered in both time and frequency axes facilitate time/frequency-domain interpolations for channel estimation. Let S_t and S_f denote the periods of pilot symbols in time and frequency, respectively. In order to keep track of the time-varying and frequency-selective channel characteristics, the pilot symbol arrangement must satisfy both Equations (6.1) and (6.2), such that

$$S_t \le \frac{1}{f_{\text{Doppler}}} \quad \text{and} \quad S_f \le \frac{1}{\sigma_{\max}}$$

where f_{Doppler} and $\sigma_{\max}$ denote the Doppler spreading and maximum delay spread, respectively.

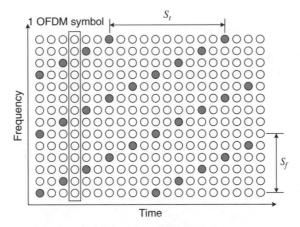

Figure 6.3 Lattice-type pilot arrangement.

6.2 Training Symbol-Based Channel Estimation

Training symbols can be used for channel estimation, usually providing a good performance. However, their transmission efficiencies are reduced due to the required overhead of training symbols such as preamble or pilot tones that are transmitted in addition to data symbols. The least-square (LS) and minimum-mean-square-error (MMSE) techniques are widely used for channel estimation when training symbols are available [127–130,132,133].

We assume that all subcarriers are orthogonal (i.e., ICI-free). Then, the training symbols for N subcarriers can be represented by the following diagonal matrix:

$$\mathbf{X} = \begin{bmatrix} X[0] & 0 & \cdots & 0 \\ 0 & X[1] & & \vdots \\ \vdots & & \ddots & 0 \\ 0 & \cdots & 0 & X[N-1] \end{bmatrix}$$

where $X[k]$ denotes a pilot tone at the kth subcarrier, with $E\{X[k]\} = 0$ and $Var\{X[k]\} = \sigma_x^2$, $k = 0, 1, 2, \cdots, N-1$. Note that $\mathbf{X}$ is given by a diagonal matrix, since we assume that all subcarriers are orthogonal. Given that the channel gain is $H[k]$ for each subcarrier k, the received training signal $Y[k]$ can be represented as

$$\mathbf{Y} \triangleq \begin{bmatrix} Y[0] \\ Y[1] \\ \vdots \\ Y[N-1] \end{bmatrix} = \begin{bmatrix} X[0] & 0 & \cdots & 0 \\ 0 & X[1] & & \vdots \\ \vdots & & \ddots & 0 \\ 0 & \cdots & 0 & X[N-1] \end{bmatrix} \begin{bmatrix} H[0] \\ H[1] \\ \vdots \\ H[N-1] \end{bmatrix} + \begin{bmatrix} Z[0] \\ Z[1] \\ \vdots \\ Z[N-1] \end{bmatrix} \quad (6.3)$$

$$= \mathbf{XH} + \mathbf{Z}$$

where $\mathbf{H}$ is a channel vector given as $\mathbf{H} = [H[0], H[1], \cdots, H[N-1]]^T$ and $\mathbf{Z}$ is a noise vector given as $\mathbf{Z} = [Z[0], Z[1], \cdots, Z[N-1]]^T$ with $E\{Z[k]\} = 0$ and $Var\{Z[k]\} = \sigma_z^2$, $k = 0, 1, 2, \cdots, N-1$. In the following discussion, let $\hat{\mathbf{H}}$ denote the estimate of channel $\mathbf{H}$.

6.2.1 LS Channel Estimation

The least-square (LS) channel estimation method finds the channel estimate $\hat{\mathbf{H}}$ in such a way that the following cost function is minimized:

$$\begin{aligned} J(\hat{\mathbf{H}}) &= \left\| \mathbf{Y} - \mathbf{X}\hat{\mathbf{H}} \right\|^2 \\ &= (\mathbf{Y} - \mathbf{X}\hat{\mathbf{H}})^H (\mathbf{Y} - \mathbf{X}\hat{\mathbf{H}}) \\ &= \mathbf{Y}^H\mathbf{Y} - \mathbf{Y}^H\mathbf{X}\hat{\mathbf{H}} - \hat{\mathbf{H}}^H\mathbf{X}^H\mathbf{Y} + \hat{\mathbf{H}}^H\mathbf{X}^H\mathbf{X}\hat{\mathbf{H}} \end{aligned} \quad (6.4)$$

By setting the derivative of the function with respect to $\hat{\mathbf{H}}$ to zero,

$$\frac{\partial J(\hat{\mathbf{H}})}{\partial \hat{\mathbf{H}}} = -2(\mathbf{X}^H\mathbf{Y})^* + 2(\mathbf{X}^H\mathbf{X}\hat{\mathbf{H}})^* = 0 \quad (6.5)$$

we have $\mathbf{X}^H\mathbf{X}\hat{\mathbf{H}} = \mathbf{X}^H\mathbf{Y}$, which gives the solution to the LS channel estimation as

$$\hat{\mathbf{H}}_{LS} = (\mathbf{X}^H\mathbf{X})^{-1}\mathbf{X}^H\mathbf{Y} = \mathbf{X}^{-1}\mathbf{Y} \tag{6.6}$$

Let us denote each component of the LS channel estimate $\hat{\mathbf{H}}_{LS}$ by $\hat{H}_{LS}[k], k = 0, 1, 2, \cdots, N-1$. Since $\mathbf{X}$ is assumed to be diagonal due to the ICI-free condition, the LS channel estimate $\hat{\mathbf{H}}_{LS}$ can be written for each subcarrier as

$$\hat{H}_{LS}[k] = \frac{Y[k]}{X[k]}, \quad k = 0, 1, 2, \cdots, N-1 \tag{6.7}$$

The mean-square error (MSE) of this LS channel estimate is given as

$$\begin{aligned}
MSE_{LS} &= E\{(\mathbf{H}-\hat{\mathbf{H}}_{LS})^H(\mathbf{H}-\mathbf{H}_{LS})\} \\
&= E\{(\mathbf{H}-\mathbf{X}^{-1}\mathbf{Y})^H(\mathbf{H}-\mathbf{X}^{-1}\mathbf{Y})\} \\
&= E\{(\mathbf{X}^{-1}\mathbf{Z})^H(\mathbf{X}^{-1}\mathbf{Z})\} \\
&= E\{\mathbf{Z}^H(\mathbf{X}\mathbf{X}^H)^{-1}\mathbf{Z}\} \\
&= \frac{\sigma_z^2}{\sigma_x^2}
\end{aligned} \tag{6.8}$$

Note that the MSE in Equation (6.8) is inversely proportional to the SNR σ_x^2/σ_z^2, which implies that it may be subject to noise enhancement, especially when the channel is in a deep null. Due to its simplicity, however, the LS method has been widely used for channel estimation.

6.2.2 MMSE Channel Estimation

Consider the LS solution in Equation (6.6), $\hat{\mathbf{H}}_{LS} = \mathbf{X}^{-1}\mathbf{Y} \triangleq \tilde{\mathbf{H}}$. Using the weight matrix $\mathbf{W}$, define $\hat{\mathbf{H}} \triangleq \mathbf{W}\tilde{\mathbf{H}}$, which corresponds to the MMSE estimate. Referring to Figure 6.4, MSE of the channel estimate $\hat{\mathbf{H}}$ is given as

$$J(\hat{\mathbf{H}}) = E\{||\mathbf{e}||^2\} = E\{||\mathbf{H}-\hat{\mathbf{H}}||^2\} \tag{6.9}$$

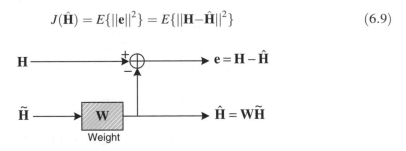

Figure 6.4 MMSE channel estimation.

Then, the MMSE channel estimation method finds a better (linear) estimate in terms of $\mathbf{W}$ in such a way that the MSE in Equation (6.9) is minimized. The orthogonality principle states that the estimation error vector $\mathbf{e} = \mathbf{H} - \hat{\mathbf{H}}$ is orthogonal to $\tilde{\mathbf{H}}$, such that

$$
\begin{aligned}
E\{\mathbf{e}\tilde{\mathbf{H}}^H\} &= E\{(\mathbf{H} - \hat{\mathbf{H}})\tilde{\mathbf{H}}^H\} \\
&= E\{(\mathbf{H} - \mathbf{W}\tilde{\mathbf{H}})\tilde{\mathbf{H}}^H\} \\
&= E\{\mathbf{H}\tilde{\mathbf{H}}^H\} - \mathbf{W}E\{\tilde{\mathbf{H}}\tilde{\mathbf{H}}^H\} \\
&= \mathbf{R}_{\mathbf{H}\tilde{\mathbf{H}}} - \mathbf{W}\mathbf{R}_{\tilde{\mathbf{H}}\tilde{\mathbf{H}}} = 0
\end{aligned}
\tag{6.10}
$$

where $\mathbf{R}_{AB}$ is the cross-correlation matrix of $N \times N$ matrices $\mathbf{A}$ and $\mathbf{B}$ (i.e., $\mathbf{R}_{AB} = E[\mathbf{A}\mathbf{B}^H]$), and $\tilde{\mathbf{H}}$ is the LS channel estimate given as

$$
\tilde{\mathbf{H}} = \mathbf{X}^{-1}\mathbf{Y} = \mathbf{H} + \mathbf{X}^{-1}\mathbf{Z}
\tag{6.11}
$$

Solving Equation (6.10) for $\mathbf{W}$ yields

$$
\mathbf{W} = \mathbf{R}_{\mathbf{H}\tilde{\mathbf{H}}}\mathbf{R}_{\tilde{\mathbf{H}}\tilde{\mathbf{H}}}^{-1}
\tag{6.12}
$$

where $\mathbf{R}_{\tilde{\mathbf{H}}\tilde{\mathbf{H}}}$ is the autocorrelation matrix of $\tilde{\mathbf{H}}$ given as

$$
\begin{aligned}
\mathbf{R}_{\tilde{\mathbf{H}}\tilde{\mathbf{H}}} &= E\{\tilde{\mathbf{H}}\tilde{\mathbf{H}}^H\} \\
&= E\{\mathbf{X}^{-1}\mathbf{Y}(\mathbf{X}^{-1}\mathbf{Y})^H\} \\
&= E\{(\mathbf{H} + \mathbf{X}^{-1}\mathbf{Z})(\mathbf{H} + \mathbf{X}^{-1}\mathbf{Z})^H\} \\
&= E\{\mathbf{H}\mathbf{H}^H + \mathbf{X}^{-1}\mathbf{Z}\mathbf{H}^H + \mathbf{H}\mathbf{Z}^H(\mathbf{X}^{-1})^H + \mathbf{X}^{-1}\mathbf{Z}\mathbf{Z}^H(\mathbf{X}^{-1})^H)\} \\
&= E\{\mathbf{H}\mathbf{H}^H\} + E\{\mathbf{X}^{-1}\mathbf{Z}\mathbf{Z}^H(\mathbf{X}^{-1})^H\} \\
&= E\{\mathbf{H}\mathbf{H}^H\} + \frac{\sigma_z^2}{\sigma_x^2}\mathbf{I}
\end{aligned}
\tag{6.13}
$$

and $\mathbf{R}_{\mathbf{H}\tilde{\mathbf{H}}}$ is the cross-correlation matrix between the true channel vector and temporary channel estimate vector in the frequency domain. Using Equation (6.13), the MMSE channel estimate follows as

$$
\begin{aligned}
\hat{\mathbf{H}} = \mathbf{W}\tilde{\mathbf{H}} &= \mathbf{R}_{\mathbf{H}\tilde{\mathbf{H}}}\mathbf{R}_{\tilde{\mathbf{H}}\tilde{\mathbf{H}}}^{-1}\tilde{\mathbf{H}} \\
&= \mathbf{R}_{\mathbf{H}\tilde{\mathbf{H}}}\left(\mathbf{R}_{\mathbf{H}\mathbf{H}} + \frac{\sigma_z^2}{\sigma_x^2}\mathbf{I}\right)^{-1}\tilde{\mathbf{H}}
\end{aligned}
\tag{6.14}
$$

The elements of $\mathbf{R}_{\mathbf{H}\tilde{\mathbf{H}}}$ and $\mathbf{R}_{\mathbf{H}\mathbf{H}}$ in Equation (6.14) are

$$
E\left\{h_{k,l}\tilde{h}_{k',l'}^*\right\} = E\left\{h_{k,l}h_{k',l'}^*\right\} = r_f[k-k']r_t[l-l']
\tag{6.15}
$$

where k and l denote the subcarrier (frequency) index and OFDM symbol (time) index, respectively. In an exponentially-decreasing multipath PDP (Power Delay Profile), the frequency-domain correlation $r_f[k]$ is given as

$$
r_f[k] = \frac{1}{1 + j2\pi\tau_{rms}k\Delta f}
\tag{6.16}
$$

where $\Delta f = 1/T_{sub}$ is the subcarrier spacing for the FFT interval length of T_{sub}. Meanwhile, for a fading channel with the maximum Doppler frequency f_{max} and Jake's spectrum, the time-domain correlation $r_t[l]$ is given as

$$r_t[l] = J_0\left(2\pi f_{max} l T_{sym}\right) \tag{6.17}$$

where $T_{sym} = T_{sub} + T_G$ for guard interval time of T_G and $J_0(x)$ is the first kind of 0th-order Bessel function. Note that $r_t[0] = J_0(0) = 1$, implying that the time-domain correlation for the same OFDM symbol is unity. Figure 6.5 shows the time-domain correlation of channel in the unit of OFDM symbol depending on the maximum Doppler frequency f_{max} when the bandwidth is 10MHz and the OFDM symbol duration is 115.2us. It can be seen from Figure 6.5 that the magnitude of $r_t[l]$ decreases as the maximum Doppler frequency f_{max} increases.

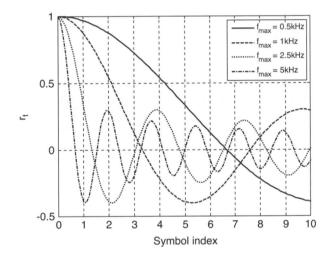

Figure 6.5   Time-domain correlation characteristic of channel (BW = 10 MHz, T_{sym} = 115.2 us).

To estimate the channel for data symbols, the pilot subcarriers must be interpolated. Popular interpolation methods include linear interpolation, second-order polynomial interpolation, and cubic spline interpolation [128–131,134].

MATLAB® Programs: For Channel Estimation and Interpolation

Three MATLAB® routines are provided. The first two, Program 6.1 ("LS_CE") and Program 6.2 ("MMSE_CE"), perform the LS and MMSE channel estimation, respectively. In "MMSE_CE," the channel impulse response is given as the sixth input argument. Also, the time-domain correlation $r_t[l]$ of channel is set to unity, such that $r_t[0] = J_0(0) = 1$, since the pilot symbol is inserted in every OFDM symbol. The third routine Program 6.3 ("interpolate") uses the MATLAB® built-in routine "interp1" to perform the linear or cubic spline interpolation, depending on whether the fourth input argument (method) is "linear" or "spline." These routines will be used in "channel_estimation.m" (Program 6.4) to test their validity.

Program 6.1 "LS_CE" for LS channel estimation method

```
function [H_LS] = LS_CE(Y,Xp,pilot_loc,Nfft,Nps,int_opt)
% LS channel estimation function
% Inputs:
%       Y        = Frequency-domain received signal
%       Xp       = Pilot signal
%       pilot_loc = Pilot location
%       N        = FFT size
%       Nps      = Pilot spacing
%       int_opt  = 'linear' or 'spline'
% output:
%       H_LS     = LS Channel estimate
Np=Nfft/Nps; k=1:Np;
LS_est(k) = Y(pilot_loc(k))./Xp(k);   % LS channel estimation
if lower(int_opt(1))=='l', method='linear'; else method='spline'; end
% Linear/Spline interpolation
H_LS = interpolate(LS_est,pilot_loc,Nfft,method);
```

Program 6.2 "MMSE_CE" for MMSE channel estimation method

```
function [H_MMSE] = MMSE_CE(Y,Xp,pilot_loc,Nfft,Nps,h,SNR)
% MMSE channel estimation function
% Inputs:
%       Y        = Frequency-domain received signal
%       Xp       = Pilot signal
%       pilot_loc = Pilot location
%       Nfft     = FFT size
%       Nps      = Pilot spacing
%       h        = Channel impulse response
%       SNR      = Signal-to-Noise Ratio[dB]
% output:
%       H_MMSE   = MMSE channel estimate
snr = 10^(SNR*0.1);  Np=Nfft/Nps; k=1:Np;
H_tilde = Y(1,pilot_loc(k))./Xp(k);  % LS estimate Eq.(6.12) or (6.8)
k=0:length(h)-1;  %k_ts = k*ts;
hh = h*h'; tmp = h.*conj(h).*k;  %tmp = h.*conj(h).*k_ts;
r = sum(tmp)/hh; r2 = tmp*k.'/hh;  %r2 = tmp*k_ts.'/hh;
tau_rms = sqrt(r2-r^2);  % rms delay
df = 1/Nfft;  %1/(ts*Nfft);
j2pi_tau_df = j*2*pi*tau_rms*df;
K1 = repmat([0:Nfft-1].',1,Np); K2 = repmat([0:Np-1],Nfft,1);
rf = 1./(1+j2pi_tau_df*Nps*(K1-K2));  % Eq.(6.17a)
K3 = repmat([0:Np-1].',1,Np); K4 = repmat([0:Np-1],Np,1);
rf2 = 1./(1+j2pi_tau_df*Nps*(K3-K4));  % Eq.(6.17a)
Rhp = rf;
Rpp = rf2 + eye(length(H_tilde),length(H_tilde))/snr;  % Eq.(6.14)
H_MMSE = transpose(Rhp*inv(Rpp)*H_tilde.');  % MMSE estimate Eq.(6.15)
```

Program 6.3 "interpolate" for channel interpolation between pilots

```
function [H_interpolated] = interpolate(H,pilot_loc,Nfft,method)
% Input:        H              = Channel estimate using pilot sequence
%            pilot_loc         = Location of pilot sequence
%               Nfft           = FFT size
%               method         = 'linear'/'spline'
% Output:  H_interpolated = interpolated channel
if pilot_loc(1)>1
  slope = (H(2)-H_est(1))/(pilot_loc(2)-pilot_loc(1));
  H = [H(1)-slope*(pilot_loc(1)-1)   H]; pilot_loc = [1 pilot_loc];
end
if pilot_loc(end)<Nfft
  slope = (H(end)-H(end-1))/(pilot_loc(end)-pilot_loc(end-1));
  H = [H  H(end)+slope*(Nfft-pilot_loc(end))];
  pilot_loc = [pilot_loc Nfft];
end
if lower(method(1))=='l', H_interpolated=interp1(pilot_loc,H,[1:Nfft]);
 else    H_interpolated = interp1(pilot_loc,H,[1:Nfft],'spline');
end
```

6.3 DFT-Based Channel Estimation

The DFT-based channel estimation technique has been derived to improve the performance of LS or MMSE channel estimation by eliminating the effect of noise outside the maximum channel delay. Let $\hat{H}[k]$ denote the estimate of channel gain at the kth subcarrier, obtained by either LS or MMSE channel estimation method. Taking the IDFT of the channel estimate $\{\hat{H}[k]\}_{k=0}^{N-1}$,

$$\text{IDFT}\{\hat{H}[k]\} = h[n] + z[n] \triangleq \hat{h}[n], \quad n = 0, 1, \ldots, N-1 \tag{6.18}$$

where $z[n]$ denotes the noise component in the time domain. Ignoring the coefficients $\{\hat{h}[n]\}$ that contain the noise only, define the coefficients for the maximum channel delay L as

$$\hat{h}_{DFT}[n] = \begin{cases} h[n] + z[n], & n = 0, 1, 2, \cdots, L-1 \\ 0, & \text{otherwise} \end{cases} \tag{6.19}$$

and transform the remaining L elements back to the frequency domain as follows [135–138]:

$$\hat{H}_{\text{DFT}}[k] = \text{DFT}\{\hat{h}_{\text{DFT}}(n)\} \tag{6.20}$$

Program 6.4 "channel_estimation.m" for DFT-based channel estimation

```
%channel_estimation.m
% for LS/DFT Channel Estimation with linear/spline interpolation
clear all; close all; clf
```

```
Nfft=32;   Ng=Nfft/8;   Nofdm=Nfft+Ng;   Nsym=100;
Nps=4; Np=Nfft/Nps; % Pilot spacing and number of pilots per OFDM symbol
Nbps=4; M=2^Nbps; % Number of bits per (modulated) symbol
mod_object = modem.qammod('M',M,'SymbolOrder','gray');
demod_object = modem.qamdemod('M',M,'SymbolOrder','gray');
Es=1; A=sqrt(3/2/(M-1)*Es); % Signal energy and QAM normalization factor
SNR = 30;   sq2=sqrt(2);   MSE = zeros(1,6); nose = 0;
for nsym=1:Nsym
   Xp = 2*(randn(1,Np)>0)-1;       % Pilot sequence generation
   msgint=randint(1,Nfft-Np,M);      % bit generation
   Data = A*modulate(mod_object,msgint);
   ip = 0;     pilot_loc = [];
   for k=1:Nfft
     if mod(k,Nps)==1
       X(k)=Xp(floor(k/Nps)+1);  pilot_loc=[pilot_loc k];  ip = ip+1;
     else      X(k) = Data(k-ip);
     end
   end
   x = ifft(X,Nfft); xt = [x(Nfft-Ng+1:Nfft) x];  % IFFT and add CP
   h = [(randn+j*randn) (randn+j*randn)/2];   % A (2-tap) channel
   H = fft(h,Nfft); ch_length=length(h); % True channel and its length
   H_power_dB = 10*log10(abs(H.*conj(H))); % True channel power in dB
   y_channel = conv(xt,h);       % Channel path (convolution)
   yt = awgn(y_channel,SNR,'measured');
   y = yt(Ng+1:Nofdm); Y = fft(y); % Remove CP and FFT
   for m=1:3
     if m==1, H_est = LS_CE(Y,Xp,pilot_loc,Nfft,Nps,'linear');
         method='LS-linear'; % LS estimation with linear interpolation
     elseif m==2, H_est = LS_CE(Y,Xp,pilot_loc,Nfft,Nps,'spline');
         method='LS-spline'; % LS estimation with spline interpolation
     else H_est = MMSE_CE(Y,Xp,pilot_loc,Nfft,Nps,h,SNR);
         method='MMSE'; % MMSE estimation
     end
     H_est_power_dB = 10*log10(abs(H_est.*conj(H_est)));
     h_est = ifft(H_est); h_DFT = h_est(1:ch_length);
     H_DFT = fft(h_DFT,Nfft); % DFT-based channel estimation
     H_DFT_power_dB = 10*log10(abs(H_DFT.*conj(H_DFT)));
     if nsym==1
       subplot(319+2*m), plot(H_power_dB,'b'); hold on;
       plot(H_est_power_dB,'r:+'); legend('True Channel',method);
       subplot(320+2*m), plot(H_power_dB,'b'); hold on;
       plot(H_DFT_power_dB,'r:+');
       legend('True Channel',[method ' with DFT']);
     end
     MSE(m) = MSE(m) + (H_H_est)*(H-H_est)';
     MSE(m+3) = MSE(m+3) + (H-H_DFT)*(H-H_DFT)';
   end
   Y_eq = Y./H_est; ip = 0;
   for k=1:Nfft
     if mod(k,Nps)==1, ip=ip+1; else Data_extracted(k-ip)=Y_eq(k); end
```

```
      end
      msg_detected = demodulate(demod_object,Data_extracted/A);
      nose = nose + sum(msg_detected~=msgint);
      MSEs = MSE/(Nfft*Nsym);
end
```

Figure 6.6 shows a block diagram of DFT-based channel estimation, given the LS channel estimation. Note that the maximum channel delay L must be known in advance. Figures 6.7(a) and (b) show the received signal constellation before and after channel compensation for the OFDM system with 16-QAM, illustrating the effect of channel estimation and compensation. Meanwhile, Figure 6.8 illustrates the channel estimates obtained by using the various types of channel estimation methods with and without DFT technique discussed in the above. Comparing Figures 6.8(a1), (b1), and (c1) with Figures 6.8(a2), (b2), and (c2) reveals that the DFT-based channel estimation method improves the performance of channel estimation.

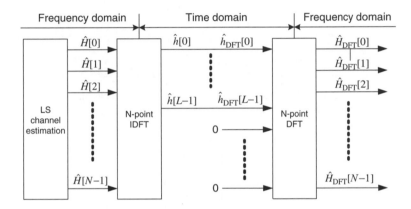

Figure 6.6   DFT-based channel estimation.

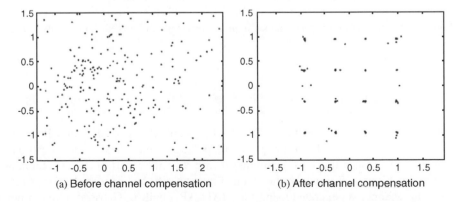

(a) Before channel compensation

(b) After channel compensation

Figure 6.7   Received signal constellation diagrams before and after channel compensation.

Also, comparing Figures 6.8(a1) and (b1) with Figure 6.8(c1), it is clear that the MMSE estimation shows better performance than the LS estimation does at the cost of requiring the additional computation and information on the channel characteristics.

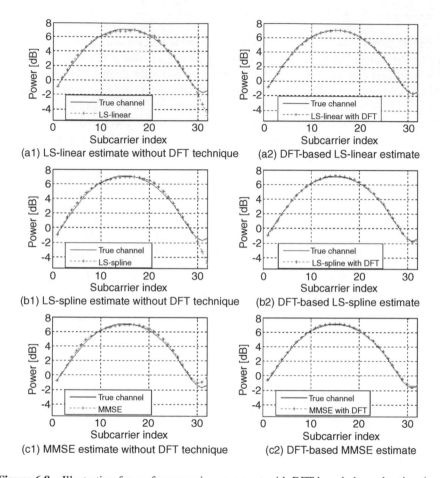

Figure 6.8   Illustration for performance improvement with DFT-based channel estimation.

MATLAB® Programs: For Channel Estimation

Program 6.4 ("channel_estimation.m") calls Program 6.1 ("LS_CE") and Program 6.2 ("MMSE_CE") for LS (with linear or spline interpolation) and MMSE channel estimations, respectively, and applies the DFT technique to improve their performances. Running the program yields Figure 6.7 in which the channel may be different every time the program is run since the channel is generated using the MATLAB® built-in Gaussian random number generator "randn."

6.4 Decision-Directed Channel Estimation

Once initial channel estimation is made with the preamble or pilots, the coefficients of channel can be updated with decision-directed (DD) channel estimation, which does not use the preamble or pilots. As depicted in Figure 6.9, the DD technique uses the detected signal feedback to track the possibly time-varying channel while subsequently using the channel estimate to detect the signal [139–141].

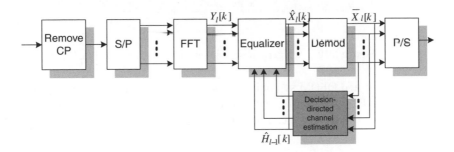

Figure 6.9 Block diagram for an OFDM receiver with decision-directed channel estimation.

Let $\hat{H}_l[k]$ denote the channel estimate made by using the lth OFDM symbol. The lth received OFDM symbol $Y_l[k]$ can be compensated by $\hat{H}_{l-1}[k]$, which is estimated by using the $(l-1)$th OFDM symbol, that is,

$$\textbf{Step 1}: \hat{X}_l[k] = \frac{Y_l[k]}{\hat{H}_{l-1}[k]}$$

Let $\overline{X}_l[k]$ denote a hard-decision value for the channel-compensated signal $\hat{X}_l[k]$. Then, the channel estimate $\hat{H}_l[k]$ is given by

$$\textbf{Step 2}: \hat{H}_l[k] = \frac{Y_l[k]}{\overline{X}_l[k]}$$

Since **Step 2** follows from the symbol decision made in **Step 1**, this particular approach is referred to as the decision-directed channel estimation technique. In this decision-directed process, any error in the detected symbol is propagated to degrade its performance. Further performance degradation is expected under fast fading, which is when channel variation is faster as compared to the OFDM symbol period. In this case, the performance degradation can be mitigated to some degree by taking the weighted average of the channel estimates between adjacent subcarriers or successive OFDM symbols.

6.5 Advanced Channel Estimation Techniques

6.5.1 Channel Estimation Using a Superimposed Signal

Consider a superimposed signal which adds a training (pilot) signal of low power to the data signal at the transmitter. Then the superimposed signal is used at the receiver for channel estimation without losing a data rate. In this approach, however, a portion of power allocated to the training signal is wasted [142,143]. Figure 6.10 illustrates a superimposed signal, which consists of a pilot signal with power $\rho \cdot P$ and a data signal with power $(1-\rho) \cdot P$, assuming that the total signal power is equal to P.

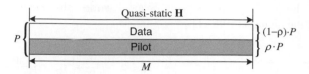

Figure 6.10   A superimposed signal for channel estimation.

For the lth transmitted OFDM symbol at the kth subcarrier, the superposed signal can be expressed as $S_l[k] + P_l[k]$ where $S_l[k]$ and $P_l[k]$ denote the data and pilot signals, respectively. The corresponding received signal can be expressed as

$$
\begin{aligned}
Y_l[k] &= H_l[k]X_l[k] + Z_l[k] \\
&= H_l[k](S_l[k] + P_l[k]) + Z_l[k]
\end{aligned} \tag{6.21}
$$

where $Y_l[k]$ and $Z_l[k]$ denote the received signal and noise at the kth subcarrier in the lth symbol period, respectively, and $H_l[k]$ the corresponding channel frequency response. Here, the coherence time or coherence bandwidth of channel is assumed to be M over which the channel response is almost constant along the time or frequency axis. Also, the pilot signals are set to be constant for the time or frequency interval of length M such that

$$
\text{Time domain:} \begin{cases} H_l[k] \approx H_{l+1}[k] \approx \cdots \approx H_{l+M-1}[k] \approx H \\ P_l[k] = P_{l+1}[k] = \cdots = P_{l+M-1}[k] = P \end{cases} \tag{6.22}
$$

or

$$
\text{Frequency domain:} \begin{cases} H_l[k] \approx H_l[k+1] \approx \cdots \approx H_l[k+M-1] \approx H \\ P_l[k] = P_l[k+1] = \cdots = P_l[k+M-1] = P \end{cases} \tag{6.23}
$$

Under the assumption that the data signal $S_l[k]$ and noise $Z_l[k]$ are the zero-mean independent and identically distributed (i.i.d.) processes, the average received signal over time or frequency interval of length M is given as

$$
\begin{aligned}
\text{Time domain: } E\{Y_l[k]\} &= \frac{1}{M}\sum_{m=0}^{M-1} Y_{l+m}[k] \\
&= \frac{1}{M}\sum_{m=0}^{M-1}(H_{l+m}[k]X_{l+m}[k] + Z_{l+m}[k]) \\
&\approx \frac{1}{M}\sum_{m=0}^{M-1}\{H(S_{l+m}[k] + P_{l+m}[k]) + Z_{l+m}[k]\} \approx HP
\end{aligned} \tag{6.24}
$$

or

$$\text{Frequency domain:} \; E\{Y_l[k]\} = \frac{1}{M} \sum_{m=0}^{M-1} Y_l[k+m]$$

$$= \frac{1}{M} \sum_{m=0}^{M-1} (H_l[k+m]X_l[k+m] + Z_l[k+m])$$

$$\approx \frac{1}{M} \sum_{m=0}^{M-1} \{II(S_l[k \mid m] + P_l[k+m]) + Z_l[k+m]\} \approx HP$$

$$(6.25)$$

In Equations (6.24) and (6.25), we have assumed that M is large enough to make the average of data signals approximately zero, that is,

$$\frac{1}{M} \sum_{m=0}^{M-1} S_l[k+m] \approx 0.$$

Using Equations (6.24) and (6.25), then, the channel estimate is obtained as

$$\hat{H} = \frac{E\{Y_l[k]\}}{P} \tag{6.26}$$

Since no additional pilot tones are required, the channel estimation technique using the superimposed signal is advantageous in the aspect of data rate. However, it requires an additional power for transmitting pilot signals and a long time interval to make the average of data signals zero. For rapidly moving terminals, the performance of channel estimation using the superposed signal in the time domain may be degraded because a short coherence time reduces M, which may be too short to make the average of data signals to zero. Similarly for channel estimation using the superposed signal in the frequency domain, its performance may be degraded because a narrow coherence bandwidth reduces M.

6.5.2 Channel Estimation in Fast Time-Varying Channels

The channel estimation methods discussed so far may be applicable only when the channel characteristic does not change within an OFDM symbol period. However, the channel for the terminals that move fast may vary with time within an OFDM symbol period, in which longer OFDM symbol period has a more severe effect on the channel estimation performance. The time-varying channel may destroy the orthogonality among subcarriers at the receiver, resulting in ICI (Inter-Channel Interference). Due to the effect of ICI, it cannot be compensated by the conventional one-tap equalizer. This section deals with the effect of ICI in the time-varying channels [144–149].

A transmitted OFDM signal can be written in the time domain as

$$x[n] = \sum_{k=0}^{N-1} X[k]e^{j2\pi kn/N}, \quad N = 0, 1, \cdots, N-1 \qquad (6.27)$$

The corresponding signal received through a wireless channel with L paths can be expressed as

$$y[n] = \sum_{i=0}^{L-1} h_i[n]x[n-\tau_i] + w[n] \qquad (6.28)$$

where $h_i[n]$ and τ_i denote the impulse response and delay time for the ith path of the time-varying channel, respectively, while $w[n]$ is an AWGN. The received signal in the frequency domain is obtained by taking the FFT of $\{y[n]\}$ as

$$\begin{aligned} Y[k] &= \frac{1}{N}\sum_{n=0}^{N-1} y[n]e^{-j2\pi kn/N} \\ &= \sum_{m=0}^{N-1}\sum_{i=0}^{L-1} X[m]H_i[k-m]e^{-j2\pi im/N} + W[k], k = 0, 1, \cdots, N-1 \end{aligned} \qquad (6.29)$$

where $W[k]$ and $H_i[k]$ denote the FFTs of $\{w[n]\}$ and the impulse response $\{h_i[n]\}$, respectively [145]. Note that $H_i[k]$ can be written as

$$H_i[k] = \frac{1}{N}\sum_{n=0}^{N-1} h_i[n]e^{-j2\pi nk/N} \qquad (6.30)$$

Define the signal vectors in the frequency domain as

$$\mathbf{Y} = \begin{bmatrix} Y[0] \\ Y[1] \\ \vdots \\ Y[N-1] \end{bmatrix}, \quad \mathbf{X} = \begin{bmatrix} X[0] \\ X[1] \\ \vdots \\ X[N-1] \end{bmatrix}, \quad \mathbf{W} = \begin{bmatrix} W[0] \\ W[1] \\ \vdots \\ W[N-1] \end{bmatrix}$$

Furthermore, the effect of channel can be represented by a channel matrix $\mathbf{H}$, which is given as

$$\mathbf{H} = \begin{bmatrix} a_{0,0} & a_{0,1} & \cdots & a_{0,N-1} \\ a_{1,0} & a_{1,1} & \cdots & a_{1,N-1} \\ \vdots & \vdots & \ddots & \vdots \\ a_{N-1,0} & a_{N-1,1} & \cdots & a_{N-1,N-1} \end{bmatrix}$$

where

$$a_{k,m} = H_0[k-m] + H_1[k-m]e^{-j2\pi m/N} + \cdots + H_{L-1}[k-m]e^{-j2\pi m(L-1)/N}, \quad m, k = 0, 1, \cdots, N-1.$$

Then, Equation (6.29) can be expressed in a matrix-vector form as

$$\mathbf{Y} = \mathbf{HX} + \mathbf{W} \qquad (6.31)$$

If the channel impulse response (CIR) $h_i[n]$ remains constant over one OFDM symbol period, the channel matrix $\mathbf{H}$ becomes diagonal with $a_{k,m} = 0, \forall\ k \neq m$. In this case, the transmitted signal can be recovered easily by a simple equalizer as follows:

$$\mathbf{X} = \mathbf{H}^{-1}\mathbf{Y} \qquad (6.32)$$

If $h_i[n]$ varies within one OFDM symbol period, $\mathbf{H}$ becomes no longer diagonal, making it difficult to solve Equation (6.31). However, if the channel is slowly time-varying, $h_i[n]$ can be approximated by a straight line. This can be illustrated from Figure 6.11(a) in which 3-path channel is simulated with Jake's model for three different Doppler frequencies [20 Hz, 50 Hz, 100 Hz]. Note that the slope of the channel impulse response increases linearly with the Doppler frequency. Figure 6.11(b) shows the corresponding magnitude responses in which most energy is concentrated around the DC component. Note that the ICI increases with the Doppler frequency as well. In cases where the channel frequency response has most of its energy near the DC component (see Figure 6.11(b)), the channel frequency response matrix $\mathbf{H}$ can be approximated by a band matrix with the size of q as

$$\mathbf{H} = \begin{bmatrix} a_{0,0} & a_{0,1} & \cdots & a_{0,q/2} & 0 & 0 & & 0 \\ a_{1,0} & a_{1,1} & \cdots & & \cdots & 0 & & \\ \vdots & \vdots & \ddots & & & & \ddots & 0 \\ a_{q/2,0} & & \cdots & & & & & a_{N-1-q/2,N-1} \\ 0 & & \cdots & & & & \ddots & \vdots \\ \vdots & \vdots & \ddots & & \vdots & a_{N-2,N-2} & & a_{N-2,N-1} \\ 0 & 0 & 0 & a_{N-1,N-1-q/2} & \cdots & a_{N-1,N-2} & & a_{N-1,N-1} \end{bmatrix}$$

$$(6.33)$$

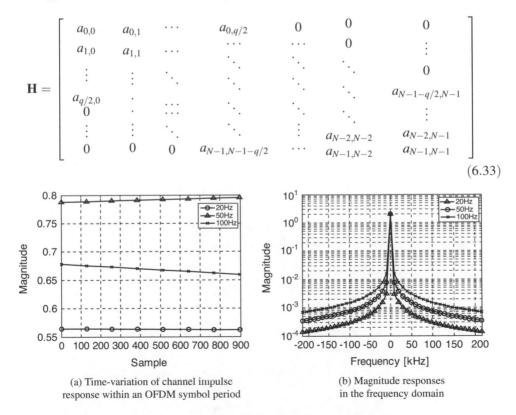

(a) Time-variation of channel impulse response within an OFDM symbol period

(b) Magnitude responses in the frequency domain

Figure 6.11   Channel characteristic in a slowly time-varying environment.

with

$$a_{k,m} = 0 \quad \text{for} \quad |k-m| \geq q/2 \tag{6.34}$$

where q denotes the number of subcarriers that incur dominant ICIs. Especially, if $\mathbf{H}$ is a sparse matrix with $q = N$, it can be transformed into a block-diagonal matrix. In this case, its inverse matrix can be obtained more easily by a block-by-block operation with less computational complexity as compared to the case of a full matrix [145].

In order to compensate for the effect of ICI under the fast-fading channels, we need an accurate estimate of the channel frequency response $\mathbf{H}$. Although extensive researches have been performed on estimation of the time-varying channels, most of the proposed techniques have been derived and verified under a limited condition [144,147–149]. Further research is expected for data transmission in very fast-fading channel environments.

6.5.3 EM Algorithm-Based Channel Estimation

The EM (Expectation-Maximization) algorithm has been widely used in a large number of areas that deal with unknown factors affecting the outcome, such as signal processing, genetics, econometric, clinical, and sociological studies. The EM-based channel estimation is an iterative technique for finding maximum likelihood (ML) estimates of a channel [150–155]. It is classified as a semi-blind method since it can be implemented when transmit symbols are not available. In this section, let us describe the channel estimation technique using the EM algorithm in [151,153,154].

Suppose that X is one of the C-ary symbols in the constellation of size C, such that $X \in \{X_1, X_2, \cdots, X_C\}$ where X_i denotes the ith symbol in the constellation. Then, the received signal for some subcarrier is given by

$$Y = H \cdot X + Z \tag{6.35}$$

The conditional probability density function of Y given H and X can be expressed as

$$f(Y|H,X) = \frac{1}{\sqrt{2\pi\sigma^2}} \exp\left(-\frac{1}{2\sigma^2}|Y - H \cdot X|^2\right) \tag{6.36}$$

Assuming that $\{X_i\}_{i=1}^{C}$ are transmitted with the same probability of $1/C$, the conditional probability density function of Y given H is

$$f(Y|H) = \sum_{i=1}^{C} \frac{1}{C} \cdot \frac{1}{\sqrt{2\pi}\sigma} \exp\left(-\frac{1}{2\sigma^2}|Y - H \cdot X_i|^2\right) \tag{6.37}$$

Assuming that the channel is not time-varying over D OFDM symbols, denote those transmitted and received symbols in a vector form as

$$\mathbf{Y} = [Y^1, Y^2, \cdots, Y^D]^T \tag{6.38}$$

$$\mathbf{X} = [X^1, X^2, \cdots, X^D]^T \tag{6.39}$$

In the context of EM algorithm, $\mathbf{Y}$ is called "*incomplete*" data since the transmitted (latent) data $\mathbf{X}$ is hidden in the observed data $\mathbf{Y}$. Meanwhile, $(\mathbf{Y}, \mathbf{X})$ is called "*complete*" data because the observed and latent data are included in the set. Since it is difficult to estimate the channel with "*incomplete*" data, the probability density function of "*incomplete*" data is converted into the probability density function of "*complete*" data. The probability density function of "*incomplete*" data is given by

$$f(\mathbf{Y}|H, \mathbf{X}) = \prod_{d=1}^{D} f(Y^d|H, X^d) \tag{6.40}$$

which can also be represented by using the log-likelihood function as

$$\log f(\mathbf{Y}|H, \mathbf{X}) = \sum_{d=1}^{D} \log f(Y^d|H, X^d) \tag{6.41}$$

Meanwhile, the probability density function of "*complete*" data can be rewritten by using the log-likelihood function as

$$\log f(\mathbf{Y}, \mathbf{X}|H) = \sum_{d=1}^{D} \log\left\{\frac{1}{C} f(Y^d|H, X^d)\right\} \tag{6.42}$$

In the conventional ML algorithm, H is estimated by maximizing the likelihood function $f(Y|H)$ in Equation (6.37). Due to the summation term of the exponential functions, however, it is not easy to derive a closed-form solution for H. In the EM algorithm, H is estimated by iteratively increasing the likelihood function in Equation (6.42). In fact, the EM algorithm consists of two iterative steps: expectation (E) step and maximization (M) step. In the E-step, the expected value of the log-likelihood function of H is computed by taking expectation over X, conditioned on Y and using the latest estimate of H, as follows:

$$\begin{aligned}
Q(H|H^{(p)}) &\triangleq E_X\left\{f(\mathbf{Y}, \mathbf{X}|H)|Y, H^{(p)}\right\} \\
&= \sum_{i=1}^{C} \sum_{d=1}^{D} \log\left\{\frac{1}{C} f(Y^d|H, X_i)\right\} \frac{f(Y^d|H^{(p)}, X_i)}{C f(Y^d|H^{(p)})}
\end{aligned} \tag{6.43}$$

where $H^{(p)}$ denotes the latest estimate of H. In the E-step, the log-likelihood functions of "*complete*" data in Equation (6.42) are averaged over D OFDM symbols. In the subsequent M-step, $H^{(p+1)}$ is determined by maximizing Equation (6.43) over all possible values of H. More specifically, it is obtained by differentiating Equation (6.43) with respect to H and setting its derivative to zero, to yield the following result [154]:

$$\begin{aligned}
H^{(p+1)} &= \arg\max_{H} Q(H|H^{(p)}) \\
&= \left[\sum_{i=1}^{C} \sum_{d=1}^{D} |X_i|^2 \frac{f(Y^d|H^{(p)}, X_i)}{C f(Y^d|H^{(p)})}\right]^{-1} \times \left[\sum_{i=1}^{C} \sum_{d=1}^{D} |Y^d X_i^*| \frac{f(Y^d|H^{(p)}, X_i)}{C f(Y^d|H^{(p)})}\right]
\end{aligned} \tag{6.44}$$

where $f(Y^d|H^{(p)}, X_i)$ and $f(Y^d|H^{(p)})$ can be obtained by Equation (6.36) and Equation (6.37), respectively. Note that Equation (6.44) can be viewed as a weighted least-square solution where an estimate of cross-correlation function is divided by an estimate of auto-correlation function, each being weighted by the corresponding probability density functions.

The EM algorithm is particularly useful for channel estimation when available data are incomplete. Incomplete data may be problematic in the situations where the information on the input (transmitted, training) signals is unavailable or insufficient. In MIMO OFDM systems, for instance, channel state information between each transmit and receive antenna pair is required for coherent decoding. However, classical channel estimation techniques cannot be used in this situation since the received signal is a superposition of signals transmitted from different antennas for each OFDM subcarrier. The EM algorithm can convert a multiple-input channel estimation problem into a number of single-input channel estimation problems. Also, the EM algorithm can be useful for channel estimation when a mobile station (MS) is located at the cell boundary subject to the inter-cell interference. In this situation, the received signal at the MS is a superposition of signals transmitted from adjacent base stations (BSs) unknown to MS. The performance of channel estimation at the cell boundary can be improved with additional received data by using the EM algorithm as long as the channel is time-invariant over D symbol periods.

Despite the advantages of the EM algorithm, its application to the channel estimation of MIMO-OFDM systems is not straightforward, simply because the computational complexity of the EM algorithm increases exponentially with the number of transmitted signals or the size of the constellation. Furthermore, EM algorithm is not applicable to a time-varying channel. There have been some attempts to reduce the computational complexity or to improve the performance of the EM algorithm. For example, a decision-directed EM (DEM) estimation technique has been proposed by combining the EM algorithm with the decision-directed (DD) channel estimation, which presents a reduced computational complexity in slowly time-varying channels [152].

6.5.4 Blind Channel Estimation

Using the statistical properties of received signals, the channel can be estimated without resorting to the preamble or pilot signals. Obviously, such a blind channel estimation technique has an advantage of not incurring an overhead with training signals. However, it often needs a large number of received symbols to extract statistical properties. Furthermore, their performance is usually worse than that of other conventional channel estimation techniques that employ the training signal. As one of the blind channel estimation techniques, Bussgang algorithm is widely used in single-carrier transmission systems [156–159]. It consists of a filter, zero-memory nonlinear estimator, and adaptive algorithm. Depending on how the zero-memory nonlinear estimator is constructed, it can be classified as Sato algorithm, CMA (Constant Modulus Algorithm), or Godard algorithm. However, Bussgang algorithm is rarely used in OFDM systems since it is not easy to find a nonlinear estimator appropriate for the received signal in the OFDM system.

The subspace-based channel estimation technique is another type of the blind channel estimation techniques developed for OFDM systems [158,160–162]. It is derived by using the second-order statistical properties and orthogonal properties of a received signal. Since the received signal space can be divided into signal subspace and noise subspace, the channel can be estimated by using the property of the noise subspace which is orthogonal to the signal subspace. The subspace-based channel estimation technique needs a high computational complexity to separate the signal subspace from noise subspace; this requires a computation of correlation from the received signal and then, eigen-decomposition. Also, a large number of received signals (i.e., a large number of equations) are required to estimate the statistical properties of received signals. Different approaches, such as increasing the number of equations by oversampling or employing a precoder matrix with full rank, have been investigated for the subspace-based channel estimation.

7

PAPR Reduction

The transmit signals in an OFDM system can have high peak values in the time domain since many subcarrier components are added via an IFFT operation. Therefore, OFDM systems are known to have a high PAPR (Peak-to-Average Power Ratio), compared with single-carrier systems. In fact, the high PAPR is one of the most detrimental aspects in the OFDM system, as it decreases the SQNR (Signal-to-Quantization Noise Ratio) of ADC (Analog-to-Digital Converter) and DAC (Digital-to-Analog Converter) while degrading the efficiency of the power amplifier in the transmitter. The PAPR problem is more important in the uplink since the efficiency of power amplifier is critical due to the limited battery power in a mobile terminal.

7.1 Introduction to PAPR

In general, even linear amplifiers impose a nonlinear distortion on their outputs due to their saturation characteristics caused by an input much larger than its nominal value. Figure 7.1 shows the input-output characteristics of high power amplifier (HPA) in terms of the input power P_{in} and the output power P_{out}. Due to the aforementioned saturation characteristic of the amplifier, the maximum possible output is limited by P_{out}^{max} when the corresponding input power is given by P_{in}^{max}. As illustrated in Figure 7.1, the input power must be backed off so as to operate in the linear region. Therefore, the nonlinear region can be described by IBO (Input Back-Off) or OBO (Output Back-Off) [163]:

$$ IBO = 10 \log_{10} \frac{P_{in}^{max}}{P_{in}}, \quad OBO = 10 \log_{10} \frac{P_{out}^{max}}{P_{out}} \tag{7.1} $$

Note that the nonlinear characteristic of HPA (High Power Amplifier), excited by a large input, causes the out-of-band radiation that affects signals in adjacent bands, and in-band distortions that result in rotation, attenuation, and offset on the received signal [164].

MIMO-OFDM Wireless Communications with MATLAB® Yong Soo Cho, Jaekwon Kim, Won Young Yang and Chung G. Kang
© 2010 John Wiley & Sons (Asia) Pte Ltd

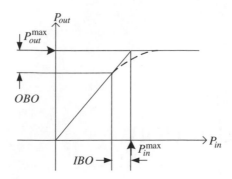

Figure 7.1   Input-output characteristic of an HPA.

7.1.1 Definition of PAPR

Consider a baseband PAM signal for a complex data sequence $\{a[n]\}$:

$$\tilde{s}(t) = \sum_k a[n]g(t-kT_s) \tag{7.2}$$

where $g(t)$ is a transmit pulse for each symbol and T_s is the symbol duration. Figure 7.2 shows a PAM transmitter for Equation (7.2), in which the output of the passband quadrature modulator is represented as

$$s(t) = \sqrt{2}\,\mathrm{Re}\big\{(\tilde{s}_I(t)+j\tilde{s}_Q(t))e^{j2\pi f_c t}\big\} \tag{7.3}$$

where $\tilde{s}_I(t)$ and $\tilde{s}_Q(t)$ denote the in-phase and quadrature components of the complex baseband PAM signal $\tilde{s}(t)$ respectively [1] (i.e., $\tilde{s}(t) = \tilde{s}_I(t)+j\tilde{s}_Q(t)$). In the sequel, we define the various terms to describe the power characteristics of Equations (7.2) and (7.3).

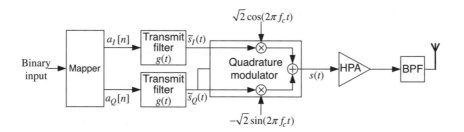

Figure 7.2   Passband PAM transmitter.

7.1.1.1 Peak-to-Mean Envelope Power Ratio (PMEPR)

PMEPR is the ratio between the maximum power and the average power for the envelope of a baseband complex signal $\tilde{s}(t)$ [165], that is,

$$PMEPR\{\tilde{s}(t)\} = \frac{\max|\tilde{s}(t)|^2}{E\{|\tilde{s}(t)|^2\}} \tag{7.4}$$

7.1.1.2 Peak Envelope Power (PEP)

PEP represents the maximum power of a complex baseband signal $\tilde{s}(t)$ [165], that is,

$$PEP\{\tilde{s}(t)\} = \max|\tilde{s}(t)|^2 \tag{7.5}$$

In the case that the average signal power is normalized (i.e., $E\{|\tilde{s}(t)|^2\} = 1$), PMEPR is equivalent to PEP.

7.1.1.3 Peak-to-Average Power Ratio (PAPR)

PAPR is the ratio between the maximum power and the average power of the complex passband signal $s(t)$ [165], that is,

$$PAPR\{\tilde{s}(t)\} = \frac{\max\left|\mathrm{Re}\left(\tilde{s}(t)e^{j2\pi f_c t}\right)\right|^2}{E\left\{\left|\mathrm{Re}(\tilde{s}(t)e^{j2\pi f_c t})\right|^2\right\}} = \frac{\max|s(t)|^2}{E\{|s(t)|^2\}} \tag{7.6}$$

The above power characteristics can also be described in terms of their magnitudes (not power) by defining the crest factor (CF) as

$$\text{Passband condition}: \quad CF = \sqrt{PAPR} \tag{7.7a}$$

$$\text{Baseband condition}: \quad CF = \sqrt{PMEPR} \tag{7.7b}$$

   In the PSK/OFDM system with N subcarriers, the maximum power occurs when all of the N subcarrier components happen to be added with identical phases. Assuming that $E\{|s(t)|^2\} = 1$, it results in $PAPR = N$, that is, the maximum power equivalent to N times the average power. We note that more PAPR is expected for M-QAM with $M > 4$ than M-ary PSK. Meanwhile, the probability of the occurrence of the maximum power signal decreases as N increases [166]. For example, suppose that there are M^2 OFDM signals with the maximum power among M^N OFDM signals in M-ary PSK/OFDM system. Accordingly, the probability of occurrence of the largest $PAPR$ is $M^2/M^N = M^{2-N}$, which turns out to be 4.7×10^{-38} in the case of QPSK/OFDM with $N = 64$ subcarriers [167]. In other words, the largest $PAPR$ rarely occurs. We are often interested in finding the probability that the signal power is out of the linear range of the HPA. Towards this end, we first consider the distribution of output signals for IFFT in the OFDM system. While the input signals of N-point IFFT have the independent and finite magnitudes which are uniformly distributed for QPSK and QAM, we can assume that the real and imaginary parts of the time-domain complex OFDM signal $s(t)$ (after IFFT at the transmitter) have asymptotically Gaussian distributions for a sufficiently large number of subcarriers by the central limit theorem. Then the amplitude of the OFDM signal $s(t)$ follows a Rayleigh distribution. Let $\{\mathbf{Z}_n\}$ be the magnitudes of complex samples $\{|s(nT_s/N)|\}_{n=0}^{N-1}$. Assuming that the average power of $s(t)$ is equal to one, that is, $E\{|s(t)|^2\} = 1$, then $\{\mathbf{Z}_n\}$ are the i.i.d. Rayleigh random variables normalized with its own average power, which has the following probability density function:

$$f_{Z_n}(z) = \frac{z}{\sigma^2} e^{-\frac{z^2}{2\sigma^2}} = 2ze^{-z^2}, \quad n = 0, 1, 2, \cdots, N-1 \tag{7.8}$$

where $E\{Z_n^2\} = 2\sigma^2 = 1$. Note that the maximum of $\mathbf{Z}_n$ is equivalent to the crest factor (CF) defined in Equation (7.7). Let $\mathbf{Z}_{\max}$ denote the crest factor (i.e., $\mathbf{Z}_{\max} = \max_{n=0,1,\cdots,N-1} \mathbf{Z}_n$). Now, the cumulative distribution function (CDF) of $\mathbf{Z}_{\max}$ is given as

$$
\begin{aligned}
F_{\mathbf{Z}_{\max}}(z) &= P(\mathbf{Z}_{\max} < z) \\
&= P(\mathbf{Z}_0 < z) \cdot P(\mathbf{Z}_1 < z) \ldots P(\mathbf{Z}_{N-1} < z) \\
&= \left(1 - e^{-z^2}\right)^N
\end{aligned}
\tag{7.9}
$$

where $P(\mathbf{Z}_n < z) = \int_0^z f_{Z_n}(x)\, dx, n = 0, 1, 2, \cdots, N-1$. In order to find the probability that the crest factor (CF) exceeds z, we consider the following complementary CDF (CCDF):

$$
\begin{aligned}
\tilde{F}_{\mathbf{Z}_{\max}}(z) &= P(\mathbf{Z}_{\max} > z) \\
&= 1 - P(\mathbf{Z}_{\max} \le z) \\
&= 1 - F_{\mathbf{Z}_{\max}}(z) \\
&= 1 - \left(1 - e^{-z^2}\right)^N
\end{aligned}
\tag{7.10}
$$

Since Equations (7.9) and (7.10) are derived under the assumption that N samples are independent and N is sufficiently large, they do not hold for the bandlimited or oversampled signals. It is due to the fact that a sampled signal does not necessarily contain the maximum point of the original continuous-time signal. However, it is difficult to derive the exact CDF for the oversampled signals and therefore, the following simplified CDF will be used:

$$F_Z(z) \approx \left(1 - e^{-z^2}\right)^{\alpha N} \tag{7.11}$$

where α has to be determined by fitting the theoretical CDF into the actual one [168]. Using simulation results, it has been shown that $\alpha = 2.8$ is appropriate for sufficiently large N. Figure 7.3 shows the theoretical and simulated CCDFs of OFDM signals with $N = 64, 128, 256, 512, 1024$ that are obtained by running Program 7.1 ("plot_CCDF.m"). Note that it calls Program 7.2 ("PAPR") for PAPR calculation and Program 7.3 ("mapper") for PSK/QAM modulation. Note that the simulation results deviate from the theoretical ones as N becomes small, which implies that Equation (7.11) is accurate only when N is sufficiently large.

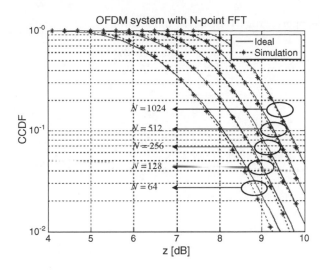

Figure 7.3   CCDFs of OFDM signals with $N = 64$, 128, 256, 512, and 1024.

MATLAB® Programs: CCDF of OFDM Signal

Program 7.1 "plot_CCDF.m" to plot CCDFs for some OFDM signals

```
% plot_CCDF.m: % Plot the CCDF curves of Fig. 7.3.
clear all; clc; clf
Ns=2.^[6:10]; b=2; M=2^b; Nblk=1e4; zdBs=[4:0.1:10];
N_zdBs=length(zdBs);
CCDF_formula=inline('1-((1-exp(-z.^2/(2*s2))).^N)','N','s2','z'); %(7.9)
for n = 1:length(Ns)
    N=Ns(n); x = zeros(Nblk,N); sqN=sqrt(N);
    for k=1:Nblk
        X=mapper(b,N); x(k,:)=ifft(X,N)*sqN; CFx(k)=PAPR(x(k,:));
    end
    s2 = mean(mean(abs(x)))^2/(pi/2);
    CCDF_theoretical=CCDF_formula(N,s2,10.^(zdBs/20));
    for i=1:N_zdBs, CCDF_simulated(i)=sum(CFx>zdBs(i))/Nblk; end
    semilogy(zdBs,CCDF_theoretical,'k-'); hold on; grid on;
    semilogy(zdBs(1:3:end),CCDF_simulated(1:3:end),'k:*');
end
axis([zdBs([1 end]) 1e-2 1]); title('OFDM system with N-point FFT');
xlabel('z[dB]'); ylabel('CCDF'); legend('Theoretical','Simulated');
```

Program 7.2 "PAPR()" to find CCDF

```
function [PAPR_dB, AvgP_dB, PeakP_dB] = PAPR(x)
% PAPR_dB = PAPR[dB], AvgP_dB = Average power[dB]
% PeakP_dB = Maximum power[dB]
Nx=length(x); xI=real(x); xQ=imag(x); Power = xI.*xI + xQ.*xQ;
AvgP = sum(Power)/Nx; AvgP_dB = 10*log10(AvgP);
PeakP = max(Power); PeakP_dB = 10*log10(PeakP);
PAPR_dB = 10*log10(PeakP/AvgP);
```

Program 7.3 "mapper()" for BPSK/QAM modulation

```
function [modulated_symbols,Mod] = mapper(b,N)
%If N is given, makes a block of N random 2^b-PSK/QAM modulated symbols
% Otherwise, a block of 2^b-PSK/QAM modulated symbols for [0:2^b-1].
M=2^b; % Modulation order or Alphabet (Symbol) size
if b==1,    Mod='BPSK';   A=1;   mod_object=modem.pskmod('M',M);
  elseif  b==2,   Mod='QPSK';   A=1;
          mod_object=modem.pskmod('M',M,'PhaseOffset',pi/4);
  else    Mod=[num2str(2^b) 'QAM'];   Es=1;   A=sqrt(3/2/(M-1)*Es);
          mod_object=modem.qammod('M',M,'SymbolOrder','gray');
end
if nargin==2 % A block of N random 2^b-PSK/QAM modulated symbols
  modulated_symbols = A*modulate(mod_object,randint(1,N,M));
 else
   modulated_symbols = A*modulate(mod_object,[0:M-1]);
end
```

Before investigating the PAPR distribution of OFDM signal, we first take a look at that of the single-carrier signal as a special case, that is, $N = 1$. Program 7.4 ("single_carrier_PAPR.m") plots Figures 7.4(a) and (b), which show a baseband QPSK-modulated signal and the corresponding passband signal with a (single) carrier frequency of $f_c = 1[\text{Hz}]$ and an over-sampling factor of 8 (i.e., $L = 8$), respectively. Since the baseband signal in Figure 7.4(a) has the same average and peak power, its PARR is 0dB. Meanwhile, the passband signal in Figure 7.4(b) shows the PAPR of 3.01dB. Note that the PAPR of the single-carrier signal can vary with the carrier frequency f_c. Therefore, the carrier frequency of the passband signal must be taken into account so as to accurately measure the PAPR of the single-carrier system. In general, the PAPR

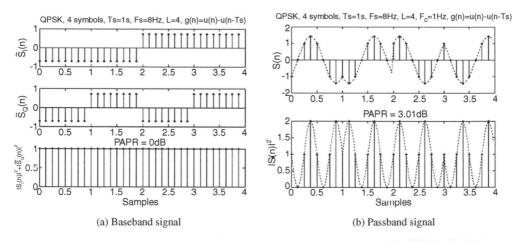

(a) Baseband signal   (b) Passband signal

Figure 7.4   Baseband/passband signals for QPSK-modulated symbols.

of the single-carrier system can be predicted directly from the modulation scheme and furthermore, it is not greatly significant, unlike the OFDM system. The PAPR characteristics of the OFDM signal will be detailed in the subsequent subsection. Program 7.4 also computes the CCDFs for both baseband and passband signals.

MATLAB® Programs: PAPR Analysis of Single-Carrier Signal

Program 7.4 "single_carrier_PAPR.m" to get PAPRs of baseband/passband single-carrier signals

```
%single_carrier_PAPR.m
clear, figure(1), clf
Ts=1; L=8; % Sampling period, Oversampling factor
Fc=1; % Carrier frequency
b=2; M=2^b; % Modulation order or Alphabet size
[X,Mod] = mapper(b); % M-PSK/QAM symbol for [0:M-1]
 L_=L*4; % Oversampling factor to make it look like continuous-time
[xt_pass_,time_] = modulation(X,Ts,L_,Fc); % Continuous-time
[xt_pass,time] = modulation(X,Ts,L,Fc); % L times oversampling
for i_s=1:M, xt_base(L*(i_s-1)+1:L*i_s) = X(i_s)*ones(1,L); end
PAPR_dB_base = PAPR(xt_base);
subplot(311), stem(time,real(xt_base),'k.'); hold on; xlabel('S_{I}');
subplot(312), stem(time,imag(xt_base),'k.'); hold on; ylabel('S_{Q}');
subplot(313), stem(time,abs(xt_base).^2,'k.'); hold on;
title(['PAPR = ' num2str(round(PAPR_dB_base*100)/100) 'dB']);
xlabel('samples'); ylabel('|S_{I}(n)|^{2}+|S_{Q}(n)|^{2}');
figure(2), clf
PAPR_dB_pass = PAPR(xt_pass);
subplot(211), stem(time,xt_pass,'k.'); hold on;
  plot(time_,xt_pass_,'k:'); ylabel('S(n)');
subplot(212), stem(time,xt_pass.*xt_pass,'r.'); hold on;
  plot(time_,xt_pass_.*xt_pass_,'k:');
  title(['PAPR = ' num2str(round(PAPR_dB_pass*100)/100) 'dB']);
  xlabel('samples'); ylabel('|S(n)|^{2}');
% PAPRs of baseband/passband signals
PAPRs_of_baseband_passband_signals=[PAPR_dB_base; PAPR_dB_pass]
```

Program 7.5 "modulation()" for modulation and oversampling

```
function [s,time] = modulation(x,Ts,Nos,Fc)
% modulates x(n*Ts) with carrier frequency Fc and
% Nos-times oversamples for time=[0:Ts/Nos:Nx*Ts-T]
% Ts: Sampling period of x[n]
% Nos: Oversampling factor
% Fc: Carrier frequency
Nx = length(x); offset = 0;
if nargin <5, % Scale and Oversampling period for Baseband
    scale=1; T=Ts/Nos;
 else % Scale and Oversampling period for Passband
    scale=sqrt(2); T=1/Fc/2/Nos;
```

```
end
t_Ts=[0:T:Ts-T]; time=[0:T:Nx*Ts-T]; % Sampling interval, Whole interval
tmp = 2*pi*Fc*t_Ts+offset;  len_Ts=length(t_Ts);
cos_wct = cos(tmp)*scale;  sin_wct = sin(tmp)*scale;
for n = 1:Nx
    s((n-1)*len_Ts+1:n*len_Ts) = real(x(n))*cos_wct-imag(x(n))*sin_wct;
end
```

7.1.2 Distribution of OFDM Signal

Figure 7.5 shows the end-to-end block diagram of an OFDM system in which the discrete-time signal $\{x[n]\}$ after IFFT at the transmitter can be expressed as

$$x[n] = \frac{1}{N}\sum_{k=0}^{N-1}X[k]e^{j\frac{2\pi}{N}kn} \tag{7.12}$$

for a sequence of PSK or QAM-modulated data symbols, $\{X[k]\}$. In other words, $x[n]$ is given by adding the N different time-domain signals $\{e^{j2\pi kn/N}\}$, each of which corresponds to the different orthogonal subcarriers, the kth one modulated with data symbol $X[k]$. Figure 7.6(a) shows the individual time-domain QPSK-modulated subcarrier signals $X[k]e^{j2\pi kn/N}$ for $N = 8$ and their sum $x[n]$, in terms of its continuous-time version $x(t)$. The PAPR characteristics of the OFDM signal is obvious from Figure 7.6(a). In general, we expect the PAPR to become significant as N increases. Meanwhile, Figure 7.6(b) shows the distributions of $|x[n]|$ including the real and imaginary parts of them for $N = 16$, which again illustrates the PAPR characteristics of OFDM signal. They show that the real and imaginary parts of $x[n]$ follow a Gaussian distribution while $|x[n]|$ or $|x(t)|$ follows a Rayleigh distribution, which is consistent with the discussion in Section 7.1.1.

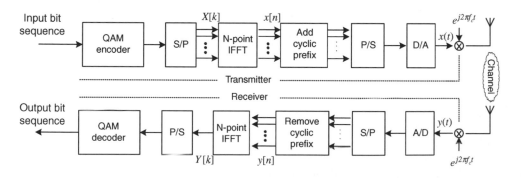

Figure 7.5 Block diagram of OFDM system.

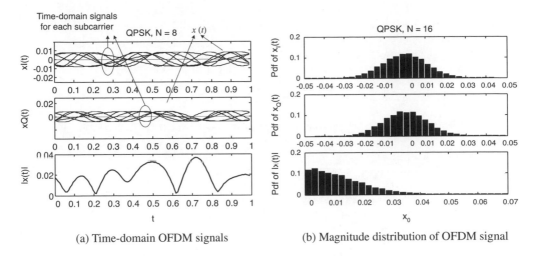

(a) Time-domain OFDM signals   (b) Magnitude distribution of OFDM signal

Figure 7.6   Characteristics of time-domain QPSK/OFDM signals: $N = 8$ and 16.

MATLAB® Program: Probabilistic Analysis of Time-Domain OFDM Signals

Program 7.6 ("OFDM_signal.m") plots Figures 7.6(a) and (b), which show the real/imaginary parts/magnitude of a baseband QPSK/OFDM signal and their distributions, respectively. Note that the oversampling factor $L = 16$ is used to approximate the continuous-time signal.

Program 7.6 "OFDM_signal.m" to get time-domain OFDM signals and their PDFs

```
% OFDM_signal.m
clear
N=8; b=2; M=2^b; L=16; NL=N*L; T=1/NL; time = [0:T:1-T];
[X,Mod] = mapper(b,N); % A block of N=8 QPSK symbols
X(1)=0+j*0; % with no DC-subcarrier
for i = 1:N
   if i<=N/2, x = ifft([zeros(1,i-1) X(i) zeros(1,NL-i+1)],NL);
     else x = ifft([zeros(1,NL-N+i-1) X(i) zeros(1,N-i)],NL);
   end
   xI(i,:) = real(x); xQ(i,:) = imag(x);
end
sum_xI = sum(xI); sum_xQ = sum(xQ);
figure(1), clf, subplot(311)
plot(time,xI,'k:'), hold on, plot(time,sum_xI,'b'),
ylabel('x_{I}(t)');
subplot(312), plot(time,xQ,'k:'); hold on, plot(time,sum_xQ,'b')
ylabel('x_{Q}(t)');
subplot(313), plot(time,abs(sum_xI+j*sum_xQ),'b'); hold on;
ylabel('|x(t)|'); xlabel('t');
clear('xI'), clear('xQ')
N=2^4; NL=N*L; T=1/NL; time=[0:T:1-T]; Nhist=1e3; N_bin=30;
for k = 1:Nhist
   [X,Mod] = mapper(b,N); % A block of N=16 QPSK symbols
```

```
X(1)=0+j*0; % with no DC-subcarrier
for i = 1:N
   if (i<= N/2) x=ifft([zeros(1,i-1) X(i) zeros(1,NL-i+1)],NL);
      else x=ifft([zeros(1,NL-N/2+i-N/2-1) X(i) zeros(1,N-i)],NL);
   end
   xI(i,:) = real(x); xQ(i,:) = imag(x);
 end
 HistI(NL*(k-1)+1:NL*k)=sum(xI); HistQ(NL*(k-1)+1:NL*k)=sum(xQ);
end
figure(2), clf
subplot(311), [xId,bin]=hist(HistI,N_bin);
 bar(bin,xId/sum(xId),'k');
title([Mod ', N=' num2str(N)]); ylabel('pdf of x_{I}(t)');
subplot(312), [xQd,bin]=hist(HistQ,N_bin);
 bar(bin,xQd/sum(xQd),'k');
ylabel('pdf of x_{Q}(t)');
subplot(313), [xAd,bin]=hist(abs(HistI+j*HistI),N_bin);
bar(bin,xAd/sum(xAd),'k'); ylabel('pdf of |x(t)|'); xlabel('x_{0}');
```

7.1.3 PAPR and Oversampling

The PAPR defined in Equation (7.6) deals with the passband signal with a carrier frequency of f_c in the continuous time domain. Since f_c in general is much higher than $1/T_s$, a continuous-time baseband OFDM signal $x(t)$ with the symbol period T_s and the corresponding passband signal $\tilde{x}(t)$ with the carrier frequency f_c have almost the same PAPR [167]. However, the PAPR for the discrete-time baseband signal $x[n]$ may not be the same as that for the continuous-time baseband signal $x(t)$. In fact, the PAPR for $x[n]$ is lower than that for $x(t)$, simply because $x[n]$ may not have all the peaks of $x(t)$ [164,169]. In practice, the PAPR for the continuous-time baseband signal can be measured only after implementing the actual hardware, including digital-to-analog convertor (DAC). In other words, measurement of the PAPR for the continuous-time baseband signal is not straightforward. Therefore, there must be some means of estimating the PAPR from the discrete-time signal $x[n]$. Fortunately, it is known that $x[n]$ can show almost the same PAPR as $x(t)$ if it is L-times interpolated (oversampled) where $L \geq 4$ [164].

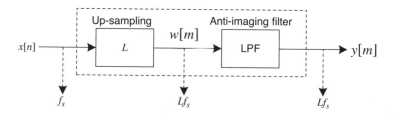

Figure 7.7   Block diagram of L-times interpolator.

Figure 7.7 shows the block diagram of interpolator with a factor of L [170]. It inserts $(L-1)$ zeros between the samples of $x[n]$ to yield $w[m]$ as follows:

$$w[m] = \begin{cases} x[m/L], & \text{for } m = 0, \pm L, \pm 2L, \ldots \\ 0, & \text{elsewhere} \end{cases}$$

A low pass filter (LPF) is used to construct the L-times-interpolated version of $x[n]$ from $w[m]$. For the LPF with an impulse response of $h[m]$, the L-times-interpolated output $y[m]$ can be represented as

$$y[m] = \sum_{k=-\infty}^{\infty} h[k]w[m-k] \tag{7.13}$$

Figures 7.8 and 7.9 illustrate the signals and their spectra appearing in the oversampling process with a sampling frequency of 2kHz to yield a result of interpolation with $L = 4$. Referring to these figures, the IFFT output signal $x[n]$ in Figure 7.5 can be expressed in terms of

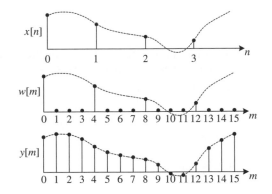

Figure 7.8   Interpolation with $L = 4$ in the time domain.

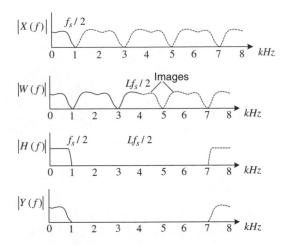

Figure 7.9   Interpolation with $L = 4$ in the frequency domain.

the L-times interpolated version as

$$x'[m] = \frac{1}{\sqrt{L \cdot N}} \sum_{k=0}^{L \cdot N-1} X'[k] \cdot e^{j2\pi m\Delta fk}/_{L \cdot N}, \quad m = 0, 1, \ldots NL-1 \tag{7.14}$$

with

$$X'[k] = \begin{cases} X[k], & \text{for} \quad 0 \leq k < N/2 \quad \text{and} \quad NL-N/2 < k < NL \\ 0, & \text{elsewhere} \end{cases} \tag{7.15}$$

where N, Δf, and $X[k]$ denote the FFT size (or the total number of subcarriers), the subcarrier spacing, and the complex symbol carried over a subcarrier k, respectively. For such an L-times-interpolated signal, the PAPR is now redefined as

$$PAPR = \frac{\max_{m=0,1,\cdots,NL} |x'[m]|^2}{E\{|x'[m]|^2\}} \tag{7.16}$$

In order to see the effect of oversampling or interpolation on PAPR, let us consider the PAPRs of the specific sequences, for example, Chu sequence and IEEE 802.16e preamble sequences. Note that Chu sequence is defined in [171,172] as

$$X_i(k) = \begin{cases} e^{j\frac{\pi}{N}i^2k} & \text{if } N \text{ is even} \\ e^{j\frac{\pi}{N}(i+1)ik} & \text{if } N \text{ is odd} \end{cases}, \quad i = 0, 1, 2, \cdots, N-1, \quad \gcd(k,N) = 1 \tag{7.17}$$

Figure 7.10(a) is obtained by running Program 7.7 ("PAPR_of_Chu.m") to show the magnitudes of $N = 16$-point IFFTs of Chu sequence without oversampling and with

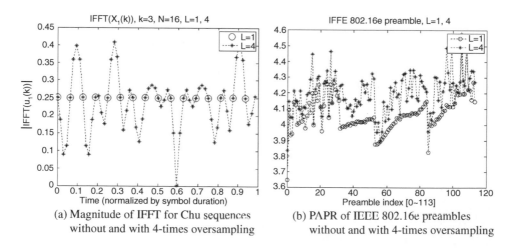

(a) Magnitude of IFFT for Chu sequences without and with 4-times oversampling

(b) PAPR of IEEE 802.16e preambles without and with 4-times oversampling

Figure 7.10 PAPR characteristics of Chu sequence and IEEE 802.16e preambles in the time domain.

$L = 4$-times oversampling. The values of PAPR without and with oversampling turn out to be 0dB and 4.27dB, respectively, which illustrate a rather significant difference in the PAPR by the sampling rate. Meanwhile, Figure 7.10(b) is obtained by running Program 7.9 ("PAPR_of_preamble.m") to show the PAPRs of the 114 preambles defined in IEEE 802.16e standard [119]. The values of PAPR with oversampling are just about 0.4dB greater than those without oversampling. In fact, these preambles are originally designed to have the low PAPR, since they are subject to power boosting in practice. That is why they do not show much difference in the PAPR by the sampling rate. As shown for the Chu sequence, however, oversampling can make a difference in general. Therefore, an oversampling process may be required to make a precise measurement of PAPR in the baseband. Furthermore, the readers must keep it in mind that the PAPR results appearing in this chapter may be different from the actual one in the RF segment, since all the simulations herein deal with the baseband signal. In fact, the PAPR of the passband signal in front of HPA is generally larger than that of the baseband signal after the RF filtering and other processing.

MATLAB® Programs: PAPRs of Chu Sequence and IEEE 802.16e Preambles

Program 7.7 "PAPR_of_Chu.m" to see the PAPR of Chu sequence in the time-domain

```
% PAPR_of_Chu.m
% Plot Fig. 7.10(a)
clear, clf
N=16; L=4; i=[0:N-1]; k=3; X = exp(j*k*pi/N*(i.*i)); % Eq.(7.17)
[x,time] = IFFT_oversampling(X,N); PAPRdB = PAPR(x);
[x_os,time_os] = IFFT_oversampling(X,N,L); PAPRdB_os = PAPR(x_os);
plot(time,abs(x),'o', time_os,abs(x_os),'k:*')
PAPRdB_without_and_with_oversampling=[PAPRdB   PAPRdB_os]
```

Program 7.8 "IFFT_oversampling()" for IFFT and oversampling (interpolation)

```
function [xt,time]=IFFT_oversampling(X,N,L)
% Zero-padding (Eq.(7.15)) & NL-point IFFT
%      => N-point IFFT & interpolation with oversampling factor L
if nargin<3, L=1; end % Virtually, no oversampling
NL=N*L; T=1/NL; time = [0:T:1-T]; X = X(:).';
xt = L*ifft([X(1:N/2) zeros(1,NL-N) X(N/2+1:end)], NL);
```

Program 7.9 "PAPR_of_preamble.m" to see the PAPR of IEEE802.16e preamble

```
% PARR_of_preamble.m
% Plot Fig. 7.10(b) (the PAPR of IEEE802.16e preamble)
clear, clf
N=1024; L=4; Npreamble=114; n=0:Npreamble-1;
for i = 1:Npreamble
    X=load(['.\\Wibro-Preamble\\Preamble_sym' num2str(i-1) '.dat']);
    X = X(:,1); X = sign(X); X = fftshift(X);
    x = IFFT_oversampling(X,N); PAPRdB(i) = PAPR(x);
    x_os = IFFT_oversampling(X,N,L); PAPRdB_os(i) = PAPR(x_os);
end
plot(n,PAPRdB,'-o', n,PAPRdB_os,':*'), title('PAPRdB with oversampling')
```

7.1.4 Clipping and SQNR

As discussed in Section 7.1.1, an OFDM signal with N subcarriers happens to exhibit the maximum power when every subcarrier component coincidently has the largest amplitude with identical phases. As N increases, it is obvious that the maximum power becomes larger and furthermore, the probability that maximum-power signal occurs decreases as N increases. This is well supported by the statistical distributions of the time-domain OFDM signal as shown in Figure 7.6(b). One simplest approach of reducing the PAPR is to clip the amplitude of the signal to a fixed level. The pseudo-maximum amplitude in this approach is referred to as the clipping level and denoted by μ. In other words, any signal whose amplitude exceeds μ will saturate its amplitude to the clipping level μ. While reducing PAPR, the clipping approach helps improve the signal-to-quantization noise ratio (SQNR) in analog-to-digital conversion (ADC).

Figure 7.11 illustrates a typical (Gaussian) distribution that the real or imaginary part of the time-domain OFDM signal $x(t)$ may have. If the clipping level is low, the signal will suffer from a clipping distortion while the PAPR and quantization noise will decrease. If the clipping level is high, a clipping distortion decreases while it suffers from the PAPR and quantization noise. This trade-off relationship between the clipping distortion and quantization noise should be taken into consideration in selecting the clipping level and the number of bits for quantization. Figure 7.12 (obtained by running Program 7.10) shows the SQNR values of OFDM signals quantized with 6, 7, 8, and 9 bits against the clipping level (normalized to the deviation σ of signal) where the maximum SQNR points are colored in black for the different quantization levels. It can be seen from Figure 7.12 that the optimal clipping level to maximize the SQNR varies with the quantization level, but it is around 4σ in most cases. Note that clipping distortion is dominant (i.e., low SQNR) on the left side of the maximum point of the SQNR since the clipping level is set to a low value. The quantization noise is dominant (i.e., low SQNR) on the right sides of the maximum point of SQNR because clipping distortion is less significant.

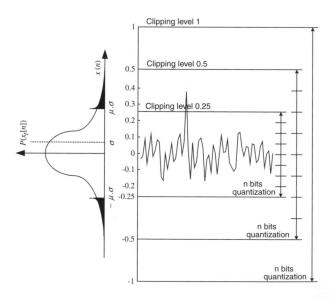

Figure 7.11   Probabilistic distribution of the real part of a time-domain OFDM signal.

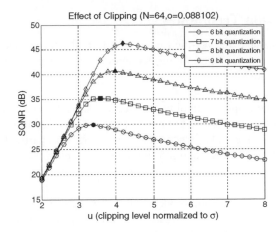

Figure 7.12 SQNR of quantized OFDM signal against the clipping level.

Program 7.10 "SQNR_with_quantization_clipping.m" to obtain SQNR vs. clipping level

```
% SQNR_with_quantization_clipping.m
% Plot Fig. 7.12
clear, clf
N=64; b=6; % FFT size, Number of bits per QAM symbol
L=8; MaxIter=10000; % Oversampling factor, Maximum number of iterations
TWLs = [6:9]; IWL = 1; % Total WordLengths and Integral WordLength
mus=[2:0.2:8]; sq2=sqrt(2); % Clipping Ratio vector
sigma=1/sqrt(N); % Variance of x
gss=['ko-';'ks-';'k^-';'kd-']; % Graphic symbols
for i = 1:length(TWLs)
   TWL = TWLs(i); FWL = TWL-IWL; % Total/Fractional WordLength
   for m = 1:length(mus)
     mu = mus(m)/sq2; % To make the real & imaginary parts of x
     % normalized to 1 (not 1/sqrt(2)) in fi() command below
     Tx = 0; Te = 0;
     for k = 1:MaxIter
       X = mapper(b,N); x = ifft(X,N);
       x = x/sigma/mu;
       xq=fi(x,1,TWL,FWL,'roundmode','round','overflowmode','saturate');
       % Note that the fi() command with TWL=FWL+1 (fraction+sign bits,
       % no integer bits) performs clipping as well as quantization
       xq=double(xq); Px = x*x'; e = x-xq; Pe = e*e';
       Tx=Tx+Px; Te=Te+Pe; % Sum of Signal power and Quantization power
     end
     SQNRdB(i,m) = 10*log10(Tx/Te);
   end
end
for i=1:size(gss,1), plot(mus,SQNRdB(i,:),gss(i,:)), hold on; end
xlabel('mus(clipping level normalized to sigma)'); ylabel('SQNR[dB]');
```

7.2 PAPR Reduction Techniques

PAPR reduction techniques are classified into the different approaches: clipping technique, coding technique, probabilistic (scrambling) technique, adaptive predistortion technique, and DFT-spreading technique.

- The clipping technique employs clipping or nonlinear saturation around the peaks to reduce the PAPR. It is simple to implement, but it may cause in-band and out-of-band interferences while destroying the orthogonality among the subcarriers. This particular approach includes block-scaling technique, clipping and filtering technique, peak windowing technique, peak cancellation technique, Fourier projection technique, and decision-aided reconstruction technique [168,173,174].
- The coding technique is to select such codewords that minimize or reduce the PAPR. It causes no distortion and creates no out-of-band radiation, but it suffers from bandwidth efficiency as the code rate is reduced. It also suffers from complexity to find the best codes and to store large lookup tables for encoding and decoding, especially for a large number of subcarriers [164]. Golay complementary sequence, Reed Muller code, M-sequence, or Hadamard code can be used in this approach [175–184].
- The probabilistic (scrambling) technique is to scramble an input data block of the OFDM symbols and transmit one of them with the minimum PAPR so that the probability of incurring high PAPR can be reduced. While it does not suffer from the out-of-band power, the spectral efficiency decreases and the complexity increases as the number of subcarriers increases. Furthermore, it cannot guarantee the PAPR below a specified level [185–192]. This approach includes SLM (SeLective Mapping), PTS (Partial Transmit Sequence), TR (Tone Reservation), and TI (Tone Injection) techniques.
- The adaptive predistortion technique can compensate the nonlinear effect of a high power amplifier (HPA) in OFDM systems [193]. It can cope with time variations of nonlinear HPA by automatically modifying the input constellation with the least hardware requirement (RAM and memory lookup encoder). The convergence time and MSE of the adaptive predistorter can be reduced by using a broadcasting technique and by designing appropriate training signals.
- The DFT-spreading technique is to spread the input signal with DFT, which can be subsequently taken into IFFT. This can reduce the PAPR of OFDM signal to the level of single-carrier transmission. This technique is particularly useful for mobile terminals in uplink transmission. It is known as the Single Carrier-FDMA (SC-FDMA), which is adopted for uplink transmission in the 3GPP LTE standard [194–197].

7.2.1 Clipping and Filtering

The clipping approach is the simplest PAPR reduction scheme, which limits the maximum of transmit signal to a pre-specified level. However, it has the following drawbacks:

- Clipping causes in-band signal distortion, resulting in BER performance degradation.
- Clipping also causes out-of-band radiation, which imposes out-of-band interference signals to adjacent channels. Although the out-of-band signals caused by clipping can be reduced by

filtering, it may affect high-frequency components of in-band signal (aliasing) when the clipping is performed with the Nyquist sampling rate in the discrete-time domain. However, if clipping is performed for the sufficiently-oversampled OFDM signals (e.g., $L \geq 4$) in the discrete-time domain before a low-pass filter (LPF) and the signal passes through a band-pass filter (BPF), the BER performance will be less degraded [169].

- Filtering the clipped signal can reduce out-of-band radiation at the cost of peak regrowth. The signal after filtering operation may exceed the clipping level specified for the clipping operation [164].

Figure 7.13 shows a block diagram of a PAPR reduction scheme using clipping and filtering where L is the oversampling factor and N is the number of subcarriers. In this scheme, the L-times oversampled discrete-time signal $x'[m]$ is generated from the IFFT of Equation (7.15) ($X'[k]$ with $N \cdot (L-1)$ zero-padding in the frequency domain) and is then modulated with carrier frequency f_c to yield a passband signal $x^p[m]$. Let $x_c^p[m]$ denote the clipped version of $x^p[m]$, which is expressed as

$$
x_c^p[m] = \begin{cases} -A & x^p[m] \leq -A \\ x^p[m] & |x^p[m]| < A \\ A & x^p[m] \geq A \end{cases}
\tag{7.18}
$$

or

$$
x_c^p[m] = \begin{cases} x^p[m] & \text{if } |x^p[m]| < A \\ \dfrac{x^p[m]}{|x^p[m]|} \cdot A & \text{otherwise} \end{cases}
\tag{7.19}
$$

where A is the pre-specified clipping level. Note that Equation (7.19) can be applied to both baseband complex-valued signals and passband real-valued signals, while Equation (7.18) can be applied only to the passband signals. Let us define the clipping ratio (CR) as the clipping level normalized by the RMS value σ of OFDM signal, such that

$$
CR = \frac{A}{\sigma}
$$

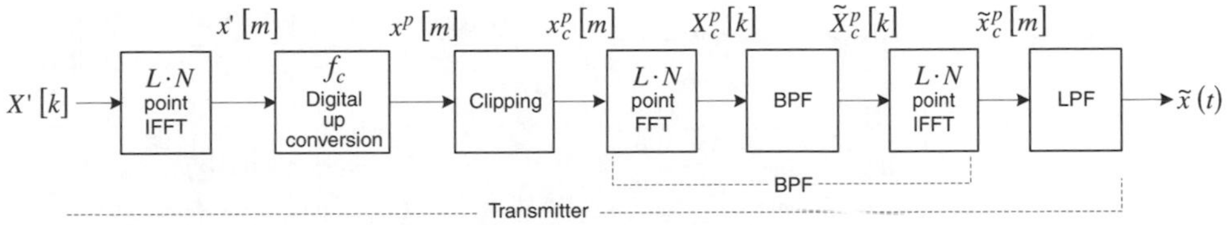

Figure 7.13 Block diagram of a PAPR reduction scheme using clipping and filtering.

Table 7.1   Parameters used for simulation of clipping and filtering.

Parameters	Value
Bandwidth, BW	1 MHz
Sampling frequency, $f_s = BW \cdot L$	
with oversampling factor, $L = 8$	8 MHz
Carrier frequency, f_c	2 MHz
FFT size, N	128
Number of guard interval samples (CP)	32
Modulation order	QPSK
Clipping ratio (CR)	0.8, 1.0, 1.2, 1.4, 1.6

It has been known that $\sigma = \sqrt{N}$ and $\sigma = \sqrt{N/2}$ in the baseband and passband OFDM signals with N subcarriers, respectively.

In general, the performance of PAPR reduction schemes can be evaluated in the following three aspects [173]:

- In-band ripple and out-of-band radiation that can be observed via the power spectral density (PSD)
- Distribution of the crest factor (CF) or PAPR, which is given by the corresponding CCDF
- Coded and uncoded BER performance.

Table 7.1 shows the values of parameters used in the QPSK/OFDM system for analyzing the performance of clipping and filtering technique with Programs 7.11 and 7.12. Figure 7.14 shows the impulse response and frequency response of the (equiripple) finite-duration impulse response (FIR) BPF used in the simulation where the sampling frequency $f_s = 8$ MHz, the stopband and passband edge frequency vectors are [1.4, 2.6][MHz] and [1.5, 2.5] [MHz], respectively, and the number of taps is set to 104 such that the stopband attenuation is about 40dB. Figure 7.15 shows the results for clipping and filtering of OFDM signals with the parameter values listed in Table 7.1. Figures 7.15(a)–(d) show the histograms as probability density functions (PDFs) and power spectra of the oversampled baseband OFDM signal $x'[m]$, the corresponding passband signal $x^p[m]$, the passband clipped signal $x_c^p[m]$, and its filtered signal $\tilde{x}_c^p[m]$. It can be seen from Figure 7.15(b) that the OFDM signal approximately follows a Gaussian distribution. Meanwhile, Figure 7.15(c) shows that the

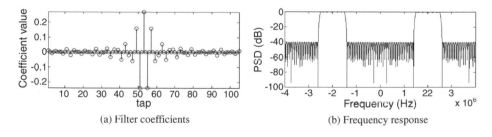

(a) Filter coefficients   (b) Frequency response

Figure 7.14   Characteristics of an equiripple passband FIR filter.

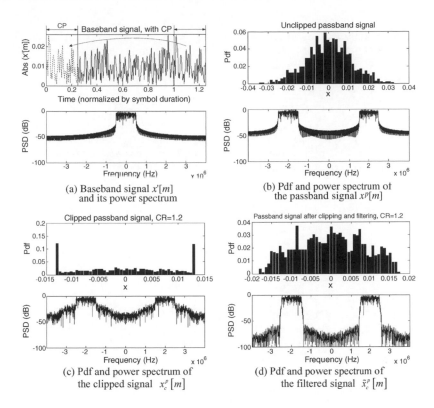

(a) Baseband signal $x'[m]$
and its power spectrum

(b) Pdf and power spectrum of
the passband signal $x^p[m]$

(c) Pdf and power spectrum of
the clipped signal $x_c^p[m]$

(d) Pdf and power spectrum of
the filtered signal $\tilde{x}_c^p[m]$

Figure 7.15   Histograms (PDFs) and power spectra of OFDM signals with clipping and filtering (CR = 1.2).

amplitude of the clipped signal is distributed below the clipping level. Finally, it can be seen from Figure 7.15(d) that the filtered signal shows its peak value beyond the clipping level. Comparing Figure 7.15(c) to Figure 7.15(d), it can also be seen that the out-of-band spectrum increases after clipping, but decreases again after filtering. Figures 7.14 and 7.15 are obtained by running Program 7.11 ("PDF_of_clipped_and_filtered_OFDM_signal.m").

MATLAB® Programs: Clipping and Filtering

Program 7.11 "PDF_of_clipped_and_filtered_OFDM_signal.m" to plot Figures 7.14 and 7.15

```
% PDF_of_clipped_and_filtered_OFDM_signal.m
% Plot Figs. 7.14 and 7.15
clear
CR = 1.2; % Clipping Ratio
b=2; N=128; Ncp=32; % Number of bits per QPSK symbol, FFT size, CP size
fs=1e6; L=8; % Sampling frequency and oversampling factor
Tsym=1/(fs/N); Ts=1/(fs*L); % Sampling frequency and sampling period
fc=2e6; wc=2*pi*fc; % Carrier frequency
t=[0:Ts:2*Tsym-Ts]/Tsym; t0=t((N/2-Ncp)*L); % Time vector
f=[0:fs/(N*2):L*fs-fs/(N*2)]-L*fs/2; % Frequency vector
Fs=8; Norder=104; dens=20; % Sampling frequency and Order of filter
```

```
dens=20; % Density factor of filter
FF=[0 1.4 1.5 2.5 2.6 Fs/2]; % Stop/Pass/Stop frequency edge vector
WW=[10 1 10]; % Stopband/Passband/Stopband weight vector
h=firpm(Norder,FF/(Fs/2),[0 0 1 1 0 0],WW,{dens}); %BPF coefficients
X = mapper(b,N); X(1) = 0; % QPSK modulation
x =IFFT_oversampling(X,N,L); % IFFT and oversampling
x_b = addcp(x,Ncp*L); % Add CP
x_b_os=[zeros(1,(N/2-Ncp)*L), x_b, zeros(1,N*L/2)]; % Oversampling
x_p = sqrt(2)*real(x_b_os.*exp(j*2*wc*t)); % From baseband to passband
x_p_c = clipping(x_p,CR); % Eq. (7.18)
X_p_c_f= fft(filter(h,1,x_p_c));
x_p_c_f = ifft(X_p_c_f);
x_b_c_f = sqrt(2)*x_p_c_f.*exp(-j*2*wc*t); % From passband to baseband

figure(1); clf
nn=(N/2-Ncp)*L+[1:N*L];
nn1=N/2*L+[-Ncp*L+1:0]; nn2=N/2*L+[0:N*L];

subplot(221)
plot(t(nn1)-t0,abs(x_b_os(nn1)),'k:'); hold on;
plot(t(nn2)-t0,abs(x_b_os(nn2)),'k-');
xlabel('t (normalized by symbol duration)'); ylabel('abs(x"[m])');
subplot(223)
XdB_p_os = 20*log10(abs(fft(x_b_os)));
plot(f,fftshift(XdB_p_os)-max(XdB_p_os),'k');
xlabel('frequency[Hz]'); ylabel('PSD[dB]'); axis([f([1 end]) -100 0]);

subplot(222)
[pdf_x_p,bin]=hist(x_p(nn),50); bar(bin,pdf_x_p/sum(pdf_x_p),'k');
xlabel('x'); ylabel('pdf'); title(['Unclipped passband signal']);
subplot(224), XdB_p = 20*log10(abs(fft(x_p)));
plot(f,fftshift(XdB_p)-max(XdB_p),'k');
xlabel('frequency[Hz]'); ylabel('PSD[dB]'); axis([f([1 end]) -100 0]);

figure(2); clf
subplot(221), [pdf_x_p_c,bin]=hist(x_p_c(nn),50);
bar(bin,pdf_x_p_c/sum(pdf_x_p_c),'k'); xlabel('x'); ylabel('pdf');
title(['Clipped passband signal, CR=' num2str(CR)]);
subplot(223), XdB_p_c = 20*log10(abs(fft(x_p_c)));
plot(f,fftshift(XdB_p_c)-max(XdB_p_c),'k');
xlabel('frequency[Hz]'); ylabel('PSD[dB]'); axis([f([1 end]) -100 0]);

subplot(222), [pdf_x_p_c_f,bin] = hist(x_p_c_f(nn),50);
bar(bin,pdf_x_p_c_f/sum(pdf_x_p_c_f),'k');
title(['Passband signal after clipping & filtering, CR=' num2str(CR)]);
xlabel('x'); ylabel('pdf');
subplot(224), XdB_p_c_f = 20*log10(abs(X_p_c_f));
plot(f,fftshift(XdB_p_c_f)-max(XdB_p_c_f),'k');
xlabel('frequency[Hz]'); ylabel('PSD[dB]'); axis([f([1 end]) -100 0]);
```

Program 7.12 "clipping" to clip a signal with CR (Clipping Ratio)

```
function [x_clipped,sigma]=clipping(x,CR,sigma)
% CR: Clipping Ratio, sigma: sqrt(variance of x)
if nargin<3
  x_mean=mean(x); x_dev=x-x_mean;
  sigma=sqrt(x_dev*x_dev'/length(x));
end
x_clipped = x; CL = CR*sigma; % Clipping level
ind = find(abs(x)>CL); % Indices to clip
x_clipped(ind)=x(ind)./abs(x(ind))*CL; % Eq. (7.18b) limitation to CL
```

Figure 7.16(a) shows the CCDFs of crest factor (CF) for the clipped and filtered OFDM signals. Recall that the CCDF of CF can be considered as the distribution of PAPR since CF is the square root of PAPR. It can be seen from this figure that the PAPR of the OFDM signal decreases significantly after clipping and increases a little after filtering. Note that the smaller the clipping ratio (CR) is, the greater the PAPR reduction effect is. Figure 7.16(b) shows the BER performance when clipping and filtering technique is used. Here, "C" and "C&F" denote the case with clipping only and the case with both clipping and filtering, respectively. It can be seen from this figure that the BER performance becomes worse as the CR decreases. Figure 7.16 has been obtained by running Program 7.13 ("CCDF_of_clipped_filtered_OFDM_signal.m".).

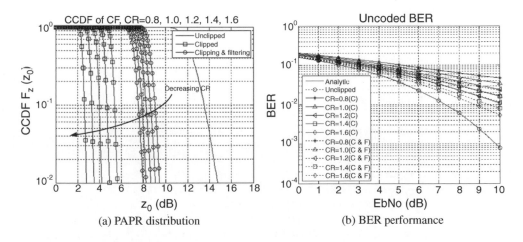

(a) PAPR distribution   (b) BER performance

Figure 7.16   PAPR distribution and BER performance with clipping and filtering.

MATLAB® Program: PAPR and BER Performance with Clipping and Filtering

Program 7.13 "CCDF_of_clipped_filtered_OFDM_signal.m" to obtain the PAPR and BER performance for the clipping and filtering technique in Figure 7.16

```
% CCDF_of_clipped_filtered_OFDM_signal.m
% Plot Fig. 7.16
clear, clf
SNRdBs=[0:10]; N_SNR=length(SNRdBs); % SNR[dB] vector
Nblk=100; CRs=[0.8:0.2:1.6]; N_CR=length(CRs); gss='*^<sd';
b = 2; M = 2^b; % Number of bits per QAM symbol and Alphabet size
N = 128; Ncp = 0; % FFT size and CP size (GI length)
fs = 1e6; L = 8; % Sampling frequency and Oversampling factor
Tsym=1/(fs/N); Ts=1/(fs*L); % OFDM symbol period and Sampling period
fc = 2e6; wc = 2*pi*fc; % Carrier frequency
t = [0:Ts:2*Tsym-Ts]/Tsym; % Time vector
A = modnorm(qammod([0:M-1],M),'avpow',1); % Normalization factor
mdmod = modem.qammod('M',M,'SymbolOrder','Gray','InputType','Bit');
mddem = modem.qamdemod('M',M,'SymbolOrder','Gray','OutputType','Bit');
Fs=8; Norder=104; % Baseband sampling frequency and Order
dens=20; % Density factor of filter
FF=[0 1.4 1.5 2.5 2.6 Fs/2]; % Stopband/Passband/Stopband frequency edge
WW=[10 1 10]; % Stopband/Passband/Stopband weight vector
h = firpm(Norder,FF/(Fs/2),[0 0 1 1 0 0],WW,{dens}); % BPF coefficients
Clipped_errCnt = zeros(size(CRs));
ClippedFiltered_errCnt = zeros(size(CRs));
CF = zeros(1,Nblk); CF_c = zeros(N_CR,Nblk); CF_cf = zeros(N_CR,Nblk);
ber_analytic = berawgn(SNRdBs-10*log10(b),'qam',M);
kk1=1:(N/2-Ncp)*L; kk2=kk1(end)+1:N/2*L+N*L; kk3=kk2(end)+[1:N*L/2];
z = [2:0.1:16]; len_z = length(z);
% -------------- Iteration with increasing SNRdB --------------%
for i = 1:N_SNR
  SNRdB = SNRdBs(i);
  for ncf = 0:2 % no/clip/clip&filter
    if ncf==2, m=ceil(length(h)/2); else m=1; end
    for cr = 1:N_CR
      if ncf==0&cr>1, break; end
      CR = CRs(cr); nobe = 0;
      for nblk = 1:Nblk %(i)
        msgbin = randint(b,N); % binary squences
        X = A*modulate(mdmod,msgbin); % 4QAM (QPSK) mapper
        X(1) = 0+j*0; % DC subcarrier not used
        x = IFFT_oversampling(X,N,L);
        x_b = addcp(x,Ncp*L);
        x_b_os = [zeros(1,(N/2-Ncp)*L), x_b, zeros(1,N*L/2)];
        x_p = sqrt(2)*real(x_b_os.*exp(j*2*wc*t));
        if ncf>0, x_p_c = clipping(x_p,CR); x_p=x_p_c; % clipping
          if ncf>1, x_p_cf = ifft(fft(h,length(x_p)).*fft(x_p)); x_p=x_p_cf;
          end
        end
        if i==N_SNR, CF(nblk) = PAPR(x_p); end
        y_p_n = [x_p(kk1) awgn(x_p(kk2),SNRdB,'measured')
        x_p(kk3)]; y_b = sqrt(2)*y_p_n.*exp(-j*2*wc*t);
        Y_b = fft(y_b);
```

```
      y_b_z = ifft(zero_pasting(Y_b));
      y_b_t = y_b_z((N/2-Ncp)*L+m+[0:L:(N+Ncp)*L-1]);
      Y_b_f = fft(y_b_t(Ncp+1:end),N)*L;
      Y_b_bin = demodulate(mddem,Y_b_f);
      nobe = nobe + biterr(msgbin(:,2:end),Y_b_bin(:,2:end));
    end % End of the nblk loop
    if ncf==0, ber_no(i) = nobe/Nblk/(N-1)/b;
      elseif ncf==1, ber_c(cr,i) = nobe/Nblk/(N-1)/b;
      else ber_cf(cr,i) = nobe/Nblk/(N-1)/b;
    end
    if i==N_SNR
      for iz=1:len_z, CCDF(iz) = sum(CF>z(iz))/Nblk; end
      if ncf==0, CCDF_no = CCDF; break;
        elseif ncf==1, CCDF_c(cr,:) = CCDF;
        else CCDF_cf(cr,:) = CCDF;
      end
    end
   end
  end
end
subplot(221), semilogy(z,CCDF_no), grid on, hold on
for cr = 1:N_CR
   gs = gss(cr);
   subplot(221), semilogy(z,CCDF_c(cr,:),[gs '-'],
     z,CCDF_cf(cr,:), [gs ':']), hold on
   subplot(222), semilogy(SNRdBs,ber_c(cr,:),[gs '-'],
     SNRdBs, ber_cf(cr,:),[gs ':']), hold on
end
semilogy(SNRdBs,ber_no,'o', SNRdBs,ber_analytic,'k'), grid on
```

```
function y=zero_pasting(x)
% Paste zeros at the center half of the input sequence x
N=length(x); M=ceil(N/4); y = [x(1:M) zeros(1,N/2) x(N-M+1:N)];
```

7.2.2 PAPR Reduction Code

It was shown in [175] that a PAPR of the maximum 3dB for the 8-carrier OFDM system can be achieved by 3/4-code rate block coding. Here, a 3-bit data word is mapped onto a 4-bit codeword. Then, the set of permissible code words with the lowest PAPRs in the time domain is chosen. The code rate must be reduced to decrease the desired level of PAPR. It was also stated in [175] that the block codes found through an exhaustive search are mostly based on Golay complementary sequence. Golay complementary sequence is defined as a pair of two sequences whose aperiodic autocorrelations sum to zero in all out-of-phase positions [176]. It is stated in [177] that Golay complementary sequences can be used for constructing OFDM signals with PAPR as low as 3dB. [178] showed the possibility of using complementary codes for both PAPR reduction and forward error correction. Meanwhile, [180] shows that a large set of binary length 2^m Golay complementary pairs can be obtained from Reed-Muller codes. However, the usefulness of these coding techniques is limited to the multicarrier systems with a small number of subcarriers. In general, the exhaustive search of a good code for OFDM systems with a large number of subcarriers is intractable, which limits the actual benefits of coding for PAPR reduction in practical OFDM systems.

First, let us consider the basic properties of complementary sequence. Two sequences $x_1[n]$ and $x_2[n]$ consisting of -1 or $+1$ with equal length N are said to be complementary if they satisfy the following condition on the sum of their autocorrelations:

$$\sum_{n=0}^{N-1}(x_1[n]x_1[n+i]+x_2[n]x_2[n+i]) = \begin{cases} 2N, & i=0 \\ 0, & i\neq 0 \end{cases} \tag{7.20}$$

Taking the Fourier transform of Equation (7.20) yields

$$|X_1[k]|^2 + |X_2[k]|^2 = 2N \tag{7.21}$$

where $X_i[k]$ is the DFT of $\{x_i[n]\}$, such that

$$X_i[k] = \sum_{n=0}^{N-1} x_i[n]e^{-2\pi nkT_s} \tag{7.22}$$

with the sampling period of T_s. The power spectral density of $X_i[k]$ is given by DFT of the autocorrelation of $x_i[n]$. Note that $|X_i[k]|^2$ is the power spectral density (PSD) of a sequence $\{x_i[n]\}$. According to Equation (7.21), the PSD $|X_i[k]|^2$ is upper-bounded by $2N$, which means

$$|X_i[k]|^2 \leq 2N \tag{7.23}$$

Since the power of $x_i[n]$ is 1, the average of $|X_i[k]|^2$ in Equation (7.22) is N and thus, the PAPR of $X_i[k]$ is upper-bounded by

$$\text{PAPR} \leq \frac{2N}{N} = 2 \text{ (or 3dB)} \tag{7.24}$$

Suppose that a sequence is applied as the input to IFFT. Since the IFFT is equivalent to taking the complex conjugate on the output of FFT and dividing it by N, we can replace $X[k]$ by the IFFT of $x[n]$ so that the PAPR can be upper-bounded by 2 (i.e., 3dB). This implies that if the complementary sequences are used as the input to IFFT for producing OFDM signals, the PAPR will not exceed 3dB. The first and second graphs in Figure 7.17 illustrate the PAPR of the uncoded OFDM signal with 16-subcarriers and that of the complementary-coded OFDM signal with 16-subcarriers, respectively. It can be seen from these figures that the complementary coding reduces the PAPR by about 9dB.

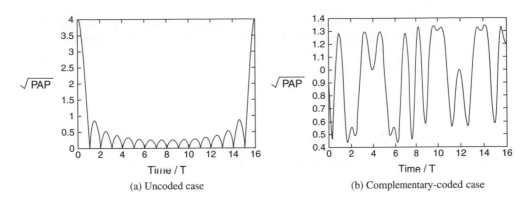

Figure 7.17 Comparison of PAPR: uncoded vs. PAPR reduction-coded OFDM system with $N = 16$.

7.2.3 Selective Mapping

Figure 7.18 shows the block diagram of selective mapping (SLM) technique for PAPR reduction. Here, the input data block $\mathbf{X} = [X[0], X[1], \ldots, X[N-1]]$ is multiplied with U different phase sequences $P^u = [P_0^u, P_1^u, \ldots, P_{N-1}^u]^T$ where $P_v^u = e^{j\varphi_v^u}$ and $\varphi_v^u \in [0, 2\pi)$ for $v = 0, 1, \cdots, N-1$ and $u = 1, 2, \cdots, U$, which produce a modified data block $\mathbf{X}^u = [X^u[1], X^u[2], \ldots, X^u[N-1]]^T$. IFFT of U independent sequences $\{\mathbf{X}^u[v]\}$ are taken to produce the sequences $\mathbf{x}^u = [x^u[0], x^u[1], \ldots, x^u[N-1]]^T$, among which the one $\tilde{\mathbf{x}} = \mathbf{x}^{\tilde{u}}$ with the lowest PAPR is selected for transmission [185], as shown as

$$\tilde{u} = \underset{u=1,2,\cdots,U}{\operatorname{argmin}} \left(\max_{n=0,1,\cdots,N-1} |x^u[n]| \right) \tag{7.25}$$

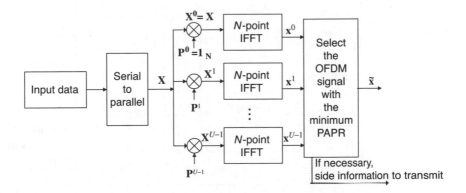

Figure 7.18 Block diagram of selective mapping (SLM) technique for PAPR reduction.

In order for the receiver to be able to recover the original data block, the information (index u) about the selected phase sequence $\mathbf{P}^u$ should be transmitted as a side information [164]. The implementation of SLM technique requires U IFFT operations. Furthermore, it requires $\lfloor \log_2 U \rfloor$ bits of side information for each data block where $\lfloor x \rfloor$ denotes the greatest integer less than x.

7.2.4 Partial Transmit Sequence

The partial transmit sequence (PTS) technique partitions an input data block of N symbols into V disjoint subblocks as follows:

$$\mathbf{X} = [\mathbf{X}^0, \mathbf{X}^1, \mathbf{X}^2, \ldots, \mathbf{X}^{V-1}]^T \tag{7.26}$$

where $\mathbf{X}^i$ are the subblocks that are consecutively located and also are of equal size. Unlike the SLM technique in which scrambling is applied to all subcarriers, scrambling (rotating its phase independently) is applied to each subblock [187] in the PTS technique (see Figure 7.19). Then each partitioned subblock is multiplied by a corresponding complex phase factor $b^v = e^{j\phi_v}$, $v = 1, 2, \ldots, V$, subsequently taking its IFFT to yield

$$\mathbf{x} = \text{IFFT}\left\{ \sum_{v=1}^{V} b^v \mathbf{X}^v \right\} = \sum_{v=1}^{V} b^v \cdot \text{IFFT}\{\mathbf{X}^v\} = \sum_{v=1}^{V} b^v \mathbf{x}^v \tag{7.27}$$

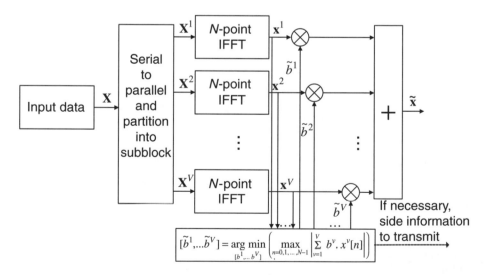

Figure 7.19 Block diagram of partial transmit sequence (PTS) technique for PAPR reduction.

where $\{\mathbf{x}^v\}$ is referred to as a partial transmit sequence (PTS). The phase vector is chosen so that the PAPR can be minimized [188], which is shown as

$$[\tilde{b}^1, \ldots \tilde{b}^V] = \arg\min_{[b^1,\ldots,b^V]} \left(\max_{n=0,1,\cdots,N-1} \left| \sum_{v=1}^{V} b^v x^v[n] \right| \right) \tag{7.28}$$

Then, the corresponding time-domain signal with the lowest PAPR vector can be expressed as

$$\tilde{\mathbf{x}} = \sum_{v=1}^{V} \tilde{b}^v \mathbf{x}^v \tag{7.29}$$

In general, the selection of the phase factors $\{b^v\}_{v=1}^{V}$ is limited to a set of elements to reduce the search complexity [164]. As the set of allowed phase factors is $\mathbf{b} = \{e^{j2\pi i/W} | i = 0, 1, \ldots, W-1\}$, W^{V-1} sets of phase factors should be searched to find the optimum set of phase vectors. Therefore, the search complexity increases exponentially with the number of subblocks.

The PTS technique requires V IFFT operations for each data block and $\lfloor \log_2 W^V \rfloor$ bits of side information. The PAPR performance of the PTS technique is affected by not only the number of subblocks, V, and the number of the allowed phase factors, W, but also the subblock partitioning. In fact, there are three different kinds of the subblock partitioning schemes: adjacent, interleaved, and pseudo-random. Among these, the pseudo-random one has been known to provide the best performance [189].

As discussed above, the PTS technique suffers from the complexity of searching for the optimum set of phase vector, especially when the number of subblock increases. In the literature [190,191], various schemes have been proposed to reduce this complexity. One particular example is a suboptimal combination algorithm, which uses the binary phase factors of $\{1, -1\}$ [190]. It is summarized as follows:

① Partition the input data block into V subblocks as in Equation (7.26).
② Set all the phase factors $b^v = 1$ for $v = 1 : V$, find PAPR of Equation (7.27), and set it as PAPR_min.
③ Set $v = 2$.
④ Find PAPR of Equation (7.27) with $b^v = -1$.
⑤ If PAPR> PAPR_min, switch b^v back to 1. Otherwise, update PAPR_min=PAPR.
⑥ If $v < V$, increment v by one and go back to Step ④. Otherwise, exit this process with the set of optimal phase factors, $\tilde{\mathbf{b}}$.

The number of computations for Equation (7.27) in this suboptimal combination algorithm is V, which is much fewer than that required by the original PTS technique (i.e., $V \ll W^V$).

Figure 7.20 shows the CCDF of PAPR for a 16-QAM/OFDM system using PTS technique as the number of subblock varies. It is seen that the PAPR performance improves as the number of subblocks increases with $V = 1, 2, 4, 8$, and 16. Figure 7.20 has been obtained by running Program 7.14 ("compare_CCDF_PTS.m"), which calls Program 7.15

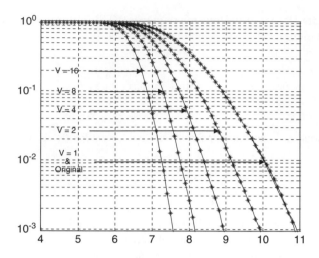

Figure 7.20   PAPR performance of a 16-QAM/OFDM system with PTS technique when the number of subblocks varies.

("CCDF_PTS") and Program 7.16 ("CCDF_OFDMA") to find the CCDF with the PTS technique.

MATLAB® Programs: Performance Analysis of PTS Technique

Program 7.14 "compare_PTS_CCDF.m" to see PAPR when the number of subblocks varies

```
% compare_CCDF_PTS.m
% Plot Fig. 7.20
clear, figure(2), clf
N=256; Nos=4; NNos=N*Nos; % FFT size and oversampling factor
b=4; M=2^b; % Number of bits per QAM symbol and Alphabet size
Nsbs=[1,2,4,8,16]; gss='*^v'; %Numbers of subblocks, Graphic symbols
dBs = [4:0.1:11]; dBcs = dBs+(dBs(2)-dBs(1))/2;
Nblk = 3000; % Number of OFDM blocks for iteration
rand('twister',5489); randn('state',0);
CCDF_OFDMa = CCDF_OFDMA(N,Nos,b,dBs,Nblk);
semilogy(dBs,CCDF_OFDMa,'k'), hold on
for k = 1:length(Nsbs)
    Nsb=Nsbs(k); str(k,:)=sprintf('No of subblocks=%2d',Nsb);
    CCDF=CCDF_PTS(N,Nos,Nsb,b,dBs,Nblk);
    semilogy(dBs,CCDF,['-' gss(k)])
end
legend(str(1,:),str(2,:),str(3,:),str(4,:),str(5,:))
axis([dBs([1 end]) 1e-3 1]); grid on;
title([num2str(M),'-QAM CCDF of OFDMA signal with PTS']);
xlabel('PAPR_0[dB]'); ylabel('Pr(PAPR>PAPR_0)');
```

Program 7.15 "CCDF_PTS()" to compute the CCDF of OFDM signal with PTS technique

```
function CCDF=CCDF_PTS(N,Nos,Nsb,b,dBs,Nblk)
% CCDF of OFDM signal with PTS (Partial Transmit Sequence) technique.
% N     : Number of Subcarriers
% Nos   : Oversampling factor
% Nsb   : Number of subblocks
% b     : Number of bits per QAM symbol
% dBs   : dB vector, Nblk: Number of OFDM blocks for iteration
NNos = N*Nos;          % FFT size
M=2^b; Es=1; A=sqrt(3/2/(M-1)*Es); % Normalization factor for M-QAM
mod_object=modem.qammod('M',M,'SymbolOrder','gray');
for nblk=1:Nblk
    w = ones(1,Nsb);   % Phase (weight) factor
    mod_sym = A*modulate(mod_object,randint(1,N,M)); % 2^b-QAM
    [Nr,Nc] = size(mod_sym);
    zero_pad_sym = zeros(Nr,Nc*Nos);
    for k=1:Nr  % zero padding for oversampling
      zero_pad_sym(k,1:Nos:Nc*Nos) = mod_sym(k,:);
    end
    sub_block=zeros(Nsb,NNos);
    for k=1:Nsb % Eq.(7.26) Disjoint Subblock Mapping
      kk = (k-1)*NNos/Nsb+1:k*NNos/Nsb;
      sub_block(k,kk) = zero_pad_sym(1,kk);
    end
    ifft_sym=ifft(sub_block.',NNos).'; % IFFT
    % -- Phase Factor Optimization - %
    for m=1:Nsb
        x = w(1:Nsb)*ifft_sym; % Eq.(7.27)
        sym_pow = abs(x).^2; PAPR = max(sym_pow)/mean(sym_pow);
        if m==1, PAPR_min = PAPR;
          else if PAPR_min<PAPR, w(m)=1; else PAPR_min = PAPR; end
        end
        w(m+1)=-1;
    end
    x_tilde = w(1:Nsb)*ifft_sym; % Eq.(7.29): The lowest PAPR symbol
    sym_pow = abs(x_tilde).^2; % Symbol power
    PAPRs(nblk) = max(sym_pow)/mean(sym_pow);
end
PAPRdBs=10*log10(PAPRs); % measure PAPR
dBcs = dBs + (dBs(2)-dBs(1))/2; % dB midpoint vector
count=0; N_bins=hist(PAPRdBs,dBcs);
for i=length(dBs):-1:1, count=count+N_bins(i); CCDF(i)=count/Nblk; end
```

Program 7.16 "CCDF_OFDMA()" to compute the CCDF of OFDM signal without PAPR reduction technique

```
function CCDF=CCDF_OFDMA(N,Nos,b,dBs,Nblk)
% CCDF of OFDM signal with no PAPR reduction technique.
```

```
% N      : Number of total subcarriers (256 by default)
% Nos    : Oversampling factor (4 by default)
% b      : Number of bits per QAM symbol
% dBs    : dB vector
% Nblk   : Number of OFDM blocks for iteration
NNos = N*Nos;
M=2^b; Es=1; A=sqrt(3/2/(M-1)*Es); % Normalization factor for M-QAM
mod_object=modem.qammod('M',M,'SymbolOrder','gray');
for nblk=1:Nblk
    mod_sym = A*modulate(mod_object,randint(1,N,M));
    [Nr,Nc]=size(mod_sym);
    zero_pad_sym=zeros(Nr,Nc*Nos);
    for k=1:Nr      % zero padding for oversampling
      zero_pad_sym(k,1:Nos:Nc*Nos)=mod_sym(k,:);
    end
    ifft_sym=ifft(zero_pad_sym,NNos);
    sym_pow=abs(ifft_sym).^2;
    mean_pow(nblk)=mean(sym_pow); max_pow(nblk)=max(sym_pow);
end
PAPR=max_pow./mean_pow; PAPRdB=10*log10(PAPR); % measure PAPR
dBcs = dBs + (dBs(2)-dBs(1))/2;   % dB   midpoint vector
count = 0; N_bins = hist(PAPRdB,dBcs);
for i=length(dBs):-1:1, count=count+N_bins(i); CCDF(i)=count/Nblk; end
```

7.2.5 Tone Reservation

A tone reservation (TR) technique partitions the N subcarriers (tones) into data tones and peak reduction tones (PRTs) [192]. Symbols in PRTs are chosen such that OFDM signal in the time domain has a lower PAPR. The positions of PRTs are known to the receiver and transmitter. Figure 7.21 shows the block diagram of the TR scheme for PAPR reduction.

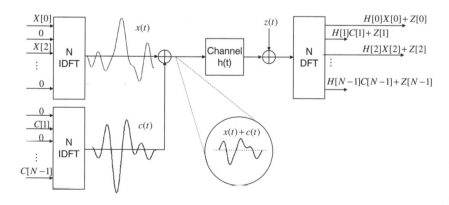

Figure 7.21   Block diagram of tone reservation (TR) technique.

Since the data tones and PRTs are exclusively assigned, the input vector to IFFT block is divided into data vector **X** and PAPR reduction vector **C**. Let $\mathbf{R} = \{i_0, \ldots, i_{R-1}\}$ and $\mathbf{R}^c$ denote the set of R PRT positions and its complement, respectively, where R denotes the number of tones reserved for peak reduction. Then the input symbols to IFFT block can be expressed as

$$X[k] + C[k] = \begin{cases} C[k], k \in \mathbf{R} \\ X[k], k \in \mathbf{R}^c \end{cases} \tag{7.30}$$

where $X[k]$ and $C[k]$ denote the data symbol and PRT symbol, respectively. By taking IFFT of the symbols given by Equation (7.30), we obtain the OFDM symbol to be transmitted as

$$x[n] + c[n] = \frac{1}{N}\sum_{k \in \mathbf{R}^c} X[n]e^{j2\pi kn/N} + \frac{1}{N}\sum_{k \in \mathbf{R}} C[n]e^{j2\pi kn/N} \tag{7.31}$$

Note that the PRT signal $c[n]$ does not cause any distortion on the data signal $x[n]$ in Equation (7.31) due to the orthogonality among subcarriers. Under the assumption that CP (Cyclic Prefix) is longer than the channel impulse response, the received OFDM symbol in the frequency domain (i.e., the output of FFT at the receiver) can be expressed as

$$H[k](X[k] + C[k]) + Z[k] = \begin{cases} H[k]C[k] + Z[k], & k \in \mathbf{R} \\ H[k]X[k] + Z[k], & k \in \mathbf{R}^c \end{cases} \tag{7.32}$$

where $H[k]$ is the channel frequency response and $Z[k]$ is the DFT of the additive noise. The receiver will decode only the data tones for $k \in \mathbf{R}^c$.

With the TR technique, additional power is required for transmitting the PRT symbols and the effective data rate decreases since the PRT tones work as an overhead.

7.2.6 Tone Injection

While the TR technique can reduce the PAPR without additional complexity, it costs the reduced data rate since the additional PRTs are required. A tone injection (TI) technique can be used to reduce the PAPR without reducing the data rate. It allows the PRTs to be overlapped with data tones [192]. Figure 7.22 shows a block diagram for the TI technique.

The basic idea of TI technique is to increase the constellation size so that each of the points in the original constellation can be mapped into several equivalent points in the expanded constellation where the extra degrees of freedom can be exploited for PAPR reduction. More specifically, the time-domain transmit signal with a reduced PAPR can be produced by combining the data signal and PAPR reduction signal as

$$\tilde{x}[n] = x[n] + c[n]$$

$$= \frac{1}{\sqrt{N}}\sum_{k=0}^{N-1}(X[k] + C[k])e^{j2\pi kn/NL} \tag{7.33}$$

where $\{C[k]\}_{k=0}^{N-1}$ and $\{c[n]\}_{n=0}^{N-1}$ denote the frequency-domain and the equivalent time-domain sequences for PAPR reduction, respectively. Since the data tones and PRTs are not

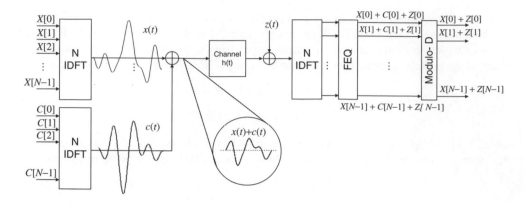

Figure 7.22   Block diagram of tone injection (TI) scheme.

separated orthogonally in the frequency domain, we need means of removing the effect of $C[k]$ at the receiver. In this technique, the PAPR reduction signal is constructed as $C[k] = p[k] \cdot D + jq[k] \cdot D$ where D is a fixed constant while $p[k]$ and $q[k]$ are chosen to minimize the PAPR. The fixed constant D is chosen as a positive real number such that $C[k]$ can be removed at the receiver by performing a modulo-D operation on the real and imaginary parts at the output of the frequency-domain equalizer (FEQ), which can be shown as $X[k] + C[k] + Z[k] = X[k] + p[k] \cdot D + q[k] \cdot D + Z[k]$.

Figure 7.23 shows an expanded 16-QAM constellation diagram used in the TI technique. Here, the black and white points denote the original QAM symbols and the expanded QAM symbols, respectively. Referring to Equation (7.33), the expanded QAM symbols can be represented as

$$\bar{X}[k] = X[k] + C[k] = X[k] + p[k] \cdot D + jq[k] \cdot D \tag{7.34}$$

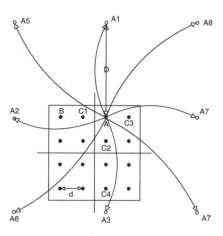

Figure 7.23   Expanded 16-QAM constellation for tone injection (TI) technique.

Figure 7.24 shows a generalized constellation with $p[k] = [-1, 0, 1]$ and $q[k] = [-1, 0, 1]$. Here, the original 16-QAM constellation points are mapped to a generalized $9 \times$ 16-QAM constellation, one of which must be chosen by the transmitter to carry the same information. These extra degrees of freedom can be used to generate OFDM symbols with a lower PAPR. Since the TI technique does not use additional subcarriers for PRTs, it does not incur any loss of data rate. However, extra signal power is required to transmit the symbols in the expanded constellation.

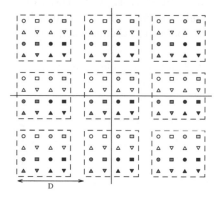

Figure 7.24 Generalized 16-QAM constellation for a given value D.

7.2.7 DFT Spreading

Before discussing the DFT-spreading technique, let us consider OFDMA (Orthogonal Frequency-Division Multiple Access) system (see Section 4.4 and [193]). As depicted in Figure 7.25, suppose that DFT of the same size as IFFT is used as a (spreading) code. Then, the OFDMA system becomes equivalent to the Single Carrier FDMA (SC-FDMA) system because the DFT and IDFT operations virtually cancel each other [195]. In this case, the transmit signal will have the same PAPR as in a single-carrier system.

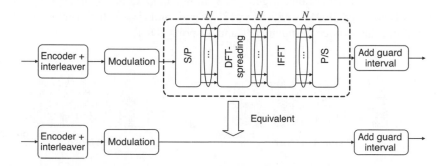

Figure 7.25 Equivalence of OFDMA system with DFT-spreading code to a single-carrier system.

In OFDMA systems, subcarriers are partitioned and assigned to multiple mobile terminals (users). Unlike the downlink transmission, each terminal in uplink uses a subset of subcarriers to transmit its own data. The rest of the subcarriers, not used for its own data transmission, will be filled with zeros. Here, it will be assumed that the number of subcarriers allocated to each user is M. In the DFT-spreading technique, M-point DFT is used for spreading, and the output of DFT is assigned to the subcarriers of IFFT. The effect of PAPR reduction depends on the way of assigning the subcarriers to each terminal [196]. As depicted in Figure 7.26, there are two different approaches of assigning subcarriers among users: DFDMA (Distributed FDMA) and LFDMA (Localized FDMA). Here, DFDMA distributes M DFT outputs over the entire band (of total N subcarriers) with zeros filled in $(N-M)$ unused subcarriers, whereas LFDMA allocates DFT outputs to M consecutive subcarriers in N subcarriers. When DFDMA distributes DFT outputs with equi-distance $N/M = S$, it is referred to as IFDMA (Interleaved FDMA) where S is called the bandwidth spreading factor.

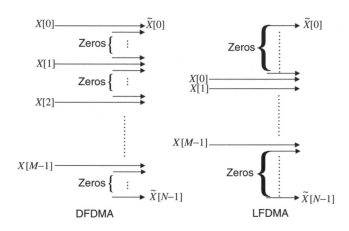

Figure 7.26 Subcarrier mapping for uplink in OFDMA systems: DFDMA and LFDMA.

Figure 7.27 illustrates the subcarriers allocated in the DFDMA and IFDMA with $M = 4$, $S = 3$, and $N = 12$. Furthermore, Figure 7.28 shows the examples of DFT spreading in DFDMA, LFDMA, and IFDMA with $N = 12$, $M = 4$, and $S = 3$. It illustrates a subcarrier mapping relationship between 4-point DFT and 12-point IDFT.

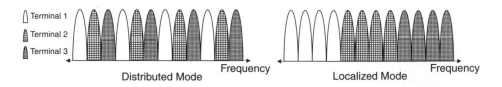

Figure 7.27 Examples of subcarrier assignment to multiple users: three users with $N = 12$, $M = 4$, and $S = 3$.

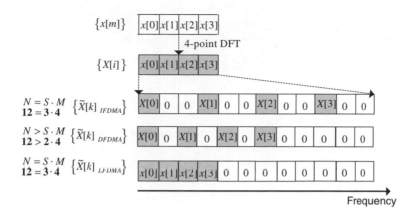

Figure 7.28 Examples of DFT spreading for IFDMA, DFDMA and LFDMA: three users with $N = 12$, $M = 4$, and $S = 3$.

Figure 7.29 shows a block diagram of the uplink transmitter with the DFT-spreading technique that employs IFDMA. Here, the input data $x[m]$ is DFT-spread to generate $X[i]$ and then, allocated as

$$\tilde{X}[k] = \begin{cases} X[k/S], & k = S \cdot m_1, \quad m_1 = 0, 1, 2, \cdots, M-1 \\ 0, & \text{otherwise} \end{cases} \tag{7.35}$$

The IFFT output sequence $\tilde{x}[n]$ with $n = M \cdot s + m$ for $s = 0, 1, 2, \cdots, S-1$ and $m = 0, 1, 2, \cdots, M-1$ can be expressed as

$$\tilde{x}[n] = \frac{1}{N} \sum_{k=0}^{N-1} \tilde{X}[k] e^{j2\pi \frac{n}{N} k}$$

$$= \frac{1}{S} \cdot \frac{1}{M} \sum_{m_1=0}^{M-1} X[m_1] e^{j2\pi \frac{n}{M} m_1}$$

$$= \frac{1}{S} \cdot \frac{1}{M} \sum_{m_1=0}^{M-1} X[m_1] e^{j2\pi \frac{Ms+m}{M} m_1}, \tag{7.36}$$

$$= \frac{1}{S} \cdot \left(\frac{1}{M} \sum_{m_1=0}^{M-1} X[m_1] e^{j2\pi \frac{m}{M} m_1} \right)$$

$$= \frac{1}{S} \cdot x[m]$$

which turns out to be a repetition of the original input signal $x[m]$ scaled by $1/S$ in the time domain [197]. In the IFDMA where the subcarrier mapping starts with the rth subcarrier

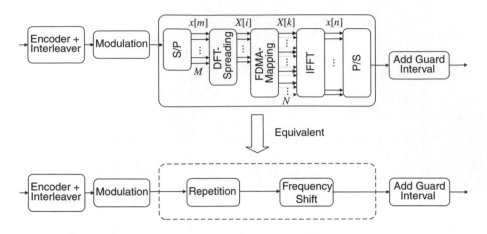

Figure 7.29   Uplink transmitter with DFT-spreading technique of IFDMA.

$(r = 0, 1, 2, \cdots, S-1)$, the DFT-spread symbol can be expressed as

$$
\tilde{X}[k] = \begin{cases} X[(k-r)/S], & k = S \cdot m_1 + r, \; m_1 = 0, 1, 2, \cdots, M-1 \\ 0, & \text{otherwise} \end{cases} \tag{7.37}
$$

Then, the corresponding IFFT output sequence, $\{\tilde{x}[n]\}$, is given by

$$
\begin{aligned}
\tilde{x}[n] &= \tilde{x}[Ms + m] \\
&= \frac{1}{N}\sum_{k=0}^{N-1}\tilde{X}[k]e^{j2\pi\frac{n}{N}k} \cdot \\
&= \frac{1}{S} \cdot \frac{1}{M}\sum_{m_1=0}^{M-1}X[m_1]e^{j2\pi\left(\frac{n}{M}m_1 + \frac{n}{N}r\right)} \\
&= \frac{1}{S} \cdot \frac{1}{M}\sum_{m_1=0}^{M-1}X[m_1]e^{j2\pi\frac{(Ms+m)}{M}m_1}e^{j2\pi\frac{n}{N}r} \\
&= \frac{1}{S} \cdot \left(\frac{1}{M}\sum_{m_1=0}^{M-1}X[m_1]e^{j2\pi\frac{m}{M}m_1}\right) \cdot e^{j2\pi\frac{n}{N}r} \\
&= \frac{1}{S}e^{j2\pi\frac{n}{N}r} \cdot x[m]
\end{aligned} \tag{7.38}
$$

Compared with Equation (7.36), one can see that the frequency shift of subcarrier allocation starting point by r subcarriers results in the phase rotation of $e^{j2\pi nr/N}$ in IFDMA.

In the DFT-spreading scheme for LFDMA, the IFFT input signal $\tilde{X}[k]$ at the transmitter can be expressed as

$$\tilde{X}[k] = \begin{cases} X[k], & k = 0, 1, 2, \cdots, M-1 \\ 0, & k = M, M+1, \cdots, N-1 \end{cases} \tag{7.39}$$

The IFFT output sequence $\tilde{x}[n]$ with $n = S \cdot m + s$ for $s = 0, 1, 2, \cdots, S-1$ can be expressed as follows [197]:

$$\tilde{x}[n] = \tilde{x}[Sm+s] = \frac{1}{N} \sum_{k=0}^{N-1} \tilde{X}[k] e^{j2\pi\frac{nk}{N}} = \frac{1}{S} \cdot \frac{1}{M} \sum_{k=0}^{M-1} X[k] e^{j2\pi\frac{Sm+sk}{SM}} \tag{7.40}$$

For $s = 0$, Equation (7.40) becomes

$$\tilde{x}[n] = \tilde{x}[Sm] = \frac{1}{S} \cdot \frac{1}{M} \sum_{k=0}^{M-1} X[k] e^{j2\pi\frac{Smk}{SM}} = \frac{1}{S} \cdot \frac{1}{M} \sum_{k=0}^{M-1} X[k] e^{j2\pi\frac{mk}{M}} = \frac{1}{S} x[m] \tag{7.41}$$

For $s \neq 0$, $X[k] = \sum_{p=0}^{M-1} x[p] e^{-j2\pi\frac{p}{N}k}$ such that Equation (7.40) becomes

$$\tilde{x}[n] = \tilde{x}[Sm+s]$$

$$= \frac{1}{S} \left(1 - e^{j2\pi\frac{s}{S}} \right) \cdot \frac{1}{M} \sum_{p=0}^{M-1} \frac{x[p]}{1 - e^{j2\pi\left\{ \frac{(m-p)}{M} + \frac{s}{SM} \right\}}}$$

$$= \frac{1}{S} e^{j\pi\frac{(M-1)s-Sm}{SM}} \cdot \sum_{p=0}^{M-1} \underbrace{\frac{\sin\left(\pi\frac{s}{S} \right)}{M \sin\left(\pi \cdot \frac{(Sm+s)}{SM} - \pi\frac{p}{M} \right)}}_{\text{Defined as } c(m,s,p)} \cdot \underbrace{\left(e^{j\pi\frac{p}{M}} x[p] \right)}_{\text{Defined as } \hat{x}[p]} \tag{7.42}$$

From Equations (7.41) and (7.42), it can be seen that the time-domain LFDMA signal becomes the $1/S$-scaled copies of the input sequence at the multiples of S in the time domain. The values in-between are obtained by summing all the input sequences with the different complex-weight factor. Figure 7.30 shows the examples of the time-domain signals when the DFT-spreading technique for IFDMA and LFDMA is applied with $N = 12$, $M = 4$, and $S = 3$, where $\tilde{x}_{\text{IFDMA}}[n]$ and $\tilde{x}_{\text{LFDMA}}[n]$ represent the signals from Equations (7.38) and (7.42), respectively.

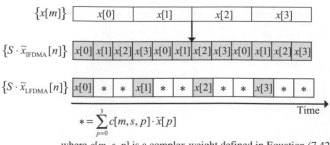

where $c[m, s, p]$ is a complex weight defined in Equation (7.42)

Figure 7.30   Time-domain signals with DFT-spreading technique: examples for IFDMA and LFDMA.

Figure 7.31 shows a comparison of PAPR performances when the DFT-spreading technique is applied to the IFDMA, LFDMA, and OFDMA. Here, QPSK, 16-QAM, and 64-QAM are used for an SC-FDMA system with $N = 256$, $M = 64$, and $S = 4$. It can be seen from Figure 7.31 that the PAPR performance of the DFT-spreading technique varies depending on the subcarrier allocation method. In the case of 16-QAM, the values of PAPRs with IFDMA, LFDMA, and LFDMA for CCDF of 1% are 3.5dB, 8.3dB, and 10.8dB, respectively. It implies that the PAPRs of IFDMA and LFDMA are lower by 7.3dB and 3.2dB, respectively, than that of OFDMA with no DFT spreading. Figure 7.31 has been obtained by running Program 7.18 ("compare_DFT_spreading.m"), which calls Program 7.17 ("CCDF_PAPR_DFTspreading") to find the PAPRs and their CCDFs for IFDMA, LFDMA, and OFDMA.

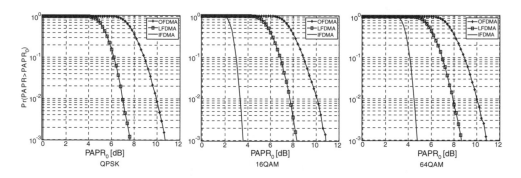

Figure 7.31   PAPR performances of DFT-spreading technique for IFDMA, LFDMA, and OFDMA.

MATLAB® Programs: PAPR Analysis of DFT Spreading

Program 7.17 "CCDF_PAPR_DFTspreading" to get CCDF and PAPR of OFDMA signals

```
function [CCDF,PAPRs]=
    CCDF_PAPR_DFTspreading(fdma_type,Ndb,b,Nfft,dBcs,Nblk,psf,Nos)
% fdma_type: 'ofdma'/'lfdma'(localized)/'ifdma'(interleaved)
% Ndb      : Data block size
% b        : Number of bits per symbol
```

```
% Nfft    : FFT size
% dBcs    : dB midpoint vector
% Nblk    : Number of OFDM blocks for iteration
% psf     : Pulse shaping filter coefficient vector
% Nos     : Oversampling factor
M=2^b; % Alphbet size
Es=1; A=sqrt(3/2/(M-1)*Es); % Normalization factor for QAM
mod_object=modem.qammod('M',M,'SymbolOrder','gray');
S=Nfft/Ndb; % Spreading factor
for iter=1:Nblk
  mod_sym = A*modulate(mod_object,randint(1,Ndb,M));
  switch upper(fdma_type(1:2))
    case 'IF', fft_sym = zero_insertion(fft(mod_sym,Ndb),S); % IFDMA
    case 'LF', fft_sym = [fft(mod_sym,Ndb) zeros(1,Nfft-Ndb)]; % LFDMA
    case 'OF', fft_sym = zero_insertion(mod_sym,S); % No DFT spreading
    otherwise fft_sym = mod_sym; % No oversampling, No DFT spraeding
  end
  ifft_sym = ifft(fft_sym,Nfft);          % IFFT
  if nargin>7, ifft_sym = zero_insertion(ifft_sym,Nos); end
  if nargin>6, ifft_sym = conv(ifft_sym,psf); end
  sym_pow = ifft_sym.*conj(ifft_sym); % measure symbol power
  PAPRs(iter) = max(sym_pow)/mean(sym_pow); % measure PAPR
end
% Find the CCDF of OFDMA signal with DFT spreading
PAPRdBs = 10*log10(PAPRs);
N_bins = hist(PAPRdBs,dBcs);      count = 0;
for i=length(dBcs):-1:1, count=count+N_bins(i); CCDF(i)=count/Nblk; end
function y=zero_insertion(x,M,N)
[Nrow,Ncol]=size(x);
if nargin<3, N=Ncol*M; end
y=zeros(Nrow,N); y(:,1:M:N) = x;
```

Program 7.18 "compare_DFT_spreading.m" to analyze PAPR of OFDMA/LFDMA/IFDMA

```
% compare_DFT_spreading.m
% Plot Fig. 7.31
clear, clf
N=256; Nd=64; % FFT size and Data block size (# of subcarriers per user)
gss='*^<sd>v.'; % Numbers of subblocks and graphic symbols
bs=[2 4 6]; N_b=length(bs);
dBs = [0:0.2:12]; dBcs = dBs+(dBs(2)-dBs(1))/2;
Nblk = 5000; % Number of OFDM blocks for iteration
for i=1:N_b
  b=bs(i); M=2^b; rand('twister',5489); randn('state',0);
  CCDF_OFDMa =
      CCDF_PAPR_DFTspreading('OF',N,b,N,dBcs,Nblk);  % CCDF of OFDMA
  CCDF_LFDMa =
      CCDF_PAPR_DFTspreading('LF',Nd,b,N,dBcs,Nblk); % CCDF of LFDMA
```

```
CCDF_IFDMa =
    CCDF_PAPR_DFTspreading('IF',Nd,b,N,dBcs,Nblk); % CCDF of IFDMA
    subplot(130+i), xlabel(['PAPR_0[dB] for ' num2str(M) '-QAM']);
    semilogy(dBs,CCDF_OFDMa,'-o',dBs,CCDF_LFDMa,'-<',dBs,CCDF_IFDMa,'-*')
    legend('OFDMA','LFDMA','IFDMA'), ylabel('Pr(PAPR>PAPR_0)'); grid on;
end
```

Now, let us consider the effect of pulse shaping on the PAPR performance of DFT-spreading technique. Figure 7.32 shows the PAPR performance of DFT-spreading technique with IFDMA and LFDMA, varying with the roll-off factor α of the RC (Raised-Cosine) filter for pulse shaping after IFFT. It can be seen from this figure that the PAPR performance of IFDMA can be significantly improved by increasing the roll-off factor from $\alpha = 0$ to 1. This is in contrast with LFDMA which is not so much affected by pulse shaping. It implies that IFDMA will have a trade-off between excess bandwidth and PAPR performance since excess bandwidth increases as the roll-off factor becomes larger. Figure 7.32 has been obtained by running Program 7.19 ("compare_DFT_spreading_w_psf.m"), which calls Program 7.17 to find the PAPRs and their CCDFs for IFDMA and LFDMA. The results here have been obtained with the simulation parameters of $N = 256$, $M = 64$, $S = 4$ (spreading factor), and $N_{os} = 8$ (oversampling factor for pulse shaping) for both QPSK and 16-QAM.

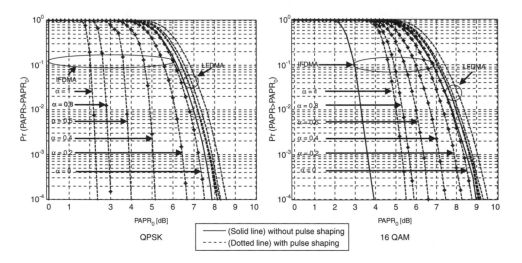

Figure 7.32   PAPR performances of DFT-spreading technique with pulse shaping.

Now, let us see how the PAPR performance of DFT-spreading technique is affected by the number of subcarriers, M, that are allocated to each user. Figure 7.33 shows that the PAPR performance of DFT-spreading technique for LFDMA with a roll-off factor of $\alpha = 0.4$ is degraded as M increases, for example, $M = 4$ to 128. Here, 64-QAM is used for the SC-FDMA system with 256-point FFT ($N = 256$). Figure 7.33 has been obtained by running Program 7.19 ("compare_DFT_spreading_w_psf.m").

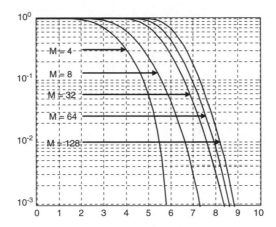

Figure 7.33   PAPR performance of DFT-spreading technique when *M* varies.

Program 7.19 "compare_DFT_spreading_w_psf.m" to see the effect of pulse shaping filter and the number of subcarriers per terminal (user)

```
% compare_DFT_spreading_w_psf.m
% To see the effect of pulse shaping using RC filter (Figs. 7.32 & 7.33)
clear, clf
N=256; Nd=64; % FFT size, # of subcarriers per user (Data block size)
S = N/Nd; % Spreading factor
Nsym=6; Nos=8; % RC filter length, Oversampling factor
rhos = [0:0.2:1]; % Roll-off factors of RC filter for pulse shaping
bs=[2 4]; % Numbers of bits per QAM symbol
dBs = [0:0.2:10]; dBcs = dBs+(dBs(2)-dBs(1))/2;
Nblk = 5000; % Number of OFDM blocks for iteration
figure(1), clf % To plot Fig. 7.32
gss='*^<sd>v.'; % Numbers of subblocks and graphic symbols
str11='IFDMA with no pulse shaping';
str12='LFDMA with no pulse shaping';
for i_b=1:length(bs)
   b=bs(i_b); M=2^b; % Number of bits per QAM symbol and Alphabet size
   rand('twister',5489); randn('state',0);
   CCDF_IF0 = CCDF_PAPR_DFTspreading('IF',Nd,b,N,dBcs,Nblk);
   subplot(220+i_b), semilogy(dBs,CCDF_IF0,'k'), hold on
   CCDF_LF0 = CCDF_PAPR_DFTspreading('LF',Nd,b,N,dBcs,Nblk);
   semilogy(dBs,CCDF_LF0,'k'), hold on
   for i=1:length(rhos)
       rho = rhos(i); % Roll-off factor
       psf = rcosfir(rho,Nsym,Nos,1,'norm')*Nos; % RC filter coeff
       CCDF_IF = CCDF_PAPR_DFTspreading('IF',Nd,b,N,dBcs,Nblk,psf,Nos);
```

```
      CCDF_LF = CCDF_PAPR_DFTspreading('LF',Nd,b,N,dBcs,Nblk,psf,Nos);
      semilogy(dBs,CCDF_IF,['-' gss(i)], dBs,CCDF_LF,[':' gss(i)])
      str1(i,:)=sprintf('IFDMA with a=%3.1f',rho);
      str2(i,:)=sprintf('LFDMA with a=%3.1f',rho);
   end
   legend(str11,str12,str1(1,:),str2(1,:),str1(2,:),str2(2,:)), grid on;
   xlabel('PAPR_0[dB]'); ylabel('Pr(PAPR>PAPR_0)');
end
figure(2), clf % To plot Fig. 7.33
Nds=[4 8 32 64 128]; N_Nds=length(Nds);
b=6; rho=0.4; % Number of bits per QAM symbol, Roll-off factor
psf = rcosfir(rho,Nsym,Nos,1,'norm')*Nos; % RC filter coeff
for i=1:N_Nds
   Nd=Nds(i); % Number of subcarriers per user (Data block size)
   rand('twister',5489); randn('state',0);
   CCDF_LFDMa = CCDF_PAPR_DFTspreading('LF',Nd,b,N,dBcs,Nblk,psf,Nos);
   semilogy(dBs,CCDF_LFDMa,['-' gss(i)]), hold on
   str(i,:)=sprintf('LFDMA with a=%3.1f for Nd=%3d',rho,Nd);
end
legend(str(1,:),str(2,:),str(3,:),str(4,:),str(5,:)), grid on;
xlabel('PAPR_0[dB]'); ylabel('Pr(PAPR>PAPR_0)');
```

In conclusion, the SC-FDMA systems with IFDMA and LFDMA have a better PAPR performance than OFDMA systems. This unique feature has been adopted for uplink transmission in 3GPP LTE, which has been evolved into one of the candidate radio interface technologies for the IMT-Advanced standards in ITU-R. Although the IFDMA has a lower PAPR than LFDMA, the LFDMA is usually preferred for implementation. It is attributed to the fact that subcarriers allocation with equi-distance over the entire band (IFDMA) is not easy to implement, since IFDMA requires additional resources such as guard band and pilots.

8

Inter-Cell Interference Mitigation Techniques

The cellular TDMA system can be designed as virtually interference-free by planning the frequency-reuse distance, which makes the same frequency channels reused sufficiently far apart. In order to maintain a sufficient frequency-reuse distance, any cell site within the same cluster cannot use the same frequency channel in the TDMA cellular network. Meanwhile, the cellular CDMA system is also virtually interference-free due to its interference-averaging capability with a wide spreading bandwidth. As long as its spreading factor is sufficiently large, the cellular CDMA system can be robust against co-channel interference, even when the same frequency channels are assigned to all neighbor cell sites, that is, frequencies are fully reused. Unlike the cellular CDMA system which has the interference-robust capability, the OFDMA-based cellular system suffers from inter-cell interference at the cell boundary, especially when all frequency channels are fully reused. In other words, some means of mitigating the inter-cell interference is required to support a full frequency-reuse operation. According to standards and literature, the inter-cell interference mitigation techniques include inter-cell interference coordination technique, inter-cell interference randomization technique, and inter-cell interference cancellation technique.

8.1 Inter-Cell Interference Coordination Technique

8.1.1 Fractional Frequency Reuse

Figure 8.1 shows the basic frequency reuse schemes for OFDMA-based cellular systems with the different frequency reuse factor (FRF), denoted by K. The FRF is defined as the number of adjacent cells which cannot use the same frequencies for transmission. Its inverse, $1/K$, corresponds to the rate at which the same frequency can be used in the network. In other words, $1/K$ is a factor to indicate how efficiently the bandwidth is used in the cellular system. When $K = 1$ as in Figure 8.1(a), the entire bandwidth available for transmission is used in all cells. In this case, the users near the cell-center will experience high signal-to-interference and noise Ratio (SINR) due to the large path loss from adjacent cells. However, the users at the cell

MIMO-OFDM Wireless Communications with MATLAB® Yong Soo Cho, Jaekwon Kim, Won Young Yang and Chung G. Kang
© 2010 John Wiley & Sons (Asia) Pte Ltd

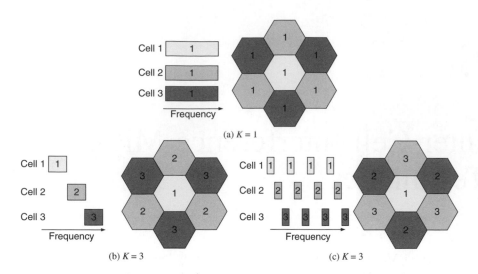

Figure 8.1 Examples of frequency reuse in an OFDMA cellular system.

boundary will suffer from a small SINR, which may increase an outage rate at the cell boundary. In order to improve the SINR throughout the cell coverage area while reducing the outage rate at the cell boundary, the whole bandwidth can be divided into three channels (Figure 8.1(b)) or subbands (Figure 8.1(c)), each of which is allocated to adjacent cells in an orthogonal manner. It corresponds to $K = 3$ and reduces the usable bandwidth for each cell. However, the users at the cell boundary will experience high SINR, reducing inter-cell interference. Note that a subband is a subset of subcarriers, which is derived from entire subcarriers of each channel in the OFDM system. Unlike the multi-channel case of Figure 8.1(b), a single channel is divided into three subbands to be assigned to each cell in Figure 8.1(c), even if both cases correspond to $K = 3$.

To improve the performance at the cell boundary, a concept of fractional frequency reuse (FFR) has been proposed [198] for the OFDMA cellular system. By definition, FFR is a subcarrier reuse scheme to allocate only a part of the total bandwidth, that is, a subset of subcarriers, to each cell such that $1 < K < 3$. In FFR schemes, the whole bandwidth is divided into subbands, some of which are allocated to a different location in the cell. Figure 8.2(a)

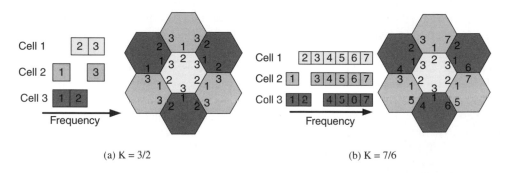

Figure 8.2 Fractional frequency reuse (FFR) in an OFDMA cellular system.

illustrates the FFR scheme with $K = 3/2$, in which two of three subbands are assigned to each cell while planning them to avoid interference from adjacent cells. As compared to the full reuse case of $K = 1$, its efficiency is reduced to 2/3. Figure 8.2(b) illustrates the FFR scheme with $K = 7/6$. In fact, this is the case when the best spectral efficiency can be achieved while ensuring the cell-edge performance in the hexagonal cellular configuration. Eliminating only one of seven subbands from each cell, orthogonal subband assignment can be still realized for any two adjacent boundaries.

Figure 8.3 shows another type of FFR scheme where a different frequency reuse is used, depending on the location in the cell. Since the users near the cell center experience a high SINR, $K = 1$ can be maintained for them. In order to avoid interference however, the higher frequency reuse factor needs to be used at the cell boundary. In Figure 8.3(a), $K = 1$ and $K = 3$

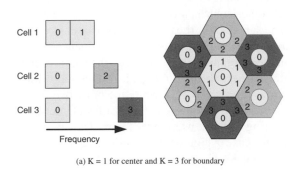

(a) K = 1 for center and K = 3 for boundary

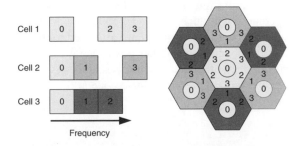

(b) K = 1 for center and K = 3/2 for boundary

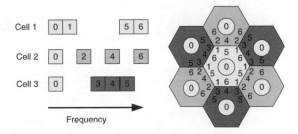

Figure 8.3 Fractional frequency reuse (FFR) with the different FRFs.

are used at the inner region and boundary, respectively. In this case, the whole bandwidth is divided into four different subbands, among which subband 0 is used by all cells (i.e., K = 1), while the rest of them are orthogonally assigned to different cells, that is, K = 3. The same idea can be extended to different configurations, for example, K = 1 for the inner region (center) and K = 3/2 for the outer region (boundary), as shown in Figure 8.3(b). In general, more than three frequency reuse factors can be used as in Figure 8.3(c).

8.1.2 Soft Frequency Reuse

In order to improve the bandwidth efficiency of the FFR schemes in Section 8.1.1, K = 1 can be still realized while reducing the inter-cell interference in OFDMA cellular systems. Toward this end, we allocate different levels of power to the subbands, depending on the user location. High power is allocated to the subbands for the users at the cell boundary, and low power is allocated to all other subbands for the users in the center (inner region), while orthogonally planning the subband for those at the cell boundary of the adjacent cells as in FFR. This particular concept is referred to as the soft frequency reuse (SFR) and it can be illustrated with an example in Figure 8.4(a), in which the whole bandwidth is divided into three subbands. In Figure 8.4(a), only one subband is orthogonally allocated to each cell for the users at the boundary as in Figure 8.1(b), while the other two subbands are allocated to each cell for those in the center. In order not to incur significant interference to the users at the boundaries of the neighboring cells, the users in the inner region must use those two subbands with lower power. Unlike the case in Figure 8.1(b), entire frequency bands can be fully reused in all cells, that is, achieving K = 1. In fact, the full frequency reuse can be enabled by allocating a lower power level to the center users. For example, Figures 8.4(a) and (b) are the soft frequency reuse versions of Figures 8.1(b) and 8.3, respectively. Figure 8.5 illustrates the soft frequency reuse scenario with nine subbands [199]. Dividing the whole bandwidth into nine subbands, six subbands with lower power are allocated to the users in the inner region and the remaining three subbands with higher power are allocated to the users at the cell boundary. The essence of the soft frequency reuse schemes is to achieve a full frequency reuse (K = 1) while reducing the inter-cell interference at the cell boundary of the cellular OFDMA system.

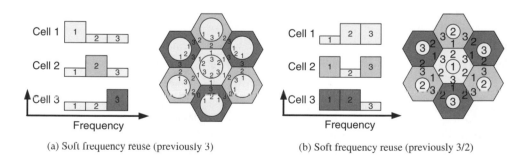

(a) Soft frequency reuse (previously 3)   (b) Soft frequency reuse (previously 3/2)

Figure 8.4 Soft frequency reuse: example with three subbands.

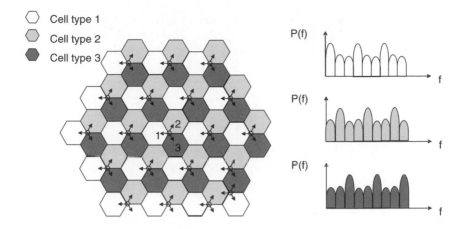

Figure 8.5   Soft frequency reuse: example with nine subbands.

8.1.3 Flexible Fractional Frequency Reuse

Assume that the whole band is divided into many subchannels in a contiguous or distributed manner. Each of these subchannels is primarily allocated to one of its adjacent cells. In other words, the whole band is divided into multiple groups, each of which is primarily reserved for one of the adjacent cells. Depending on the traffic demand in each cell, each cell can borrow some of the subchannels reserved for the adjacent cells. In the case that resource allocation information on adjacent cells is fully known, the subchannels that are not used in their primary cell can be immediately borrowed. Furthermore, those borrowed subchannels can be allocated to the users with better channel conditions, since they can use a lower power so as to reduce the inter-cell interference. The whole band is fully reused in every cell by reducing the power of the subchannels that are allocated to the users at the boundary of the adjacent cells. This approach is similar to the soft frequency reuse (SFR) scheme, but more flexible than the existing FFR and SFR schemes, and thus referred to as a flexible FFR. However, CQI information of the user and resource allocation information in adjacent cells must be known for resource borrowing and power allocation in this approach [200]. In fact, a special interface between the neighbor base stations (known as X-interface in 3GPP LTE standard) has been specified to support the current aspect of inter-cell interference coordination (ICIC) by sharing information among them.

Figure 8.6 illustrates the flexible FFR scheme for three cells, each with its own primary group of subchannels. Here, cell A, B, and C are reserved with the primary group of five subchannels, denoted as $\{A_m\}_{m=1}^{5}$, $\{B_m\}_{m=1}^{5}$, and $\{C_m\}_{m=1}^{5}$, respectively. Suppose that cell A requests more resources than $|\{A_m\}_{m=1}^{5}|$, which requires additional subchannels, B_5 and C_5, to be borrowed from other groups. The subchannels $\{A_m\}_{m=1}^{5}$ are allocated with high power to the users in low SINRs while the borrowed subchannels, B_5 and C_5, are allocated with low power to the users in high SINRs. On the other hand, suppose that cell B requests less resource than $|\{B_m\}_{m=1}^{5}|$. In this example, high power is used for the subchannels in $\{B_m\}_{m=1}^{5}$ allocated to the user in a low SINR region (i.e., a weak user). Meanwhile, low power is used for the subchannels in $\{B_m\}_{m=1}^{5}$

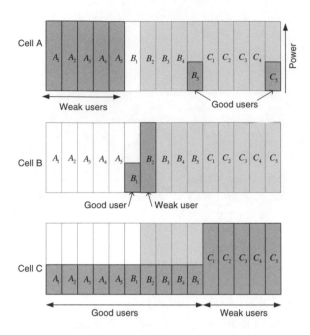

Figure 8.6   Flexible FFR with five primary subchannels in each cell.

allocated to the user in a high SINR region (i.e., a good user). Finally, it is illustrated that cell C is fully overloaded; and thus, it is using all the subchannels with lower power, that belong to the primary group of the adjacent cells, in addition to its own primary group $\{C_m\}_{m=1}^{5}$.

8.1.4 Dynamic Channel Allocation

In general, a fixed number of channels are allocated to each cell. Depending on the traffic load and time-varying channel environment, the number of channels allocated to each cell can be varied by dynamic channel allocation (DCA). DCA is known to perform better under light to moderate traffic in a time-varying channel environment. However, the implementation cost for DCA is high because it incurs a high signaling load among BSs and significant computational effort. DCA schemes can be divided into two categories: centralized DCA and distributed DCA. Although the centralized DCA provides optimal performance, a large amount of communication among BSs leads to large system latencies, which is impractical. The distributed DCA schemes, relying on the limited information shared among BSs, allocate channels by using the results on signal strength or SNR.

In OFDM-based cellular systems, DCA schemes can be used more efficiently while increasing spectral efficiency. Note that wideband channel estimation and rapid channel re-allocation are required in DCA. Since the channel information and interference signal levels on each subcarrier can be measured easily in OFDM systems, DCA schemes can be efficiently applied to the OFDM-based cellular systems. A measurement-based DCA scheme has been applied to an OFDMA-based cellular system, known as Advanced Cellular Internet Service (ACIS) [201]. In this system, interference sensing is performed at both MS and BS to avoid selecting a channel that is already in use in the adjacent cells. Also, fast channel reassignment can be implemented to avoid rapidly changing interference.

As another inter-cell interference coordination technique, Hybrid Division Duplex scheme (HDD) was proposed [202, 203]. By exploiting the advantages of both TDD and FDD, the HDD scheme is more flexible and efficient in providing asymmetric data service as well as managing inter-cell interference. The HDD scheme has a pair of frequency bands as in the FDD, performing a TDD operation using one of the bands in such a manner that allows for simultaneous FDD and TDD operations.

8.2 Inter-Cell Interference Randomization Technique

8.2.1 Cell-Specific Scrambling

As long as intra-cell and inter-cell synchronization can be maintained in the OFDM-based cellular systems, each subchannel can be considered independent due to the orthogonality among subcarriers. However, the interferences from adjacent cells may cause significant performance degradation; therefore, the interference signal can be randomized for enabling the averaging effect of the inter-cell interference. More specifically, a cell-specific scrambling code or cell-specific interleaver can be used for randomizing the interference signal [204]. Let $X^{(m)}[k]$ and $C^{(m)}[k]$ denote the transmitted signal and a scrambling code of the mth cell for subcarrier k, $m = 0, 1, 2, \ldots, M-1$. The received OFDM signal in the frequency-domain can be expressed by

$$Y[k] \approx \sum_{m=0}^{M-1} H^{(m)}[k]\, C^{(m)}[k]\, X^{(m)}[k] + Z[k] \qquad (8.1)$$

where $H^{(m)}[k]$ is the channel gain and $Z[k]$ is the additive noise for subcarrier k. In Equation (8.1), OFDM symbol index is omitted for simplicity. Assuming that the cell index $m = 0$ denotes the serving cell, Equation (8.1) can be decomposed into the desired signal and inter-cell interference components as follows:

$$Y[k] \approx H^{(0)}[k]\, C^{(0)}[k]\, X^{(0)}[k] + \sum_{m=1}^{M-1} H^{(m)}[k]\, C^{(m)}[k]\, X^{(m)}[k] + Z[k] \qquad (8.2)$$

Descrambling the received signal $Y[k]$ by the descrambling code $\left(C^{(0)}[k]\right)^*$ of the serving cell yields

$$
\begin{aligned}
Y^{(0)}[k] &\approx \left(C^{(0)}[k]\right)^* Y[k] \\
&\approx H^{(0)}[k]X^{(0)}[k] + \sum_{m=1}^{M-1} \left(C^{(0)}[k]\right)^* H^{(m)}[k]C^{(m)}[k]X^{(m)}[k] + Z[k]
\end{aligned} \qquad (8.3)
$$

where $\left|C^{(0)}[k]\right|^2 = 1$. Let us assume that $X^{(m)}[k]$ is an i.i.d. random signal with uniform distribution while $H^{(m)}[k]$ has Gaussian distribution in real and imaginary parts (i.e., the channel is Rayleigh-faded). As long as the scrambling codes are orthogonal, the second term can be approximated by an additive white Gaussian noise (AWGN). In other words, the interferences from adjacent cells have been whitened by the scrambling codes.

Figure 8.7 shows a block diagram of the cell-specific scrambling technique [205]. In this technique, the transmitted signal from each cell is multiplied by the scrambling code that is uniquely assigned to the cell. The signal may have been encoded and interleaved by a FEC block. By multiplying the received signal with the same scrambling code as the one in the transmitter, the cell-specific scrambling technique allows us to whiten the interferences from adjacent cells in unicast transmission. Unlike spreading techniques, the scrambling technique does not require to expand the bandwidth, since its spreading factor corresponds to one.

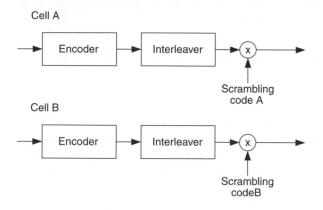

Figure 8.7   Cell-specific scrambling technique.

8.2.2 Cell-Specific Interleaving

A cell-specific interleaving technique is often referred to as Interleaved Division Multiple Access (IDMA) technique [206]. The IDMA technique is similar to the cell-specific scrambling technique for the case of single-user detection, in that it whitens the interferences from adjacent cells. The IDMA technique whitens the inter-cell interference by using a specific interleaver at each cell, while the cell-specific scrambling technique performs the same job by using a specific scrambling code. Especially when the multi-user detection technique is employed in IDMA, it can reduce inter-cell interference more effectively than the cell-specific scrambling technique, by canceling interference iteratively with multiuser detector.

Figure 8.8 shows an example of a cell-specific interleaving technique applied to a downlink of the OFDMA-based cellular system, in which we assume that MS1 in BS1 and MS2 in BS 2 share the same subchannel. BS1 and BS2 use interleaver pattern 1 and interleaver pattern 2, respectively. Each MS decodes the signal by using its own interleaver pattern of the serving BS. In this case, the interference from adjacent BSs can be approximated by AWGN.

8.2.3 Frequency-Hopping OFDMA

Consider a situation that transmits a data burst over many time slots, using a subband, which is defined as a subset of contiguous subcarriers in the OFDMA system. Using different subbands for transmitting the same data burst in different time slots, the burst error can be avoided, or

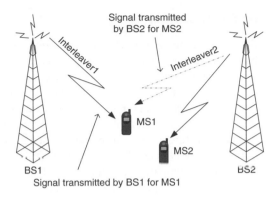

Signal transmitted
by BS2 for MS2

Interleaver1

Interleaver2

MS1

MS2

BS1   BS2

Signal transmitted by BS1 for MS1

Figure 8.8   Cell-specific interleaving technique.

frequency diversity can be obtained in frequency-selective fading channels. It is the frequency-hopping technique in the OFDMA system that makes the subband hop in the frequency domain, in accordance with a predefined frequency hopping pattern [207]. In the cellular OFDMA system, frequency hopping (FH) is a useful technique to average out inter-cell interferences when a different hopping pattern is used for each cell. In other words, it can randomize the collision between the subbands that are used in all adjacent cells. It is referred to as a frequency-hopping OFDMA (FH-OFDMA) technique. From an implementation viewpoint, FH-OFDMA technique usually requires a large memory, since all the data over a period of the hopping pattern should be buffered for decoding at the receiver.

Figure 8.9(a) illustrates two different hopping patterns for the data bursts in two adjacent cells over the consecutive time slots, denoting the burst A in one cell by a solid line and interference from the cell by a dotted line. Note that a subband for the burst A falls into deep fading at time slot 5, while it is interfered with that of the neighbor cell at time slot 3. However, it is not interfered with the subband of the neighbor cell at other time slots due to the different hopping patterns. This subband-by-subband frequency-hopping technique allows us to perform both frequency diversity and inter-cell interference randomization in the OFDMA cellular system. Figure 8.9(b) shows an example of different hopping patterns for different bursts.

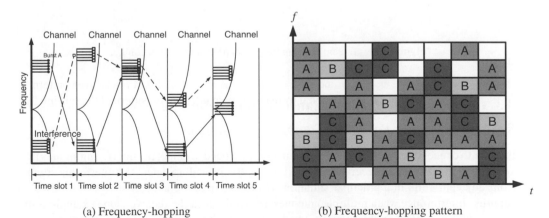

(a) Frequency-hopping   (b) Frequency-hopping pattern

Figure 8.9   Frequency-hopping OFDMA technique: an illustration.

8.2.4 Random Subcarrier Allocation

In a random subcarrier allocation technique, we define a *subchannel* as a subset of subcarriers that are spread out randomly over the whole band. Diversity gain can be achieved by randomly allocating the subcarriers in each subchannel. The subcarriers can be distributed in such a way that the probability of subcarrier collision among adjacent cells is minimal, randomizing inter-cell interference in the cellular OFDMA system. Figure 8.10 illustrates an idea of the random subcarrier allocation technique. Here, collisions occur only at one subcarrier in any two adjacent cells chosen from eight different cells, minimizing inter-cell interference. In this figure, only a few subcarriers subject to collision between cells (cell 1 and cell 4, cell 0 and cell 2, cell 3 and cell 6, etc.) are shown. Compared with the FH-OFDMA technique, the random subcarrier allocation technique shows a comparable performance, as long as the number of total subcarriers used in the random allocation technique is the same as the one used within a period of the hopping pattern in the FH technique. The random subcarrier allocation technique has an advantage of not requiring a large memory (or delay) for buffering, since data can be decoded on the unit of the OFDM symbol at the receiver. Note that subchannelization in the Mobile WiMAX system is a specific example of the random sub-carrier allocation, as detailed in Section 4.5.1 [208].

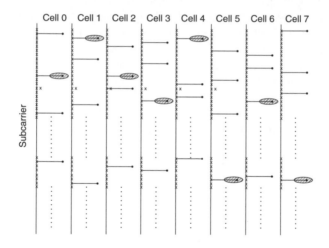

Figure 8.10   Example of random subcarrier allocation technique.

8.3 Inter-Cell Interference Cancellation Technique

8.3.1 Interference Rejection Combining Technique

In order to cancel interferences from adjacent cells, we need to detect the interfering signals first and cancel them from the received signal. It is usually difficult to detect the interfering signals from adjacent cells in a practical situation. However, spatial characteristics can be used to suppress interference when multiple antennas are available at the receiver. One example is the interference rejection combining (IRC) technique, which takes advantage of the interference statistics (correlation property of co-channel interference) received at multiple antennas [209]. The IRC technique can be viewed as a generalization of the maximum ratio combining (MRC) technique that incorporates the spatial characteristics of the received signal for interference rejection combining at the receiver.

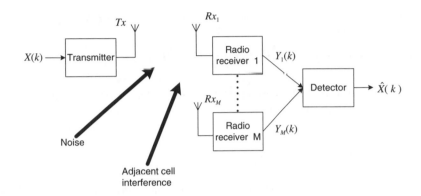

Figure 8.11   System model for interference rejection combining (IRC) technique.

As illustrated in Figure 8.11, consider the receiver with M antennas subject to the adjacent-cell interference as well as noise. Let $H_i[k]$ and $Z_i[k]$ denote the channel gain and additive noise/interference for the kth subcarrier of the ith antenna in the receiver, respectively ($i = 1, 2, \ldots, M$). We assume that $Z_i[k]$ is modeled as a zero-mean Gaussian random variable. For the transmitted signal $X[k]$, the received signal at the ith antenna is expressed as

$$Y_i[k] = H_i[k]X[k] + Z_i[k], \quad i = 1, 2, \ldots, M \tag{8.4}$$

or in a vector form,

$$\mathbf{Y}[k] = \mathbf{H}[k]X[k] + \mathbf{Z}[k] \tag{8.5}$$

where

$$\mathbf{Y}[k] = [Y_1[k]\, Y_2[k] \cdots Y_M[k]]^T$$

$$\mathbf{H}[k] = [H_1[k]\, H_2[k] \cdots H_M[k]]^T$$

$$\mathbf{Z}[k] = [Z_1[k]\, Z_2[k] \cdots Z_M[k]]^T.$$

Let $\mathbf{Q}[k]$ denote the covariance matrix of $\mathbf{Z}[k]$ (i.e., $\mathbf{Q} = E\{\mathbf{Z}[k]\mathbf{Z}^H[k]\}$). For K subcarriers within the coherence bandwidth, the covariance matrix can be approximated as

$$\hat{\mathbf{Q}} = \sum_{k=1}^{K} (\mathbf{Y}[k] - \hat{\mathbf{H}}[k]X[k]) \cdot (\mathbf{Y}[k] - \hat{\mathbf{H}}[k]X[k])^H \tag{8.6}$$

Here, the estimated channels $\hat{\mathbf{H}}[k]$ are assumed correct. The covariance matrix, estimated by using a preamble (or pilot) at the initial stage, can be used for signal detection at data transmission period as long as the channel does not vary. Assuming that all the received signals have identical probability distributions, the maximum likelihood (ML) solution for the transmitted signal is given by

$$\hat{X}[k] = \underset{X[k]}{\arg\max} \ \frac{1}{\pi^M |\hat{\mathbf{Q}}|} \ \exp\left\{ -\left(\mathbf{Y}[k] - \hat{\mathbf{H}}[k]X[k]\right)^H \hat{\mathbf{Q}}^{-1} \left(\mathbf{Y}[k] - \hat{\mathbf{H}}[k]X[k]\right) \right\}$$

$$= \underset{X[k]}{\arg\min} \ \left(\mathbf{Y}[k] - \hat{\mathbf{H}}[k]X[k]\right)^H \hat{\mathbf{Q}}^{-1} \left(\mathbf{Y}[k] - \hat{\mathbf{H}}[k]X[k]\right) \tag{8.7}$$

Practical techniques for intercell interference cancellation can be developed by approximating the ideal ML solution in Equation (8.7).

8.3.2 IDMA Multiuser Detection

In Section 8.2.3, we have discussed the IDMA technique for a single user, in which the interference at the receiver is regarded as noise. In the IDMA technique with multi-user receivers, the performance is improved by demodulating the interference signals as well as a desired signal, and detecting iteratively with *a posteriori* probability decoder [206]. Figure 8.12 shows a block diagram of the iterative multiuser detector in the OFDM-IDMA receiver.

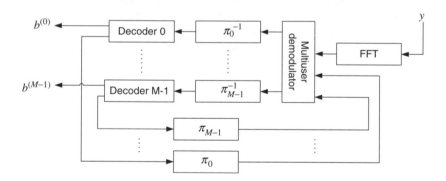

Figure 8.12   Block diagram for iterative multiuser detector in OFDM-IDMA receiver.

In the transmitter side, the coded signals are first interleaved by the cell-specific interleaver π_m, and then modulated onto subcarriers via IFFT. In the receiver side, the received signal y passes through the FFT block and then, a multiuser demodulator. In the multiuser demodulator, the soft output, channel information, and extrinsic information for each user are calculated and passed to a user decoder block, where decisions on the transmitted bit are made. The extrinsic information with a decision bit is fed back to the multiuser demodulator. The same cell-specific interleaver π_m as in the transmitter, and the corresponding cell-specific de-interleaver π_m^{-1} are used to reduce correlation between the multiuser demodulator and user decoder block. The decoded bits become more reliable as the number of iterations increases.

9

MIMO: Channel Capacity

Compared to a conventional single antenna system, the channel capacity of a multiple antenna system with N_T transmit and N_R receive antennas can be increased by the factor of $\min(N_T, N_R)$, without using additional transmit power or spectral bandwidth. Due to the ever increasing demand of faster data transmission speed in the recent or future telecommunication systems, the multiple antenna systems have been actively investigated [210, 211] and successfully deployed for the emerging broadband wireless access networks (e.g., Mobile WiMAX) [212].

Even when a wireless channel with high channel capacity is given, we still need to find good techniques to achieve high-speed data transmission or high reliability. Multiple antenna techniques can be broadly classified into two categories: diversity techniques and spatial-multiplexing techniques [213]. The diversity techniques intend to receive the same information-bearing signals in the multiple antennas or to transmit them from multiple antennas, thereby improving the transmission reliability [214, 215]. A basic idea of the diversity techniques is to convert Rayleigh fading wireless channel into more stable AWGN-like channel without any catastrophic signal fading. We will address the diversity techniques in Chapter 10. In the spatial–multiplexing techniques, on the other hand, the multiple independent data streams are simultaneously transmitted by the multiple transmit antennas, thereby achieving a higher transmission speed. We will address the spatial-multiplexing techniques in Chapter 11. When the spatial-multiplexing techniques are used, the maximum achievable transmission speed can be the same as the capacity of the MIMO channel; however, when the diversity techniques are used, the achievable transmission speed can be much lower than the capacity of the MIMO channel [216].

In this chapter, we discuss the capacity of the MIMO wireless channel. First, we address useful matrix identities that are frequently used in the expression of the corresponding capacity. In the subsequent sections, we derive the MIMO system capacities for deterministic and random channels.

9.1 Useful Matrix Theory

The matrix $\mathbf{H} \in \mathbb{C}^{N_R \times N_T}$ has a singular value decomposition (SVD), represented as

$$\mathbf{H} = \mathbf{U}\mathbf{\Sigma}\mathbf{V}^H \tag{9.1}$$

MIMO-OFDM Wireless Communications with MATLAB® Yong Soo Cho, Jaekwon Kim, Won Young Yang and Chung G. Kang
© 2010 John Wiley & Sons (Asia) Pte Ltd

where $\mathbf{U} \in \mathbb{C}^{N_R \times N_R}$ and $\mathbf{V} \in \mathbb{C}^{N_T \times N_T}$ are unitary matrices[1], and $\mathbf{\Sigma} \in \mathbb{C}^{N_R \times N_T}$ is a rectangular matrix, whose diagonal elements are non-negative real numbers and whose off-diagonal elements are zero. The diagonal elements of $\mathbf{\Sigma}$ are the singular values of the matrix $\mathbf{H}$, denoting them by $\sigma_1, \sigma_2, \cdots, \sigma_{N_{min}}$, where $N_{min} \triangleq \min(N_T, N_R)$. In fact, assume that $\sigma_1 \geq \sigma_2 \geq \ldots \geq \sigma_{N_{min}}$, that is, the diagonal elements of $\mathbf{\Sigma}$, are the ordered singular values of the matrix $\mathbf{H}$. The rank of $\mathbf{H}$ corresponds to the number of non-zero singular values (i.e., $\mathrm{rank}(\mathbf{H}) \leq N_{min}$). In case of $N_{min} = N_T$, SVD in Equation (9.1) can also be expressed as

$$\mathbf{H} = \mathbf{U}\mathbf{\Sigma}\mathbf{V}^H$$

$$= \underbrace{[\mathbf{U}_{N_{min}} \ \mathbf{U}_{N_R - N_{min}}]}_{\mathbf{U}} \underbrace{\begin{bmatrix} \mathbf{\Sigma}_{N_{min}} \\ \mathbf{0}_{N_R - N_{min}} \end{bmatrix}}_{\mathbf{\Sigma}} \mathbf{V}^H \tag{9.2}$$

$$= \mathbf{U}_{N_{min}} \mathbf{\Sigma}_{N_{min}} \mathbf{V}^H$$

where $\mathbf{U}_{N_{min}} \in \mathbb{C}^{N_R \times N_{min}}$ is composed of N_{min} left-singular vectors corresponding to the maximum possible nonzero singular values, and $\mathbf{\Sigma}_{N_{min}} \in \mathbb{C}^{N_{min} \times N_{min}}$ is now a square matrix. Since N_{min} singular vectors in $\mathbf{U}_{N_{min}}$ are of length N_R, there always exist $(N_R - N_{min})$ singular vectors such that $[\mathbf{U}_{N_{min}} \ \mathbf{U}_{N_R - N_{min}}]$ is unitary. In case of $N_{min} = N_R$, SVD in Equation (9.1) can be expressed as

$$\mathbf{H} = \mathbf{U}\underbrace{[\mathbf{\Sigma}_{N_{min}} \ \mathbf{0}_{N_T - N_{min}}]}_{\mathbf{\Sigma}} \underbrace{\begin{bmatrix} \mathbf{V}_{N_{min}}^H \\ \mathbf{V}_{N_T - N_{min}}^H \end{bmatrix}}_{\mathbf{V}^H} \tag{9.3}$$

$$= \mathbf{U}\mathbf{\Sigma}_{N_{min}} \mathbf{V}_{N_{min}}^H$$

where $\mathbf{V}_{N_{min}} \in \mathbb{C}^{N_T \times N_{min}}$ is composed of N_{min} right-singular vectors. Given SVD of $\mathbf{H}$, the following eigen-decomposition holds:

$$\mathbf{H}\mathbf{H}^H = \mathbf{U}\mathbf{\Sigma}\mathbf{\Sigma}^H\mathbf{U}^H = \mathbf{Q}\mathbf{\Lambda}\mathbf{Q}^H \tag{9.4}$$

where $\mathbf{Q} = \mathbf{U}$ such that $\mathbf{Q}^H\mathbf{Q} = \mathbf{I}_{N_R}$, and $\mathbf{\Lambda} \in \mathbb{C}^{N_R \times N_R}$ is a diagonal matrix with its diagonal elements given as

$$\lambda_i = \begin{cases} \sigma_i^2, & \text{if } i = 1, 2, \cdots, N_{min} \\ 0, & \text{if } i = N_{min} + 1, \cdots, N_R. \end{cases} \tag{9.5}$$

As the diagonal elements of $\mathbf{\Lambda}$ in Equation (9.4) are eigenvalues $\{\lambda_i\}_{i=1}^{N_R}$, Equation (9.5) indicates that the squared singular values $\{\sigma_i^2\}$ for $\mathbf{H}$ are the eigenvalues of the Hermitian symmetric matrix $\mathbf{H}\mathbf{H}^H$, or similarly, of $\mathbf{H}^H\mathbf{H}$.

---

[1] Recall that a unitary matrix $\mathbf{U}$ satisfies $\mathbf{U}^H\mathbf{U} = \mathbf{I}_{N_R}$ where $\mathbf{I}_{N_R}$ is an $N_R \times N_R$ identity matrix.

For a non-Hermitian square matrix $\mathbf{H} \in \mathbb{C}^{n \times n}$ (or non-symmetric real matrix), the eigen-decomposition is expressed as

$$\mathbf{H} \underbrace{[\mathbf{x}_1 \ \mathbf{x}_2 \cdots \mathbf{x}_n]}_{\mathbf{X}} = \underbrace{[\mathbf{x}_1 \ \mathbf{x}_2 \cdots \mathbf{x}_n]}_{\mathbf{X}} \mathbf{\Lambda}_{\text{non-}H} \tag{9.6}$$

or equivalently,

$$\mathbf{H} = \mathbf{X}\mathbf{\Lambda}_{\text{non-}H}\mathbf{X}^{-1} \tag{9.7}$$

where $\{\mathbf{x}_i\}_{i=1}^n \in \mathbb{C}^{n \times 1}$ are the right-side eigenvectors corresponding to eigenvalues in $\mathbf{\Lambda}_{\text{non-}H} \in \mathbb{C}^{n \times n}$. In Equation (9.7), linear independence of the eigenvectors is assumed. Comparing Equation (9.4) to Equation (9.7), it can be seen that the eigenvectors of a non-Hermitian matrix $\mathbf{H} \in \mathbb{C}^{n \times n}$ are not orthogonal, while those of a Hermitian matrix $\mathbf{H}\mathbf{H}^H$ are orthonormal (i.e., $\mathbf{Q}^{-1} = \mathbf{Q}^H$).

Meanwhile, the squared Frobenius norm of the MIMO channel is interpreted as a total power gain of the channel, that is,

$$\|\mathbf{H}\|_F^2 = \text{Tr}(\mathbf{H}\mathbf{H}^H) = \sum_{i=1}^{N_R} \sum_{j=1}^{N_T} |h_{i,j}|^2. \tag{9.8}$$

Using Equation (9.4), the squared Frobenius norm in Equation (9.8) can also be represented in various ways as follows:

$$\begin{aligned}
\|\mathbf{H}\|_F^2 &= \|\mathbf{Q}^H\mathbf{H}\|_F^2 \\
&= \text{Tr}(\mathbf{Q}^H\mathbf{H}\mathbf{H}^H\mathbf{Q}) \\
&= \text{Tr}(\mathbf{Q}^H\mathbf{Q}\mathbf{\Lambda}\mathbf{Q}^H\mathbf{Q}) \\
&= \text{Tr}(\mathbf{\Lambda}) \\
&= \sum_{i=1}^{N_{\min}} \lambda_i \\
&= \sum_{i=1}^{N_{\min}} \sigma_i^2
\end{aligned} \tag{9.9}$$

In deriving Equation (9.9), we have used the fact that the Frobenious norm of a matrix does not change by multiplication with a unitary matrix.

9.2 Deterministic MIMO Channel Capacity

For a MIMO system with N_T transmit and N_R receive antennas, as shown in Figure 9.1, a narrowband time-invariant wireless channel can be represented by $N_R \times N_T$ deterministic matrix $\mathbf{H} \in \mathbb{C}^{N_R \times N_T}$. Consider a transmitted symbol vector $\mathbf{x} \in \mathbb{C}^{N_T \times 1}$, which is composed of N_T independent input symbols $x_1, x_2, \cdots, x_{N_T}$. Then, the received signal $\mathbf{y} \in \mathbb{C}^{N_R \times 1}$ can be rewritten in a matrix form as follows:

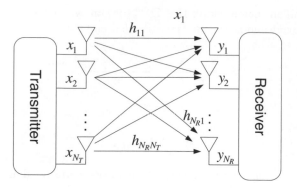

Figure 9.1 $N_R \times N_T$ MIMO system.

$$\mathbf{y} = \sqrt{\frac{E_x}{N_T}}\mathbf{H}\mathbf{x} + \mathbf{z} \qquad (9.10)$$

where $\mathbf{z} = (z_1, z_2, \cdots, z_{N_R})^T \in \mathbb{C}^{N_R \times 1}$ is a noise vector, which is assumed to be zero-mean *circular symmetric* complex Gaussian (ZMCSCG). Note that the noise vector $\mathbf{z}$ is referred to as *circular symmetric* when $e^{j\theta}\mathbf{z}$ has the same distribution as $\mathbf{z}$ for any θ. The autocorrelation of transmitted signal vector is defined as

$$\mathbf{R}_{xx} = E\{\mathbf{x}\mathbf{x}^H\}. \qquad (9.11)$$

Note that $\mathrm{Tr}(\mathbf{R}_{xx}) = N_T$ when the transmission power for each transmit antenna is assumed to be 1.

9.2.1 Channel Capacity when CSI is Known to the Transmitter Side

The capacity of a deterministic channel is defined as

$$C = \max_{f(\mathbf{x})} I(\mathbf{x}; \mathbf{y}) \text{ bits/channel use} \qquad (9.12)$$

in which $f(\mathbf{x})$ is the probability density function (PDF) of the transmit signal vector $\mathbf{x}$, and $I(\mathbf{x}; \mathbf{y})$ is the mutual information of random vectors $\mathbf{x}$ and $\mathbf{y}$. Namely, the channel capacity is the maximum mutual information that can be achieved by varying the PDF of the transmit signal vector. From the fundamental principle of the information theory, the mutual information of the two continuous random vectors, $\mathbf{x}$ and $\mathbf{y}$, is given as

$$I(\mathbf{x}; \mathbf{y}) = H(\mathbf{y}) - H(\mathbf{y}|\mathbf{x}) \qquad (9.13)$$

in which $H(\mathbf{y})$ is the differential entropy of $\mathbf{y}$ and $H(\mathbf{y}|\mathbf{x})$ is the conditional differential entropy of $\mathbf{y}$ when $\mathbf{x}$ is given. Using the statistical independence of the two random vectors $\mathbf{z}$ and $\mathbf{x}$ in Equation (9.10), we can show the following relationship:

$$H(\mathbf{y}|\mathbf{x}) = H(\mathbf{z}) \qquad (9.14)$$

Using Equation (9.14), we can express Equation (9.13) as

$$I(\mathbf{x};\mathbf{y}) = H(\mathbf{y}) - H(\mathbf{z}) \qquad (9.15)$$

From Equation (9.15), given that $H(\mathbf{z})$ is a constant, we can see that the mutual information is maximized when $H(\mathbf{y})$ is maximized. Using Equation (9.10), meanwhile, the auto-correlation matrix of $\mathbf{y}$ is given as

$$
\begin{aligned}
\mathbf{R}_{yy} = E\{\mathbf{y}\mathbf{y}^H\} &= E\left\{\left(\sqrt{\frac{E_x}{N_T}}\mathbf{H}\mathbf{x} + \mathbf{z}\right)\left(\sqrt{\frac{E_x}{N_T}}\mathbf{x}^H\mathbf{H}^H + \mathbf{z}^H\right)\right\} \\
&= E\left\{\left(\frac{E_x}{N_T}\mathbf{H}\mathbf{x}\mathbf{x}^H\mathbf{H}^H + \mathbf{z}\mathbf{z}^H\right)\right\} \\
&= \frac{E_x}{N_T}E\left\{\mathbf{H}\mathbf{x}\mathbf{x}^H\mathbf{H}^H + \mathbf{z}\mathbf{z}^H\right\} \\
&= \frac{E_x}{N_T}\mathbf{H}E\left\{\mathbf{x}\mathbf{x}^H\right\}\mathbf{H}^H + E\left\{\mathbf{z}\mathbf{z}^H\right\} \\
&= \frac{E_x}{N_T}\mathbf{H}\mathbf{R}_{xx}\mathbf{H}^H + N_0\mathbf{I}_{N_R}
\end{aligned}
\qquad (9.16)
$$

where E_x is the energy of the transmitted signals, and N_0 is the power spectral density of the additive noise $\{z_i\}_{i=1}^{N_R}$. The differential entropy $H(\mathbf{y})$ is maximized when $\mathbf{y}$ is ZMCSCG, which consequently requires $\mathbf{x}$ to be ZMCSCG as well. Then, the mutual information of $\mathbf{y}$ and $\mathbf{z}$ is respectively given as

$$
\begin{aligned}
H(\mathbf{y}) &= \log_2\left\{\det\left(\pi e\mathbf{R}_{yy}\right)\right\} \\
H(\mathbf{z}) &= \log_2\left\{\det(\pi e N_0\mathbf{I}_{N_R})\right\}
\end{aligned}
\qquad (9.17)
$$

In [217], it has been shown that using Equation (9.17), the mutual information of Equation (9.15) is expressed as

$$I(\mathbf{x};\mathbf{y}) = \log_2\det\left(\mathbf{I}_{N_R} + \frac{E_x}{N_T N_0}\mathbf{H}\mathbf{R}_{xx}\mathbf{H}^H\right) \text{ bps/Hz.} \qquad (9.18)$$

Then, the channel capacity of deterministic MIMO channel is expressed as

$$C = \max_{Tr(\mathbf{R}_{xx})=N_T} \log_2 \det\left(\mathbf{I}_{N_R} + \frac{\mathsf{E}_x}{N_T\mathsf{N}_0}\mathbf{H}\mathbf{R}_{xx}\mathbf{H}^H\right) \text{ bps/Hz.} \qquad (9.19)$$

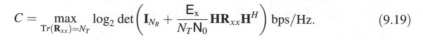

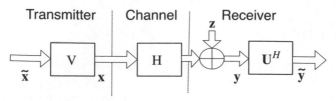

Figure 9.2 Modal decomposition when CSI is available at the transmitter side.

When channel state information (CSI) is available at the transmitter side, modal decomposition can be performed as shown in Figure 9.2, in which a transmitted signal is pre-processed with $\mathbf{V}$ in the transmitter and then, a received signal is post-processed with $\mathbf{U}^H$ in the receiver. Referring to the notations in Figure 9.2, the output signal in the receiver can be written as

$$\tilde{\mathbf{y}} = \sqrt{\frac{\mathsf{E}_x}{N_T}}\mathbf{U}^H\mathbf{H}\mathbf{V}\tilde{\mathbf{x}} + \tilde{\mathbf{z}} \qquad (9.20)$$

where $\tilde{\mathbf{z}} = \mathbf{U}^H\mathbf{z}$. Using the singular value decomposition in Equation (9.1), we can rewrite Equation (9.20) as

$$\tilde{\mathbf{y}} = \sqrt{\frac{\mathsf{E}_x}{N_T}}\mathbf{\Sigma}\tilde{\mathbf{x}} + \tilde{\mathbf{z}}$$

which is equivalent to the following r virtual SISO channels, that is,

$$\tilde{y}_i = \sqrt{\frac{\mathsf{E}_x}{N_T}}\sqrt{\lambda_i}\tilde{x}_i + \tilde{z}_i, \quad i = 1, 2, \cdots, r. \qquad (9.21)$$

The above equivalent representation can be illustrated as in Figure 9.3. If the transmit power for the ith transmit antenna is given by $\gamma_i = E\{|x_i|^2\}$, the capacity of the ith virtual SISO channel is

$$C_i(\gamma_i) = \log_2\left(1 + \frac{\mathsf{E}_x\gamma_i}{N_T\mathsf{N}_0}\lambda_i\right), \quad i = 1, 2, \cdots, r. \qquad (9.22)$$

Assume that total available power at the transmitter is limited to

$$E\{\mathbf{x}^H\mathbf{x}\} = \sum_{i=1}^{N_T} E\{|x_i|^2\} = N_T. \qquad (9.23)$$

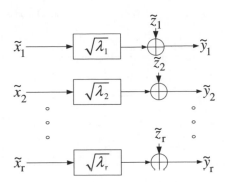

Figure 9.3 The r virtual SISO channels obtained from the modal decomposition of a MIMO channel.

The MIMO channel capacity is now given by a sum of the capacities of the virtual SISO channels, that is,

$$C = \sum_{i=1}^{r} C_i(\gamma_i) = \sum_{i=1}^{r} \log_2\left(1 + \frac{E_x\gamma_i}{N_T N_0}\lambda_i\right) \qquad (9.24)$$

where the total power constraint in Equation (9.23) must be satisfied. The capacity in Equation (9.24) can be maximized by solving the following power allocation problem:

$$C = \max_{\{\gamma_i\}} \sum_{i=1}^{r}\log_2\left(1 + \frac{E_x\gamma_i}{N_T N_0}\lambda_i\right) \qquad (9.25)$$

subject to $\sum_{i=1}^{r}\gamma_i = N_T$.

It can be shown that a solution to the optimization problem in Equation (9.25) is given as

$$\gamma_i^{opt} = \left(\mu - \frac{N_T N_0}{E_x\lambda_i}\right)^+, \quad i = 1, \cdots, r \qquad (9.26)$$

$$\sum_{i=1}^{r}\gamma_i^{opt} = N_T. \qquad (9.27)$$

where μ is a constant and $(x)^+$ is defined as

$$(x)^+ = \begin{cases} x & \text{if } x \geq 0 \\ 0 & \text{if } x < 0 \end{cases}. \qquad (9.28)$$

The above solution in Equation (9.26) satisfying the constraint in Equation (9.27) is the well-known *water-pouring* power allocation algorithm, which is illustrated in Figure 9.4 (also, refer to Section 4.2.5). It addresses the fact that more power must be allocated to the mode with

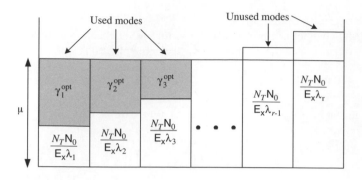

Figure 9.4   Water-pouring power allocation algorithm.

higher SNR. Furthermore, if an SNR is below the threshold given in terms of μ, the corresponding modes must not be used, that is, no power allocated to them.

9.2.2 Channel Capacity when CSI is Not Available at the Transmitter Side

When $\mathbf{H}$ is not known at the transmitter side, one can spread the energy equally among all the transmit antennas, that is, the autocorrelation function of the transmit signal vector $\mathbf{x}$ is given as

$$\mathbf{R}_{xx} = \mathbf{I}_{N_T} \tag{9.29}$$

In this case, the channel capacity is given as

$$C = \log_2 \det \left(\mathbf{I}_{N_R} + \frac{E_x}{N_T N_0} \mathbf{H}\mathbf{H}^H \right). \tag{9.30}$$

Using the eigen-decomposition $\mathbf{H}\mathbf{H}^H = \mathbf{Q}\mathbf{\Lambda}\mathbf{Q}^H$ and the identity $\det(\mathbf{I}_m + \mathbf{A}\mathbf{B}) = \det(\mathbf{I}_n + \mathbf{B}\mathbf{A})$, where $\mathbf{A} \in \mathbb{C}^{m\times n}$ and $\mathbf{B} \in \mathbb{C}^{n\times m}$, the channel capacity in Equation (9.30) is expressed as

$$C = \log_2 \det \left(\mathbf{I}_{N_R} + \frac{E_x}{N_T N_0} \mathbf{Q}\mathbf{\Lambda}\mathbf{Q}^H \right) = \log_2 \det \left(\mathbf{I}_{N_R} + \frac{E_x}{N_T N_0} \mathbf{\Lambda} \right)$$

$$= \sum_{i=1}^{r} \log_2 \left(1 + \frac{E_x}{N_T N_0} \lambda_i \right) \tag{9.31}$$

where r denotes the rank of $\mathbf{H}$, that is, $r = N_{\min} \triangleq \min(N_T, N_R)$. From Equation (9.31), we can see that a MIMO channel is converted into r virtual SISO channels with the transmit power E_x/N_T for each channel and the channel gain of λ_i for the ith SISO channel. Note that the result in Equation (9.31) is a special case of Equation (9.23) with $\gamma_i = 1$, $i = 1, 2, \ldots, r$, when CSI is not available at the transmitter and thus, the total power is equally allocated to all transmit antennas.

If we assume that the total channel gain is fixed, for example, $\|\mathbf{H}\|_F^2 = \sum_{i=1}^{r} \lambda_i = \zeta$, $\mathbf{H}$ has a full rank, $N_T = N_R = N$, and $r = N$, then the channel capacity Equation (9.31) is maximized when the singular values of $\mathbf{H}$ are the same for all (SISO) parallel channels, that is,

$$\lambda_i = \frac{\zeta}{N}, \quad i = 1, 2, \cdots, N. \tag{9.32}$$

Equation (9.32) implies that the MIMO capacity is maximized when the channel is orthogonal, that is,

$$\mathbf{H}\mathbf{H}^H = \mathbf{H}^H\mathbf{H} = \frac{\zeta}{N}\mathbf{I}_N \tag{9.33}$$

which leads its capacity to N times that of each parallel channel, that is,

$$C = N \log_2\left(1 + \frac{\zeta E_x}{N_0 N}\right). \tag{9.34}$$

9.2.3 Channel Capacity of SIMO and MISO Channels

For the case of a SIMO channel with one transmit antenna and N_R receive antennas, the channel gain is given as $\mathbf{h} \in \mathbb{C}^{N_R \times 1}$, and thus $r = 1$ and $\lambda_1 = \|\mathbf{h}\|_F^2$. Consequently, regardless of the availability of CSI at the transmitter side, the channel capacity is given as

$$C_{SIMO} = \log_2\left(1 + \frac{E_x}{N_0}\|\mathbf{h}\|_F^2\right). \tag{9.35}$$

If $|h_i|^2 = 1$, $i = 1, 2, \cdots, N_R$, and consequently $\|\mathbf{h}\|_F^2 = N_R$, the capacity is given as

$$C_{SIMO} = \log_2\left(1 + \frac{E_x}{N_0}N_R\right). \tag{9.36}$$

From Equation (9.36), we can see that the channel capacity increases logarithmically as the number of antennas increases. We can also see that only a single data stream can be transmitted and that the availability of CSI at the transmitter side does not improve the channel capacity at all.

For the case of a MISO channel, the channel gain is given as $\mathbf{h} \in \mathbb{C}^{1 \times N_T}$, thus $r = 1$ and $\lambda_1 = \|\mathbf{h}\|_F^2$. When CSI is not available at the transmitter side, the channel capacity is given as

$$C_{MISO} = \log_2 \left(1 + \frac{\mathsf{E_x}}{N_T \mathsf{N_0}} \|\mathbf{h}\|_F^2 \right). \tag{9.37}$$

If $|h_i|^2 = 1$, $i = 1, 2, \cdots, N_T$, and consequently $\|\mathbf{h}\|_F^2 = N_T$, Equation (9.37) reduces to

$$C_{MISO} = \log_2 \left(1 + \frac{\mathsf{E_x}}{\mathsf{N_0}} \right). \tag{9.38}$$

From Equation (9.38), we can see that the capacity is the same as that of a SISO channel. One might ask what the benefit of multiple transmit antennas is when the capacity is the same as that of a single transmit antenna system. Although the maximum achievable transmission speeds of the two systems are the same, there are various ways to utilize the multiple antennas, for example, the space-time coding technique, which improves the transmission reliability as will be addressed in Chapter 10.

When CSI is available at the transmitter side (i.e., $\mathbf{h}$ is known), the transmit power can be concentrated on that particular mode of the current channel. In other words, $(\mathbf{h}^H/\|\mathbf{h}\|)x$ is transmitted instead of x directly. Then the received signal can be expressed as

$$y = \sqrt{\mathsf{E_x}} \mathbf{h} \cdot \frac{\mathbf{h}^H}{\|\mathbf{h}\|} x + z = \sqrt{\mathsf{E_x}} \|\mathbf{h}\| x + z \tag{9.39}$$

Note that the received signal power has been increased by N_T times in Equation (9.39) and thus, the channel capacity is given as

$$C_{MISO} = \log_2 \left(1 + \frac{\mathsf{E_x}}{\mathsf{N_0}} \|\mathbf{h}\|_F^2 \right) = \log_2 \left(1 + \frac{\mathsf{E_x}}{\mathsf{N_0}} N_T \right). \tag{9.40}$$

9.3 Channel Capacity of Random MIMO Channels

In Section 9.2, we have assumed that MIMO channels are deterministic. In general, however, MIMO channels change randomly. Therefore, $\mathbf{H}$ is a random matrix, which means that its channel capacity is also randomly time-varying. In other words, the MIMO channel capacity can be given by its time average. In practice, we assume that the random channel is an *ergodic*[2] process. Then, we should consider the following statistical notion of the MIMO channel capacity:

$$\overline{C} = E\{C(\mathbf{H})\} = E\left\{ \max_{\mathrm{Tr}(\mathbf{R}_{xx})=N_T} \log_2 \det \left(\mathbf{I}_{N_R} + \frac{\mathsf{E_x}}{N_T \mathsf{N_0}} \mathbf{H} \mathbf{R}_{xx} \mathbf{H}^H \right) \right\} \tag{9.41}$$

which is frequently known as an ergodic channel capacity. For example, the ergodic channel capacity for the open-loop system without using CSI at the transmitter side, from Equation (9.31), is given as

---

[2] A random process is ergodic if its time average converges to the same limit for almost all realizations of the process, for example, for a discrete random process $X[n]$, $1N \sum n = 1 N X[n] \to EX[n]$ as $N \to \infty$.

$$\overline{C_{OL}} = E\left\{\sum_{i=1}^{r} \log_2\left(1 + \frac{E_x}{N_T N_0}\lambda_i\right)\right\}. \tag{9.42}$$

Similarly, the ergodic channel capacity for the closed-loop (CL) system using CSI at the transmitter side, from Equation (9.24), is given as

$$\overline{C}_{CL} = E\left\{\max_{\sum_{i=1}^{r}\gamma_i=N_T}\sum_{i=1}^{r}\log_2\left(1 + \frac{E_x}{N_T N_0}\gamma_i\lambda_i\right)\right\} \tag{9.43}$$

$$= E\left\{\sum_{i=1}^{r}\log_2\left(1 + \frac{E_x}{N_T N_0}\gamma_i^{opt}\lambda_i\right)\right\}. \tag{9.44}$$

Another statistical notion of the channel capacity is the outage channel capacity. Define the outage probability as

$$P_{out}(R) = \Pr(C(\mathbf{H}) < R) \tag{9.45}$$

In other words, the system is said to be in outage if the decoding error probability cannot be made arbitrarily small with the transmission rate of R bps/Hz. Then, the ε-outage channel capacity is defined as the largest possible data rate such that the outage probability in Equation (9.45) is less than ε. In other words, it is corresponding to C_ε such that $P(C(\mathbf{H}) \leq C_\varepsilon) = \varepsilon$.

Using Program 9.1 ("Ergodic_Capacity_CDF.m"), we can produce the cumulative distribution function (CDF) of the capacity for the random MIMO channel when CSI is not available at the transmitter side. Figure 9.5 shows the CDFs of the random 2×2 and 4×4 MIMO channel capacities when SNR is 10dB, in which $\varepsilon = 0.01$-outage capacity is indicated. It is clear

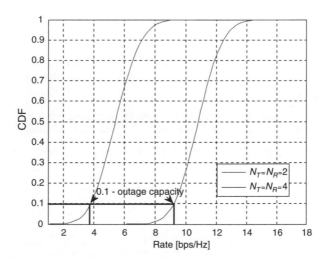

Figure 9.5 Distribution of MIMO channel capacity (SNR = 10dB; CSI is not available at the transmitter side).

from Figure 9.5 that the MIMO channel capacity improves with increasing the number of transmit and receive antennas.

MATLAB® Program: Ergodic Channel Capacity

Program 9.1 "Ergodic_Capacity_CDF.m" for ergodic capacity of MIMO channel

```
% Ergodic_Capacity_CDF.m
clear all, close all
SNR_dB=10; SNR_linear=10.^(SNR_dB/10.);
N_iter=50000; sq2=sqrt(0.5); grps = ['b:'; 'b-'];
for Icase=1:2
   if Icase==1, nT=2; nR=2; % 2x2
   else nT=4; nR=4; % 4x4
   end
   n=min(nT,nR); I = eye(n);
   for iter=1:N_iter
      H = sq2*(randn(nR,nT)+j*randn(nR,nT));
      C(iter) = log2(real(det(I+SNR_linear/nT*H'*H)));
   end
   [PDF,Rate] = hist(C,50);
   PDF = PDF/N_iter;
   for i=1:50
      CDF(Icase,i) = sum(PDF([1:i]));
   end
   plot(Rate,CDF(Icase,:),grps(Icase,:)); hold on
end
xlabel('Rate[bps/Hz]'); ylabel('CDF')
axis([1 18 0 1]); grid on; set(gca,'fontsize',10);
legend('{\it N_T}={\it N_R}=2','{\it N_T}={\it N_R}=4');
```

Using Program 9.2 ("Ergodic_Capacity_vs_SNR.m"), we can compute the ergodic capacity of the MIMO channel as SNR is varied, when CSI is not known at the transmitter side. Figure 9.6 shows the ergodic channel capacity as varying the number of antennas, under the same conditions as for Figure 9.5.

MATLAB® Program: Ergodic Channel Capacity for Various Antenna Configurations

Program 9.2 "Ergodic_Capacity_vs_SNR.m" for ergodic channel capacity vs. SNR in Figure 9.6.

```
% Ergodic_Capacity_vs_SNR.m
clear all, close all
SNR_dB=[0:5:20]; SNR_linear=10.^(SNR_dB/10);
N_iter=1000; sq2 = sqrt(0.5);
for Icase=1:5
   if Icase==1, nT=1; nR=1; % 1x1
```

```
    elseif Icase==2, nT=1; nR=2; % 1x2
    elseif Icase==3, nT=2; nR=1; % 2x1
    elseif Icase==4, nT=2; nR=2; % 2x2
    else nT=4; nR=4; % 4x4
  end
  n=min(nT,nR); I = eye(n);
  C(Icase,:) = zeros(1,length(SNR_dB));
  for iter=1:N_iter
    H = sq2*(randn(nR,nT)+j*randn(nR,nT));
    if nR>=nT, HH = H'*H; else HH = H*H'; end
    for i=1:length(SNR_dB) % Random channel generation
        C(Icase,i) = C(Icase,i)+log2(real(det(I+SNR_linear(i)/nT*HH)));
    end
  end
end
C = C/N_iter;
plot(SNR_dB,C(1,:),'b-o', SNR_dB,C(2,:),'b-', SNR_dB,C(3,:),'b-s');
hold on, plot(SNR_dB,C(4,:),'b->', SNR_dB,C(5,:),'b-^');
xlabel('SNR[dB]'); ylabel('bps/Hz');
```

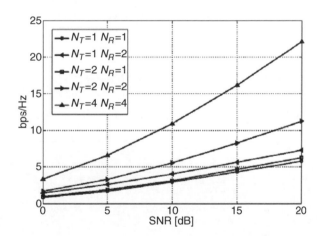

Figure 9.6 Ergodic MIMO channel capacity when CSI is not available at the transmitter.

Using Programs 9.3 ("OL_CL_Comparison.m") and Program 9.4 ("Water_Pouring"), the ergodic capacities for the closed-loop and open-loop systems are computed and compared. Figure 9.7 compares the ergodic capacities for 4×4 MIMO channels with and without using CSI at the transmitter side. It shows that the closed-loop system provides more capacity than the

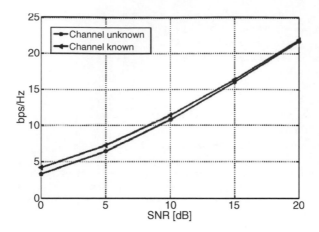

Figure 9.7 Ergodic channel capacity: $N_T = N_R = 4$.

open-loop system. However, we can see that the CSI availability does not help to improve the channel capacity when the average SNR is extremely high. It implies that even the lowest SNR mode is good enough to get almost the same transmit power allocated as the highest SNR mode, when the average SNR is extremely high.

MATLAB® Programs: Open-Loop vs. Closed-Loop MIMO Channel Capacity

Program 9.3 "OL_CL_Comparison.m" for Ergodic channel capacity: open-loop vs. closed-loop

```
%OL_CL_Comparison.m
clear all, close all;
SNR_dB=[0:5:20]; SNR_linear=10.^(SNR_dB/10.);
rho=0.2;
Rtx=[1       rho     rho^2   rho^3;
     rho     1       rho     rho^2;
     rho^2   rho     1       rho;
     rho^3   rho^2   rho     1];
Rrx=[1       rho     rho^2   rho^3;
     rho     1       rho     rho^2;
     rho^2   rho     1       rho;
     rho^3   rho^2   rho     1];
N_iter=1000;
nT=4; nR=4; n=min(nT,nR);
I = eye(n); % 4x4
```

```
sq2 = sqrt(0.5);
C_44_OL=zeros(1,length(SNR_dB));
C_44_CL=zeros(1,length(SNR_dB));
for iter=1:N_iter
    Hw = sq2*(randn(4,4) + j*randn(4,4));
    H = Rrx^(1/2)*Hw*Rtx^(1/2);
    tmp = H'*H/nT;
    SV = svd(H'*H);
    for i=1:length(SNR_dB) %random channel generation
        C_44_OL(i) = C_44_OL(i)+log2(det(I+SNR_linear(i)*tmp)); %Eq.(9.41)
        Gamma = Water_Pouring(SV,SNR_linear(i),nT);
        C_44_CL(i) = C_44_CL(i) + log2(det(I+SNR_linear(i)/nT*diag(Gamma)
                              *diag(SV))); %Eq.(9.44)
    end
end
C_44_OL = real(C_44_OL)/N_iter;
C_44_CL = real(C_44_CL)/N_iter;
plot(SNR_dB, C_44_OL,'-o', SNR_dB, C_44_CL,'-');
```

Program 9.4 "Water_Pouring" for water-pouring algorithm

```
function [Gamma]=Water_Pouring(Lamda,SNR,nT)
Gamma=zeros(1,length(Lamda));
r=length(Lamda); index=[1:r]; index_temp=index;
p=1;
while p<r
    irp=1:r-p+1; temp = sum(1./Lamda(index_temp(irp)));
    mu = nT/(r-p+1)*(1+1/SNR*temp);
    Gamma(index_temp(irp)) = mu - nT./(SNR*Lamda(index_temp(irp)));
    if min(Gamma(index_temp))<0
        i=find(Gamma==min(Gamma));
        ii=find(index_temp==i);
        index_temp2=[index_temp([1:ii-1]) index_temp([ii+1:end])];
        clear index_temp;
        index_temp=index_temp2;
        p=p+1;
        clear Gamma;
    else
        p=r;
    end
end
Gamma_t=zeros(1,length(Lamda));
Gamma_t(index_temp)=Gamma(index_temp);
Gamma=Gamma_t;
```

In general, the MIMO channel gains are not independent and identically distributed (i.i.d.). The channel correlation is closely related to the capacity of the MIMO channel. In the sequel, we consider the capacity of the MIMO channel when the channel gains between transmit and received antennas are correlated. When the SNR is high, the deterministic channel capacity can be approximated as

$$C \approx \max_{\mathrm{Tr}(\mathbf{R}_{xx})=N} \log_2 \det(\mathbf{R}_{xx}) + \log_2 \det\left(\frac{\mathsf{E}_x}{N N_0} \mathbf{H}_w \mathbf{H}_w^H\right) \tag{9.46}$$

From Equation (9.46), we can see that the second term is constant, while the first term involving $\det(\mathbf{R}_{xx})$ is maximized when $\mathbf{R}_{xx} = I_N$. Consider the following correlated channel model:

$$\mathbf{H} = \mathbf{R}_r^{1/2} \mathbf{H}_w \mathbf{R}_t^{1/2} \tag{9.47}$$

where $\mathbf{R}_t$ is the correlation matrix, reflecting the correlations between the transmit antennas (i.e., the correlations between the column vectors of $\mathbf{H}$), $\mathbf{R}_r$ is the correlation matrix reflecting the correlations between the receive antennas (i.e., the correlations between the row vectors of $\mathbf{H}$), and $\mathbf{H}_w$ denotes the i.i.d. Rayleigh fading channel gain matrix. The diagonal entries of $\mathbf{R}_t$ and $\mathbf{R}_r$ are constrained to be a unity. From Equation (9.30), then, the MIMO channel is given as

$$C = \log_2 \det\left(\mathbf{I}_{N_R} + \frac{\mathsf{E}_x}{N_T N_0} \mathbf{R}_r^{1/2} \mathbf{H}_w \mathbf{R}_t \mathbf{H}_w^H \mathbf{R}_r^{H/2}\right). \tag{9.48}$$

If $N_T = N_R = N$, $\mathbf{R}_r$ and $\mathbf{R}_t$ are of full rank, and SNR is high, Equation (9.48) can be approximated as

$$C \approx \log_2 \det\left(\frac{\mathsf{E}_x}{N_T N_0} \mathbf{H}_w \mathbf{H}_w^H\right) + \log_2 \det(\mathbf{R}_r) + \log_2 \det(\mathbf{R}_t). \tag{9.49}$$

We find from Equation (9.49) that the MIMO channel capacity has been reduced, and the amount of capacity reduction (in bps) due to the correlation between the transmit and receive antennas is

$$\log_2 \det(\mathbf{R}_r) + \log_2 \det(\mathbf{R}_t). \tag{9.50}$$

In the sequel, it is shown that the value in Equation (9.50) is always negative by the fact that $\log_2 \det(\mathbf{R}) \leq 0$ for any correlation matrix $\mathbf{R}$. Since $\mathbf{R}$ is a symmetric matrix, eigen-decomposition in Equation (9.2) is applicable, that is, $\mathbf{R} = \mathbf{Q}\mathbf{\Lambda}\mathbf{Q}^H$. Since the determinant of a unitary matrix is unity, the determinant of a correlation matrix can be expressed as

$$\det(\mathbf{R}) = \prod_{i=1}^{N} \lambda_i. \tag{9.51}$$

Note that the geometric mean is bounded by the arithmetic mean, that is,

$$\left(\Pi_{i=1}^{N}\lambda_i\right)^{\frac{1}{N}} \leq \frac{1}{N}\sum_{i=1}^{N}\lambda_i = 1. \tag{9.52}$$

From Equations (9.51) and (9.52), it is obvious that

$$\log_2 \det(\mathbf{R}) \leq 0 \tag{9.53}$$

The equality in Equation (9.53) holds when the correlation matrix is the identity matrix. Therefore, the quantities in Equation (9.50) are all negative.

Program 9.5 computes the ergodic MIMO channel capacity when there exists a correlation between the transmit and receive antennas, with the following channel correlation matrices: $\mathbf{R}_r = \mathbf{I}_4$ and

$$\mathbf{R}_t = \begin{bmatrix} 1 & 0.76e^{j0.17\pi} & 0.43e^{j0.35\pi} & 0.25e^{j0.53\pi} \\ 0.76e^{-j0.17\pi} & 1 & 0.76e^{j0.17\pi} & 0.43e^{j0.35\pi} \\ 0.43e^{-j0.35\pi} & 0.76e^{-j0.17\pi} & 1 & 0.76e^{j0.17\pi} \\ 0.25e^{-j0.53\pi} & 0.43e^{-j0.35\pi} & 0.76e^{-j0.17\pi} & 1 \end{bmatrix} \tag{9.54}$$

$\mathbf{R}_r = \mathbf{I}_4$ states that no correlation exists between the receive antennas. Figure 9.8 has been generated by Program 9.5, from which it can be shown that a capacity of 3.3 bps/Hz is lost due to the channel correlation when SNR is 18dB.

MATLAB® Program: Ergodic MIMO Capacity for Correlated Channel

Program 9.5 "Ergodic_Capacity_Correlation.m:" Channel capacity reduction due to correlation

```
% Ergodic_Capacity_Correlation.m
% Capacity reduction due to correlation of MIMO channels (Fig. 9.8)
clear all, close all;
SNR_dB=[0:5:20]; SNR_linear=10.^(SNR_dB/10.);
N_iter=1000; N_SNR=length(SNR_dB);
nT=4; nR=4; n=min(nT,nR); I = eye(n); sq2=sqrt(0.5); % 4x
R=[1 0.76*exp(0.17j*pi) 0.43*exp(0.35j*pi) 0.25*exp(0.53j*pi);
    0.76*exp(-0.17j*pi) 1 0.76*exp(0.17j*pi) 0.43*exp(0.35j*pi);
    0.43*exp(-0.35j*pi) 0.76*cxp(-0.17j*pi) 1 0.76*exp(0.17j*pi);
    0.25*exp(-0.53j*pi) 0.43*exp(-0.35j*pi) 0.76*exp(-0.17j*pi) 1]; %(9.54)
C_44_iid=zeros(1,N_SNR); C_44_corr=zeros(1,N_SNR);
```

```
for iter=1:N_iter
    H_iid = sq2*(randn(nR,nT)+j*randn(nR,nT));
    H_corr = H_iid*R^(1/2);
    tmp1 = H_iid'*H_iid/nT; tmp2 = H_corr'*H_corr/nT;
    for i=1:N_SNR % Eq.(9.48)
        C_44_iid(i) = C_44_iid(i) + log2(det(I+SNR_linear(i)*tmp1));
        C_44_corr(i) = C_44_corr(i) + log2(det(I+SNR_linear(i)*tmp2));
    end
end
C_44_iid = real(C_44_iid)/N_iter;
C_44_corr = real(C_44_corr)/N_iter;
plot(SNR_dB,C_44_iid, SNR_dB,C_44_corr,':');
```

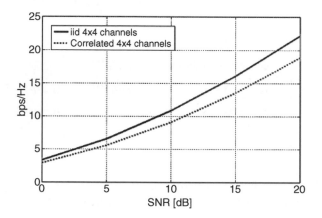

Figure 9.8   Capacity reduction due to the channel correlation.

10

Antenna Diversity and Space-Time Coding Techniques

In Chapter 9, we have mentioned that the multiple antenna techniques can be broadly classified into two categories: spatial multiplexing techniques or diversity techniques. In this chapter, we will study the basic concepts of antenna diversity techniques. For AWGN channel, the slope of BER versus SNR curve for AWGN channel goes to infinity as SNR becomes large, that is, showing a water-falling type of BER performance as SNR increases. For a Rayleigh fading wireless channel, however, the corresponding slope is linear in the log-log scale. It implies that the transmission performance over the Rayleigh fading wireless channel is significantly degraded, even at high SNR. The fundamental goal of the antenna diversity techniques is to convert an unstable time-varying wireless fading channel into a stable AWGN-like channel without significant instantaneous fading, thereby steepening the BER versus SNR curve.

Among many different types of antenna diversity techniques, transmit diversity techniques have been widely adopted in practice, since it is useful in reducing the processing complexity of the receiver. Furthermore, it requires multiple antennas only on the transmitter side. In this chapter, we will focus on the space-time coding techniques that are used for achieving the antenna diversity gain.

10.1 Antenna Diversity

Diversity techniques are used to mitigate degradation in the error performance due to unstable wireless fading channels, for example, subject to the multipath fading [218, 219]. Diversity in data transmission is based on the following idea: The probability that multiple statistically independent fading channels simultaneously experience deep fading is very low. There are various ways of realizing diversity gain, including the following ones:

- **Space diversity:** sufficiently separated (more than 10λ) multiple antennas are used to implement independent wireless channels.

MIMO-OFDM Wireless Communications with MATLAB® Yong Soo Cho, Jaekwon Kim, Won Young Yang and Chung G. Kang
© 2010 John Wiley & Sons (Asia) Pte Ltd

- **Polarization diversity:** independent channels are implemented using the fact that vertically and horizontally polarized paths are independent.
- **Time diversity:** same information is repeatedly transmitted at sufficiently separated (more than coherence time) time instances.
- **Frequency diversity:** same information is repeatedly transmitted at sufficiently separated (more than coherence bandwidth) frequency bands.
- **Angle diversity:** multiple receive antennas with different directivity are used to receive the same information-bearing signal at different angles.

Time, frequency and spatial diversity techniques are illustrated in Figure 10.1. In time diversity, data is transmitted over multiple time slots. In frequency diversity, the same data is transmitted at multiple spectral bands to achieve diversity gain. As shown in Figure 10.1(a) and (b), time diversity and frequency diversity techniques require additional time resource and frequency resource, respectively. However, antenna or space diversity techniques do not require any additional time or frequency resource. Figure 10.1(c) illustrates a concept of the space-time diversity that employs multiple transmit antennas, not requiring additional time

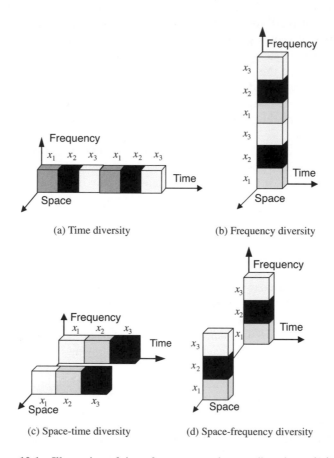

Figure 10.1   Illustration of time, frequency, and space diversity techniques.

resource as opposed to one in Figure 10.1(a). Similarly, Figure 10.1(d) illustrates a concept of the space-frequency diversity that employs multiple transmit antennas, which do not require additional frequency resource as opposed to the one in Figure 10.1(b). Although two transmit antennas are illustrated for the antenna diversity in Figure 10.1, the concept can be extended to various antenna configurations. Some examples of single input multiple output (SIMO), multiple input single output (MISO), and multiple input multiple output (MIMO) antenna configurations are illustrated in Figure 10.2.

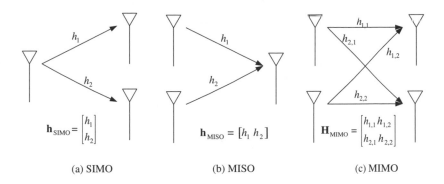

(a) SIMO   (b) MISO   (c) MIMO

Figure 10.2   Examples of various antenna configurations.

10.1.1 Receive Diversity

Consider a receive diversity system with N_R receiver antennas. Assuming a single transmit antenna as in the single input multiple output (SIMO) channel of Figure 10.2, the channel is expressed as

$$\mathbf{h} = [h_1 h_2 \cdots h_{N_R}]^T \tag{10.1}$$

for N_R independent Rayleigh fading channels. Let x denote the transmitted signal with the unit variance in the SIMO channel. The received signal $\mathbf{y} \in \mathbb{C}^{N_R \times 1}$ is expressed as

$$\mathbf{y} = \sqrt{\frac{E_x}{N_0}}\mathbf{h}x + \mathbf{z} \tag{10.2}$$

where $\mathbf{z}$ is ZMCSCG noise with $E\{\mathbf{z}\mathbf{z}^H\} = \mathbf{I}_{N_R}$. The received signals in the different antennas can be combined by various techniques. These combining techniques include selection combining (SC), maximal ratio combining (MRC), and equal gain combing (EGC). In SC, the received signal with the highest SNR among N_R branches is selected for decoding. Let γ_i be the instantaneous SNR for the ith branch, which is given as

$$\gamma_i = |h_i|^2 \frac{E_x}{N_0}, \quad i = 1, 2, \ldots, N_R. \tag{10.3}$$

Then the average SNR for SC is given as

$$\rho_{SC} = E\left\{ \max_i \left(|h_i|^2 \right) \right\} \cdot \frac{E_x}{N_0}, \quad i = 1, 2, \ldots, N_R \tag{10.4}$$

In MRC, all N_R branches are combined by the following weighted sum:

$$y_{MRC} = \underbrace{\left[w_1^{(MRC)} w_2^{(MRC)} \cdots w_{N_R}^{(MRC)} \right]}_{\mathbf{w}_{MRC}^T} \mathbf{y} = \sum_{i=1}^{N_R} w_i^{(MRC)} y_i \tag{10.5}$$

where $\mathbf{y}$ is the received signal in Equation (10.2) and $\mathbf{w}_{MRC}$ is the weight vector. As $y_i = \sqrt{E_s/N_0}\, h_i x + z_i$ from Equation (10.2), the combined signal can be decomposed into the signal and noise parts, that is,

$$
\begin{aligned}
y_{MRC} &= \mathbf{w}_{MRC}^T \left(\sqrt{\frac{E_x}{N_0}} \mathbf{h} x + \mathbf{z} \right) \\
&= \sqrt{\frac{E_x}{N_0}} \mathbf{w}_{MRC}^T \mathbf{h} x + \mathbf{w}_{MRC}^T \mathbf{z}
\end{aligned}
\tag{10.6}
$$

Average power of the instantaneous signal part and that of the noise part in Equation (10.6) are respectively given as

$$P_s = E\left\{ \left| \sqrt{\frac{E_x}{N_0}} \mathbf{w}_{MRC}^T \mathbf{h} x \right|^2 \right\} = \frac{E_x}{N_0} E\left\{ |\mathbf{w}_{MRC}^T \mathbf{h} x|^2 \right\} = \frac{E_x}{N_0} |\mathbf{w}_{MRC}^T \mathbf{h}|^2 \tag{10.7}$$

and

$$P_z = E\left\{ |\mathbf{w}_{MRC}^T \mathbf{z}|^2 \right\} = \left\| \mathbf{w}_{MRC}^T \right\|_2^2 \tag{10.8}$$

From Equations (10.7) and (10.8), the average SNR for the MRC is given as

$$\rho_{MRC} = \frac{P_s}{P_z} = \frac{E_x}{N_0} \frac{|\mathbf{w}_{MRC}^T \mathbf{h}|^2}{\left\| \mathbf{w}_{MRC}^T \right\|_2^2} \tag{10.9}$$

Invoking the Cauchy-Schwartz inequality,

$$\left| \mathbf{w}_{MRC}^T \mathbf{h} \right|^2 \leq \left\| \mathbf{w}_{MRC}^T \right\|_2^2 \|\mathbf{h}\|_2^2 \tag{10.10}$$

Equation (10.9) is upper-bounded as

$$\rho_{MRC} = \frac{E_x}{N_0} \frac{|\mathbf{w}_{MRC}^T \mathbf{h}|^2}{\|\mathbf{w}_{MRC}^T\|_2^2}$$

$$\leq \frac{E_x}{N_0} \frac{\|\mathbf{w}_{MRC}^T\|_2^2 \|\mathbf{h}\|_2^2}{\|\mathbf{w}_{MRC}^T\|_2^2} = \frac{E_x}{N_0} \|\mathbf{h}\|_2^2 \tag{10.11}$$

Note that the SNR in Equation (10.11) is maximized at $\mathbf{w}_{MRC} = \mathbf{h}^*$, which yields $\rho_{MRC} = E_x \|\mathbf{h}\|_2^2 / N_0$. In other words, the weight factor of each branch in Equation (10.5) must be matched to the corresponding channel for maximal ratio combining (MRC). Equal gain combining (EGC) is a special case of MRC in the sense that all signals from multiple branches are combined with equal weights. In fact, MRC achieves the best performance, maximizing the post-combining SNR. Running Program 10.1 ("MRC_scheme.m") yields Figure 10.3, which shows that the performance improves with the number of receiving antennas.

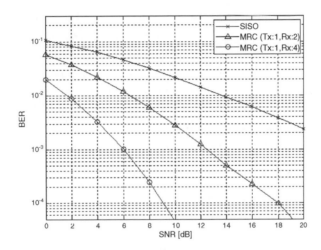

Figure 10.3 The performance of MRC for Rayleigh fading channels.

MATLAB® Programs: MRC Performance for a Rayleigh Fading Channel

Program 10.1 "MRC_scheme.m" for performance of MRC for Rayleigh fading channels

```
% MRC_scheme.m
clear, clf
N_frame=130; N_packet=4000;
b=2;    % Set to 1/2/3/4 for BPSK/QPSK/16QAM/64QAM
NT=1;   NR=2;   % Numbers of Tx/Rx antennas
SNRdBs=[0:2:20]; sq_NT=sqrt(NT);   sq2-sqrt(2);
for i_SNR=1:length(SNRdBs)
  SNRdB=SNRdBs(i_SNR);   sigma=sqrt(0.5/(10^(SNRdB/10)));
```

```
for i_packet=1:N_packet
  msg_symbol=randint(N_frame*b,NT);
  [temp,sym_tab,P]=modulator(msg_symbol.',b);
  X=temp.';
  Hr = (randn(N_frame,NR)+j*randn(N_frame,NR))/sq2;
  H= reshape(Hr,N_frame,NR); Habs=sum(abs(H).^2,2); Z=0;
  for i=1:NR
      R(:,i) = sum(H(:,i).*X,2)/sq_NT ...
                + sigma*(randn(N_frame,1)+j*randn(N_frame,1));
      Z = Z + R(:,i).*conj(H(:,i));
  end
  for m=1:P
    d1(:,m) = abs(sum(Z,2)-sym_tab(m)).^2 ...
                + (-1+sum(Habs,2))*abs(sym_tab(m))^2;
  end
  [y1,i1] = min(d1,[],2);   Xd=sym_tab(i1).';
  temp1 = X>0;    temp2 = Xd>0;
  noeb_p(i_packet) = sum(sum(temp1~=temp2));
end
BER(i_SNR) = sum(noeb_p)/(N_packet*N_frame*b);
end % end of FOR loop for SNR
semilogy(SNRdBs,BER), grid on, axis([SNRdBs([1 end]) 1e-6 1e0])
```

Program 10.2 "modulator" for BPSK, QPSK, 8-PSK, 16-QAM mapping function

```
function [mod_symbols,sym_table,M]=modulator(bitseq,b)
N_bits=length(bitseq);sq10=sqrt(10);
if b==1        % BPSK modulation
  sym_table=exp(j*[0 -pi]); sym_table=sym_table([1 0]+1);
  inp=bitseq; mod_symbols=sym_table(inp+1); M=2;
elseif b==2   % QPSK modulation
  sym_table=exp(j*pi/4*[-3 3 1 -1]);sym_table=sym_table([0 1 3 2]+1);
inp=reshape(bitseq,b,N_bits/b);
  mod_symbols=sym_table([2 1]*inp+1);   M=4;
elseif b==3   % generates 8-PSK symbols
  sym_table=exp(j*pi/4*[0:7]);
sym_table=sym_table([0 1 3 2 6 7 5 4]+1);
  inp=reshape(bitseq,b,N_bits/b);
mod_symbols=sym_table([4 2 1]*inp+1);   M=8;
elseif b==4 % 16-QAM modulation
  m=0;
  for k=-3:2:3 % Power normalization
    for l=-3:2:3,   m=m+1; sym_table(m)=(k+j*l)/sq10;   end
  end
sym_table=sym_table([0 1 3 2 4 5 7 6 12 13 15 14 8 9 11 10]+1); inp=reshape
  (bitseq,b,N_bits/b);
mod_symbols=sym_table([8 4 2 1]*inp+1); M=16; %16-ary symbol sequence
else   error('Unimplemented modulation');end
```

10.1.2 Transmit Diversity

A critical drawback of receive diversity is that most of computational burden is on the receiver side, which may incur high power consumption for mobile units in the case of downlink. Diversity gain can also be achieved by space-time coding (STC) at the transmit side, which requires only simple linear processing in the receiver side for decoding. In order to further reduce the computational complexity in mobile units, differential space-time codes can be used, which do not require CSI estimation at the receiver side [219–221].

10.2 Space-Time Coding (STC): Overview

In this section, we provide a mathematical description of space-time coded systems. Based on the mathematical model, a pairwise error probability is derived. Finally, a space-time code design criterion is described by using a pairwise error probability. Specific examples of STC are provided in the subsequent discussion.

10.2.1 System Model

Figure 10.4 illustrates space-time-coded MIMO systems with N_T transmit antennas and N_R receive antennas. In the space-time coded MIMO systems, bit stream is mapped into symbol stream $\{\tilde{x}_i\}_{i=1}^{N}$. As depicted in Figure 10.4, a symbol stream of size N is space-time-encoded into $\{x_i^{(t)}\}_{i=1}^{N_T}$, $t = 1, 2, \ldots, T$, where i is the antenna index and t is the symbol time index. Note that the number of symbols in a space-time codeword is $N_T \cdot T$ (i.e., $N = N_T \cdot T$). In other words, $\{x_i^{(t)}\}_{i=1}^{N_T}$, $t = 1, 2, \ldots, T$, forms a space-time codeword. As N symbols are transmitted by a codeword over T symbol times, the symbol rate of the space-time-coded system example in Figure 10.4 is given as

$$R = \frac{N}{T} \quad \text{[symbols/channel use]} \tag{10.12}$$

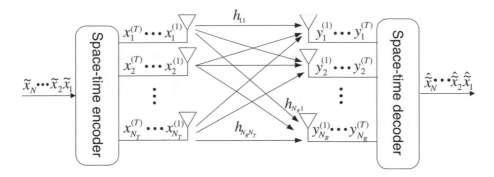

Figure 10.4 Space-time coded MIMO systems.

At the receiver side, the symbol stream $\{\tilde{x}_i\}_{i=1}^{N}$ is estimated by using the received signals $\{y_j^{(t)}\}_{j=1}^{N_R}$, $t = 1, 2, \ldots, T$. Let $h_{ji}^{(t)}$ denote the Rayleigh-distributed channel gain from the ith transmit antenna to the jth receive antenna over the tth symbol period ($i = 1, 2, \ldots, N_T$,

$j = 1, 2, \ldots, N_R$, and $t = 1, 2, \ldots, T$). If we assume that the channel gains do not change during T symbol periods, the symbol time index can be omitted. Furthermore, as long as the transmit antennas and receive antennas are spaced sufficiently apart, $N_R \times N_T$ fading gains $\{h_{ji}^{(t)}\}$ can be assumed to be statistically independent. If $x_i^{(t)}$ is the transmitted signal from the ith transmit antenna during tth symbol period, the received signal at the jth receive antenna during tth symbol period is

$$
y_j^{(t)} = \sqrt{\frac{E_x}{N_0 N_T}} \left[h_{j1}^{(t)} \; h_{j2}^{(t)} \cdots h_{jN_T}^{(t)} \right] \underbrace{\begin{bmatrix} x_1^{(t)} \\ x_2^{(t)} \\ \vdots \\ x_{N_T}^{(t)} \end{bmatrix}}_{\mathbf{x}^{(t)}} + z_j^{(t)} \tag{10.13}
$$

where $z_j^{(t)}$ is the noise process at the jth receive antenna during tth symbol period, which is modeled as the ZMCSCG noise of unit variance, and E_x is the average energy of each transmitted signal. Meanwhile, the total transmitted power is constrained as

$$
\sum_{i=1}^{N_T} E\left\{ \left| x_i^{(t)} \right|^2 \right\} = N_T, \quad t = 1, 2, \ldots, T \tag{10.14}
$$

Note that when variance is assumed to be 0.5 for real and imaginary parts of h_{ji}, the PDF of each channel gain is given as

$$
\begin{aligned}
f_{H_{ji}}(h_{ji}) &= f_{H_{ji}}\left(\mathrm{Re}\{h_{ji}\}, \mathrm{Im}\{h_{ji}\}\right) \\
&= \frac{1}{\sqrt{\pi}} \exp\left(-\left|\mathrm{Re}\{h_{ji}\}\right|^2\right) \cdot \frac{1}{\sqrt{\pi}} \exp\left(-\left|\mathrm{Im}\{h_{ji}\}\right|^2\right) \\
&= \frac{1}{\pi} \exp\left(-\left|\mathrm{Re}\{h_{ji}\}\right|^2 - \left|\mathrm{Im}\{h_{ji}\}\right|^2\right) \\
&= \frac{1}{\pi} \exp\left(-\left|h_{ji}\right|^2\right).
\end{aligned} \tag{10.15}
$$

In a similar manner, the PDF of the additive noise can be expressed as

$$
f_{Z_j^{(t)}}\left(z_j^{(t)}\right) = \frac{1}{\pi} \exp\left(-\left|z_j^{(t)}\right|^2\right) \tag{10.16}
$$

Considering the relationship in Equation (10.13) during a period of T symbols for the jth receive antenna, we have

$$
\left[y_1^{(t)} \; y_j^{(t)} \cdots y_j^{(t)} \right] = \sqrt{\frac{E_x}{N_0 N_T}} \left[h_{j1}^{(t)} \; h_{j2}^{(t)} \cdots h_{jN_T}^{(t)} \right] \underbrace{\begin{bmatrix} x_1^{(1)} & x_1^{(2)} & & x_1^{(T)} \\ x_2^{(1)} & x_2^{(2)} & \cdots & x_2^{(2)} \\ \vdots & \vdots & & \vdots \\ x_{N_T}^{(1)} & x_{N_T}^{(2)} & & x_{N_T}^{(T)} \end{bmatrix}}_{\mathbf{X}} + \left[z_j^{(1)} \; z^{(2)} \cdots z_j^{(T)} \right]
$$

$$\tag{10.17}$$

Again considering the relationship in Equation (10.17) for N_R receive antennas, while assuming quasi-static channel gains (i.e., $h_{ji}^{(t)} = h_{ji}, t = 1, 2, \cdots, T$), the system equation is given as

$$
\underbrace{\begin{bmatrix} y_1^{(1)} & y_1^{(2)} & \cdots & y_1^{(T)} \\ y_2^{(1)} & y_2^{(2)} & \cdots & y_1^{(T)} \\ \vdots & \vdots & \ddots & \vdots \\ y_{N_R}^{(1)} & y_{N_R}^{(2)} & \cdots & y_{N_R}^{(T)} \end{bmatrix}}_{\mathbf{Y}} = \sqrt{\frac{E_x}{N_0 N_T}} \underbrace{\begin{bmatrix} h_{11} & h_{12} & \cdots & h_{1N_T} \\ h_{21} & h_{22} & \cdots & h_{2N_T} \\ \vdots & \vdots & \ddots & \vdots \\ h_{N_R 1} & h_{N_R 2} & \cdots & h_{N_R N_T} \end{bmatrix}}_{\mathbf{H}} \underbrace{\begin{bmatrix} x_1^{(1)} & x_1^{(2)} & & x_1^{(T)} \\ x_2^{(1)} & x_2^{(2)} & \cdots & x_2^{(T)} \\ \vdots & \vdots & & \vdots \\ x_{N_T}^{(1)} & x_{N_T}^{(2)} & & x_{N_T}^{(T)} \end{bmatrix}}_{\mathbf{X}}
$$

$$
+ \underbrace{\begin{bmatrix} z_1^{(1)} & z_1^{(2)} & \cdots & z_1^{(T)} \\ z_2^{(1)} & z_2^{(2)} & \cdots & z_2^{(T)} \\ \vdots & \vdots & \ddots & \vdots \\ z_{N_R}^{(1)} & z_{N_R}^{(2)} & \cdots & z_{N_R}^{(T)} \end{bmatrix}}_{\mathbf{Z}} \tag{10.18}
$$

Based on the above relationship, a pairwise error probability is derived in the next section.

10.2.2 Pairwise Error Probability

Assuming that CSI is exactly known at the receiver side and the noise components are independent, the conditional PDF of the received signal in Equation (10.18) is given as

$$
\begin{aligned}
f_Y(\mathbf{Y}|\mathbf{H}, \mathbf{X}) &= f_Z(\mathbf{Z}) \\
&= \prod_{j=1}^{N_R} \prod_{t=1}^{T} \frac{1}{\pi} \exp\left(-\left|z_j^{(t)}\right|^2\right) \\
&= \frac{1}{\pi^{N_R T}} \exp\left(-\sum_{j=1}^{N_R} \sum_{t-1}^{T} \left|z_j^{(t)}\right|^2\right) \\
&= \frac{1}{\pi^{N_R T}} \exp\left(-\|\mathbf{Z}\|_F^2\right) \\
&= \frac{1}{\pi^{N_R T}} \exp\left(-\mathrm{tr}\left(\mathbf{Z}\mathbf{Z}^H\right)\right).
\end{aligned} \tag{10.19}
$$

Using the above conditional PDF, the ML codeword $\mathbf{X}_{ML}$ can be found by maximizing Equation (10.19), that is,

$$
\begin{aligned}
\mathbf{X}_{ML} &= \arg\max_{\mathbf{X}} f_Y(\mathbf{Y}|\mathbf{H}, \mathbf{X}) \\
&= \arg\max_{\mathbf{X}} \frac{1}{\pi^{N_R T}} \exp\left(-\mathrm{tr}\left[\left(\mathbf{Y} - \sqrt{\frac{E_x}{N_0 N_T}}\mathbf{H}\mathbf{X}\right)\left(\mathbf{Y} - \sqrt{\frac{E_x}{N_0 N_T}}\mathbf{H}\mathbf{X}\right)^H\right]\right) \\
&= \arg\min_{\mathbf{X}} \mathrm{tr}\left[\left(\mathbf{Y} - \sqrt{\frac{E_x}{N_0 N_T}}\mathbf{H}\mathbf{X}\right)\left(\mathbf{Y} - \sqrt{\frac{E_x}{N_0 N_T}}\mathbf{H}\mathbf{X}\right)^H\right]
\end{aligned} \tag{10.20}
$$

Note that the detected symbol $\mathbf{X}$ is erroneous (i.e., $\mathbf{X}_{ML} \neq \mathbf{X}$) when the following condition is satisfied:

$$\mathrm{tr}\left[\left(\mathbf{Y}-\sqrt{\frac{E_x}{N_0 N_T}}\mathbf{H}\mathbf{X}\right)\left(\mathbf{Y}-\sqrt{\frac{E_x}{N_0 N_T}}\mathbf{H}\mathbf{X}\right)^H\right]$$
$$\geq \mathrm{tr}\left[\left(\mathbf{Y}-\sqrt{\frac{E_x}{N_0 N_T}}\mathbf{H}\mathbf{X}_{ML}\right)\left(\mathbf{Y}-\sqrt{\frac{E_x}{N_0 N_T}}\mathbf{H}\mathbf{X}_{ML}\right)^H\right] \quad (10.21)$$

The above condition of error can be re-expressed as

$$\mathrm{tr}[\mathbf{Z}\mathbf{Z}^H] \geq \mathrm{tr}\left[\left(\mathbf{Y}-\sqrt{\frac{E_x}{N_0 N_T}}\mathbf{H}\mathbf{X}_{ML}\right)\left(\mathbf{Y}-\sqrt{\frac{E_x}{N_0 N_T}}\mathbf{H}\mathbf{X}_{ML}\right)^H\right]$$
$$= \mathrm{tr}\left[\left(\sqrt{\frac{E_x}{N_0 N_T}}\mathbf{H}(\mathbf{X}-\mathbf{X}_{ML})+\mathbf{Z}\right)\left(\sqrt{\frac{E_x}{N_0 N_T}}\mathbf{H}(\mathbf{X}-\mathbf{X}_{ML})+\mathbf{Z}\right)^H\right]$$
$$= \mathrm{tr}\left[\frac{E_x}{N_0 N_T}\mathbf{H}(\mathbf{X}-\mathbf{X}_{ML})(\mathbf{X}-\mathbf{X}_{ML})^H\mathbf{H}^H\right] \quad (10.22)$$
$$+ \mathrm{tr}\left[2\mathrm{Re}\left\{\sqrt{\frac{E_x}{N_0 N_T}}\mathbf{H}(\mathbf{X}-\mathbf{X}_{ML})\mathbf{Z}^H\right\}\right] + \mathrm{tr}\left[\mathbf{Z}\mathbf{Z}^H\right]$$

from which we have the following inequality:

$$\mathrm{tr}\left[-2\mathrm{Re}\left\{\sqrt{\frac{E_x}{N_0 N_T}}\mathbf{H}(\mathbf{X}-\mathbf{X}_{ML})\mathbf{Z}^H\right\}\right] \geq \mathrm{tr}\left[\frac{E_x}{N_0 N_T}\mathbf{H}(\mathbf{X}-\mathbf{X}_{ML})(\mathbf{X}-\mathbf{X}_{ML})^H\mathbf{H}^H\right]$$
$$= \left\|\sqrt{\frac{E_x}{N_0 N_T}}\mathbf{H}(\mathbf{X}-\mathbf{X}_{ML})\right\|_F^2 \quad (10.23)$$

The left-hand side of Equation (10.23) is again expanded as

$$W \triangleq \mathrm{tr}\left[-2\mathrm{Re}\left\{\sqrt{\frac{E_x}{N_0 N_T}}\mathbf{H}(\mathbf{X}-\mathbf{X}_{ML})\mathbf{Z}^H\right\}\right]$$
$$= \sum_{l=1}^{N_R}\mathbf{e}_l^T 2\mathrm{Re}\left\{-\sqrt{\frac{E_x}{N_0 N_T}}\mathbf{H}(\mathbf{X}-\mathbf{X}_{ML})\mathbf{Z}^H\right\}\mathbf{e}_l$$
$$= \sqrt{\frac{E_x}{N_0 N_T}}\sum_{l=1}^{N_R}2\mathrm{Re}\left\{-\mathbf{e}_l^T\mathbf{H}(\mathbf{X}-\mathbf{X}_{ML})\mathbf{Z}^H\mathbf{e}_l\right\} \quad (10.24)$$
$$= 2\sqrt{\frac{E_x}{N_0 N_T}}\sum_{l=1}^{N_R}[\mathrm{Im}\{\mathbf{h}_l(\mathbf{X}-\mathbf{X}_{ML})\}\mathrm{Im}\{\mathbf{z}_l\}-\mathrm{Re}\{\mathbf{h}_l(\mathbf{X}-\mathbf{X}_{ML})\}\mathrm{Re}\{\mathbf{z}_l\}]$$

where $\mathbf{e}_l = [0_1 \; \cdots \; 0_{l-1} \; 1 \; 0_{l+1} \cdots 0_{N_R}]^T$, $\mathbf{h}_l$ and $\mathbf{z}_l$ are the lth row of $\mathbf{H}$ and the lth column of $\mathbf{Z}^H$, respectively. Since $\text{Re}\{\mathbf{z}_l\}$ and $\text{Im}\{\mathbf{z}_l\}$ are the zero-mean independent Gaussian random vectors, each term of summation in Equation (10.24), $\text{Im}\{\mathbf{h}_l(\mathbf{X}-\mathbf{X}_{ML})\}\text{Im}\{\mathbf{z}_l\} - \text{Re}\{\mathbf{h}_l(\mathbf{X}-\mathbf{X}_{ML})\}\text{Re}\{\mathbf{z}_l\}$, is also a zero-mean Gaussian random variable with its variance given as

$$\frac{1}{2}\left[\|\text{Re}\{\mathbf{h}_l(\mathbf{X}-\mathbf{X}_{ML})\}\|^2 + \|\text{Im}\{\mathbf{h}_l(\mathbf{X}-\mathbf{X}_{ML})\}\|^2\right] = \frac{1}{2}\|\mathbf{h}_l(\mathbf{X}-\mathbf{X}_{ML})\|^2 \qquad (10.25)$$

Also, since $\mathbf{z}_l$ and $\mathbf{z}_m$ are statistically independent for $l \neq m$, the left-hand side of Equation (10.23) can be shown as a zero-mean Gaussian random variable with its variance given as

$$\begin{aligned}
\sigma_0^2 = Var\{W\} &= \left(2\sqrt{\frac{E_x}{N_0 N_T}}\right)^2 \sum_{l=1}^{N_R} \frac{1}{2}\|\mathbf{h}_l(\mathbf{X}-\mathbf{X}_{ML})\|^2 \\
&= \frac{2E_x}{N_0 N_T} \sum_{l=1}^{N_R} \mathbf{h}_l(\mathbf{X}-\mathbf{X}_{ML})(\mathbf{X}-\mathbf{X}_{ML})^H \mathbf{h}_l^H \\
&= \frac{2E_x}{N_0 N_T} \sum_{l=1}^{N_R} \mathbf{e}_l^H \mathbf{H}(\mathbf{X}-\mathbf{X}_{ML})(\mathbf{X}-\mathbf{X}_{ML})^H \mathbf{H}^H \mathbf{e}_l \qquad (10.26) \\
&= \frac{2E_x}{N_0 N_T} \text{tr}\left(\mathbf{H}(\mathbf{X}-\mathbf{X}_{ML})(\mathbf{X}-\mathbf{X}_{ML})^H \mathbf{H}^H\right) \\
&= \left\|\sqrt{\frac{2E_x}{N_0 N_T}}\mathbf{H}(\mathbf{X}-\mathbf{X}_{ML})\right\|_F^2
\end{aligned}$$

Normalizing the left-hand side of Equation (10.23) by $\sqrt{\sigma_0^2}$, the inequality in Equation (10.23) can be re-expressed as

$$z_{\text{unit}} \triangleq \frac{W}{\sqrt{\sigma_0^2}} \geq \sqrt{\frac{E_x}{2N_0 N_T}}\|\mathbf{H}(\mathbf{X}-\mathbf{X}_{ML})\|_F \qquad (10.27)$$

where z_{unit} is a zero-mean real Gaussian random variable with unit variance. Therefore, the probability that $\mathbf{X}$ is transmitted but $\mathbf{X}_{ML} \neq \mathbf{X}$ is given as

$$\text{Pr}(\mathbf{X} \rightarrow \mathbf{X}_{ML}) = Q\left(\sqrt{\frac{E_x}{2N_0 N_T}}\|\mathbf{H}(\mathbf{X}-\mathbf{X}_{ML})\|_F\right) \qquad (10.28)$$

Invoking the Chernoff bound $Q(x) \leq (1/2)\exp(-x^2/2)$, the pairwise error probability of Equation (10.28) is upper-bounded as

$$\text{Pr}(\mathbf{X} \rightarrow \mathbf{X}_{ML}) \leq \frac{1}{2}\exp\left(-\frac{E_x}{N_0 N_T}\frac{\|\mathbf{H}(\mathbf{X}-\mathbf{X}_{ML})\|_F^2}{4}\right) \qquad (10.29)$$

10.2.3 Space-Time Code Design

Starting from the pairwise space-time codeword error probability in Equation (10.29), we discuss design principles of space-time codes. The matrix norm on the right-hand side of Equation (10.29) can be expressed as

$$\left\| \mathbf{H}(\mathbf{X}-\hat{\mathbf{X}}) \right\|_F^2 = \mathrm{tr}\left(\mathbf{H}(\mathbf{X}-\hat{\mathbf{X}})(\mathbf{X}-\hat{\mathbf{X}})^H \mathbf{H}^H \right)$$
$$= \sum_{l=1}^{N_R} \mathbf{h}_l (\mathbf{X}-\hat{\mathbf{X}})(\mathbf{X}-\hat{\mathbf{X}})^H \mathbf{h}_l^H \qquad (10.30)$$

where $\mathbf{h}_l$ is the lth row of $\mathbf{H}$, $l = 1, 2, \ldots, N_T$. Using eigen-decomposition that we have discussed in Section 9.1, Equation (10.30) can be modified as

$$\sum_{l=1}^{N_R} \mathbf{h}_l (\mathbf{X}-\hat{\mathbf{X}})(\mathbf{X}-\hat{\mathbf{X}})^H \mathbf{h}_l^H = \sum_{l=1}^{N_R} \mathbf{h}_l \mathbf{V} \mathbf{\Lambda} \mathbf{V}^H \mathbf{h}_l^H \qquad (10.31)$$

where $\mathbf{V}$ is a unitary matrix with orthonormal eigenvector columns $\{\mathbf{v}_i\}_{i=1}^{N_T}$, and $\mathbf{\Lambda}$ is a diagonal matrix defined as

$$\mathbf{\Lambda} = \mathrm{Diag}\{\lambda_1 \ \lambda_2 \ \cdots \ \lambda_{N_T}\} \qquad (10.32)$$

where $\{\lambda_i\}_{i=1}^{N_T}$ are the eigenvalues of $(\mathbf{X}-\hat{\mathbf{X}})(\mathbf{X}-\hat{\mathbf{X}})^H$. Using Equation (10.32), the right-hand side of Equation (10.31) is expressed as

$$\sum_{l=1}^{N_R} \mathbf{h}_l \mathbf{V} \mathbf{\Lambda} \mathbf{V}^H \mathbf{h}_l^H = \sum_{l=1}^{N_R} \sum_{i=1}^{N_T} \lambda_i |\mathbf{h}_l \mathbf{v}_i|^2 \qquad (10.33)$$

Note that $\mathbf{h}_l$ is a row vector and $\mathbf{v}_i$ is a column vector of the same dimension. From Equation (10.33), we can see that space-time codewords need to be designed such that Equation (10.33) is maximized. Also note that the design parameters are the eigenvalues $\{\lambda_i\}_{i=1}^{N_T}$ and the corresponding eigenvectors $\{\mathbf{v}_i\}_{i=1}^{N_T}$. Using Equations (10.30)–(10.33), the upper bound in Equation (10.29) can be expressed as

$$\Pr\left(\mathbf{X} \to \hat{\mathbf{X}} \middle| \mathbf{H} \right) \leq \frac{1}{2} \exp\left(-\frac{E_x}{4N_0 N_T} \sum_{l=1}^{N_R} \sum_{i=1}^{N_T} \lambda_i |\beta_{l,i}|^2 \right) \qquad (10.34)$$

where $\beta_{l,i} = \mathbf{h}_l \mathbf{v}_i$. We assume that a channel gain matrix $\mathbf{H}$ is given and that if each gain $\mathbf{h}_{j,i}$ is a complex Gaussian random variable with zero-mean unit variance (i.e., corresponding to the Rayleigh fading channel), then the PDF of random variable $|\beta_{l,i}|$ is given as

$$f_{|B_{l,i}|}(|\beta_{l,i}|) = 2|\beta_{l,i}| \exp\left(-|\beta_{l,i}|^2 \right) \qquad (10.35)$$

Using the conditional probability of codeword error in Equation (10.34) and the distribution in Equation (10.35), the unconditional upper bound of codeword error probability can be obtained as

$$\Pr(\mathbf{X} \to \hat{\mathbf{X}}) \leq \left(\prod_{i=1}^{N_T} \frac{1}{1 + \frac{\mathsf{E_x}}{4\mathsf{N_0}} \lambda_i} \right)^{N_R} \tag{10.36}$$

If $\mathsf{E_x}/\mathsf{N_0}$ is large enough, the right-hand side of Equation (10.36) can be further bounded by

$$\left(\prod_{i=1}^{N_T} \frac{1}{1 + \frac{\mathsf{E_x}}{4\mathsf{N_0}} \lambda_i} \right)^{N_R} \leq \left\{ \prod_{i=1}^{r} \left(\frac{\mathsf{E_x}}{4\mathsf{N_0}} \lambda_i \right)^{-1} \right\}^{N_R} = \left(\prod_{i=1}^{r} \lambda_i \right)^{-N_R} \left(\frac{\mathsf{E_x}}{4\mathsf{N_0}} \right)^{rN_R}. \tag{10.37}$$

where $r = \mathrm{rank}\{(\mathbf{X}-\hat{\mathbf{X}})(\mathbf{X}-\hat{\mathbf{X}})^H\}$, which corresponds to the number of nonzero eigenvalues. The error probability in Equation (10.37) is expressed in the log scale as follows:

$$\log(\Pr(\mathbf{X} \to \hat{\mathbf{X}})) \leq -N_R \log \left(\prod_{i=1}^{r} \lambda_i \right) - rN_R \log \left(\frac{\mathsf{E_x}}{4\mathsf{N_0}} \right) \tag{10.38}$$

Therefore, r and $\prod_{i=1}^{r} \lambda_i$ must be maximized so as to minimize the codeword error probability in Equation (10.36) with respect to every pair of $\mathbf{X}_p$ and $\mathbf{X}_q, q \neq p$. Consider the minimum rank of r, that is,

$$v = \min_{p \neq q} \mathrm{rank}\{(\mathbf{X}_p - \mathbf{X}_q)(\mathbf{X}_p - \mathbf{X}_q)^H\} \tag{10.39}$$

From Equation (10.38), the diversity order is given by $N_R v$. If $T \geq N_T$, the maximum possible value of v is N_T, and the maximum achievable diversity gain of $N_R N_T$ is obtained when $v = N_T$. This is the so-called rank criterion that critically governs the error performance. Assuming the maximum diversity order, we have the following relationship:

$$\prod_{i=1}^{N_T} \lambda_i(\mathbf{X}_p, \mathbf{X}_q) = \left| (\mathbf{X}_p - \mathbf{X}_q)(\mathbf{X}_p - \mathbf{X}_q)^H \right|. \tag{10.40}$$

Note that Equation (10.40) is another factor to provide additional gain beyond the diversity gain, which is referred to as a coding gain.

Considering all possible pairs of $\mathbf{X}_p$ and $\mathbf{X}_q$, let $\Lambda_{\min}$ represent the minimum value of $\prod_{i=1}^{r} \lambda_i(\mathbf{X}_p, \mathbf{X}_q)$, that is,

$$\Lambda_{\min} = \min_{p \neq q} \prod_{i=1}^{N_T} \lambda_i(\mathbf{X}_p, \mathbf{X}_q). \tag{10.41}$$

Therefore, in order to further improve the performance, the minimum coding gain can be achieved by maximizing Equation (10.41) in the course of space-time code design. Figure 10.5 illustrates the coding gain as well as diversity gain that can be achieved with space-time codes. As illustrated here, the diversity gain stands out by the slope of the error curves while coding gain is measured by the amount of parallel shift in the log-log scale BER curves [222–224].

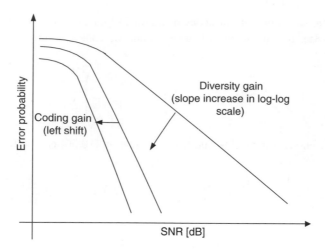

Figure 10.5   Diversity gain vs. coding gain.

In general, two different criteria for space-time code design can be considered: rank criterion and determinant criterion [215, 219]. With emphasis on the rank criterion, more advanced space-time codes were developed in [225, 226]. Furthermore, the performance of space-time coding techniques was analyzed in [227, 222]. Space-time codes can be classified into space-time block codes (STBC) and space-time trellis codes (STTC). Each of these codes is addressed in Sections 10.3 and 10.4, respectively. In general, STTCs achieve better performance than STBCs, at the expense of more complexity associated with the maximum likelihood (ML) decoder in the receiver.

10.3 Space-Time Block Code (STBC)

The very first and well-known STBC is the Alamouti code, which is a complex orthogonal space-time code specialized for the case of two transmit antennas [214]. In this section, we first consider the Alamouti space-time coding technique and then, its generalization to the case of three antennas or more [228].

10.3.1 Alamouti Space-Time Code

A complex orthogonal space-time block code for two transmit antennas was developed by Alamouti [214]. In the Alamouti encoder, two consecutive symbols x_1 and x_2 are encoded with the following space-time codeword matrix:

$$\mathbf{X} = \begin{bmatrix} x_1 & -x_2^* \\ x_2 & x_1^* \end{bmatrix} \tag{10.42}$$

As depicted in Figure 10.6, Alamouti encoded signal is transmitted from the two transmit antennas over two symbol periods. During the first symbol period, two symbols x_1 and x_2 are simultaneously transmitted from the two transmit antennas. During the second symbol period, these symbols are transmitted again, where $-x_2^*$ is transmitted from the first transmit antenna and x_1^* transmitted from the second transmit antenna.

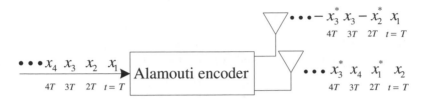

Figure 10.6   Alamouti encoder.

Note that the Alamouti codeword **X** in Equation (10.42) is a complex-orthogonal matrix, that is,

$$\mathbf{X}\mathbf{X}^H = \begin{bmatrix} |x_1|^2 + |x_2|^2 & 0 \\ 0 & |x_1|^2 + |x_2|^2 \end{bmatrix} = \left(|x_1|^2 + |x_2|^2 \right) \mathbf{I}_2 \qquad (10.43)$$

where $\mathbf{I}_2$ denotes the 2×2 identity matrix. Since $N = 2$ and $T = 2$, the transmission rate of Alamouti code is shown to be unity by Equation (10.12). Consider two different Alamouti codes,

$$\mathbf{X}_p = \begin{bmatrix} x_{1,p} & -x_{2,p}^* \\ x_{2,p} & x_{1,p}^* \end{bmatrix} \quad \text{and} \quad \mathbf{X}_q = \begin{bmatrix} x_{1,q} & -x_{2,q}^* \\ x_{2,q} & x_{1,q}^* \end{bmatrix} \qquad (10.44)$$

where $[x_{1,p}\, x_{2,p}]^T \neq [x_{1,q}\, x_{2,q}]^T$. Then the minimum rank in Equation (10.39) is evaluated as

$$
\begin{aligned}
v &= \min_{p \neq q} \mathrm{rank} \left\{ \begin{bmatrix} x_{1,p}-x_{1,q} & -x_{2,p}^*+x_{2,q}^* \\ x_{2,p}-x_{2,q} & x_{1,p}^*-x_{1,q}^* \end{bmatrix} \begin{bmatrix} x_{1,p}-x_{1,q} & -x_{2,p}^*+x_{2,q}^* \\ x_{2,p}-x_{2,q} & x_{1,p}^*-x_{1,q}^* \end{bmatrix}^H \right\} \\
&= \min_{p \neq q} \mathrm{rank} \left\{ \begin{bmatrix} e_1 & -e_2^* \\ e_2 & e_1^* \end{bmatrix} \begin{bmatrix} e_1^* & e_2^* \\ -e_2 & e_1 \end{bmatrix} \right\} \\
&= \min_{p \neq q} \mathrm{rank} \left\{ \left(|e_1|^2 + |e_2|^2 \right) \mathbf{I}_2 \right\} \\
&= 2
\end{aligned}
\qquad (10.45)
$$

where $e_1 = x_{1,p}-x_{1,q}$ and $e_2 = x_{2,p}-x_{2,q}$. Note that e_1 and e_2 cannot be zeros simultaneously. From Equation (10.44), the Alamouti code has been shown to have a diversity gain of 2. Note that the diversity analysis is based on ML signal detection at the receiver side. We now discuss ML signal detection for Alamouti space-time coding scheme. Here, we assume that two channel gains, $h_1(t)$ and $h_2(t)$, are time-invariant over two consecutive symbol periods, that is,

$$h_1(t) = h_1(t + T_s) = h_1 = |h_1| e^{j\theta_1}$$

$$\text{and} \quad h_2(t) = h_2(t + T_s) = h_2 = |h_2| e^{j\theta_2} \qquad (10.46)$$

where $|h_i|$ and θ_i denote the amplitude gain and phase rotation over the two symbol periods, $i = 1, 2$. Let y_1 and y_2 denote the received signals at time t and $t + T_s$, respectively, then

$$
\begin{aligned}
y_1 &= h_1 x_1 + h_2 x_2 + z_1 \\
y_2 &= -h_1 x_2^* + h_2 x_1^* + z_2
\end{aligned}
\tag{10.47}
$$

where z_1 and z_2 are the additive noise at time t and $t + T_s$, respectively. Taking complex conjugation of the second received signal, we have the following matrix vector equation:

$$
\begin{bmatrix} y_1 \\ y_2^* \end{bmatrix} = \begin{bmatrix} h_1 & h_2 \\ h_2^* & -h_1^* \end{bmatrix} \begin{bmatrix} x_1 \\ x_2 \end{bmatrix} + \begin{bmatrix} z_1 \\ z_2^* \end{bmatrix}
\tag{10.48}
$$

In the course of time, from time t to $t + T$, the estimates for channels, $\hat{h}_1$ and $\hat{h}_2$, are provided by the channel estimator. In the following discussion, however, we assume an ideal situation in which the channel gains, h_1 and h_2, are exactly known to the receiver. Then the transmit symbols are now two unknown variables in the matrix of Equation (10.48). Multiplying both sides of Equation (10.48) by the Hermitian transpose of the channel matrix, that is,

$$
\begin{aligned}
\begin{bmatrix} h_1^* & h_2 \\ h_2^* & -h_1 \end{bmatrix} \begin{bmatrix} y_1 \\ y_2^* \end{bmatrix} &= \begin{bmatrix} h_1^* & h_2 \\ h_2^* & -h_1 \end{bmatrix} \begin{bmatrix} h_1 & h_2 \\ h_2^* & -h_1^* \end{bmatrix} \begin{bmatrix} x_1 \\ x_2 \end{bmatrix} + \begin{bmatrix} h_1^* & h_2 \\ h_2^* & -h_1 \end{bmatrix} \begin{bmatrix} z_1 \\ z_1^* \end{bmatrix} \\
&= \left(|h_1|^2 + |h_2|^2 \right) \begin{bmatrix} x_1 \\ x_2 \end{bmatrix} + \begin{bmatrix} h_1^* z_1 + h_2 z_1^* \\ h_2^* z_1 - h_1 z_1^* \end{bmatrix}
\end{aligned}
\tag{10.49}
$$

we obtain the following input-output relations:

$$
\begin{bmatrix} \tilde{y}_1 \\ \tilde{y}_2 \end{bmatrix} = \left(|h_1|^2 + |h_2|^2 \right) \begin{bmatrix} x_1 \\ x_2 \end{bmatrix} + \begin{bmatrix} \tilde{z}_1 \\ \tilde{z}_2 \end{bmatrix}
\tag{10.50}
$$

where

$$
\begin{bmatrix} \tilde{y}_1 \\ \tilde{y}_2 \end{bmatrix} \triangleq \begin{bmatrix} h_1^* & h_2 \\ h_2^* & -h_1 \end{bmatrix} \begin{bmatrix} y_1 \\ y_2^* \end{bmatrix}
$$

and

$$
\begin{bmatrix} \tilde{z}_1 \\ \tilde{z}_2 \end{bmatrix} \triangleq \begin{bmatrix} h_1^* & h_2 \\ h_2^* & -h_1 \end{bmatrix} \begin{bmatrix} z_1 \\ z_1^* \end{bmatrix}
\tag{10.51}
$$

In Equation (10.50), we note that other antenna interference does not exist anymore, that is, the unwanted symbol x_2 dropped out of y_1, while the unwanted symbol x_1 dropped out of y_2. This is attributed to complex orthogonality of the Alamouti code in Equation (10.42). This particular feature allows for simplification of the ML receiver structure as follows:

$$
\hat{x}_{i,ML} = Q\left(\frac{\tilde{y}_i}{|h_1|^2 + |h_2|^2} \right), \quad i = 1, 2.
\tag{10.52}
$$

where $Q(\cdot)$ denotes a slicing function that determines a transmit symbol for the given constellation set. The above equation implies that x_1 and x_2 can be decided separately, which

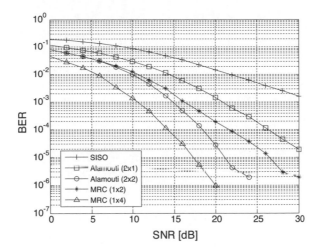

Figure 10.7   Error performance of Alamouti encoding scheme.

reduces the decoding complexity of original ML-decoding algorithm from $|\mathbf{C}|^2$ to $2|\mathbf{C}|$ where $\mathbf{C}$ represents a constellation for the modulation symbols, x_1 and x_2. Furthermore, the scaling factor $(|h_1|^2 + |h_2|^2)$ in Equation (10.50) warrants the second-order spatial diversity, which is one of the main features of the Alamouti code.

Figure 10.7 compares the Alamouti coding and MRC in terms of BER performance that is obtained by using Program 10.3 ("Alamouti_scheme.m"). Here, we assume the independent Rayleigh fading channels and perfect channel estimation at the receiver. Note that the Alamouti coding achieves the same diversity order as 1×2 MRC technique (implied by the same slope of the BER curves). Due to a total transmit power constraint (i.e., total transmit power split into each antenna by one half in the Alamouti coding), however, MRC technique outperforms Alamouti technique in providing a power combining gain in the receiver. Also shown is the 2×2 Alamouti technique which achieves the same diversity order as 1×4 MRC technique.

MATLAB® Program: Alamouti Space-Time Coding

Program 10.3 "Alamouti_scheme.m" for Alamouti space-time block coding

```
% Alamouti_scheme.m
N_frame=130; N_packet=4000;NT=2; NR=1; b=2;
SNRdBs=[0:2:30];   sq_NT=sqrt(NT);   sq2=sqrt(2);
for i_SNR=1:length(SNRdBs)
  SNRdB=SNRdBs(i_SNR);   sigma=sqrt(0.5/(10^(SNRdB/10)));
  for i_packet=1:N_packet
    msg_symbol=randint(N_frame*b,NT);
    tx_bits=msg_symbol.';   tmp=[];   tmp1=[];
    for i=1:NT
      [tmp1,sym_tab,P]=modulator(tx_bits(i,:),b); tmp=[tmp; tmp1];
    end
    X=tmp.'; X1=X; X2=[-conj(X(:,2)) conj(X(:,1))];
    for n=1:NT
```

```
    Hr(n,:,:)=(randn(N_frame,NT)+j*randn(N_frame,NT))/sq2;
  end
  H=reshape(Hr(n,:,:),N_frame,NT); Habs(:,n)=sum(abs(H).^2,2);
  R1 = sum(H.*X1,2)/sq_NT+sigma*(randn(N_frame,1)+j*randn(N_frame,1));
  R2 = sum(H.*X2,2)/sq_NT+sigma*(randn(N_frame,1)+j*randn(N_frame,1));
  Z1 = R1.*conj(H(:,1)) + conj(R2).*H(:,2);
  Z2 = R1.*conj(H(:,2)) - conj(R2).*H(:,1);
  for m=1:P
    tmp = (-1+sum(Habs,2))*abs(sym_tab(m))^2;
    d1(:,m) = abs(sum(Z1,2)-sym_tab(m)).^2 + tmp;
    d2(:,m) = abs(sum(Z2,2)-sym_tab(m)).^2 + tmp;
  end
  [y1,i1]=min(d1,[],2);   S1d=sym_tab(i1).';   clear d1
  [y2,i2]=min(d2,[],2);   S2d=sym_tab(i2).';   clear d2
  Xd = [S1d S2d];   tmp1=X>0  ; tmp2=Xd>0;
  noeb_p(i_packet) = sum(sum(tmp1~=tmp2)); % for coded
 end % end of FOR loop for i_packet
 BER(i_SNR) = sum(noeb_p)/(N_packet*N_frame*b);
end   % end of FOR loop for i_SNR
semilogy(SNRdBs,BER), axis([SNRdBs([1 end]) 1e-6 1e0]);
grid on; xlabel('SNR[dB]'); ylabel('BER');
```

10.3.2 Generalization of Space-Time Block Coding

In the previous section, we have shown that thanks to the orthogonality of the Alamouti space-time code for two transmit antenna cases, ML decoding at the receiver can be implemented by simple linear processing. This idea was generalized for an arbitrary number of transmit antennas using the general orthogonal design method in [215]. Two main objectives of orthogonal space-time code design are to achieve the diversity order of $N_T N_R$ and to implement computationally-efficient per-symbol detection at the receiver that achieves the ML performance.

Figure 10.8 shows the general structure of space-time block encoder. The output of the space-time block encoder is a codeword matrix $\mathbf{X}$ with dimension of $N_T \times T$, where N_T is the number

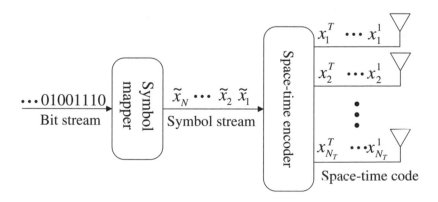

Figure 10.8   Space-time block encoder.

of transmit antennas and T is the number of symbols for each block. Let a row vector $\mathbf{x}_i$ denote the ith row of the codeword matrix $\mathbf{X}$ (i.e., $\mathbf{x}_i = [x_i^1 x_i^2 \ldots x_i^T], i = 1, 2, \ldots, N_T$). Then, $\mathbf{x}_i$ will be transmitted by the ith transmit antenna over the period of T symbols. In order to facilitate computationally-efficient ML detection at the receiver, the following property is required:

$$
\begin{aligned}
\mathbf{X}\mathbf{X}^H &= c\left(|x_i^1|^2 + |x_i^2|^2 + \cdots + |x_i^T|^2\right)\mathbf{I}_{N_T} \\
&= c\|\mathbf{x}_i\|^2 \mathbf{I}_{N_T}
\end{aligned}
\tag{10.53}
$$

where c is a constant. The above property implies that the row vectors of the codeword matrix $\mathbf{X}$ are orthogonal to each other, that is,

$$
\mathbf{x}_i \mathbf{x}_j^H = \sum_{t=1}^{T} x_i^t (x_j^t)^* = 0, \quad i \neq j, \quad i, j \in \{1, 2, \ldots, N_T\}
\tag{10.54}
$$

In the following sections, some examples of the space-time block codes will be provided.

10.3.2.1 Real Space-Time Block Codes

This section provides the examples of the space-time block codes with real entries. We consider square space-time codes with the coding rate of 1 and diversity order of $N_T = 2, 4, 8$. Following are the examples of those real space-time codes:

$$
\mathbf{X}_{2,\text{real}} = \begin{bmatrix} x_1 & -x_2 \\ x_2 & x_1 \end{bmatrix}
\tag{10.55}
$$

$$
\mathbf{X}_{4,\text{real}} = \begin{bmatrix} x_1 & -x_2 & -x_3 & -x_4 \\ x_2 & x_1 & x_4 & -x_3 \\ x_3 & -x_4 & x_1 & x_2 \\ x_4 & x_3 & -x_2 & x_1 \end{bmatrix}
\tag{10.56}
$$

$$
\mathbf{X}_{8,\text{real}} = \begin{bmatrix}
x_1 & -x_2 & -x_3 & -x_4 & -x_5 & -x_6 & -x_7 & -x_8 \\
x_2 & x_1 & -x_4 & x_3 & -x_6 & x_5 & x_8 & -x_7 \\
x_3 & x_4 & x_1 & -x_2 & -x_7 & -x_8 & x_5 & x_6 \\
x_4 & -x_3 & x_2 & x_1 & -x_8 & x_7 & -x_6 & x_5 \\
x_5 & x_6 & x_7 & x_8 & x_1 & -x_2 & -x_3 & -x_4 \\
x_6 & -x_5 & x_8 & -x_7 & x_2 & x_1 & x_4 & -x_3 \\
x_7 & -x_8 & -x_5 & x_6 & x_3 & -x_4 & x_1 & x_2 \\
x_8 & x_7 & -x_6 & -x_5 & x_4 & x_3 & -x_2 & x_1
\end{bmatrix}
\tag{10.57}
$$

Note that all the codes in the above have the coding rate of 1, because $N = T = N_T$. For example, the space-time code in Equation (10.56), $T = N = 4$, which yields the coding rate of 1 as follows:

$$
R = \frac{N}{T} = \frac{4}{4} = 1
\tag{10.58}
$$

There exists non-square space-time block codes with the coding rate of $R = 1$, that is, $N = T \neq N_T$. Following are the examples of the space-time block codes with real entries for $N_T = 3, 5, 6$, achieving both the maximum diversity and the maximum coding rate simultaneously:

$$\mathbf{X}_{3,\text{real}} = \begin{bmatrix} x_1 & -x_2 & -x_3 & -x_4 \\ x_2 & x_1 & x_4 & -x_3 \\ x_3 & -x_4 & x_1 & x_2 \end{bmatrix} \tag{10.59}$$

$$\mathbf{X}_{5,\text{real}} = \begin{bmatrix} x_1 & -x_2 & -x_3 & -x_4 & -x_5 & -x_6 & -x_7 & -x_8 \\ x_2 & x_1 & -x_4 & x_3 & -x_6 & x_5 & x_8 & -x_7 \\ x_3 & x_4 & x_1 & -x_2 & -x_7 & -x_8 & x_5 & x_6 \\ x_4 & -x_3 & x_2 & x_1 & -x_8 & x_7 & -x_6 & x_5 \\ x_5 & x_6 & x_7 & x_8 & x_1 & -x_2 & -x_3 & -x_4 \end{bmatrix} \tag{10.60}$$

$$\mathbf{X}_{6,\text{real}} = \begin{bmatrix} x_1 & -x_2 & -x_3 & -x_4 & -x_5 & -x_6 & -x_7 & -x_8 \\ x_2 & x_1 & -x_4 & x_3 & -x_6 & x_5 & x_8 & -x_7 \\ x_3 & x_4 & x_1 & -x_2 & -x_7 & -x_8 & x_5 & x_6 \\ x_4 & -x_3 & x_2 & x_1 & -x_8 & x_7 & -x_6 & x_5 \\ x_5 & x_6 & x_7 & x_8 & x_1 & -x_2 & -x_3 & -x_4 \\ x_6 & -x_5 & x_8 & -x_7 & x_2 & x_1 & x_4 & -x_3 \end{bmatrix} \tag{10.61}$$

Let us check the diversity gain of $\mathbf{x}_{3,\text{real}}$ in Equation (10.59) as an example. For two different space-time codewords, we have the following difference matrix:

$$\mathbf{X}_{3,\text{real},p} - \mathbf{X}_{3,\text{real},q} = \begin{bmatrix} x_{1,p} - x_{1,q} & -x_{2,p} + x_{2,q} & -x_{3,p} + x_{3,q} & -x_{4,p} + x_{4,q} \\ x_{2,p} - x_{2,q} & x_{1,p} - x_{1,p} & x_{4,p} - x_{4,q} & -x_{3,p} + x_{3,q} \\ x_{3,p} - x_{3,q} & -x_{4,p} + x_{4,q} & x_{1,p} - x_{1,q} & x_{2,p} - x_{2,q} \end{bmatrix}$$
$$= \begin{bmatrix} e_1 & -e_2 & -e_3 & -e_4 \\ e_2 & e_1 & e_4 & -e_3 \\ e_3 & -e_4 & e_1 & e_2 \end{bmatrix} \tag{10.62}$$

where $e_i = x_{i,p} - x_{i,q}, i = 1, 2, 3, 4$. Then the minimum rank in Equation (10.39) is evaluated as

$$v = \min_{p \neq q} \text{rank} \left\{ (\mathbf{X}_{3,\text{real},p} - \mathbf{X}_{3,\text{real},q})(\mathbf{X}_{3,\text{real},p} - \mathbf{X}_{3,\text{real},q})^T \right\}$$
$$= \min_{p \neq q} \text{rank} \left\{ \begin{bmatrix} e_1 & -e_2 & -e_3 & -e_4 \\ e_2 & e_1 & e_4 & -e_3 \\ e_3 & -e_4 & e_1 & e_2 \end{bmatrix} \begin{bmatrix} e_1 & e_2 & e_3 \\ -e_2 & e_1 & -e_4 \\ -e_3 & e_4 & e_1 \\ -e_4 & -e_3 & e_2 \end{bmatrix} \right\} \tag{10.63}$$
$$= \min_{p \neq q} \text{rank} \left\{ \left(|e_1|^2 + |e_2|^2 + |e_3|^2 + |e_4|^2 \right) \mathbf{I}_3 \right\} = 3.$$

From Equation (10.63), it is clear that $\mathbf{x}_{3,\text{real}}$ in Equation (10.59) achieves the maximum diversity gain of 3. The maximum diversity gain of the other real space-time codewords can be shown in a similar way.

10.3.2.2 Complex Space-Time Block Codes

Recall that the Alamouti code is a complex space-time block code with $N_T = 2$, which achieves the maximum diversity order of 2 with the maximum possible coding rate (i.e., $R = 1$), given as

$$\mathbf{X}_{2,\text{complex}} = \begin{bmatrix} x_1 & -x_2^* \\ x_2 & x_1^* \end{bmatrix} \tag{10.64}$$

When $N_T \geq 3$, however, it has been known that there does not exist a complex space-time code that satisfies two design goals of achieving the maximum diversity gain and the maximum coding rate at the same time. Consider the following examples for $N_T = 3$ and $N_T = 4$:

$$\mathbf{X}_{3,\text{complex}} = \begin{bmatrix} x_1 & -x_2 & -x_3 & -x_4 & x_1^* & -x_2^* & -x_3^* & -x_4^* \\ x_2 & x_1 & x_4 & -x_3 & x_2^* & x_1^* & x_4^* & -x_3^* \\ x_3 & -x_4 & x_1 & x_2 & x_3^* & -x_4^* & x_1^* & x_2^* \end{bmatrix} \tag{10.65}$$

$$\mathbf{X}_{4,\text{complex}} = \begin{bmatrix} x_1 & -x_2 & -x_3 & -x_4 & x_1^* & -x_2^* & -x_3^* & -x_4^* \\ x_2 & x_1 & x_4 & -x_3 & x_2^* & x_1^* & x_4^* & -x_3^* \\ x_3 & -x_4 & x_1 & x_2 & x_3^* & -x_4^* & x_1^* & x_2^* \\ x_4 & x_3 & -x_2 & x_1 & x_4^* & x_3^* & -x_2^* & x_1^* \end{bmatrix} \tag{10.66}$$

Both of the two space-time block codes in the above have a coding rate of 1/2, while satisfying a full-rank condition. For example, the space-time code in Equation (10.66) transmits $N = 4$ symbols (x_1, x_2, x_3, x_4) over $T = 8$ symbol periods, thus yielding a coding rate of $R = 1/2$.

If a decoding complexity at the receiver is compromised, however, higher coding rates can be achieved by the following codes:

$$\mathbf{X}_{3,\text{complex}}^{\text{high rate}} = \begin{bmatrix} x_1 & -x_2^* & \dfrac{x_3^*}{\sqrt{2}} & \dfrac{x_3^*}{\sqrt{2}} \\[2mm] x_2 & x_1^* & \dfrac{x_3}{\sqrt{2}} & \dfrac{-x_3}{\sqrt{2}} \\[2mm] \dfrac{x_3}{\sqrt{2}} & \dfrac{x_3}{\sqrt{2}} & \dfrac{(-x_1-x_1^*+x_2-x_2^*)}{2} & \dfrac{(x_2+x_2^*+x_1-x_1^*)}{2} \end{bmatrix} \tag{10.67}$$

$$\mathbf{X}_{4,\text{complex}}^{\text{high rate}} = \begin{bmatrix} x_1 & -x_2 & \dfrac{x_3^*}{\sqrt{2}} & \dfrac{x_3^*}{\sqrt{2}} \\[2mm] x_2 & x_1 & \dfrac{x_3^*}{\sqrt{2}} & \dfrac{-x_3^*}{\sqrt{2}} \\[2mm] \dfrac{x_3}{\sqrt{2}} & \dfrac{x_3}{\sqrt{2}} & \dfrac{(-x_1-x_1^*+x_2-x_2^*)}{2} & \dfrac{(x_2+x_2^*+x_1-x_1^*)}{2} \\[2mm] \dfrac{x_3}{\sqrt{2}} & \dfrac{-x_3}{\sqrt{2}} & \dfrac{(-x_2-x_2^*+x_1-x_1^*)}{2} & \dfrac{-(x_1+x_1^*+x_2-x_2^*)}{2} \end{bmatrix} \tag{10.68}$$

Note that the above two space-time block codes provide a coding rate of $R = 3/4$.

10.3.3 Decoding for Space-Time Block Codes

In this section, we consider some examples of space-time block decoding for the various codewords in the previous section. The space-time block codes can be used for various numbers of receive antennas. However, only a single receive antenna is assumed in this section. Let us first consider a real space-time block code $\mathbf{X}_{4,\text{real}}$ in Equation (10.56). We express the received signals as

$$[y_1\ y_2\ y_3\ y_4] = \sqrt{\frac{E_x}{4N_0}}[h_1\ h_2\ h_3\ h_4]\begin{bmatrix} x_1 & -x_2 & -x_3 & -x_4 \\ x_2 & x_1 & x_4 & -x_3 \\ x_3 & -x_4 & x_1 & x_2 \\ x_4 & x_3 & -x_2 & x_1 \end{bmatrix} + [z_1\ z_2\ z_3\ z_4], \qquad (10.69)$$

from which the following input-output relationship can be obtained:

$$\underbrace{\begin{bmatrix} y_1 \\ y_2 \\ y_3 \\ y_4 \end{bmatrix}}_{\mathbf{y}_{\text{eff}}} = \sqrt{\frac{E_x}{4N_0}}\underbrace{\begin{bmatrix} h_1 & h_2 & h_3 & h_4 \\ h_2 & -h_1 & h_4 & -h_3 \\ h_3 & -h_4 & -h_1 & h_2 \\ h_4 & h_3 & -h_2 & -h_1 \end{bmatrix}}_{\mathbf{H}_{\text{eff}}}\underbrace{\begin{bmatrix} x_1 \\ x_2 \\ x_3 \\ x_4 \end{bmatrix}}_{\mathbf{x}_{\text{eff}}} + \underbrace{\begin{bmatrix} z_1 \\ z_2 \\ z_3 \\ z_3 \end{bmatrix}}_{\mathbf{z}_{\text{eff}}} \qquad (10.70)$$

Note that the columns of the effective channel matrix $\mathbf{H}_{\text{eff}}$ in Equation (10.70) are orthogonal to each other. Using the orthogonality of the effective channel, we can decode Equation (10.70) as

$$\begin{aligned} \tilde{\mathbf{y}}_{\text{eff}} &= \mathbf{H}_{\text{eff}}^T \mathbf{y}_{\text{eff}} \\ &= \sqrt{\frac{E_x}{N_0 4}}\mathbf{H}_{\text{eff}}^T \mathbf{H}_{\text{eff}} \mathbf{x}_{\text{eff}} + \mathbf{H}_{\text{eff}}^T \mathbf{z}_{\text{eff}} \\ &= \sqrt{\frac{E_x}{N_0 4}}\sum_{i=1}^{4}|h_i|^2 \mathbf{I}_4 \mathbf{x}_{\text{eff}} + \tilde{\mathbf{z}}_{\text{eff}}. \end{aligned} \qquad (10.71)$$

Using the above result, the ML signal detection is performed as

$$\hat{x}_{i,ML} = Q\left(\frac{\tilde{y}_{\text{eff},i}}{\sqrt{\frac{E_x}{4N_0}\sum_{j=1}^{4}|h_j|^2}}\right), \qquad i = 1,2,3,4. \qquad (10.72)$$

where $\tilde{y}_{\text{eff},i}$ is the ith entry of $\tilde{\mathbf{y}}_{\text{eff}}$. Now, let us consider space-time decoding for the complex space-time block code $\mathbf{X}_{3,\text{complex}}$ in Equation (10.65). Then we express the received signals from a single receive antenna as

$$\begin{aligned} [y_1\ y_2\ y_3\ y_4 y_5\ y_6\ y_7\ y_8] &= \sqrt{\frac{E_x}{3N_0}}[h_1\ h_2\ h_3]\begin{bmatrix} x_1 & -x_2 & -x_3 & -x_4 & x_1^* & -x_2^* & -x_3^* & -x_4^* \\ x_2 & x_1 & x_4 & -x_3 & x_2^* & x_1^* & x_4^* & -x_3^* \\ x_3 & -x_4 & x_1 & x_2 & x_3^* & -x_4^* & x_1^* & x_2^* \end{bmatrix} \\ &\quad + [z_1\ z_2\ z_3\ z_4\ z_5\ z_6\ z_7\ z_8] \end{aligned}$$

$$(10.73)$$

The above input-output relation can be also expressed as

$$
\underbrace{\begin{bmatrix} y_1 \\ y_2 \\ y_3 \\ y_4 \\ y_5^* \\ y_6^* \\ y_7^* \\ y_8^* \end{bmatrix}}_{\mathbf{y}_{\text{eff}}} = \sqrt{\frac{E_x}{3N_0}} \underbrace{\begin{bmatrix} h_1 & h_2 & h_3 & 0 \\ h_2 & -h_1 & 0 & -h_3 \\ h_3 & 0 & -h_1 & h_2 \\ 0 & h_3 & -h_2 & -h_1 \\ h_1^* & h_2^* & h_3^* & 0 \\ h_2^* & -h_1^* & 0 & -h_3^* \\ h_3^* & 0 & -h_1^* & h_2^* \\ 0 & h_3^* & -h_2^* & -h_1^* \end{bmatrix}}_{\mathbf{H}_{\text{eff}}} \underbrace{\begin{bmatrix} x_1 \\ x_2 \\ x_3 \\ x_4 \end{bmatrix}}_{\mathbf{x}_{\text{eff}}} + \underbrace{\begin{bmatrix} z_1 \\ z_2 \\ z_3 \\ z_4 \\ z_5^* \\ z_6^* \\ z_7^* \\ z_8^* \end{bmatrix}}_{\mathbf{z}_{\text{eff}}}
\tag{10.74}
$$

Again, using the orthogonality of the above effective channel matrix, the received signal is modified as

$$
\tilde{\mathbf{y}}_{\text{eff}} = \mathbf{H}_{\text{eff}}^H \mathbf{y}_{\text{eff}}
$$

$$
= \sqrt{\frac{E_x}{3N_0}} \mathbf{H}_{\text{eff}}^H \mathbf{H}_{\text{eff}} \mathbf{x}_{\text{eff}} + \mathbf{H}_{\text{eff}}^H \mathbf{z}_{\text{eff}}
$$

$$
= 2\sqrt{\frac{E_x}{3N_0}} \sum_{j=1}^{3} |h_j|^2 \mathbf{I}_4 \mathbf{x}_{\text{eff}} + \tilde{\mathbf{z}}_{\text{eff}}
\tag{10.75}
$$

Using the above result, the ML signal detection is performed as

$$
\hat{x}_{i,ML} = Q\left(\frac{\tilde{y}_{\text{eff},i}}{2\sqrt{\dfrac{E_x}{3N_0} \displaystyle\sum_{j=1}^{3} |h_j|^2}} \right), \quad i = 1,2,3,4
\tag{10.76}
$$

We now consider the high rate space-time block code $\mathbf{X}_{3,\text{complex}}^{\text{high rate}}$ in Equation (10.67). Then we express the received signals as

$$
[y_1 \; y_2 \; y_3 \; y_4] = \sqrt{\frac{E_x}{3N_0}}[h_1 \; h_2 \; h_3] \begin{bmatrix} x_1 & -x_2^* & \dfrac{x_3^*}{\sqrt{2}} & \dfrac{x_3^*}{\sqrt{2}} \\[2ex] x_2 & x_1^* & \dfrac{x_3^*}{\sqrt{2}} & \dfrac{-x_3^*}{\sqrt{2}} \\[2ex] \dfrac{x_3}{\sqrt{2}} & \dfrac{x_3}{\sqrt{2}} & \dfrac{(-x_1-x_1^*+x_2-x_2^*)}{2} & \dfrac{(x_2+x_2^*+x_1-x_1^*)}{2} \end{bmatrix}
$$

$$
+ [z_1 \; z_2 \; z_3 \; z_4]
\tag{10.77}
$$

from which the following input-output relation can be obtained:

$$
\begin{bmatrix} y_1 \\ y_2 \\ y_3 \\ y_4 \end{bmatrix} = \sqrt{\frac{E_x}{3N_0}} \begin{bmatrix} h_1 & h_2 & \dfrac{h_3}{\sqrt{2}} & 0 & 0 & 0 \\[2mm] 0 & 0 & \dfrac{h_3}{\sqrt{2}} & h_2 & -h_1 & 0 \\[2mm] -\dfrac{h_3}{2} & \dfrac{h_3}{2} & 0 & -\dfrac{h_3}{2} & -\dfrac{h_3}{2} & \dfrac{h_1+h_2}{\sqrt{2}} \\[2mm] \dfrac{h_3}{2} & \dfrac{h_3}{2} & 0 & -\dfrac{h_3}{2} & \dfrac{h_3}{2} & \dfrac{h_1-h_2}{\sqrt{2}} \end{bmatrix} \begin{bmatrix} x_1 \\ x_2 \\ x_3 \\ x_1^* \\ x_2^* \\ x_3^* \end{bmatrix} + \begin{bmatrix} z_1 \\ z_2 \\ z_3 \\ z_4 \end{bmatrix}
\tag{10.78}
$$

From Equation (10.78), we can derive the following three equations:

$$
\underbrace{\begin{bmatrix} y_1 \\ y_2^* \\ \dfrac{y_4-y_3}{\sqrt{2}} \\ \dfrac{y_4^*+y_3^*}{\sqrt{2}} \end{bmatrix}}_{\mathbf{y}_{\text{eff}}^1} = \sqrt{\frac{E_x}{3N_0}} \underbrace{\begin{bmatrix} h_1 & h_2 & \dfrac{h_3}{\sqrt{2}} & 0 & 0 & 0 \\[2mm] h_2^* & -h_1^* & 0 & 0 & 0 & \dfrac{h_3^*}{\sqrt{2}} \\[2mm] \dfrac{h_3}{\sqrt{2}} & 0 & 0 & 0 & \dfrac{h_3}{\sqrt{2}} & -h_2 \\[2mm] -\dfrac{h_3^*}{\sqrt{2}} & 0 & h_1^* & 0 & \dfrac{h_3^*}{\sqrt{2}} & 0 \end{bmatrix}}_{\mathbf{H}_{\text{eff}}^1} \underbrace{\begin{bmatrix} x_1 \\ x_2 \\ x_3 \\ x_1^* \\ x_2^* \\ x_3^* \end{bmatrix}}_{\mathbf{x}_{\text{eff}}} + \underbrace{\begin{bmatrix} z_1 \\ z_2^* \\ \dfrac{z_4-z_3}{\sqrt{2}} \\ \dfrac{z_4^*+z_3^*}{\sqrt{2}} \end{bmatrix}}_{\mathbf{z}_{\text{eff}}^1}
\tag{10.79}
$$

$$
\underbrace{\begin{bmatrix} y_1 \\ y_2^* \\ \dfrac{y_3+y_4}{\sqrt{2}} \\ \dfrac{y_4^*-y_3^*}{\sqrt{2}} \end{bmatrix}}_{\mathbf{y}_{\text{eff}}^2} = \sqrt{\frac{E_x}{3N_0}} \underbrace{\begin{bmatrix} h_1 & h_2 & \dfrac{h_3}{\sqrt{2}} & 0 & 0 & 0 \\[2mm] h_2^* & -h_1^* & 0 & 0 & 0 & \dfrac{h_3^*}{\sqrt{2}} \\[2mm] 0 & \dfrac{h_3}{\sqrt{2}} & 0 & -\dfrac{h_3}{\sqrt{2}} & 0 & h_1 \\[2mm] 0 & \dfrac{h_3^*}{\sqrt{2}} & -h_2^* & \dfrac{h_3^*}{\sqrt{2}} & 0 & 0 \end{bmatrix}}_{\mathbf{H}_{\text{eff}}^2} \underbrace{\begin{bmatrix} x_1 \\ x_2 \\ x_3 \\ x_1^* \\ x_2^* \\ x_3^* \end{bmatrix}}_{\mathbf{x}_{\text{eff}}} + \underbrace{\begin{bmatrix} z_1 \\ z_2^* \\ \dfrac{z_3+z_4}{\sqrt{2}} \\ \dfrac{z_4^*-z_3^*}{\sqrt{2}} \end{bmatrix}}_{\mathbf{z}_{\text{eff}}^2}
\tag{10.80}
$$

$$
\underbrace{\begin{bmatrix} \dfrac{y_1 + y_2}{\sqrt{2}} \\ y_3^* \\ y_4^* \end{bmatrix}}_{\mathbf{y}_{\text{eff}}^3} = \sqrt{\dfrac{E_x}{3N_0}} \underbrace{\begin{bmatrix} \dfrac{h_1}{\sqrt{2}} & \dfrac{h_2}{\sqrt{2}} & h_3 & \dfrac{h_2}{\sqrt{2}} & -\dfrac{h_2}{\sqrt{2}} & 0 \\ -\dfrac{h_3^*}{\sqrt{2}} & -\dfrac{h_3^*}{\sqrt{2}} & \dfrac{h_1^* + h_2^*}{\sqrt{2}} & -\dfrac{h_3^*}{\sqrt{2}} & \dfrac{h_3^*}{\sqrt{2}} & 0 \\ -\dfrac{h_3^*}{\sqrt{2}} & \dfrac{h_3^*}{\sqrt{2}} & \dfrac{h_1^* - h_2^*}{\sqrt{2}} & \dfrac{h_3^*}{\sqrt{2}} & -\dfrac{h_3^*}{\sqrt{2}} & 0 \end{bmatrix}}_{\mathbf{H}_{\text{eff}}^3} \underbrace{\begin{bmatrix} x_1 \\ x_2 \\ x_3 \\ x_1^* \\ x_2^* \\ x_3^* \end{bmatrix}}_{\mathbf{x}_{\text{eff}}} + \underbrace{\begin{bmatrix} \dfrac{z_1 + z_2}{\sqrt{2}} \\ z_3^* \\ z_4^* \end{bmatrix}}_{\mathbf{z}_{\text{eff}}^3}
$$

$$(10.81)$$

From Equations (10.79)–(10.81), the following decision statistics are derived:

$$
\tilde{y}_{\text{eff},1} = \left(\mathbf{h}_{\text{eff},1}^1\right)^H \mathbf{y}_{\text{eff}}^1 = \sqrt{\dfrac{E_x}{3N_0}}\left(|h_1|^2 + |h_2|^2 + |h_3|^2\right)x_1 + \left(\mathbf{h}_{\text{eff},1}^1\right)^H \mathbf{z}_{\text{eff}}^1 \tag{10.82}
$$

$$
\tilde{y}_{\text{eff},2} = \left(\mathbf{h}_{\text{eff},2}^2\right)^H \mathbf{y}_{\text{eff}}^2 = \sqrt{\dfrac{E_x}{3N_0}}\left(|h_1|^2 + |h_2|^2 + |h_3|^2\right)x_2 + \left(\mathbf{h}_{\text{eff},2}^2\right)^H \mathbf{z}_{\text{eff}}^2 \tag{10.83}
$$

$$
\tilde{y}_{\text{eff},3} = \left(\mathbf{h}_{\text{eff},3}^3\right)^H \mathbf{y}_{\text{eff}}^3 = \sqrt{\dfrac{E_x}{3N_0}}\left(|h_1|^2 + |h_2|^2 + |h_3|^2\right)x_3 + \left(\mathbf{h}_{\text{eff},3}^3\right)^H \mathbf{z}_{\text{eff}}^3 \tag{10.84}
$$

where $\mathbf{h}_{\text{eff},i}^i$ is the ith column of $\mathbf{H}_{\text{eff}}^i$, $i = 1, 2, 3$. Using the above results in Equations (10.82)–(10.84), the ML signal detection is performed as

$$
\hat{x}_{i,ML} = Q\left(\dfrac{\tilde{y}_{\text{eff},i}}{\sqrt{\dfrac{E_x}{3N_0}\displaystyle\sum_{j=1}^{3}|h_j|^2}}\right), \quad i = 1, 2, 3 \tag{10.85}
$$

Although construction of the effective channel construction for $\mathbf{X}_{3,\text{complex}}^{\text{high rate}}$ is rather more complex than the previous examples, the detection processes in Equations (10.82)–(10.85) still have all simple linear processing structures.

Program 10.4 ("STBC_3x4_simulation.m") can be used to simulate the error performance of various space-time block codes for the quasi-static Rayleigh fading channels. The simulation results are depicted in Figure 10.9. As expected, a higher-order diversity is obtained with a larger number of transmit antennas, that is, steepening the slope of BER curves as the number of transmit antenna increases. In fact, it confirms that all space-time block codes achieve the maximum diversity order of N_T.

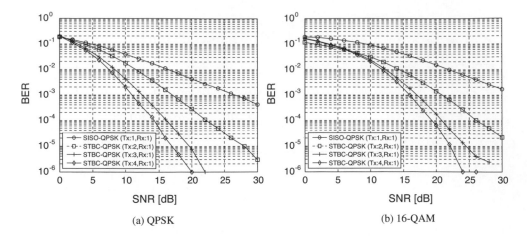

Figure 10.9   BER performance of various space-time block codes.

Program 10.4 "STBC_3x4_simulation.m" to simulate 3x4 Space-Time Block Coding

```
% STBC_3x4_simulation.m
clear; clf
Lfr=130; N_packet=4000;  NT=3; NR=4; b=2; M=2^b;
SNRdBs=[0:2:30]; sq_NT=sqrt(NT); sq2=sqrt(2);
for i_SNR=1:length(SNRdBs)
   SNRdB=SNRdBs(i_SNR); sigma=sqrt(0.5/(10^(SNRdB/10)));
   for i_packet=1:N_packet
     msg_symbol = randint(Lfr*b,M);
     tx_bits = msg_symbol.'; temp=[]; temp1=[];
     for i=1:4
       [temp1,sym_tab,P]=modulator(tx_bits(i,:),b); temp=[temp; temp1];
     end
     X=temp.';
     % Block signals in the l-th time slot % Block coding for G3 STBC
     X1=X(:,1:3); X5=conj(X1);
     X2=[-X(:,2) X(:,1) -X(:,4)]; X6=conj(X2);
     X3=[-X(:,3) X(:,4) X(:,1)]; X7=conj(X3);
     X4=[-X(:,4) -X(:,3) X(:,2)]; X8=conj(X4);
     for n=1:NT
        Hr(n,:,:)=(randn(Lfr,NT)+j*randn(Lfr,NT))/sq2;
     end
     for n=1:NT
        H = reshape(Hr(n,:,:),Lfr,NT); Hc=conj(H);
        Habs(:,n) = sum(abs(H).^2,2);
        R1n = sum(H.*X1,2)/sq_NT +sigma*(randn(Lfr,1)+j*randn(Lfr,1));
        R2n = sum(H.*X2,2)/sq_NT +sigma*(randn(Lfr,1)+j*randn(Lfr,1));
```

```
        R3n = sum(H.*X3,2)/sq_NT +sigma*(randn(Lfr,1)+j*randn(Lfr,1));
        R4n = sum(H.*X4,2)/sq_NT +sigma*(randn(Lfr,1)+j*randn(Lfr,1));
        R5n = sum(H.*X5,2)/sq_NT +sigma*(randn(Lfr,1)+j*randn(Lfr,1));
        R6n = sum(H.*X6,2)/sq_NT +sigma*(randn(Lfr,1)+j*randn(Lfr,1));
        R7n = sum(H.*X7,2)/sq_NT +sigma*(randn(Lfr,1)+j*randn(Lfr,1));
        R8n = sum(H.*X8,2)/sq_NT +sigma*(randn(Lfr,1)+j*randn(Lfr,1));
        Z1_1 = R1n.*Hc(:,1) + R2n.*Hc(:,2) + R3n.*Hc(:,3);
        Z1_2 = conj(R5n).*H(:,1) + conj(R6n).*H(:,2) + conj(R7n).*H(:,3);
        Z(:,n,1) = Z1_1 + Z1_2;
        Z2_1 = R1n.*Hc(:,2) - R2n.*Hc(:,1) + R4n.*Hc(:,3);
        Z2_2 = conj(R5n).*H(:,2) - conj(R6n).*H(:,1) + conj(R8n).*H(:,3);
        Z(:,n,2) = Z2_1 + Z2_2;
        Z3_1 = R1n.*Hc(:,3) - R3n.*Hc(:,1) -R4n.*Hc(:,2);
        Z3_2 = conj(R5n).*H(:,3) -conj(R7n).*H(:,1) -conj(R8n).*H(:,2);
        Z(:,n,3) = Z3_1 + Z3_2;
        Z4_1 = -R2n.*Hc(:,3) + R3n.*Hc(:,2) - R4n.*Hc(:,1);
        Z4_2 = -conj(R6n).*H(:,3) + conj(R7n).*H(:,2) - conj(R8n).*H(:,1);
        Z(:,n,4) = Z4_1 + Z4_2;
    end
    for m=1:P
        tmp = (-1+sum(Habs,2))*abs(sym_tab(m))^2;
        for i=1:4
          d(:,m,i) = abs(sum(Z(:,:,i),2)-sym_tab(m)).^2 + tmp;
        end
    end
    Xd = [];
    for n=1:4, [yn,in]=min(d(:,:,n),[],2); Xd=[Xd sym_tab(in).']; end
    temp1=X>0; temp2=Xd>0;
    noeb_p(i_packet) = sum(sum(temp1~=temp2));
  end % End of FOR loop for i_packet
  BER(i_SNR) = sum(noeb_p)/(N_packet*Lfr*b);
end % End of FOR loop for i_SNR
semilogy(SNRdBs,BER), axis([SNRdBs([1 end]) 1e-6 1e0])
```

10.3.4 Space-Time Trellis Code

The main advantage of space-time block codes is that a maximum diversity gain can be achieved with a relatively simple linear-processing receiver. In general, however, its coding gain can be further improved with another type of STC, known as a space-time trellis code (STTC). It was first introduced in [219], and its performance in terms of coding gain, spectral efficiency, and diversity gain was confirmed in [224, 229, 230]. In this section, we present an example of the space-time trellis code using M-PSK.

10.3.4.1 Space-Time Trellis Encoder

Figure 10.10 shows a simplified encoder part of the space-time trellis-coded system that employs M-PSK modulation with N_T transmit antennas. Let $\mathbf{b}_t = \begin{bmatrix} b_t^1 \ b_t^2 \ \cdots \ b_t^m \end{bmatrix}^T$ denote input

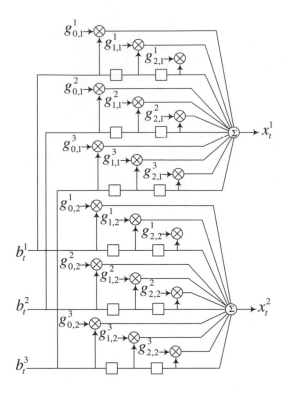

Figure 10.10 Example of space-time trellis encoder: $N_T = 2$, $m = 3$, $\nu_1 = \nu_2 = \nu_3 = 2$.

data symbol of $m = \log_2 M$ bits, which is input to the encoder at time t, $t = 0, 1, 2, \ldots$. Then, a sequence of input data symbols is represented as

$$
\mathbf{B} = [\mathbf{b}_0 \; \mathbf{b}_1 \; \cdots \; \mathbf{b}_t \; \cdots] =
\begin{bmatrix}
b_0^1 & b_1^1 & \cdots & b_t^1 & \cdots \\
b_0^2 & b_1^2 & \cdots & b_t^2 & \cdots \\
\vdots & \vdots & \ddots & \vdots & \\
b_0^m & b_1^m & \cdots & b_t^m & \cdots
\end{bmatrix}.
\tag{10.86}
$$

The STTC encoder can be considered as a convolutional encoder with the memory size of ν_k delay units for the kth branch for each output symbol. Let $\{\nu_k\}_{k=1}^{m}$ denote the size of memory used to store the kth branch metrics that is calculated as

$$
\nu_k = \left\lfloor \frac{\nu + k - 1}{\log_2 M} \right\rfloor
\tag{10.87}
$$

where ν is the size of total required memory for the space-time trellis code, that is,

$$\nu = \sum_{k=1}^{m} v_k. \tag{10.88}$$

Then, the output of the STTC encoder is specified by the following generator polynomials:

$$\mathbf{g}^1 = \left[\left(g_{0,1}^1, g_{0,2}^1, \ldots, g_{0,N_T}^1\right), \left(g_{1,1}^1, g_{1,2}^1, \ldots, g_{1,N_T}^1\right), \ldots, \left(g_{v_1,1}^1, g_{v_1,2}^1, \ldots, g_{v_1,N_T}^1\right)\right]$$

$$\mathbf{g}^2 = \left[\left(g_{0,1}^2, g_{0,2}^2, \ldots, g_{0,N_T}^2\right), \left(g_{1,1}^2, g_{1,2}^2, \ldots, g_{1,N_T}^2\right), \ldots, \left(g_{v_2,1}^2, g_{v_2,2}^2, \ldots, g_{v_2,N_T}^2\right)\right] \tag{10.89}$$

$$\vdots$$

$$\mathbf{g}^m = \left[\left(g_{0,1}^m, g_{0,2}^m, \ldots, g_{0,N_T}^m\right), \left(g_{1,1}^m, g_{1,2}^m, \ldots, g_{1,N_T}^m\right), \ldots, \left(g_{v_m,1}^m, g_{v_m,2}^m, \ldots, g_{v_m,N_T}^m\right)\right]$$

where $g_{j,i}^k$ denotes M-PSK symbols, $k = 1, 2, \ldots, m$, $j = 1, 2, \ldots, v_k$, $i = 1, 2, \ldots, N_T$.

Let x_t^i denote the outputs of the STTC encoder for the ith transmit antenna at time t, $i = 1, 2, \ldots, N_T$, which are given as

$$x_t^i = \sum_{k=1}^{m} \sum_{j=0}^{v_k} g_{j,i}^k b_{t-j}^k \bmod M \tag{10.90}$$

Space-time trellis-encoded M-PSK symbols are now expressed as

$$\mathbf{X} = [\mathbf{x}_0 \, \mathbf{x}_1 \, \cdots \, \mathbf{x}_t \, \cdots] = \begin{bmatrix} x_0^1 & x_1^1 & \cdots & x_t^1 & \cdots \\ x_0^2 & x_1^2 & \cdots & x_t^2 & \cdots \\ \vdots & \vdots & \ddots & \vdots & \\ x_0^{N_T} & x_1^{N_T} & \cdots & x_t^{N_T} & \cdots \end{bmatrix} \tag{10.91}$$

where $\mathbf{x}_t = \begin{bmatrix} x_t^1 & x_t^2 & \cdots & x_t^{N_T} \end{bmatrix}^T$ is the output of the encoder that is composed of N_T M-PSK symbols, $t = 0, 1, 2, \ldots$. Figure 10.10 shows an example of the STTC encoder for $N_T = 2$, $m = 3$, and $\nu = 6$, that is, $v_1 = v_2 = v_3 = 2$ by Equation (10.88).

The Viterbi algorithm can be used for decoding the space-time trellis-coded systems. In the Viterbi algorithm, the branch metric is given by the following squared Euclidian distance:

$$\sum_{t=1}^{T} \sum_{j=1}^{N_R} \left| y_t^j - \sum_{i=1}^{N_T} h_{j,i} x_t^i \right|^2 \tag{10.92}$$

where y_t^j is the received signal at the jth receive antenna during tth symbol period, and $h_{j,i}$ is the channel gain between the ith transmit antenna and jth receive antenna. Using the branch metric in Equation (10.92), a path with the minimum accumulated Euclidian distance is selected for the detected sequence of transmitted symbols. Generator polynomials for STTC can be designed with either rank-determinant or trace criterion [231]. They are summarized in Tables 10.1 and 10.2, respectively.

Table 10.1   Generator polynomials designed by rank-determinant criteria [231].

Modulation	v	N_T	Generator sequences	Rank	Det.	Trace
QPSK	2	2	$\mathbf{g}^1 = [(0,2)\quad(2,0)]$ $\mathbf{g}^2 = [(0,1)\quad(1,0)]$	2	4.0	
QPSK	4	2	$\mathbf{g}^1 = [(0,2)\quad(2,0)\quad(0,2)]$ $\mathbf{g}^2 = [(0,1)\quad(1,2)\quad(2,0)]$	2	12.0	
QPSK	4	3	$\mathbf{g}^1 = [(0,0,2)\quad(0,1,2)\quad(2,3,1)]$ $\mathbf{g}^2 = [(2,0,0)\quad(1,2,0)\quad(2,3,3)]$	3	32	16
8PSK	3	2	$\mathbf{g}^1 = [(0,4)\quad(4,0)]$ $\mathbf{g}^2 = [(0,2)\quad(2,0)]$ $\mathbf{g}^3 = [(0,1)\quad(5,0)]$	2	2	4
8PSK	4	2	$\mathbf{g}^1 = [(0,4)\quad(4,4)]$ $\mathbf{g}^2 = [(0,2)\quad(2,2)]$ $\mathbf{g}^3 = [(0,1)\quad(5,1)\quad(1,5)]$	2	3.515	6

Table 10.2   Generator polynomials designed with trace criterion.

Modulation	v	N_T	Generator sequences	Rank	Det.	Trace
QPSK	2	2	$\mathbf{g}^1 = [(0,2)\quad(1,0)]$ $\mathbf{g}^2 = [(2,3)\quad(2,0)]$	2	4.0	10.0
QPSK	4	2	$\mathbf{g}^1 = [(1,2)\quad(1,3)\quad(3,2)]$ $\mathbf{g}^2 = [(2,0)\quad(2,2)\quad(2,0)]$	2	8.0	16.0
QPSK	4	3	$\mathbf{g}^1 = [(0,2,2,0)\quad(1,2,3,2)]$ $\mathbf{g}^2 = [(2,3,3,2)\quad(2,0,2,1)]$	2	—	20.0
8PSK	3	2	$\mathbf{g}^1 = [(2,4)\quad(3,7)]$ $\mathbf{g}^2 = [(4,0)\quad(6,6)]$ $\mathbf{g}^3 = [(7,2)\quad(0,7)\quad(4,4)]$	2	0.686	8.0
8PSK	4	2	$\mathbf{g}^1 = [(2,4,2,2)\quad(3,7,2,4)]$ $\mathbf{g}^2 = [(4,0,4,4)\quad(6,6,4,0)]$ $\mathbf{g}^3 = [(7,2,2,0)\quad(0,7,6,3)\quad(4,4,0,2)]$	2	—	20.0

10.3.4.2 Space-Time Trellis Code: Illustrative Example

In this section, an example is provided for the space-time trellis code with two transmit antennas ($N_T = 2$), 4-state ($v = 2$) trellis, QPSK modulation, and the following generator polynomials:

$$\begin{aligned}
\mathbf{g}^1 &= \left[\left(g_{0,1}^1, g_{0,2}^1\right)\left(g_{1,1}^1, g_{1,2}^1\right)\right] = [(0,2)\,(2,0)] \\
\mathbf{g}^2 &= \left[\left(g_{0,1}^2, g_{0,2}^2\right)\left(g_{1,1}^2, g_{1,2}^2\right)\right] = [(0,1)\,(1,0)]
\end{aligned} \tag{10.93}$$

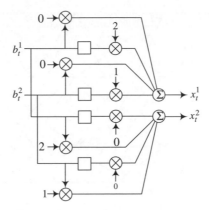

Figure 10.11   Example of space-time trellis encoder: $N_T = 2$, $m = 2$, $\nu_1 = \nu_2 = 2$.

Figure 10.11 shows a structure of the STTC encoder for this example. The encoder state at time t is $\left(b_{t-1}^1\, b_{t-1}^2\right)$ or $2b_{t-1}^1 + b_{t-1}^2$. The output for the ith transmit antenna at time t is calculated as

$$
\begin{aligned}
x_t^1 &= \left(g_{0,1}^1 b_t^1 + g_{1,1}^1 b_{t-1}^1 + g_{0,1}^2 b_t^2 + g_{1,1}^2 b_{t-1}^2\right) \bmod 4 \\
&= \left(2b_{t-1}^1 + b_{t-1}^2\right) \bmod 4 \qquad\qquad\qquad\qquad\qquad (10.94)\\
&= 2b_{t-1}^1 + b_{t-1}^2
\end{aligned}
$$

and

$$
\begin{aligned}
x_t^2 &= \left(g_{0,2}^1 b_t^1 + g_{1,2}^1 b_{t-1}^1 + g_{0,2}^2 b_t^2 + g_{1,2}^2 b_{t-1}^2\right) \bmod 4 \\
&= \left(2b_t^1 + b_t^2\right) \bmod 4 \qquad\qquad\qquad\qquad\qquad\quad (10.95)\\
&= 2b_t^1 + b_t^2
\end{aligned}
$$

From Equations (10.94) and (10.95), it can be seen that $x_t^1 = x_{t-1}^1$, that is, the signal transmitted from the first antenna is a delayed version of the transmitted signal from the second transmit antenna. Note that the output x_t^2 at time t becomes the encoder state at time $(t+1)$ in this particular example. Figure 10.12 shows the corresponding trellis diagram, in which the branch labels indicate two output symbols, x_t^1 and x_t^2.

For example, consider the following input bit sequence:

$$
\mathbf{B} = \begin{bmatrix} b_0^1 & b_1^1 & b_2^1 & b_3^1 & b_4^1 & \cdots \\ b_0^2 & b_1^2 & b_2^2 & b_3^2 & b_4^2 & \cdots \end{bmatrix} = \begin{bmatrix} 1 & 0 & 1 & 0 & 0 & \cdots \\ 0 & 1 & 1 & 0 & 1 & \cdots \end{bmatrix} \qquad (10.96)
$$

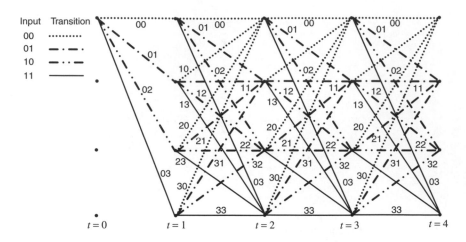

Figure 10.12   Trellis diagram for STTC encoder in Figure 10.11: $N_T = 2$, $m = 2$, $\nu_1 = \nu_2 = 2$.

Equations (10.94) and (10.95) are used to generate the encoded symbols, represented equivalently by the trellis diagram in Figure 10.12. Assuming that the initial state at $t = 0$ is 0, the input of $\left(b_0^1 \ b_0^2\right) = (1 \ 0)$ produces the output of $\left(x_0^1 \ x_0^2\right) = (0 \ 2)$ by Equations (10.93) and (10.94), leading to the state of 2 at $t = 1$. At $t = 1$, the input of $\left(b_1^1 \ b_1^2\right) = (0 \ 1)$ produces the output of $\left(x_0^1 \ x_0^2\right) = (2 \ 1)$, leading to the state of 1 at $t = 2$. At $t = 2$, the input of $\left(b_2^1 \ b_2^2\right) = (1 \ 1)$ produces the output of $\left(x_0^1 \ x_0^2\right) = (1 \ 3)$, leading to the state of 3 at $t = 3$. Continuing the same encoding process, the trellis-encoded symbol stream is represented as

$$\mathbf{X} = \begin{bmatrix} x_0^1 & x_1^1 & x_2^1 & x_3^1 & x_4^1 & \cdots \\ x_0^2 & x_1^2 & x_2^2 & x_3^2 & x_4^2 & \cdots \end{bmatrix} = \begin{bmatrix} 0 & 2 & 1 & 3 & 0 & \cdots \\ 2 & 1 & 3 & 0 & 1 & \cdots \end{bmatrix} \qquad (10.97)$$

Figure 10.13 shows the error performance of the space-time trellis codes with various numbers of states and antenna configurations under the quasi-static fading channels. Figure 10.13(a) shows the performance for 2×1 antenna configuration. We observe that all the curves have the same slope, which implies that they have the same diversity gain, and that the performance improves with the number of these states. Figure 10.13(b) shows the performance for 2×2 antenna systems. It is obvious that 2×2 antenna systems outperform 2×1 antenna systems. Program 10.5 ("STTC_simulation.m") can be used to obtain Figure 10.13.

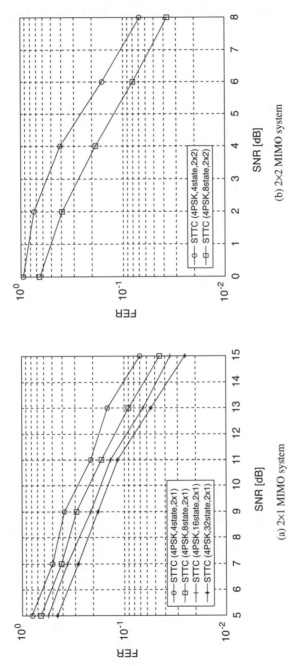

(a) 2×1 MIMO system

(b) 2×2 MIMO system

Figure 10.13   Performance of space-time trellis code with various numbers of states.

MATLAB® Programs: Space-Time Trellis Code

Program 10.5 "STTC_simulation.m" to simulate the STTC (Space-Time Trellis Coding)

```
% STTC_simulation.m
clear, clf
N_frame=130; N_packets=1000;
state='4_State_4PSK';
NT=1; NR=2; zf=3;
[dlt,slt,M] = STTC_stage_modulator(state,NR);
data_source = data_generator(N_frame,N_packet,M,zf);
data_encoded = trellis_encoder(data_source,dlt,slt);
mod_sig = STTC_modulator(data_encoded,M);
SNRdBs=5:2:15;
for i_SNR=1:length(SNRdBs)
  [signal,ch_coefs] = channel1(mod_sig,SNRdBs(i_SNR),NR);
  [data_est,state_est] = STTC_detector(signal,dlt,slt,ch_coefs);
  [N_frame1,space_dim,N_packets] = size(data_est);
  FER(i_SNR) = sum(sum(data_source~=data_est)>0)/N_packets;
end
```

Program 10.6 "STTC_stage_modulator"

```
function [dlt,slt,M]=STTC_stage_modulator(state,NR)
switch state
  case '4_State_4PSK', M=4; Ns=4; % Modulation order & Number of states
  case '8_State_4PSK', M=4; Ns=8;
  case '16_State_4PSK', M=4; Ns=16;
  case '32_State_4PSK', M=4; Ns=32;
  case '8_State_8PSK', M=8; Ns=8;
  case '16_State_8PSK', M=8; Ns=16;
  case '32_State_8PSK', M=8; Ns=32;
  case 'DelayDiv_8PSK', M=8; Ns=8;
  case '16_State_16qam', M=16; Ns=16;
  case 'DelayDiv_16qam', M=16; Ns=16;
  otherwise, disp('Wrong option!!');
end
base = reshape(1:Ns,M,Ns/M)';
slt = repmat(base,M,1);
stc_bc16=[ 0 0;11 1; 2 2; 9 3; 4 4;15 5; 6 6;13 7; 8 8; 3 9;
           10 10; 1 11;12 12;  7 13;14 14; 5 15];
for n=1:M
 l=n-1;
 ak=bitget(l,1); bk=bitget(l,2);dk=bitget(l,3);ek=bitget(l,4);
 switch M
   case 4 % 4 PSK
   for m=1:Ns
     k=m-1;ak_1=bitget(k,1); ak_2=bitget(k,3); ak_3=bitget(k,5);
     bk_1=bitget(k,2);bk_2=bitget(k,4);bk_3=bitget(k,6);
     switch Ns
```

```
      case 4 %4state_4psk
        if NR~=2
          dlt(m,n,1) = mod(2*bk_1+ak_1,M);% Rank and Determinant
          dlt(m,n,2) = mod(2*bk+ak,M);
        else
          dlt(m,n,1)=mod(2*bk_1+ak_1,M);dlt(m,n,2)=Smod(2*bk+ak,M);
        end
      case 8 %8state_4psk
        if NR~=2
          dlt(m,n,1) = mod(2*ak_2+2*bk_1+ak_1,M);
          dlt(m,n,2) = mod(2*ak_2+2*bk+ak,M);
        else
          dlt(m,n,1) = mod(2*bk_1+2*bk+ak_1+2*ak,M);
          dlt(m,n,2) = mod(bk_1+2*ak_2+2*bk+2*ak_1,M);
        end
      case 16 %16state_4psk
        if NR~=2
          dlt(m,n,1)=mod(2*ak_2+2*bk_1+ak_1,M);
          dlt(m,n,2)=mod(2*bk_2+2*ak_1+2*bk,M);
        else
          dlt(m,n,1)=mod(2*bk_3+2*ak_3+3*bk_2+3*bk_1+2*ak_1+2*ak,M);
          dlt(m,n,2)=mod(2*bk_3+3*bk_2+bk_1+2*ak_1+2*bk+2*ak,M);
        end
      case 32 %32state_4psk
        dlt(m,n,1)=mod(2*ak_3+3*bk_2+2*ak_2+2*bk_1+ak_1,M);
        dlt(m,n,2)=mod(2*ak_3+3*bk_2+2*bk_1+ak_1+2*bk+ak,M);
    end% End of switch Ns
  end% End of for m loop
case 8 % 'rank & determinant' criteria only
 for m=1:Ns
   k=m-1;ak_1=bitget(k,1); bk_1=bitget(k,2);  dk_1=bitget(k,3);
   ak_2 = bitget    (k,4);bk_2 = bitget(k,5);
   switch Ns
     case 8
       switch state
         case '8_State_8PSK'
           dlt(m,n,1) = mod(4*dk_1+2*bk_1+5*ak_1,M);
           dlt(m,n,2) = mod(4*dk+2*bk+ak,M);
         case 'DelayDiv_8PSK'
           dlt(m,n,1)=mod(4*dk_1+2*bk_1+ak_1,M);
           dlt(m,n,2)=mod(4*dk+2*bk+ak,M);
       end
     case 16 %16state_8psk
       dlt(m,n,1)=mod(ak_2+4*dk_1+2*bk_1+5*ak_1,M);
       dlt(m,n,2)=mod(5*ak_2+4*dk_1+2*bk_1+ak_1+4*dk+2*bk+ak,M);
     case 32 %32state_8psk
       dlt(m,n,1)=mod(2*bk_2+3*ak_2+4*dk_1+2*bk_1+5*ak_1,M);
       dlt(m,n,2)=mod(2*bk_2+7*ak_2+4*dk_1+2*bk_1+ak_1+4*dk+ ...
         2*bk+ak,M);
```

```
     end% End of switch Ns
   end% End of for m loop
 case 16 % 16 QAM 'rank & determinant' criteria only
  for m=1:Ns
    k = m-1;
    ak_1=bitget(k,1); bk_1=bitget(k,2);
    dk_1=bitget(k,3); ek_1=bitget(k,4);
    switch Ns
      case 16
        switch state
          case '16_State_16qam'
            dlt(m,n,1)=stc_bc16(k+1,1);
            dlt(m,n,2)=stc_bc16(k+1,2)-m+n;
          case 'DelayDiv_16qam'
            dlt(m,n,1)=mod(8*ek_1+4*dk_1+2*bk_1+ak_1,M);
            dlt(m,n,2)=mod(8*ek+4*dk+2*bk+ak,M);
        end% End of switch state
    end% End of switch Ns
  end% End of for m loop
 end% End of switch M
end% End of for n loop
```

Program 10.7 "data_generator"

```
function ]data]=data_generator(N_frame,N_packet,M,zf)
data = round((M-1)*rand(N_frame,1,N_packet));
[m,n,o] = size(data);
data(m+1:m+zf,:,1:o) = 0;
```

Program 10.8 "trellis_encoder"

```
function [enc_data]=trellis_encoder(data,dlt,slt)
[N_frame,1,N_packet] = size(data);
n_state = 1;
for k=1:N_packet
  for i=1:N_frame
    d = data(i,1,k) + 1;
    enc_data(i,:,k) = dlt(n_state,d,:);
    n_state = slt(n_state,d,:);
  end
end
```

Program 10.9 "STTC_modulator"

```
function [sig_mod]=STTC_modulator(data,M,sim_options)
qam16=[1 1; 2 1; 3 1; 4 1; 4 2; 3 2; 2 2; 1 2; 1 3; 2 3; 3 3; 4 3; 4 4; 3 4; 2 4; 1 4];
[N_frame,space_dim,N_packets]=size(data); j2piM=j*2*pi/M;
for k=1:N_packets
  switch M
    case 16    % 16QAM
```

```
      for l=1:space_dim
        k1(:,l) = qam16(data(:,l,k)+1,1);
        k2(:,l) = qam16(data(:,l,k)+1,2);
      end
        q(:,:,k) = 2*k1-M-1 -j*(2*k2-M-1);
    otherwise
        q(:,:,k) = exp(j2piM*data(:,:,k));
  end
  sig_mod=q;
end
```

Program 10.10 "channel1": channel generation for STTC (Space-Time Trellis Code)

```
function [varargout]=channel1(sig,SNRdB,NR)
ch_conf=[2 NR];
[N_frame,space_dim,N_packets]=size(sig);
spowr=sum(abs(sig(:,1,1)))/N_frame;
sigma =sqrt(0.5*spowr*(10^(-SNRdB/10)));
sq2 = sqrt(2);
ch_coefs=(randn(ch_conf(1),ch_conf(2),N_packets) +...
j*randn(ch_conf(1),ch_conf(2),N_packets))/sq2;
ch_noise= sigma*(randn(N_frame,ch_conf(2),N_packets) +...
j*randn(N_frame,ch_conf(2),N_packets));
for k=1:N_packets
  sig_add(:,:,k) = sig(:,:,k)*ch_coefs(:,:,k);
end
sig_corr = sig_add + ch_noise;
varargout = {sig_corr,ch_coefs};
```

Program 10.11 "STTC_detector": STTC (Space-Time Trellis Code) detector

```
function [data_est,state_est] = STTC_detector(sig,dlt,slt,ch_coefs)
[step_final,space_dim,N_packets] = size(sig);
[s,md,foo] = size(dlt);
qam16=[1 1;2 1;3 1;4 1;4 2;3 2;2 2;1 2;1 3;2 3;3 3;4 3;4 4;3 4;2 4;1 4];
for k=1:N_packets
  metric(1,2:s) = realmax;
  for l=1:step_final
    for m=1:s % current m
      [s_pre,foo] = find(slt==m);
      pos = mod(m-1,md) + 1;
      data_test = dlt(s_pre,pos,:);
      data_test = reshape(data_test,[md 2]);
      if md==16% 16QAM
        for r=1:2
          k1(:,r) = qam16(data_test(:,r)+1,1);
          k2(:,r) = qam16(data_test(:,r)+1,2);
        end
        q_test = (2*k1-md-1) - j*(2*k2-md-1);
      else % 4,8PSK
```

```
        expr = j*2*pi/md;
        q_test = exp(expr*data_test);
      end
      metric_d=branch_metric(sig(l,:,k),q_test,ch_coefs(:,:,k));
      metric_md = metric(l,s_pre)' + metric_d;
      [metric_min,metric_pos] = min(metric_md);
      metric(l+1,m) = metric_min;
      vit_state(l+1,m) = s_pre(metric_pos);
      vit_data(l+1,m) = pos-1;
    end
  end
end
[foo,state_best] = min(metric(end,:));
state_est(step_final+1) = state_best;
for l=step_final:-1:1
  state_est(l) = vit_state(l+1,state_est(l+1));
  data_est(l,:,k) = vit_data(l+1,state_est(l+1));
end
end
```

11

Signal Detection for Spatially Multiplexed MIMO Systems

Spatially multiplexed MIMO (SM-MIMO) systems can transmit data at a higher speed than MIMO systems using antenna diversity techniques in Chapter 10. However, spatial de-multiplexing or signal detection at the receiver side is a challenging task for SM MIMO systems. This chapter addresses signal detection techniques for SM MIMO systems. Consider the $N_R \times N_T$ MIMO system in Figure 11.1. Let $\mathbf{H}$ denote a channel matrix with it (j, i)th entry h_{ji} for the channel gain between the ith transmit antenna and the jth receive antenna, $j = 1, 2, \ldots, N_R$ and $i = 1, 2, \ldots, N_T$. The spatially-multiplexed user data and the correspond-ing received signals are represented by $\mathbf{x} = [x_1, x_2, \ldots, x_{N_T}]^T$ and $\mathbf{y} = [y_1, y_2, \ldots, y_{N_R}]^T$, respectively, where x_i and y_j denote the transmit signal from the ith transmit antenna and the received signal at the jth receive antenna, respectively. Let z_j denote the white Gaussian noise with a variance of σ_z^2 at the jth receive antenna, and $\mathbf{h}_i$ denote the ith column vector of the channel matrix $\mathbf{H}$. Now, the $N_R \times N_T$ MIMO system is represented as

$$
\begin{aligned}
\mathbf{y} &= \mathbf{H}\mathbf{x} + \mathbf{z} \\
&= \mathbf{h}_1 x_1 + \mathbf{h}_2 x_2 + \cdots + \mathbf{h}_{N_T} x_{N_T} + \mathbf{z}
\end{aligned}
\tag{11.1}
$$

where $\mathbf{z} = [z_1, z_2, \cdots, z_{N_R}]^T$.

11.1 Linear Signal Detection

Linear signal detection method treats all transmitted signals as interferences except for the desired stream from the target transmit antenna. Therefore, interference signals from other transmit antennas are minimized or nullified in the course of detecting the desired signal from the target transmit antenna. To facilitate the detection of desired signals from each antenna, the effect of the channel is inverted by a weight matrix $\mathbf{W}$ such that

$$
\tilde{\mathbf{x}} = [\tilde{x}_1 \tilde{x}_2 \cdots \tilde{x}_{N_T}]^T = \mathbf{W}\mathbf{y},
\tag{11.2}
$$

MIMO-OFDM Wireless Communications with MATLAB®   Yong Soo Cho, Jaekwon Kim, Won Young Yang and Chung G. Kang
© 2010 John Wiley & Sons (Asia) Pte Ltd

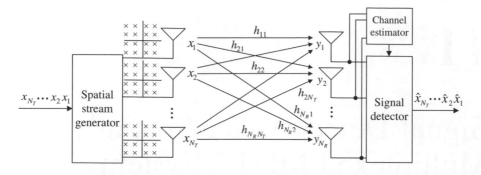

Figure 11.1 Spatially multiplexed MIMO systems.

that is, detection of each symbol is given by a linear combination of the received signals. The standard linear detection methods include the zero-forcing (ZF) technique and the minimum mean square error (MMSE) technique.

11.1.1 ZF Signal Detection

The zero-forcing (ZF) technique nullifies the interference by the following weight matrix:

$$\mathbf{W}_{ZF} = \left(\mathbf{H}^H \mathbf{H}\right)^{-1} \mathbf{H}^H. \tag{11.3}$$

where $(\,\cdot\,)^H$ denotes the Hermitian transpose operation. In other words, it inverts the effect of channel as

$$\begin{aligned} \tilde{\mathbf{x}}_{ZF} &= \mathbf{W}_{ZF}\mathbf{y} \\ &= \mathbf{x} + \left(\mathbf{H}^H \mathbf{H}\right)^{-1} \mathbf{H}^H \mathbf{z} \\ &= \mathbf{x} + \tilde{\mathbf{z}}_{ZF} \end{aligned} \tag{11.4}$$

where $\tilde{\mathbf{z}}_{ZF} = \mathbf{W}_{ZF}\mathbf{z} = \left(\mathbf{H}^H \mathbf{H}\right)^{-1} \mathbf{H}^H \mathbf{z}$. Note that the error performance is directly connected to the power of $\tilde{\mathbf{z}}_{ZF}$ (i.e., $\|\tilde{\mathbf{z}}_{ZF}\|_2^2$). Using the SVD in Section 9.1, the post-detection noise power can be evaluated as

$$\begin{aligned} \|\tilde{\mathbf{z}}_{ZF}\|_2^2 &= \left\| \left(\mathbf{H}^H \mathbf{H}\right)^{-1} \mathbf{H}^H \mathbf{z} \right\|^2 \\ &= \left\| \left(\mathbf{V}\boldsymbol{\Sigma}^2 \mathbf{V}^H\right)^{-1} \mathbf{V}\boldsymbol{\Sigma}\mathbf{U}^H \mathbf{z} \right\|^2 \\ &= \left\| \mathbf{V}\boldsymbol{\Sigma}^{-2} \mathbf{V}^H \mathbf{V}\boldsymbol{\Sigma}\mathbf{U}^H \mathbf{z} \right\|^2 \\ &= \left\| \mathbf{V}\boldsymbol{\Sigma}^{-1} \mathbf{U}^H \mathbf{z} \right\|^2 \end{aligned} \tag{11.5}$$

Since $\|\mathbf{Q}\mathbf{x}\|^2 = \mathbf{x}^H \mathbf{Q}^H \mathbf{Q}\mathbf{x} = \mathbf{x}^H \mathbf{x} = \|\mathbf{x}\|^2$ for a unitary matrix $\mathbf{Q}$, the expected value of the noise power is given as

$$
\begin{aligned}
E\left\{\|\tilde{\mathbf{z}}_{ZF}\|_2^2\right\} &= E\left\{\left\|\mathbf{\Sigma}^{-1}\mathbf{U}^H\mathbf{z}\right\|_2^2\right\} \\
&= E\left\{\mathrm{tr}\left(\mathbf{\Sigma}^{-1}\mathbf{U}^H\mathbf{z}\mathbf{z}^H\mathbf{U}\mathbf{\Sigma}^{-1}\right)\right\} \\
&= \mathrm{tr}\left(\mathbf{\Sigma}^{-1}\mathbf{U}^H E\{\mathbf{z}\mathbf{z}^H\}\mathbf{U}\mathbf{\Sigma}^{-1}\right) \\
&= \mathrm{tr}\left(\sigma_z^2\mathbf{\Sigma}^{-1}\mathbf{U}^H\mathbf{U}\mathbf{\Sigma}^{-1}\right) \\
&= \sigma_z^2\mathrm{tr}\left(\mathbf{\Sigma}^{-2}\right) \\
&= \sum_{i=1}^{N_T}\frac{\sigma_z^2}{\sigma_i^2}.
\end{aligned}
\tag{11.6}
$$

11.1.2 MMSE Signal Detection

In order to maximize the post-detection signal-to-interference plus noise ratio (SINR), the MMSE weight matrix is given as

$$
\mathbf{W}_{MMSE} = \left(\mathbf{H}^H\mathbf{H} + \sigma_z^2\mathbf{I}\right)^{-1}\mathbf{H}^H.
\tag{11.7}
$$

Note that the MMSE receiver requires the statistical information of noise σ_z^2. Note that the ith row vector $\mathbf{w}_{i,MMSE}$ of the weight matrix in Equation (11.7) is given by solving the following optimization equation:

$$
\mathbf{w}_{i,MMSE} = \underset{\mathbf{w}=(w_1,w_2,\ldots,w_{N_T})}{\arg\max}\ \frac{|\mathbf{w}\mathbf{h}_i|^2 \mathsf{E}_x}{\mathsf{E}_x\displaystyle\sum_{j=1,j\neq i}^{N_T}|\mathbf{w}\mathbf{h}_j|^2 + \|\mathbf{w}\|^2\sigma_z^2}
\tag{11.8}
$$

Using the MMSE weight in Equation (11.7), we obtain the following relationship:

$$
\begin{aligned}
\tilde{\mathbf{x}}_{MMSE} &= \mathbf{W}_{MMSE}\mathbf{y} \\
&= \left(\mathbf{H}^H\mathbf{H} + \sigma_z^2\mathbf{I}\right)^{-1}\mathbf{H}^H\mathbf{y} \\
&= \tilde{\mathbf{x}} + \left(\mathbf{H}^H\mathbf{H} + \sigma_z^2\mathbf{I}\right)^{-1}\mathbf{H}^H\mathbf{z} \\
&= \tilde{\mathbf{x}} + \tilde{\mathbf{z}}_{MMSE}
\end{aligned}
\tag{11.9}
$$

where $\tilde{\mathbf{z}}_{MMSE} = \left(\mathbf{H}^H\mathbf{H} + \sigma_z^2\mathbf{I}\right)^{-1}\mathbf{H}^H\mathbf{z}$. Using SVD again as in Section 9.1, the post-detection noise power is expressed as

$$
\begin{aligned}
\|\tilde{\mathbf{z}}_{MMSE}\|_2^2 &= \left\|\left(\mathbf{H}^H\mathbf{H} + \sigma_z^2\mathbf{I}\right)^{-1}\mathbf{H}^H\mathbf{z}\right\|^2 \\
&= \left\|\left(\mathbf{V}\mathbf{\Sigma}^2\mathbf{V}^H + \sigma_z^2\mathbf{I}\right)^{-1}\mathbf{V}\mathbf{\Sigma}\mathbf{U}^H\mathbf{z}\right\|^2.
\end{aligned}
\tag{11.10}
$$

Because $\left(\mathbf{V}\mathbf{\Sigma}^2\mathbf{V}^H + \sigma_z^2\mathbf{I}\right)^{-1}\mathbf{V}\mathbf{\Sigma} = \left(\mathbf{V}\mathbf{\Sigma}^2\mathbf{V}^H + \sigma_z^2\mathbf{I}\right)^{-1}\left(\mathbf{\Sigma}^{-1}\mathbf{V}^H\right)^{-1} = \left(\mathbf{\Sigma}\mathbf{V}^H + \sigma_z^2\mathbf{\Sigma}^{-1}\mathbf{V}^H\right)^{-1}$, the noise power in Equation (11.10) can be expressed as

$$
\|\tilde{\mathbf{z}}_{MMSE}\|_2^2 = \left\|\left(\mathbf{\Sigma}\mathbf{V}^H + \sigma_z^2\mathbf{\Sigma}^{-1}\mathbf{V}^H\right)^{-1}\mathbf{U}^H\mathbf{z}\right\|^2 = \left\|\mathbf{V}\left(\mathbf{\Sigma} + \sigma_z^2\mathbf{\Sigma}^{-1}\right)^{-1}\mathbf{U}^H\mathbf{z}\right\|^2
\tag{11.11}
$$

Again by the fact that multiplication with a unitary matrix does not change the vector norm, that is, $\|\mathbf{V}\mathbf{x}\|^2 = \|\mathbf{x}\|^2$, the expected value of Equation (11.11) is given as

$$
\begin{aligned}
E\left\{\|\tilde{\mathbf{z}}_{MMSE}\|_2^2\right\} &= E\left\{\left\|\left(\mathbf{\Sigma} + \sigma_z^2\mathbf{\Sigma}^{-1}\right)^{-1}\mathbf{U}^H\mathbf{z}\right\|^2\right\} \\
&= E\left\{\mathrm{tr}\left(\left(\mathbf{\Sigma} + \sigma_z^2\mathbf{\Sigma}^{-1}\right)^{-1}\mathbf{U}^H\mathbf{z}\mathbf{z}^H\mathbf{U}\left(\mathbf{\Sigma} + \sigma_z^2\mathbf{\Sigma}^{-1}\right)^{-1}\right)\right\} \\
&= \mathrm{tr}\left(\left(\mathbf{\Sigma} + \sigma_z^2\mathbf{\Sigma}^{-1}\right)^{-1}\mathbf{U}^H E\{\mathbf{z}\mathbf{z}^H\}\mathbf{U}\left(\mathbf{\Sigma} + \sigma_z^2\mathbf{\Sigma}^{-1}\right)^{-1}\right) \\
&= \mathrm{tr}\left(\sigma_z^2\left(\mathbf{\Sigma} + \sigma_z^2\mathbf{\Sigma}^{-1}\right)^{-2}\right) \\
&= \sum_{i=1}^{N_T}\sigma_z^2\left(\sigma_i + \frac{\sigma_z^2}{\sigma_i}\right)^{-2} \\
&= \sum_{i=1}^{N_T}\frac{\sigma_z^2\sigma_i^2}{\left(\sigma_i^2 + \sigma_z^2\right)^2}.
\end{aligned}
\tag{11.12}
$$

Noise enhancement effect in the course of linear filtering is significant when the condition number of the channel matrix is large, that is, the minimum singular value is very small. Referring to Equations (11.6) and (11.12), the noise enhancement effects due to the minimum singular value for the ZF and MMSE linear detectors are respectively given as

$$
E\left\{\|\tilde{\mathbf{z}}_{ZF}\|_2^2\right\} = \sum_{i=1}^{N_T}\frac{\sigma_z^2}{\sigma_i^2} \approx \frac{\sigma_z^2}{\sigma_{\min}^2} \quad \text{for ZF}
\tag{11.13a}
$$

$$
E\left\{\|\tilde{\mathbf{z}}_{MMSE}\|_2^2\right\} = \sum_{i=1}^{N_T}\frac{\sigma_z^2\sigma_i^2}{\left(\sigma_i^2 + \sigma_z^2\right)^2} \approx \frac{\sigma_z^2\sigma_{\min}^2}{\left(\sigma_{\min}^2 + \sigma_z^2\right)^2} \quad \text{for MMSE}
\tag{11.13b}
$$

where $\sigma_{\min}^2 = \min\{\sigma_1^2, \sigma_2^2, ..., \sigma_{N_T}^2\}$. Comparing Equation (11.13b) to Equation (11.13a), it is clear that the effect of noise enhancement in MMSE filtering is less critical than that in ZF filtering. Note that if $\sigma_{\min}^2 \gg \sigma_z^2$ and thus, $\sigma_{\min}^2 + \sigma_z^2 \approx \sigma_{\min}^2$, then the noise enhancement effect of the two linear filters becomes the same. The diversity order achieved by the ZF technique is $N_R - N_T + 1$. In the case of the single transmit antenna and multiple receive antennas, a ZF receiver is equivalent to a maximal ratio combining (MRC) receiver that achieves the diversity order of N_R.

11.2 OSIC Signal Detection

In general, the performance of the linear detection methods is worse than that of other nonlinear receiver techniques. However, linear detection methods require a low complexity of hardware implementation. We can improve their performance without increasing the complexity significantly by an ordered successive interference cancellation (OSIC) method. It is a bank of linear receivers, each of which detects one of the parallel data streams, with the detected signal components successively canceled from the received signal at each stage. More specifically, the detected signal in each stage is subtracted from the received signal so that the remaining signal with the reduced interference can be used in the subsequent stage [211, 213, 232].

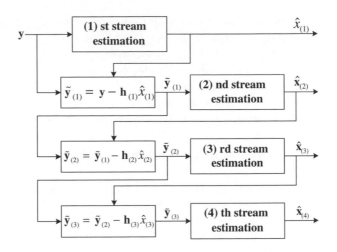

Figure 11.2 Illustration of OSIC signal detection for four spatial streams (i.e., $N_T = 4$).

Figure 11.2 illustrates the OSIC signal detection process for four spatial streams. Let $x_{(i)}$ denote the symbol to be detected in the ith order, which may be different from the transmit signal at the ith antenna, since $x_{(i)}$ depends on the order of detection. Let $\hat{x}_{(i)}$ denote a sliced value of $x_{(i)}$. In the course of OSIC, either ZF method in Equation (11.3) or MMSE method in Equation (11.7) can be used for symbol estimation. Suppose that the MMSE method is used in the following discussion. The (1)st stream is estimated with the (1)st row vector of the MMSE weight matrix in Equation (11.7). After estimation and slicing to produce $\hat{x}_{(1)}$, the remaining signal in the first stage is formed by subtracting it from the received signal, that is,

$$\begin{aligned} \tilde{\mathbf{y}}_{(1)} &= \mathbf{y} - \mathbf{h}_{(1)}\hat{x}_{(1)} \\ &= \mathbf{h}_{(1)}\left(x_{(1)} - \hat{x}_{(1)}\right) + \mathbf{h}_{(2)}x_{(2)} + \cdots + \mathbf{h}_{(N_T)}x_{(N_T)} + \mathbf{z}. \end{aligned} \tag{11.14}$$

If $x_{(1)} = \hat{x}_{(1)}$, then the interference is successfully canceled in the course of estimating $x_{(2)}$; however, if $x_{(1)} \neq \hat{x}_{(1)}$, then error propagation is incurred because the MMSE weight that has been designed under the condition of $x_{(1)} = \hat{x}_{(1)}$ is used for estimating $x_{(2)}$.

Due to the error propagation caused by erroneous decision in the previous stages, the order of detection has significant influence on the overall performance of OSIC detection. In the sequel, we describe the different methods of detection ordering.

- **Method 1 (SINR-Based Ordering):** Signals with a higher post-detection signal-to-interference-plus-noise-ratio (SINR) are detected first. Consider the linear MMSE detection with the following post-detection SINR:

$$\text{SINR}_i = \frac{\mathsf{E}_x \left| \mathbf{w}_{i,MMSE}\mathbf{h}_i \right|^2}{\mathsf{E}_x \displaystyle\sum_{l \neq i} \left| \mathbf{w}_{i,MMSE}\mathbf{h}_l \right| + \sigma_z^2 \left\| \mathbf{w}_{i,MMSE} \right\|^2}, \quad i = 1, 2, \ldots, N_T \tag{11.15}$$

where E_x is the energy of the transmitted signals, $\mathbf{w}_{i,MMSE}$ is the ith row of the MMSE weight matrix in Equation (11.7), and $\mathbf{h}_i$ is the ith column vector of the channel matrix. Note that the mean-square error (MSE) is minimized and furthermore, the post-detection SINR is maximized by the MMSE detection. Once N_T SINR values are calculated by using the MMSE weight matrix of Equation (11.7), we choose the corresponding layer with the highest SINR. In the course of choosing the second-detected symbol, the interference due to the first-detected symbol is canceled from the received signals. Suppose that $(1) = l$ (i.e., the lth symbol has been canceled first). Then, the channel matrix in Equation (11.7) is modified by deleting the channel gain vector corresponding to the lth symbol as follows:

$$\mathbf{H}^{(1)} = [\mathbf{h}_1 \, \mathbf{h}_2 \, \cdots \, \mathbf{h}_{l-1} \, \mathbf{h}_{l+1} \, \cdots \, \mathbf{h}_{N_T}] \qquad (11.16)$$

Using Equation (11.16) in place of $\mathbf{H}$ in Equation (11.7), the MMSE weight matrix is recalculated. Now, (N_T-1) SINR values, $\{\text{SINR}_i\}_{i=1, i \neq l}^{N_T}$, are calculated to choose the symbol with the highest SINR. The same process is repeated with the remaining signal after canceling the next symbol with the highest SINR. In MMSE-OSIC, the total number of SINR values to be calculated is $\sum_{j=1}^{N_T} j = N_T(N_T + 1)/2$.

- **Method 2 (SNR-Based Ordering):** When ZF weight in Equation (11.3) is used, the interference term in Equation (11.15) disappears, and the signal power $|\mathbf{w}_i \, \mathbf{h}_l|^2 = 1$, which reduces the post-detection SINR to

$$\text{SNR}_i = \frac{\mathsf{E}_{\mathsf{x}}}{\sigma_z^2 \|\mathbf{w}_i\|^2}, \quad i = 1, 2, \ldots, N_T. \qquad (11.17)$$

The same procedure of detection ordering as in Method 1 can be used, now using the SNR in Equation (11.17), instead of SINR in Equation (11.15). In this method, the number of SNR values to be calculated is also given by $\sum_{j=1}^{N_T} j = N_T(N_T + 1)/2$.

- **Method 3 (Column Norm-Based Ordering):** Note that both Methods 1 and 2 involve rather complex computation of a large number of SINR and SNR values, respectively, for detection ordering. In order to reduce the ordering complexity, we can use the norm of the column vectors in a channel matrix. Consider the following representation of the received signal:

$$\mathbf{y} = \mathbf{Hx} + \mathbf{z} = \mathbf{h}_1 x_1 + \mathbf{h}_2 x_2 + \cdots + \mathbf{h}_{N_T} x_{N_T} + \mathbf{z} \qquad (11.18)$$

from which we observe that the received signal strength of the ith transmitted signal is proportional to the norm of the ith column in the channel matrix. Therefore, we can detect the signal in the order of the norms $\|\mathbf{h}_i\|$. In this method, we need to compute N_T norms and then, order them only once. Detection is performed in the decreasing order of norms. Since ordering is required only once, complexity is significantly reduced as compared to the previous two methods.

- **Method 4 (Received Signal-Based Ordering):** In Methods 1, 2, and 3, the channel gains and noise characteristics were used for deciding an order of detection. However, the received signals can also be incorporated into deciding a detection order [233]. This method achieves the better performance than the above three methods. As opposed to the Methods 1, 2, and 3, in which ordering required only once as long as the channel matrix is fixed, this method involves the highest complexity as detection ordering is required every time a signal is received.

Using the OSIC method, the diversity order can be larger than $N_R - N_T + 1$ for all symbols. Thanks to ordering, the diversity order of the first-detected symbol is also larger than $N_R - N_T + 1$. However, the rest of symbols have the diversity order that depends on whether the previously-detected symbols are correct or not. If they are all correct, the diversity order of the ith-detected symbol becomes $N_R - N_T + i$. We note that the (i)th-detected symbol is different from the one transmitted from the ith transmit antenna.

Program 11.1 ("OSIC_detector"), with Program 11.2 ("QAM16_slicer") as a subroutine, implements the OSIC detection methods. In this program, OSIC type can be selected by assigning appropriate value to OSIC_type argument. Figure 11.3 compares the error performance of OSIC detection for the different ordering methods for $N_R = N_T = 4$. It is observed that the post-detection SINR-based ordering method achieves the best performance among these three methods.

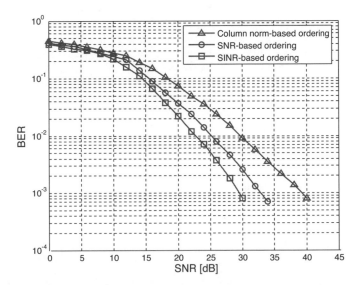

Figure 11.3 Performance of OSIC methods with different detection ordering.

MATLAB® Programs: OSIC Detection Methods

Program 11.1 "OSIC_detector" implementing the various OSIC signal detection methods

```
function [X_hat]=OSIC_detector(y,H,sigma2,NT,OSIC_type)
% Input
%   y     : Received signal, NRx1
%   H     : Channel matrix, NRxNT
%   sigma2: Noise variance
```

```
%    NT      : Number of Tx antennas
%    OSIC_type=1/2/3 for Post_detection_SINR/Column_max/Post_detection_SNR
% Output
%   X_hat : Estimated signal, NTx1
if OSIC_type==1  % Post_detection_SINR
   Order=[]; % Detection order
   index_array=[1:NT]; % yet to be detected signal index
   % V-BLAST
   for stage=1:NT
      Wmmse=inv(H'*H+sigma2*eye(NT+1-stage))*H'; % MMSE filter (Eq.(11.7))
      WmmseH=Wmmse*H;
      SINR=[];
      for i=1:NT-(stage-1)
         tmp= norm(WmmseH(i,[1:i-1 i+1:NT-(stage-1)]))^2 ...
               + sigma2*norm(Wmmse(i,:))^2;
         SINR(i) = abs(Wmmse(i,i))^2/tmp; % SINR calculation Eq.(11.15)
      end
      [val_max,index_tmp] = max(SINR);     % Ordering using SINR
      Order = [Order index_array(index_tmp)];
      index_array = index_array([1:index_tmp-1 index_tmp+1:end]);
      x_tmp(stage) = Wmmse(index_tmp,:)*y;     % MMSE filtering (Eq.(11.9))
      X_hat(stage) = QAM16_slicer(x_tmp(stage),1); % Slicing
      y_tilde = y-H(:,index_tmp)*X_hat(stage); % Interference subtraction
      H_tilde = H(:,[1:index_tmp-1 index_tmp+1:NT-(stage-1)]); % new H
      H = H_tilde;     y = y_tilde;
   end
   X_hat(Order) = X_hat;
elseif OSIC_type==2 % Column_norm ordering detection
   X_hat=zeros(NT,1);
   G = inv(H);                % Inverse of H
   for i=1:NT                 % Column_norm calculation
       norm_array(i) = norm(H(:,i));
   end
   [sorted_norm_array,Order_tmp] = sort(norm_array);
   Order = wrev(Order_tmp);
   % V-BLAST
   for stage=1:NT
      x_tmp = G(Order(stage),:)*y;     % Tx signal estimation
      X_hat(Order(stage)) = QAM16_slicer(x_tmp,1); % Slicing
      y_tilde = y-H(:,Order(stage))*X_hat(Order(stage));
   end
else % OSIC with Post_detection_SNR ordering
   Order=[];
   index_array=[1:NT]; % Set of indices of signals to be detected
   % V-BLAST
   for stage=1:NT
      G = inv(H'*H)*H';
```

```
        norm_array=[];
        for i=1:NT-(stage-1) % Detection ordering
        norm_array(i) = norm(G(i,:));
      end
      [val_min,index_min] = min(norm_array); % Ordering in SNR
      Order = [Order index_array(index_min)];
      index_array = index_array([1:index_min-1 index_min+1:end]);
      x_tmp(stage) = G(index_min,:)*y;    % Tx signal estimation
      X_hat(stage) = QAM16_slicer(x_tmp(stage),1);    % Slicing
      y_tilde = y-H(:,index_min)*X_hat(stage); % Interference subtraction
      H_tilde = H(:,[1:index_min-1 index_min+1:NT-(stage-1)]); % New H
      H = H_tilde;    y = y_tilde;
   end
   X_hat(Order) = X_hat;
end
```

Program 11.2 "QAM16_slicer"

```
function [X_hat] = QAM16_slicer(X,N)
if nargin<2,   N = length(X);   end
sq10=sqrt(10); b = [-2 0 2]/sq10; c = [-3 -1 1 3]/sq10;
Xr = real(X);   Xi = imag(X);
for i=1:N
   R(find(Xr<b(1))) = c(1);    I(find(Xi<b(1))) = c(1);
   R(find(b(1)<=Xr&Xr<b(2))) = c(2);    I(find(b(1)<=Xi&Xi<b(2))) = c(2);
   R(find(b(2)<=Xr&Xr<b(3))) = c(3);    I(find(b(2)<=Xi&Xi<b(3))) = c(3);
   R(find(b(3)<=Xr)) = c(4);    I(find(b(3)<=Xi)) = c(4);
end
X_hat = R + j*I;
```

11.3 ML Signal Detection

Maximum likelihood (ML) detection calculates the Euclidean distance between the received signal vector and the product of all possible transmitted signal vectors with the given channel $\mathbf{H}$, and finds the one with the minimum distance. Let $\mathbf{C}$ and N_T denote a set of signal constellation symbol points and a number of transmit antennas, respectively. Then, ML detection determines the estimate of the transmitted signal vector $\mathbf{x}$ as

$$\hat{\mathbf{x}}_{ML} = \underset{\mathbf{x} \in C^{N_T}}{\text{argmin}} \|\mathbf{y} - \mathbf{Hx}\|^2 \tag{11.19}$$

where $\|\mathbf{y} - \mathbf{Hx}\|^2$ corresponds to the ML metric. The ML method achieves the optimal performance as the maximum a posteriori (MAP) detection when all the transmitted vectors are equally likely. However, its complexity increases exponentially as modulation order and/or the number of transmit antennas increases [234]. The required number of ML metric calculation is $|C|^{N_T}$, that is, the complexity of metric calculation exponentially increases with the number of antennas. Even if this particular method suffers from computational complexity, its performance serves as a reference to other detection methods since it corresponds to the best possible performance. It has been shown that the number of ML metric calculations can be reduced from $|C|^{N_T}$ to $|C|^{N_T-1}$ by the modified ML (MML) detection method [234]. In other words, it will be useful for reducing the complexity when $N_T = 2$. However, its complexity is still too much for $N_T \geq 3$.

Note that the linear detection methods in Section 11.1 and OSIC detection methods in Section 11.2 require much lower complexity than the optimal ML detection, but their performance is significantly inferior to the ML detection. The performances of the OSIC detection method with post-detection SINR-based ordering and the ML detection are compared in Figure 11.4, which has been obtained by using Program 11.3. It is obvious that the ML detection outperforms the OSIC detection. Therefore, there have been active researches to develop the detection methods that still consider the ML detection criterion in Equation (11.19) while still achieving a near-optimal performance with less complexity.

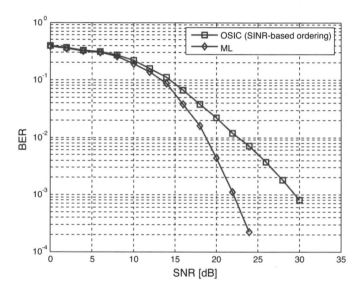

Figure 11.4   Performance comparison: OSIC vs. ML detection methods.

MATLAB® Program: ML Signal Detection

Program 11.3 "ML_detector" for ML signal detection

```
function [X_hat]=ML_detector(y,H)
QAM_table = [-3-3j, -3-j, -3+3j, -3+j, -1-3j, -1-j, -1+3j, -1+j,
             3-3j, 3-j, 3+3j, 3+j, 1-3j, 1-j, 1+3j, 1+j]/sqrt(10);
metric = 100000;
for l = 1:16
    x_tmp(1) = QAM_table(l);    Esti_y1(:,1) = y - H(:,1).*x1_tmp;
    for m = 1:16
        x_tmp(2) = QAM_table(m);
        Esti_y2(:,1) = Esti_y1(:,1) - H(:,2).*x2_tmp;
        for n = 1:16
```

```
            x_tmp(3) = QAM_table(n);
            Esti_y3(:,1) = Esti_y2(:,1) - H(:,3).*x3_tmp;
            for o = 1:16
                x_tmp(4) = QAM_table(o);
                Esti_y4(:,1) = Esti_y3(:,1) - H(:,4).*x4_tmp;
                metric_tmp = sqrt(Esti_y4(:,1)'*Esti_y4(:,1));
                if metric_tmp<metric
                   X_hat = x_tmp;    metric = metric_tmp;
                end
            end
        end
    end
end
```

11.4 Sphere Decoding Method

Sphere decoding (SD) method intends to find the transmitted signal vector with minimum ML metric, that is, to find the ML solution vector. However, it considers only a small set of vectors within a given sphere rather than all possible transmitted signal vectors [235, 236]. SD adjusts the sphere radius until there exists a single vector (ML solution vector) within a sphere. It increases the radius when there exists no vector within a sphere, and decreases the radius when there exist multiple vectors within the sphere.

In the sequel, we sketch the idea of SD through an example. Consider a square QAM in a 2×2 complex MIMO channel. The underlying complex system can be converted into an equivalent real system. Let y_{jR} and y_{jI} denote the real and imaginary parts of the received signal at the jth receive antenna, that is, $y_{jR} = \text{Re}\{y_j\}$ and $y_{jI} = \text{Im}\{y_j\}$. Similarly, the input signal x_i from the ith antenna can be represented by $x_{iR} = \text{Re}\{x_i\}$ and $x_{iI} = \text{Im}\{x_i\}$. For the 2×2 MIMO channel, the received signal can be expressed in terms of its real and imaginary parts as follows:

$$\begin{bmatrix} y_{1R} + jy_{1I} \\ y_{2R} + jy_{2I} \end{bmatrix} = \begin{bmatrix} h_{11R} + jh_{11I} & h_{12R} + jh_{12I} \\ h_{21R} + jh_{21I} & h_{22R} + jh_{22I} \end{bmatrix} \begin{bmatrix} x_{1R} + jx_{1I} \\ x_{2R} + jx_{2I} \end{bmatrix} + \begin{bmatrix} z_{1R} + jz_{1I} \\ z_{2R} + jz_{2I} \end{bmatrix} \quad (11.20)$$

where $h_{ij} = \text{Re}\{h_{ij}\}$, $h_{ij} = \text{Im}\{h_{ij}\}$, $z_i = \text{Re}\{z_i\}$, and $z_i = \text{Im}\{z_i\}$. The real and imaginary parts of Equation (11.20) can be respectively expressed as

$$\begin{bmatrix} y_{1R} \\ y_{2R} \end{bmatrix} = \begin{bmatrix} h_{11R} & h_{12R} \\ h_{21R} & h_{22R} \end{bmatrix} \begin{bmatrix} x_{1R} \\ x_{2R} \end{bmatrix} - \begin{bmatrix} h_{11I} & h_{12I} \\ h_{21I} & h_{22I} \end{bmatrix} \begin{bmatrix} x_{1I} \\ x_{2I} \end{bmatrix} + \begin{bmatrix} z_{1R} \\ z_{2R} \end{bmatrix}$$

$$= \begin{bmatrix} h_{11R} & h_{12R} & -h_{11I} & -h_{12I} \\ h_{21R} & h_{22R} & -h_{21I} & -h_{22I} \end{bmatrix} \begin{bmatrix} x_{1R} \\ x_{2R} \\ x_{1I} \\ x_{2I} \end{bmatrix} + \begin{bmatrix} z_{1R} \\ z_{2R} \end{bmatrix} \quad (11.21a)$$

and

$$
\begin{bmatrix} y_{1I} \\ y_{2I} \end{bmatrix} = \begin{bmatrix} h_{11I} & h_{12I} & h_{11R} & h_{12R} \\ h_{21I} & h_{22I} & h_{21R} & h_{22R} \end{bmatrix} \begin{bmatrix} x_{1R} \\ x_{2R} \\ x_{1I} \\ x_{2I} \end{bmatrix} + \begin{bmatrix} z_{1I} \\ z_{2I} \end{bmatrix}.
\tag{11.21b}
$$

The above two Equations (11.21a) and (11.21b) can be combined to yield the following expression:

$$
\underbrace{\begin{bmatrix} y_{1R} \\ y_{2R} \\ y_{1I} \\ y_{2I} \end{bmatrix}}_{\bar{\mathbf{y}}} = \underbrace{\begin{bmatrix} h_{11R} & h_{12R} & -h_{11I} & -h_{12I} \\ h_{21R} & h_{22R} & -h_{21I} & -h_{22I} \\ h_{11I} & h_{12I} & h_{11R} & h_{12R} \\ h_{21I} & h_{22I} & h_{21R} & h_{22R} \end{bmatrix}}_{\bar{\mathbf{H}}} \underbrace{\begin{bmatrix} x_{1R} \\ x_{2R} \\ x_{1I} \\ x_{2I} \end{bmatrix}}_{\bar{\mathbf{x}}} + \underbrace{\begin{bmatrix} z_{1R} \\ z_{2R} \\ z_{1I} \\ z_{2I} \end{bmatrix}}_{\bar{\mathbf{z}}}
\tag{11.22}
$$

For $\bar{\mathbf{y}}$, $\bar{\mathbf{H}}$, $\bar{\mathbf{x}}$, and $\bar{\mathbf{z}}$ defined in Equation (11.22), the SD method exploits the following relation:

$$
\arg \min_{\bar{\mathbf{x}}} \|\bar{\mathbf{y}} - \bar{\mathbf{H}}\bar{\mathbf{x}}\|^2 = \arg \min_{\bar{\mathbf{x}}} (\bar{\mathbf{x}} - \hat{\bar{\mathbf{x}}})^T \bar{\mathbf{H}}^T \bar{\mathbf{H}} (\bar{\mathbf{x}} - \hat{\bar{\mathbf{x}}})
\tag{11.23}
$$

where $\hat{\bar{\mathbf{x}}} = (\bar{\mathbf{H}}^H \bar{\mathbf{H}})^{-1} \bar{\mathbf{H}}^H \bar{\mathbf{y}}$, which is the unconstrained solution[1] of the real system shown in Equation (11.22). Equation (11.23) is proved in Appendix 11.A. It shows that the ML solution can be determined by the different metric $(\bar{\mathbf{x}} - \hat{\bar{\mathbf{x}}})^T \bar{\mathbf{H}}^T \bar{\mathbf{H}} (\bar{\mathbf{x}} - \hat{\bar{\mathbf{x}}})$. Consider the following sphere with the radius of R_{SD}:

$$
(\bar{\mathbf{x}} - \hat{\bar{\mathbf{x}}})^T \bar{\mathbf{H}}^T \bar{\mathbf{H}} (\bar{\mathbf{x}} - \hat{\bar{\mathbf{x}}}) \le R_{SD}^2.
\tag{11.24}
$$

The SD method considers only the vectors inside a sphere defined by Equation (11.24). Figure 11.5 illustrates a sphere with the center of $\hat{\bar{\mathbf{x}}} = (\bar{\mathbf{H}}^H \bar{\mathbf{H}})^{-1} \bar{\mathbf{H}}^H \bar{\mathbf{y}}$ and radius of R_{SD}. In this example, this sphere includes four candidate vectors, one of which is the ML solution vector. We note that no vector outside the sphere can be the ML solution vector because their ML metric values are bigger than the ones inside the sphere. If we were fortunate to choose the closest one among the four candidate vectors, we can reduce the radius in Equation (11.24) so that we may have a sphere within which a single vector remains. In other words, the ML solution vector is now contained in this sphere with a reduced radius, as illustrated in Figure 11.5(b).

Note that the new metric in Equation (11.23) is also expressed as

$$
(\bar{\mathbf{x}} - \hat{\bar{\mathbf{x}}})^T \bar{\mathbf{H}}^T \bar{\mathbf{H}} (\bar{\mathbf{x}} - \hat{\bar{\mathbf{x}}}) = (\bar{\mathbf{x}} - \hat{\bar{\mathbf{x}}})^T \mathbf{R}^T \mathbf{R} (\bar{\mathbf{x}} - \hat{\bar{\mathbf{x}}}) = \|\mathbf{R}(\bar{\mathbf{x}} - \hat{\bar{\mathbf{x}}})\|^2
\tag{11.25}
$$

[1] Being unconstrained means that the entries of $\bar{\mathbf{x}}$ in Equation (11.23) are not constrained to be one of the symbol points in signal constellation.

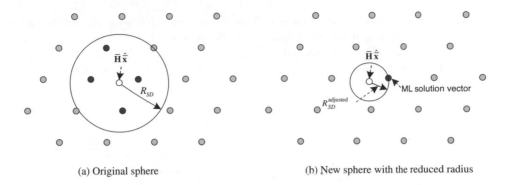

(a) Original sphere   (b) New sphere with the reduced radius

Figure 11.5   Illustration of the sphere in sphere decoding.

where $\mathbf{R}$ is obtained from QR decomposition of the real channel matrix $\bar{\mathbf{H}} = \mathbf{QR}$. When $N_T = N_R = 2$, the metric in Equation (11.25) is given as

$$
\|\mathbf{R}(\bar{\mathbf{x}}-\hat{\bar{\mathbf{x}}})\|^2 = \left\| \begin{bmatrix} r_{11} & r_{12} & r_{13} & r_{14} \\ 0 & r_{22} & r_{23} & r_{24} \\ 0 & 0 & r_{33} & r_{34} \\ 0 & 0 & 0 & r_{44} \end{bmatrix} \begin{bmatrix} \bar{x}_1-\hat{\bar{x}}_1 \\ \bar{x}_2-\hat{\bar{x}}_2 \\ \bar{x}_3-\hat{\bar{x}}_3 \\ \bar{x}_4-\hat{\bar{x}}_4 \end{bmatrix} \right\|^2
$$

$$
= \left| r_{44}\left(\bar{x}_4-\hat{\bar{x}}_4\right) \right|^2 + \left| r_{33}\left(\bar{x}_3-\hat{\bar{x}}_3\right) + r_{34}\left(\bar{x}_4-\hat{\bar{x}}_4\right) \right|^2
$$
$$
+ \left| r_{22}\left(\bar{x}_2-\hat{\bar{x}}_2\right) + r_{23}\left(\bar{x}_3-\hat{\bar{x}}_3\right) + r_{24}\left(\bar{x}_4-\hat{\bar{x}}_4\right) \right|^2
$$
$$
+ \left| r_{11}\left(\bar{x}_1-\hat{\bar{x}}_1\right) + r_{12}\left(\bar{x}_2-\hat{\bar{x}}_2\right) + r_{13}\left(\bar{x}_3-\hat{\bar{x}}_3\right) + r_{14}\left(\bar{x}_4-\hat{\bar{x}}_4\right) \right|^2 \qquad (11.26)
$$

From Equations (11.25) and (11.26), the sphere in Equation (11.24) can be expressed as

$$
\left| r_{44}\left(\bar{x}_4-\hat{\bar{x}}_4\right) \right|^2 + \left| r_{33}\left(\bar{x}_3-\hat{\bar{x}}_3\right) + r_{34}\left(\bar{x}_4-\hat{\bar{x}}_4\right) \right|^2 + \left| r_{22}\left(\bar{x}_2-\hat{\bar{x}}_2\right) + r_{23}\left(\bar{x}_3-\hat{\bar{x}}_3\right) + r_{24}\left(\bar{x}_4-\hat{\bar{x}}_4\right) \right|^2
$$
$$
+ \left| r_{11}\left(\bar{x}_1-\hat{\bar{x}}_1\right) + r_{12}\left(\bar{x}_2-\hat{\bar{x}}_2\right) + r_{13}\left(\bar{x}_3-\hat{\bar{x}}_3\right) + r_{14}\left(\bar{x}_4-\hat{\bar{x}}_4\right) \right|^2 \le R_{SD}^2
$$

$$(11.27)$$

Using the sphere in Equation (11.27), the details of SD method are now described with the following four steps:

Step 1: Referring to Equation (11.27), we first consider a candidate value for $\bar{x}_4$ in its own single dimension, that is, which is arbitrarily chosen from the points in the sphere $\left| r_{44}\left(\bar{x}_4-\hat{\bar{x}}_4\right) \right|^2 \le R_{SD}^2$. In other words, this point must be chosen in the following range:

$$
\hat{\bar{x}}_4 - \frac{R_{SD}}{r_{44}} \le \bar{x}_4 \le \hat{\bar{x}}_4 + \frac{R_{SD}}{r_{44}} \qquad (11.28)
$$

Let $\tilde{\tilde{x}}_4$ denote the point chosen in **Step 1**. If there exists no candidate point satisfying the inequalities, the radius needs to be increased. We assume that a candidate value was successfully chosen. Then we proceed to **Step 2**.

Step 2: Referring to Equation (11.27) again, a candidate value for $\bar{x}_3$ is chosen from the points in the following sphere:

$$\left|r_{44}\left(\tilde{\tilde{x}}_4-\hat{\hat{x}}_4\right)\right|^2 + \left|r_{33}\left(\bar{x}_3-\hat{\hat{x}}_3\right)+r_{34}\left(\tilde{\tilde{x}}_4-\hat{\hat{x}}_4\right)\right|^2 \le R_{SD}^2 \qquad (11.29)$$

which is equivalent to

$$\hat{\hat{x}}_3 - \frac{\sqrt{R_{SD}^2-\left|r_{44}(\tilde{\tilde{x}}_4-\hat{\hat{x}}_4)\right|^2}-r_{34}(\tilde{\tilde{x}}_4-\hat{\hat{x}}_4)}{r_{33}} \le \bar{x}_3 \le \hat{\hat{x}}_3 + \frac{\sqrt{R_{SD}^2-\left|r_{44}(\tilde{\tilde{x}}_4-\hat{\hat{x}}_4)\right|^2}-r_{34}(\tilde{\tilde{x}}_4-\hat{\hat{x}}_4)}{r_{33}}$$

$$(11.30)$$

Note that $\tilde{\tilde{x}}_4$ in Equation (11.30) is the one already chosen in **Step 1**. If a candidate value for $\bar{x}_3$ does not exist, we go back to **Step 1** and choose other candidate value $\tilde{\tilde{x}}_4$. Then search for $\bar{x}_3$ that meets the inequalities in Equation (11.30) for the given $\tilde{\tilde{x}}_4$. In case that no candidate value $\bar{x}_3$ exists with all possible values of $\tilde{\tilde{x}}_4$, we increase the radius of sphere, R_{SD}, and repeat the **Step 1**. Let $\tilde{\tilde{x}}_4$ and $\tilde{\tilde{x}}_3$ denote the final points chosen from **Step 1** and **Step 2**, respectively.

Step 3: Given $\tilde{\tilde{x}}_4$ and $\tilde{\tilde{x}}_3$, a candidate value for $\bar{x}_2$ is chosen from the points in the following sphere:

$$\begin{aligned}\left|r_{44}\left(\tilde{\tilde{x}}_4-\hat{\hat{x}}_4\right)\right|^2 &+ \left|r_{33}\left(\tilde{\tilde{x}}_3-\hat{\hat{x}}_3\right)+r_{34}\left(\tilde{\tilde{x}}_4-\hat{\hat{x}}_4\right)\right|^2 \\ &+ \left|r_{22}\left(\bar{x}_2-\hat{\hat{x}}_2\right)+r_{23}\left(\tilde{\tilde{x}}_3-\hat{\hat{x}}_3\right)+r_{24}\left(\tilde{\tilde{x}}_4-\hat{\hat{x}}_4\right)\right|^2 \le R_{SD}^2.\end{aligned} \qquad (11.31)$$

Arbitrary value is chosen for $\bar{x}_2$ inside the sphere of Equation (11.31). In choosing a point, the inequality in Equation (11.31) is used as in the previous steps. If no candidate value for $\bar{x}_2$ exists, we go back to **Step 2** and choose another candidate value $\tilde{\tilde{x}}_3$. In case that no candidate value for $\bar{x}_2$ exists after trying all possible candidate values for $\tilde{\tilde{x}}_3$, we go back to **Step 1** and choose another candidate value for $\tilde{\tilde{x}}_4$. The final points chosen from **Step 1** through **Step 3** are denoted as $\tilde{\tilde{x}}_4$, $\tilde{\tilde{x}}_3$, and $\tilde{\tilde{x}}_2$, respectively.

Step 4: Now, a candidate value for $\bar{x}_1$ is chosen from the points in the following sphere:

$$\begin{aligned}\left|r_{44}\left(\tilde{\tilde{x}}_4-\hat{\hat{x}}_4\right)\right|^2 &+ \left|r_{33}\left(\tilde{\tilde{x}}_3-\hat{\hat{x}}_3\right)+r_{34}\left(\tilde{\tilde{x}}_4-\hat{\hat{x}}_4\right)\right|^2 + \left|r_{22}\left(\tilde{\tilde{x}}_2-\hat{\hat{x}}_2\right)+r_{23}\left(\tilde{\tilde{x}}_3-\hat{\hat{x}}_3\right)+r_{24}\left(\tilde{\tilde{x}}_4-\hat{\hat{x}}_4\right)\right|^2 \\ &+ \left|r_{11}\left(\bar{x}_1-\hat{\hat{x}}_1\right)+r_{12}\left(\tilde{\tilde{x}}_2-\hat{\hat{x}}_2\right)+r_{13}\left(\tilde{\tilde{x}}_3-\hat{\hat{x}}_3\right)+r_{14}\left(\tilde{\tilde{x}}_4-\hat{\hat{x}}_4\right)\right|^2 \le R_{SD}^2.\end{aligned}$$

$$(11.32)$$

An arbitrary value satisfying Equation (11.32) is chosen for $\bar{x}_1$. If no candidate value for $\bar{x}_1$ exists, we go back to **Step 3** to choose other candidate value $\tilde{\tilde{x}}_2$. In case that no candidate value for $\bar{x}_1$ exists after trying with all possible candidate values for $\tilde{\tilde{x}}_2$, we go back to **Step 2** to choose another value for $\bar{x}_3$. Let $\tilde{\tilde{x}}_1$ denote the candidate value for $\bar{x}_1$. Once we find all candidate values, $\tilde{\tilde{x}}_4$, $\tilde{\tilde{x}}_3$, $\tilde{\tilde{x}}_2$, and $\tilde{\tilde{x}}_1$, then the corresponding radius is calculated by using Equation (11.32). Using the new reduced radius, **Step 1** is repeated. If $\left[\tilde{\tilde{x}}_1 \ \tilde{\tilde{x}}_2 \ \tilde{\tilde{x}}_3 \ \tilde{\tilde{x}}_4\right]$ turns out to be a single point inside a sphere with that radius, it is declared as the ML solution vector and our searching procedure stops.

Program 11.4 ("SD_detector") implements the SD method. Note that the performance of SD is the same as that of ML, but the complexity is significantly reduced. Now we present a simple example of sphere decoding to illustrate the above steps for 2×2 MIMO system.

MATLAB® Programs: Sphere Decoding Method

Program 11.4 "SD_detector"

```
function [X_hat]=SD_detector(y,H,NT)
% Input parameters
%      y : received signal, NRx1
%      H : Channel matrix, NRxNT
%     NT : number of Tx antennas
% Output parameter
%      X_hat : estimated signal, NTx1
global x_list;            % candidate symbols in real constellations
global x_now;             % temporary x_vector elements
global x_hat;             % inv(H)*y
global x_sliced;          % sliced x_hat
global x_pre;             % x vectors obtained in the previous stage
global real_constellation; % real constellation
global R;                 % R in the QR decomposition
global radius_squared;    % radius^2
global x_metric;          % ML metrics of previous stage candidates
global len;               % NT*2
QAM_table2 = [-3-3j, -3-j, -3+3j, -3+j, -1-3j, -1-j, -1+3j, -1+j,3-3j,
              3-j, 3+3j, 3+j, 1-3j, 1-j, 1+3j, 1+j]/sqrt(10); % 16-QAM
real_constellation = [-3 -1 1 3]/sqrt(10);
y =[real(y); imag(y)];   % y : complex vector  -> real vector
H =[real(H)    -(imag(H)) ; imag(H)    real(H)];
                            % H : complex vector -> real vector
len = NT*2; % complex -> real
x_list = zeros(len,4); % 4 : real constellation length, 16-QAM
x_now = zeros(len,1);
x_hat = zeros(len,1);
x_pre = zeros(len,1);
x_metric = 0;
[Q,R] = qr(H);       % NR x NT QR decomposition
x_hat = inv(H)*y;    % zero forcing equalization
x_sliced = QAM16_real_slicer(x_hat,len)';   % slicing
radius_squared = norm(R*(x_sliced-x_hat))^2;     % Radious^2
transition = 1; % meaning of transition
             % 0 : radius*2, 1~len : stage number
             % len+1 : compare two vectors in terms of norm values
             % len+2 : finish
flag = 1;        % transition tracing
          % 0 : stage index increases by +1
          % 1 : stage index decreases by -1
          % 2 : 1->len+2 or len+1->1
while(transition<len+2)
```

```
   if transition == 0     % radius_squared*2
       [flag,transition,radius_squared,x_list]
              = radius_control(radius_squared,transition);
   elseif transition <= len
       [flag,transition] = stage_processing(flag,transition);
   elseif transition == len+1 %
       [flag,transition] = compare_vector_norm(transition);
   end
end
ML = x_pre;
for i=1:len/2
   X_hat(i) = ML(i)+j*ML(i+len/2);
end
```

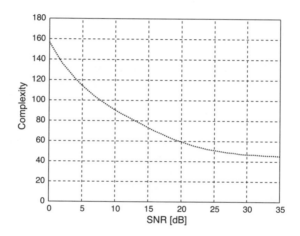

Figure 11.6   The complexity of SD for 16-QAM, 2 × 2 MIMO channel, and ZF method for initial radius calculation.

Let us illustrate the complexity of SD with the example of 2 × 2 MIMO system. As can be observed in Figure 11.6, the complexity of SD depends on how well the initial radius is chosen. Although an initial radius can be determined in various ways, suppose that the initial radius is determined as

$$R_{SD}^2 = \sum_{i=1}^{4} \left| \sum_{k=i}^{4} r_{ik} \left(\bar{\bar{x}}_k - \hat{\bar{x}}_k \right) \right|^2 \tag{11.33}$$

where $\hat{\bar{x}} = \begin{bmatrix} \hat{\bar{x}}_1 & \hat{\bar{x}}_2 & \hat{\bar{x}}_3 & \hat{\bar{x}}_4 \end{bmatrix}^T$ is the unconstrained LS solution, and $\bar{\bar{x}}_i = Q(\hat{\bar{x}}_i)$, $i = 1, 2, 3, 4$. Note that computation of the initial radius in Equation (11.33) requires 14 real multiplications. Using this initial radius, the inequality conditions for candidate selection for $\bar{x}_i$,

$i = 4-s+1$ in **Step** s, $s = 1, 2, 3, 4$, can be generalized as

$$\hat{x}_{i,LS} + \frac{-\alpha_i - \beta_i}{r_{ii}} \leq x_i \leq \hat{x}_{i,LS} + \frac{\alpha_i - \beta_i}{r_{ii}} \tag{11.34}$$

where

$$\alpha_i = \sqrt{R_{SD}^2 - \sum_{k=i+1}^{4} \left| \sum_{p=k}^{4} r_{kp} \left(\tilde{x}_p - \hat{x}_{p,LS} \right) \right|^2} \quad \text{and} \quad \beta_i = \sum_{k=i+1}^{4} r_{ik} \left(\bar{\tilde{x}}_k - \hat{\tilde{x}}_k \right).$$

In Equation (11.34), $\{\tilde{x}_k\}_{k=i+1}^4$ are the selected symbols in the previous steps. Using the fact that each symbol $\bar{x}_k$ is an integer, and reusing the results in the previous steps, one multiplication, two divisions, and one square-root operations are required for calculating the Equation (11.34). In the first step $(s = 1)$, $\beta_4 = 0$ and thus, one division and one square-root operation are required. In order to calculate the new radius using a new vector of length $2 \times N_R (= 4)$, only one multiplication is required, because the results in the previous steps can be reused. Table 11.1 summarizes the complexity of SD. In Table 11.1, the calculation $\hat{\tilde{x}} = (\bar{H})^{-1} \bar{y}$ refers to the multiplication of $\bar{H}^{-1}$ and $\bar{y}$, excluding the calculation of $\bar{H}^{-1}$. Program 11.4 ("SD_detector") can be used to evaluate the complexity of SD. Figure 11.6 shows the complexity SD in terms of the sum of multiplication, division, and square-root operations, as SNR varies. As the SNR increases, the ZF solution $\hat{\tilde{x}}$ becomes more likely to coincide with the ML solution vector. Therefore, the initial radius is properly chosen as in Figure 11.4b, eliminating the necessity of adjusting the radius. Note that the complexity of the ML signal detection corresponds to the ML metric calculation of $16^2 = 256$ times. Assuming that four real multiplications are required for each ML metric calculation, $256 \times 4 = 1024$ real multiplications are required in total. The main drawback of SD is that its complexity depends on SNR. Furthermore, the worst-case complexity is the same as that of ML detection, although the average complexity is significantly reduced as shown in Figure 11.6.

Table 11.1 Complexity of sphere decoding in each step.

	Multiplications	Divisions	Square roots
$\hat{\tilde{x}} = (\bar{H})^{-1} \bar{y}$	16	0	0
R_{SD}^2 in Equation (11.33)	14	0	0
Step 1	0	1	1
Step 2–4 each	1	2	1
R_{SD}^2 update	1	0	0

Program 11.5 "state_processing"

```
function [flag, transition] = stage_processing(flag, transition)
% Input
%     flag : previous stage index
%         flag = 0 : stage index decreased
%             -> x_now empty -> new x_now
```

```
%          flag = 1 : stage index decreased -> new x_now
%          flag = 2 : previous stage index =len+1
%              -> If R>R' → start from the first stage
%       transition : stage number
% Output
%       flag : stage number is calculated from flag
%       transition : next stage number, 0 : R*2, 1: next stage,
%                   len+2: finish
global x_list   x_metric   x_now   x_hat   R
global real_constellation   radius_squared   x_sliced
stage_index = length(R(1,:))-(transition-1);
if flag==2 % previous stage=len+1 : recalculate radius R'
   radius_squared  = norm(R*(x_sliced-x_hat))^2;
end
if flag~=0 % previous stage=len+1 or 0
   % -> upper and lower bound calculation, x_list(stage_index,:)
   [bound_lower bound_upper] = bound(transition);
   for i =1:4     % search for a candidate in x_now(stage_index),
                % 4=size(real_constellation), 16-QAM assumed
      if bound_lower <= real_constellation(i)
         && real_constellation(i) <= bound_upper
            list_len = list_length(x_list(stage_index,:));
            x_list(stage_index,list_len+1) = real_constellation(i);
      end
   end
end
list_len = list_length(x_list(stage_index,:));
if list_len==0      % no candidate in x_now
   if x_metric == 0 || transition ~= 1
   % transition >=2 → if no candidate → decrease stage index
      flag = 0;
      transition = transition-1;
   elseif x_metric ~= 0 && transition == 1
   % above two conditions are met→ ML solution found
      transition = length(R(1,:))+2;   % finish stage
   end
else                % candidate exist in x_now → increase stage index
   flag = 1;
   transition = transition+1;
   x_now(stage_index) = x_list(stage_index,1);
   x_list(stage_index,:) = [x_list(stage_index,[2:4]) 0];
end
```

Program 11.6 "list_length"

```
function [len]=list_length(list)
% Input
%       list : vector type
% Output
```

```
%      len : index number
len = 0;
for i=1:4
    if list(i)==0,   break;    else   len = len+1;    end
end
```

Program 11.7 "bound"

```
function [bound_lower,bound_upper]=bound(transition)
% Input
%     R : [Q R] = qr(H)
%     radius_squared : R^2
%     transition      : stage number
%     x_hat           : inv(H)*y
%     x_now           : slicing x_hat
% Output
%     bound_lower     : bound lower
%     bound_upper     : bound upper
global R radius_squared x_now x_hat
len = length(x_hat);
temp_sqrt = radius_squared;
temp_k=0;
for i=1:1:transition-1
    temp_abs=0;
    for k=1:1:i
        index_1 = len-(i-1);
        index_2 = index_1+ (k-1);
        temp_k = R(index_1,index_2)*(x_now(index_2)-x_hat(index_2));
        temp_abs=temp_abs+temp_k;
    end
    temp_sqrt = temp_sqrt - abs(temp_abs)^2;
end
temp_sqrt = sqrt(temp_sqrt);
temp_no_sqrt = 0;
index_1 = len-(transition-1);
index_2 = index_1;
for i=1:1:transition-1
    index_2 = index_2+1;
    temp_i = R(index_1,index_2)*(x_now(index_2)-x_hat(index_2));
    temp_no_sqrt = temp_no_sqrt - temp_i;
end
temp_lower = -temp_sqrt + temp_no_sqrt;
temp_upper = temp_sqrt + temp_no_sqrt;
index = len-(transition-1);
bound_lower = temp_lower/R(index,index) + x_hat(index);
bound_upper = temp_upper/R(index,index) + x_hat(index);
bound_upper = fix(bound_upper*sqrt(10))/sqrt(10);
bound_lower = ceil(bound_lower*sqrt(10))/sqrt(10);
```

Program 11.8 "radius_control"

```
function [flag,transition,radius_squared,x_list]
                       =radius_control(radius_squared,transition)
% Input parameters
%      radius_squared : current radius
%      transition : current stage number
% Output parameters
%      radius_squared : doubled radius
%      transition : next stage number
%      flag : next stage number is calculated from flag
global len;
radius_squared = radius_squared*2;
transition = transition+1;
flag = 1;
x_list(len,:)=zeros(1,4);
```

Program 11.9 "vector_comparison"

```
function [check]=vector_comparison(vector_1,vector_2)
% To check if the two vectors are the same
% Input
%   pre_x : vector 1
%   now_x : vector 2
% Output
%   check : 1-> same vectors, 0-> different vectors
check = 0;
len1 = length(vector_1);   len2 = length(vector_2);
if len1~=len2,   error('Vector size is different');   end
for column_num = 1:len1
    if vector_1(column_num,1)==vector_2(column_num,1)
      check = check + 1;
    end
end
if check == len1,   check = 1;   else   check = 0;   end
```

Program 11.10 "compare_vector_norm"

```
function [flag,transition]=compare_vector_norm(transition)
% stage index increased(flag = 1) : recalculate x_list(index,:)
% stage index decreased(flag = 0) : in the previous stage,
                             no candidate x_now in x_list
% Input
%     flag : previous stage
%     transition : stage number
% Output
```

```
%      flag : next stage number is calculated from flag
%      transition : next stage number
global x_list x_pre x_metric x_now x_hat R len radius_squared x_sliced
vector_identity = vector_comparison(x_pre,x_now);
        % check if the new candidate is among the ones we found before
if vector_identity==1   % if 1 → ML solution found
   len_total = 0;
   for i=1:len   % if the vector is unique → len_total = 0
      len_total = len_total + list_length(x_list(i,:));
   end
   if len_total==0       % ML solution vector found
      transition = len+2;  % finish
      flag = 1;
   else                          % more than one candidates
      transition = transition-1;  % go back to the previous stage
      flag =0;
   end
else   % if 0 → new candidate vector is different from the previous candidate
        vector and norm is smaller → restart
     x_sliced_temp = x_now;
     metric_temp   = norm(R*(x_sliced_temp-x_hat))^2;
      if metric_temp<=radius_squared
       % new candidate vector has smaller metric → restart
       x_pre = x_now;   x_metric = metric_temp;   x_sliced = x_now;
       transition = 1;      % restart
       flag = 2;
       x_list=zeros(len,4);   x_now=zeros(len,1);   % initialization
     else % new candidate vector has a larger ML metric
       transition = transition-1;   % go back to the previous stage
       flag =0;
   end
 end
end
```

11.5 QRM-MLD Method

Assuming that the numbers of the transmit and receive antennas are equal, consider QR decomposition of the channel matrix, that is, $\mathbf{H} = \mathbf{QR}$. Then, the ML metric in Equation (11.19) can be equivalently expressed as

$$
\begin{aligned}
\|\mathbf{y}-\mathbf{Hx}\| &= \|\mathbf{y}-\mathbf{QRx}\| \\
&= \left\|\mathbf{Q}^{H}(\mathbf{y}-\mathbf{QRx})\right\| \\
&= \|\tilde{\mathbf{y}}-\mathbf{Rx}\|.
\end{aligned}
\tag{11.35}
$$

We note that QR decomposition also has been applied to sphere decoding in Equation (11.25) where it is for an equivalent real system. In Equation (11.35), however, it is for a complex channel matrix. In the following discussion, we will illustrate the QRM-MLD method [237,

238] for $N_T = N_R = 4$, in which case Equation (11.35) can be expanded as

$$\|\tilde{\mathbf{y}} - \mathbf{R}\mathbf{x}\|^2 = \left\| \begin{bmatrix} \tilde{y}_1 \\ \tilde{y}_2 \\ \tilde{y}_3 \\ \tilde{y}_4 \end{bmatrix} - \begin{bmatrix} r_{11} & r_{12} & r_{13} & r_{14} \\ 0 & r_{22} & r_{23} & r_{24} \\ 0 & 0 & r_{33} & r_{34} \\ 0 & 0 & 0 & r_{44} \end{bmatrix} \begin{bmatrix} x_1 \\ x_2 \\ x_3 \\ x_4 \end{bmatrix} \right\|^2 \tag{11.36}$$

$$= |\tilde{y}_4 - r_{44}x_4|^2 + |\tilde{y}_3 - r_{33}x_3 - r_{34}x_4|^2 + |\tilde{y}_2 - r_{22}x_2 - r_{23}x_3 - r_{24}x_4|^2$$
$$+ |\tilde{y}_1 - r_{11}x_1 - r_{12}x_2 - r_{13}x_3 - r_{14}x_4|^2$$

Let $\mathbf{C}$ denote a signal constellation. To facilitate the following discussion, let $\arg\min_M f(x_1, x_2, \ldots, x_n)$ operation to select M of candidate vector $(x_1, x_2, \ldots, x_n)$ that $(x_1, x_2, \cdots, x_n)$ correspond to M smallest values of $f(x_1, x_2, \ldots, x_n)$. Referring to Equation (11.36), the QRM-MLD method is detailed with the following four steps, one for each symbol:

Step 1: Among $|\mathbf{C}|$ candidates for $x_4 \in \mathbf{C}$, select M of them that correspond to M smallest values of $f_1(x_4) = |\tilde{y}_4 - r_{44}x_4|^2$. Let $\tilde{x}_{4,c,1}$ denote one of those M selected symbols, $c = 1, 2, \cdots, M$, that is,

$$\{\tilde{x}_{4,c,1}\}_{c=1}^{M} = \arg\min_{M} f_1(x_4) \tag{11.37}$$
$$\substack{x_4 \in \mathbf{C}}$$

Step 2: Among $M \times |\mathbf{C}|$ candidate vectors for $\{[x_3 \ \tilde{x}_{4,c,1}]\}_{c=1}^{M}$ where $x_3 \in \mathbf{C}$, select M of them that corresponds to M smallest values of $f_2(x_3, \tilde{x}_{4,c,1}) = |\tilde{y}_4 - r_{44}\tilde{x}_{4,c,1}|^2 + |\tilde{y}_3 - r_{33}x_3 - r_{34}\tilde{x}_{4,c,1}|^2$ for $\{\tilde{x}_{4,c,1}\}_{c=1}^{M}$ given from **Step 1**. Let $[\tilde{x}_{3,c,2} \ \tilde{x}_{4,c,2}]$ denote one of those M selected vectors, that is,

$$\{[\tilde{x}_{3,c,2} \ \tilde{x}_{4,c,2}]\}_{c=1}^{M} = \arg\min_{M} f_2(x_3, \tilde{x}_{4,c,1}). \tag{11.38}$$
$$\substack{x_3 \in \mathbf{C}, \{\tilde{x}_{4,c,1}\}_{c=1}^{M}}$$

Step 3: Among $M \times |\mathbf{C}|$ candidate vectors for $\{[x_2 \ \tilde{x}_{3,c,2} \ \tilde{x}_{4,c,2}]\}_{c=1}^{M}$ where $x_2 \in \mathbf{C}$, select M of them that corresponds to M smallest values of $f_3(x_2, \tilde{x}_{3,c,2}, \tilde{x}_{4,c,2}) = |\tilde{y}_4 - r_{44}\tilde{x}_{4,c,2}|^2 + |\tilde{y}_3 - r_{33}\tilde{x}_{3,c,2} - r_{34}\tilde{x}_{4,c,2}|^2 + |\tilde{y}_2 - r_{22}x_2 - r_{23}\tilde{x}_{3,c,2} - r_{24}\tilde{x}_{4,c,2}|^2$. Let $\{[\tilde{x}_{2,c,3} \ \tilde{x}_{3,c,3} \ \tilde{x}_{4,c,3}]\}_{c=1}^{M}$ denote those M selected vectors, that is,

$$\{[\tilde{x}_{2,c,3} \ \tilde{x}_{3,c,3} \ \tilde{x}_{4,c,3}]\}_{c=1}^{M} = \arg\min_{M} f_3(x_2, \tilde{x}_{3,c,2}, \tilde{x}_{4,c,2}). \tag{11.39}$$
$$\substack{x_2 \in \mathbf{C}, \{[\tilde{x}_{3,c,2}, \tilde{x}_{4,c,2}]\}_{c=1}^{M}}$$

Step 4: Among $M \times |\mathbf{C}|$ candidate vectors for $\{[x_1 \tilde{x}_{2,c,3} \ \tilde{x}_{3,c,3} \ \tilde{x}_{4,c,3}]\}_{c=1}^{M}$ where $x_1 \in \mathbf{C}$, select M of them that corresponds to M smallest values of $f_4(x_1, \tilde{x}_{2,c,3}, \tilde{x}_{3,c,3}, \tilde{x}_{4,c,3}) = |\tilde{y}_4 - r_{44}\tilde{x}_{4,c,3}|^2 + |\tilde{y}_3 - r_{33}\tilde{x}_{3,c,3} - r_{34}\tilde{x}_{4,c,3}|^2 + |\tilde{y}_2 - r_{22}\tilde{x}_{2,c,3} - r_{23}\tilde{x}_{3,c,3} - r_{24}\tilde{x}_{4,c,3}|^2 + |\tilde{y}_1 - r_{11}x_1 - r_{12}\tilde{x}_{2,c,3} - r_{13}\tilde{x}_{3,c,3} - r_{14}\tilde{x}_{4,c,3}|^2$. Let $\{[\tilde{x}_{1,c,4} \ \tilde{x}_{2,c,4} \ \tilde{x}_{3,c,4} \ \tilde{x}_{4,c,4}]\}_{c=1}^{M}$ denote those M

selected vectors, that is,

$$\left\{\left[\tilde{x}_{1,c,4}\ \tilde{x}_{2,c,4}\ \tilde{x}_{3,c,4}\ \tilde{x}_{4,c,4}\right]\right\}_{c=1}^{M} = \underset{x_1 \in \mathbf{C}, \left\{\left[\tilde{x}_{2,c,3}\ \tilde{x}_{3,c,3}\ \tilde{x}_{4,c,3}\right]\right\}_{c=1}^{M}}{\arg\min_{M}}\ f_4\left(x_1, \tilde{x}_{2,c,3}, \tilde{x}_{3,c,3}, \tilde{x}_{4,c,3}\right) \quad (11.40)$$

Among M candidate vectors found in the above, only one that minimizes the metric in Equation (11.36) will be selected as the final detected symbol in the hard decision receiver, while the rest of $(M-1)$ vectors are discarded. In case we intend to provide soft decision values, however, $M \times |\mathbf{C}|$ candidate vectors, $\left\{\left[x_1 \in \mathbf{C}\ \tilde{x}_{2,c,3}\ \tilde{x}_{3,c,3}\ \tilde{x}_{4,c,3}\right]\right\}_{c=1}^{M}$, and their metric values are stored and used. This soft decision approach will be extensively treated in Section 11.7.4.

The performance of QRM-MLD depends on the parameter M. As M increases, its performance approaches ML performance at the sacrifice of the complexity. Program 11.11 ("QRM_MLD_detector") can be used to perform the QRD-MLD method. Figure 11.7 shows the performance of QRM-MLD for $M = 4$ and 16. When $M = 16$, QRM-MLD achieves the same performance as SD which corresponds to the ML performance. When $M = 4$, however, QRM-MLD performance degrades at the benefit of the reduced complexity. It implies that the optimal performance of QRM-MLD is warranted only with a sufficiently large value of M. Note that the complexity of SD in Section 11.4 depends on SNR, channel condition number, and method of initial radius calculation. Regardless of SNR and channel condition number, however, the complexity of QRM-MLD is fixed for the given value of M, which makes its hardware implementation simpler.

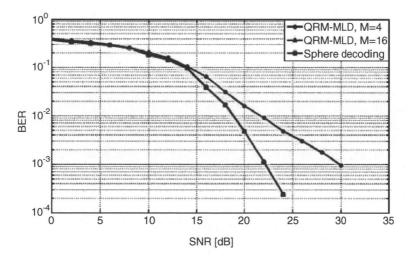

Figure 11.7 Performance comparison: SD vs. QRM-MLD.

MATLAB® Programs: QRM-MLD Method

Program 11.11 "QRM_MLD_detector"

```
function [X_hat]=QRM_MLD_detector(Y,H)
% Input
%     Y : Received signal, NRx1
%     H : Channel matrix, NTxNT
% Output
%     X_hat : Estimated signal, NTx1
global NT  M;   % NT=# of Tx antenna, M=M-algorithm parameter
QAM_table = [-3-3j, -3-j, -3+3j, -3+j, -1-3j, -1-j, -1+3j, -1+j, 3-3j,
             3-j, 3+3j, 3+j, 1-3j, 1-j, 1+3j, 1+j]/sqrt(10); % QAM table
[Q,R] = qr(H);    % QR-decomposition
Y_tilde = Q'*Y;
symbol_replica = zeros(NT,M,NT); % QAM table index
for stage = 1:NT
    symbol_replica = stage_processing(symbol_replica,stage);
end
X_hat = QAM_table(symbol_replica(:,1));
```

Program 11.12 "stage_processing1"

```
function [symbol_replica] = stage_processing(symbol_replica,stage)
% Input
%      symbol_replica : M candidate vectors
%      stage : Stage number
% Output
%      symbol_replica : M candidate vectors
global NT M; % NT=Number of Tx antennas, M=M-algorithm parameter
if stage==1; m = 1;   else m = M;   end
symbol_replica_norm = calculate_norm(symbol_replica,stage);
[symbol_replica_norm_sorted, symbol_replica_sorted]
       = sort_matrix(symbol_replica_norm);
       % sort in norm order, data is in a matrix form
symbol_replica_norm_sorted = symbol_replica_norm_sorted(1:M);
symbol_replica_sorted = symbol_replica_sorted(:,[1:M]);
if stage>=2
   for i=1:m
     symbol_replica_sorted([2:stage],i) = ...
     symbol_replica([1:stage-1],symbol_replica_sorted(2,i),(NT+2)-stage);
   end
end
if stage==1 % In stage 1, size of symbol_replica_sorted is 2xM,
              the second row is not necessary
   symbol_replica([1:stage],:,(NT+1)-stage) =
              symbol_replica_sorted(1,:);
else
   symbol_replica([1:stage],:,(NT+1)-stage) = symbol_replica_sorted;
end
```

Program 11.13 "calculate_norm"

```
function [symbol_replica_norm]=calculate_norm(symbol_replica,stage)
% Input
%     symbol_replica : M candidate vectors
%     stage : stage number
% Output
%     symbol_replica_norm : Norm values of M candidate vectors
global  QAM_table  R  Y_tilde  NT  M;
%[Q,R]=qr(H), Y_tilde=Q'*Y, NT=# of Tx antenna, M=M-algorithm parameter
if stage==1;   m = 1;   else   m = M;   end
stage_index = (NT+1)-stage;
for i=1:m
    X_temp = zeros(NT,1);
    for a=NT:-1:(NT+2)-stage
        X_temp(a) = QAM_table(symbol_replica((NT+1)-a,i,stage_index+1));
    end
    X_temp([(NT+2)-stage:(NT)]) = wrev(X_temp([(NT+2)-stage:(NT)]));
       % reordering
    Y_tilde_now = Y_tilde([(NT+1)-stage:(NT)]);
       % Y_tilde used in the current stage
    R_now = R([(NT+1)-stage:(NT)],[(NT+1)-stage:(NT)]);
       % R used in the current stage
    for k=1:length(QAM_table) % norm calculation,
       % the norm values in the previous stages can be used, however,
       % we recalculate them in an effort to simplify the MATLAB code
       X_temp(stage_index) = QAM_table(k);
       X_now = X_temp([(NT+1)-stage:(NT)]);
       symbol_replica_norm(i,k) = norm(Y_tilde_now - R_now*X_now)^2;
    end
end
end
```

Program 11.14 "sort_matrix"

```
function [entry_sorted,entry_index_sorted]=sort_matrix(matrix)
% Input
%     matrix : A matrix to be sorted
% Output
%     entry_sorted : increasingly ordered norm
%     entry_index_sorted : ordered QAM_table index
[Nrow,Ncol] = size(matrix);
flag=0; % flag = 1 → the least norm is found
matrix_T=matrix.'; vector=matrix_T(:).'; % matrix  → vector form
entry_index_sorted =[];
for m=1:Nrow*Ncol
    entry_min = min(vector);   flag=0;
    for i=1:Nrow
        if flag==1,   break;   end
        for k=1:Ncol
            if flag==1,   break;   end
```

```
            entry_temp = matrix(i,k);
            if entry_min==entry_temp
              entry_index_sorted = [entry_index_sorted [k; i]];
              entry_sorted(m) = entry_temp;
              vector((i-1)*Ncol+k) = 10000000;
              flag=1;
            end
        end
    end
end
```

11.6 Lattice Reduction-Aided Detection

In general, the linear detection and OSIC methods may increase the noise component in the course of linear filtering, thereby degrading the performance. Such noise enhancement problem becomes critical, especially when the condition number of channel matrix increases as discussed in Section 11.1. Lattice reduction method can be useful for reducing the condition numbers of channel matrices [239, 240]. Figure 11.8 illustrates two different sets of basis vectors that span the same space for two transmit antenna cases. Each vector corresponds to one of two columns in the channel matrix.

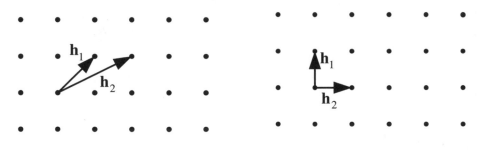

(a) A basis vector set with a large condition number   (b) An orthogonal basis vector set

Figure 11.8   Two sets of basis vectors that span the same space.

The basis vector set in Figure 11.8(a) has a larger condition number than that in Figure 11.8(b). A basis vector set with a small condition number reduces the noise enhancement in the linear detection and OSIC methods. When the basis vectors are orthogonal as in Figure 11.8(b), there is no noise enhancement at all in the process of linear filtering; thus ZF linear detection in Section 11.1 achieves the same error performance as ML or SD. The simplicity of Alamouti decoding in Section 10.3.1 is attributed to the fact that the virtual channel matrix is composed of orthogonal column vectors.

Using QR decomposition $\bar{\mathbf{H}} = \mathbf{QR}$, the system of Equation (11.22) is expressed as

$$\bar{\mathbf{y}} = \bar{\mathbf{H}}\bar{\mathbf{x}} + \bar{\mathbf{z}} = \mathbf{QR}\bar{\mathbf{x}} + \bar{\mathbf{z}}. \qquad (11.41)$$

Multiplying both sides with $\mathbf{Q}^H$, we have

$$\tilde{\tilde{\mathbf{y}}} = \mathbf{Q}^H \bar{\mathbf{y}} = \mathbf{R}\bar{\mathbf{x}} + \tilde{\mathbf{z}} \tag{11.42}$$

where $\tilde{\mathbf{z}} = \mathbf{Q}^H \bar{\mathbf{z}}$. Since $\mathbf{Q}$ is unitary, the statistical characteristics of noise components does not change, that is, $\tilde{\mathbf{z}}$ and $\bar{\mathbf{z}}$ have the same statistics, while $\bar{\mathbf{H}}$ has the same condition number as $\mathbf{R}$. Since $\|\mathbf{QR}\|_2 = \|\mathbf{RQ}\|_2 = \|\mathbf{R}\|_2$ for a unitary matrix $\mathbf{Q}$ where $\|\cdot\|_2$ denotes the matrix-induced norm[2], the condition numbers of $\bar{\mathbf{H}}$ and $\mathbf{R}$ are equal, that is, $k_2(\bar{\mathbf{H}}) = \|\bar{\mathbf{H}}\|\|\bar{\mathbf{H}}^{-1}\| = \|\mathbf{QR}\|\|\mathbf{R}^{-1}\mathbf{Q}^H\| = \|\mathbf{R}\|\|\mathbf{R}^{-1}\| = k_2(\mathbf{R})$. We note that lattice reduction does not necessarily require QR decomposition, but it will be much simpler with the triangular matrix from the implementation perspective. In the following subsections, we show how the condition number can be reduced for a triangular matrix by a lattice reduction technique.

11.6.1 Lenstra-Lenstra-Lovasz (LLL) Algorithm

Lenstra-Lenstra-Lovasz (LLL) algorithm can be used to reduce the condition number of a triangular matrix [239, 240]. We consider a real system described in Equation (11.41). Our objective is to build an equivalent system equation that is better conditioned than Equation (11.41) whose condition number depends on $\mathbf{R}$. In the sequel, let us present the LLL algorithm for 4×4 matrix $\mathbf{R} = [\mathbf{r}_1\ \mathbf{r}_2\ \mathbf{r}_3\ \mathbf{r}_4]$ where $\{\mathbf{r}_i\}_{i=1}^4$ are the ith column vectors of $\mathbf{R}$. Let $r_{j,i}$ denote the (j,i)th entry of matrix $\mathbf{R}$. Consider another 4×4 matrix $\mathbf{T} = [\mathbf{t}_1\ \mathbf{t}_2\ \mathbf{t}_3\ \mathbf{t}_4]$ where $\{\mathbf{t}_i\}_{i=1}^4$ is the ith column vector. At the time of initialization, $\mathbf{T}$ is set to an identity matrix (i.e., $\mathbf{T} = \mathbf{I}_{4\times4}$). In general, it takes $(N\text{-}1)$ steps in this algorithm for lattice reduction with an $N \times N$ matrix. In our 4×4 matrix example, we follow the three steps shown below, each of which reduces the length of the second, third, and fourth column vectors of $\mathbf{R}$.

Step 1 (Length reduction of the second column vector $\mathbf{r}_2$): Define $\mu_{1,2}$ as

$$\mu_{1,2} = \left\langle \frac{r_{1,2}}{r_{1,1}} \right\rangle \tag{11.43}$$

where $\langle \cdot \rangle$ denotes the integer closest to the argument. Using $\mu_{1,2}$ in Equation (11.43), the second column vectors of matrices $\mathbf{R}$ and $\mathbf{T}$ are modified as

$$\begin{aligned} \mathbf{r}_2 &\leftarrow \mathbf{r}_2 - \mu_{1,2}\mathbf{r}_1 \\ \mathbf{t}_2 &\leftarrow \mathbf{t}_2 - \mu_{1,2}\mathbf{t}_1 \end{aligned} \tag{11.44}$$

If $\mu_{1,2} = 0$, the above two modifications must be skipped. If the modified matrix $\mathbf{R}$ does not satisfy the following condition:

$$\delta r_{1,1}^2 \le r_{1,2}^2 + r_{2,2}^2 \tag{11.45}$$

where $1/4 < \delta \le 1$, the corresponding two column vectors of $\mathbf{R}$ and $\mathbf{T}$ are exchanged as

$$\begin{aligned} \mathbf{R} &\leftarrow [\mathbf{r}_2\ \mathbf{r}_1\ \mathbf{r}_3\ \mathbf{r}_4] \\ \mathbf{T} &\leftarrow [\mathbf{t}_2\ \mathbf{t}_1\ \mathbf{t}_3\ \mathbf{t}_4] \end{aligned} \tag{11.46}$$

---

[2] For a matrix $\mathbf{A} \in \mathbb{C}^{m \times n}$, the matrix-induced norm is defined as $\|\mathbf{A}\|_2 = \arg\max_{\mathbf{x}} \dfrac{\|\mathbf{Ax}\|_2}{\|\mathbf{x}\|_2}$.

The above updated matrix $\mathbf{R}$ is not upper-triangular, requiring the following Givens rotation so that the new matrix may still be upper-triangular:

$$\mathbf{R} \leftarrow \Theta_1 \mathbf{R} \tag{11.47}$$

The modification is reflected on the unitary matrix $\mathbf{Q}$ as follows:

$$\mathbf{Q} \leftarrow \mathbf{Q}\,\Theta_1^T \tag{11.48}$$

where Θ_1 is an orthogonal rotation matrix that is defined as

$$\Theta_1 = \begin{bmatrix} \alpha_1 & \beta_1 & 0 & 0 \\ -\beta_1 & \alpha_1 & 0 & 0 \\ 0 & 0 & 1 & 0 \\ 0 & 0 & 0 & 1 \end{bmatrix} \tag{11.49}$$

with α_1 and β_1 given as

$$\alpha_1 = \frac{r_{1,1}}{\sqrt{r_{1,1}^2 + r_{2,1}^2}} \quad \text{and} \quad \beta_1 = \frac{r_{2,1}}{\sqrt{r_{1,1}^2 + r_{2,1}^2}} \tag{11.50}$$

Then we go back to the beginning of **Step 1** and repeat the same process until the condition in (11.45) is satisfied. Once the condition in (11.45) is satisfied, operations in (11.47) and (11.48) are skipped on to **Step 2**.

Step 2 (Length reduction of the third column vector $\mathbf{r}_3$): Define $\mu_{2,3}$ as

$$\mu_{2,3} = \left\langle \frac{r_{2,3}}{r_{2,2}} \right\rangle \tag{11.51}$$

Using $\mu_{2,3}$ in Equation (11.51), the third column vectors of $\mathbf{R}$ and $\mathbf{T}$ are modified as

$$\begin{aligned} \mathbf{r}_3 &\leftarrow \mathbf{r}_3 - \mu_{2,3}\,\mathbf{r}_2 \\ \mathbf{t}_3 &\leftarrow \mathbf{t}_3 - \mu_{2,3}\,\mathbf{t}_2 \end{aligned} \tag{11.52}$$

Subsequently, define $\mu_{1,3}$ as

$$\mu_{1,3} = \left\langle \frac{r_{1,3}}{r_{1,1}} \right\rangle \tag{11.53}$$

Using $\mu_{1,3}$, the third column vectors of $\mathbf{R}$ and $\mathbf{T}$ are again modified as

$$\begin{aligned} \mathbf{r}_3 &\leftarrow \mathbf{r}_3 - \mu_{1,3}\,\mathbf{r}_1 \\ \mathbf{t}_3 &\leftarrow \mathbf{t}_3 - \mu_{1,3}\,\mathbf{t}_1 \end{aligned} \tag{11.54}$$

Note that the calculation of $\mu_{2,3}$ and the corresponding reduction in (11.52) are performed prior to the calculation of $\mu_{1,3}$ and the corresponding reduction in (11.54). The following condition is then checked to decide whether the second and third column vectors need to be exchanged:

$$\delta r_{2,2}^2 \leq r_{2,3}^2 + r_{3,3}^2 \tag{11.55}$$

In other words, if the modified matrix $\mathbf{R}$ with (11.54) does not satisfy the condition in (11.55), the second and third column vectors of $\mathbf{R}$ and $\mathbf{T}$ are exchanged as

$$\mathbf{R} \leftarrow [\mathbf{r}_1 \; \mathbf{r}_3 \; \mathbf{r}_2 \; \mathbf{r}_4]$$
$$\mathbf{T} \leftarrow [\mathbf{t}_1 \; \mathbf{t}_3 \; \mathbf{t}_2 \; \mathbf{t}_4] \tag{11.56}$$

The above updated matrix $\mathbf{R}$ is not upper-triangular and requires the following Givens rotation so that the new matrix may still be upper-triangular:

$$\mathbf{R} \leftarrow \Theta_2 \mathbf{R} \tag{11.57}$$

The modification is reflected on the unitary matrix $\mathbf{Q}$ as follows:

$$\mathbf{Q} \leftarrow \mathbf{Q} \Theta_2^T \tag{11.58}$$

where Θ_2 is an orthogonal rotation matrix that is defined as

$$\Theta_2 = \begin{bmatrix} 1 & 0 & 0 & 0 \\ 0 & \alpha_2 & \beta_2 & 0 \\ 0 & -\beta_2 & \alpha_2 & 0 \\ 0 & 0 & 0 & 1 \end{bmatrix} \tag{11.59}$$

with α_2 and β_2 given as

$$\alpha_2 = \frac{r_{2,2}}{\sqrt{r_{2,2}^2 + r_{3,2}^2}} \quad \text{and} \quad \beta_2 = \frac{r_{3,2}}{\sqrt{r_{2,2}^2 + r_{3,2}^2}}. \tag{11.60}$$

Note that the length of the second column was reduced in **Step 1**. In the case where the condition in (11.55) is not satisfied, in which consequently the second and third columns are exchanged, it requires reduction of the length of the new second column vector. Thus, we go back to **Step 1** and repeat the same process until the condition in (11.55) is satisfied. Once the condition in (11.55) is satisfied, we go on to the **Step 3**. In this example, the above condition is assumed to be met, thus no column vector exchange is necessary.

Step 3 (Length reduction of the fourth column vector $\mathbf{r}_4$): Define $\mu_{3,4}$ as

$$\mu_{3,4} = \left\langle \frac{r_{3,4}}{r_{3,3}} \right\rangle \tag{11.61}$$

Using the above $\mu_{3,4}$, the fourth column vectors of $\mathbf{R}$ and $\mathbf{T}$ are modified as

$$
\begin{aligned}
\mathbf{r}_4 &\leftarrow \mathbf{r}_4 - \mu_{3,4}\,\mathbf{r}_3 \\
\mathbf{t}_4 &\leftarrow \mathbf{t}_4 - \mu_{3,4}\,\mathbf{t}_3
\end{aligned}
\tag{11.62}
$$

Subsequently, we define $\mu_{2,4}$ as

$$
\mu_{2,4} = \left\langle \frac{r_{2,4}}{r_{2,2}} \right\rangle
\tag{11.63}
$$

Note that the above $r_{2,4}$ has been modified in Equation (11.62). Using the above $\mu_{2,4}$, the fourth column vectors of $\mathbf{R}$ and $\mathbf{T}$ are again modified as

$$
\begin{aligned}
\mathbf{r}_4 &\leftarrow \mathbf{r}_4 - \mu_{2,4}\,\mathbf{r}_2 \\
\mathbf{t}_4 &\leftarrow \mathbf{t}_4 - \mu_{2,4}\,\mathbf{t}_2
\end{aligned}
\tag{11.64}
$$

Finally, we define $\mu_{1,4}$ as

$$
\mu_{1,4} = \left\langle \frac{r_{1,4}}{r_{1,1}} \right\rangle
\tag{11.65}
$$

which will be used to modify the fourth column vectors as

$$
\begin{aligned}
\mathbf{r}_4 &\leftarrow \mathbf{r}_4 - \mu_{1,4}\,\mathbf{r}_1 \\
\mathbf{t}_4 &\leftarrow \mathbf{t}_4 - \mu_{1,4}\,\mathbf{t}_1
\end{aligned}
\tag{11.66}
$$

Then the following condition is checked:

$$
\delta r_{3,3}^2 \le r_{3,4}^2 + r_{4,4}^2
\tag{11.67}
$$

If the condition (11.67) is satisfied, a procedure of the LLL lattice reduction is complete. Otherwise, the two column vectors of $\mathbf{R}$ and $\mathbf{T}$ are exchanged as

$$
\begin{aligned}
\mathbf{R} &\leftarrow \begin{bmatrix} \mathbf{r}_1 & \mathbf{r}_2 & \mathbf{r}_4 & \mathbf{r}_3 \end{bmatrix} \\
\mathbf{T} &\leftarrow \begin{bmatrix} \mathbf{t}_1 & \mathbf{t}_2 & \mathbf{t}_4 & \mathbf{t}_3 \end{bmatrix}
\end{aligned}
\tag{11.68}
$$

We convert the matrix $\mathbf{R}$ into an upper-triangular matrix via the following Givens rotation:

$$
\begin{aligned}
\mathbf{R} &\leftarrow \Theta_3 \mathbf{R} \\
\mathbf{Q} &\leftarrow \mathbf{Q}\,\Theta_3^T
\end{aligned}
\tag{11.69}
$$

where Θ is the orthogonal rotation matrix that is defined as

$$\Theta_3 = \begin{bmatrix} 1 & 1 & 0 & 0 \\ 1 & 1 & 0 & 0 \\ 0 & 0 & \alpha_3 & \beta_3 \\ 0 & 0 & -\beta_3 & \alpha_3 \end{bmatrix} \tag{11.70}$$

with α_3 and β_3 given as

$$\alpha_3 = \frac{r_{3,3}}{\sqrt{r_{3,3}^2 + r_{4,3}^2}} \quad \text{and} \quad \beta_3 = \frac{r_{4,3}}{\sqrt{r_{3,3}^2 + r_{4,3}^2}} \tag{11.71}$$

In case that the condition in (11.67) is not satisfied, in which consequently the third and fourth columns are exchanged, it requires the reduction of the length of the new third column vector. Thus, we go back to **Step 2** and repeat the same procedure until the condition in (11.67) is satisfied.

In the above procedure, we started with $\mathbf{Q}$, $\mathbf{R}$, and $\mathbf{T} = \mathbf{I}$ as inputs to the LLL algorithm. According to the LLL algorithm, these matrices have been modified to yield a new set of matrices, $\mathbf{Q}_{LLL}$, $\mathbf{R}_{LLL}$, and $\mathbf{T}_{LLL}$. Note that the condition number of $\mathbf{R}_{LLL}$ is less than or equal to that of $\mathbf{R}$. Using the new matrix set, the system of Equation (11.41) can be re-written as

$$\begin{aligned} \bar{\mathbf{y}} &= \bar{\mathbf{H}}\bar{\mathbf{x}} + \bar{\mathbf{z}} \\ &= \mathbf{Q}\mathbf{R}\bar{\mathbf{x}} + \bar{\mathbf{z}} \\ &= \mathbf{Q}_{LLL}\mathbf{R}_{LLL}\mathbf{T}_{LLL}^{-1}\bar{\mathbf{x}} + \bar{\mathbf{z}} \end{aligned} \tag{11.72}$$

where $\mathbf{T}_{LLL}^{-1}$ is used to recover $\mathbf{Q}$ and $\mathbf{R}$ from their modifications. Also note that the matrix $\mathbf{Q}_{LLL}$ is still orthogonal, because the input matrix $\mathbf{Q}$ has been modified only by multiplications with orthogonal rotation matrices whenever necessary.

11.6.2 Application of Lattice Reduction

The above lattice reduction can be combined with various signal detection methods. If the lattice reduction is combined with the linear detection methods in Section 11.1, noise enhancement can be mitigated, especially when the original condition number is too large. The same is true for OSIC signal detection in Section 11.2. If the lattice reduction is made for SD in Section 11.4, the initial guess becomes more likely to be an ML solution, thus providing a smaller sphere radius. In this subsection, for instance, we consider the lattice reduction associated with the linear MMSE detection method.

Multiplying both sides of Equation (11.72) with $\mathbf{Q}_{LLL}^H$, we have

$$\mathbf{Q}_{LLL}^H \bar{\mathbf{y}} = \mathbf{R}_{LLL}\mathbf{T}_{LLL}^{-1}\bar{\mathbf{x}} + \mathbf{Q}_{LLL}^H \bar{\mathbf{z}} \tag{11.73}$$

If we let $\tilde{\mathbf{y}} = \mathbf{Q}_{LLL}^H \bar{\mathbf{y}}$, $\tilde{\mathbf{x}} = \mathbf{T}_{LLL}^{-1}\bar{\mathbf{x}}$, and $\tilde{\mathbf{z}} = \mathbf{Q}_{LLL}^H \bar{\mathbf{z}}$, Equation (11.73) is represented in a new form of system equations:

$$\tilde{\mathbf{y}} = \mathbf{R}_{LLL}\tilde{\mathbf{x}} + \tilde{\mathbf{z}} \tag{11.74}$$

which is expected to be *well-conditioned*[3]. Suppose that the linear MMSE signal detection technique is applied to the above well-conditioned system equation to yield the estimate of $\tilde{\mathbf{x}}$,

---

[3] When the system matrix has a small condition number, the corresponding system is said to be well-conditioned.

denoted as $\hat{\bar{\mathbf{x}}}$, that is,

$$\hat{\bar{\mathbf{x}}}_{MMSE} = \left(\mathbf{R}_{LLL}^H \mathbf{R}_{LLL} + \sigma_z^2 \mathbf{I}\right)^{-1} \mathbf{R}_{LLL}^H \tilde{\bar{\mathbf{y}}} \qquad (11.75)$$

Note that $\hat{\bar{\mathbf{x}}}_{MMSE}$ is the estimate of $\tilde{\bar{\mathbf{x}}} = \mathbf{T}_{LLL}^{-1}\bar{\mathbf{x}}$. Since the entries of the matrix $\mathbf{T}_{LLL}$ are integers and its determinant is unity, the entries of $\mathbf{T}_{LLL}^{-1}$ are also integers. Thus, if the entries of $\hat{\bar{\mathbf{x}}}_{MMSE}$ in Equation (11.75) have non-integer parts, the entries can be sliced so that the estimated values can be composed of the closest integer values. Let $\tilde{\bar{\mathbf{x}}}_{\text{sliced}}$ denote the corresponding sliced value. Then the estimate of $\bar{\mathbf{x}}$ is obtained as

$$\hat{\bar{\mathbf{x}}} = \mathbf{T}_{LLL}^{-1}\tilde{\bar{\mathbf{x}}}_{\text{sliced}} \qquad (11.76)$$

Program 11.15 ("LRAD_MMSE") implements the MMSE detection method subject to the lattice reduction using Lenstrat-Lenstra-Lovasz (LLL) algorithm. Its BER performance is shown in Figure 11.9, which has been obtained by running Program 11.15. It is clear that the lattice reduction significantly improves the performance of the linear MMSE signal detection method.

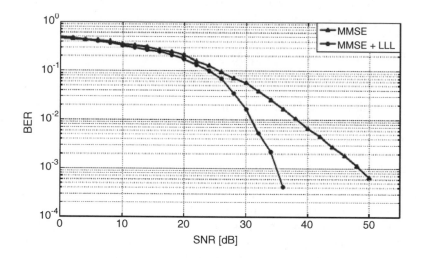

Figure 11.9   Performance improvement with lattice reduction: $N_T = N_R = 4$.

MATLAB® Programs: Lattice Reduction-Aided MMSE Detection

Program 11.15 "LRAD_MMSE" for Lattice Reduction-Aided Detector with MMSE detection

```
function [X_estimate] = LRAD_MMSE(H_complex,y,sigma2,delta)
% Lattice Reduction-Aided Detector with MMSE
% Input
%    H_complex : Complex channel matrix, NRxNT
%    y         : Complex received signal, NRx1
```

```
%     sigma2    : Noise variance
%     delta : Scaling variable
% Output
%     X_estimate : Estimated signal, NTx1
Nt = 4; Nr = 4;  N = 2*Nt;
H_real=[[real(H_complex) -imag(H_complex)];
        [imag(H_complex) real(H_complex)]]; % Complex to Real channel
H=[H_real; sqrt(sigma2)*eye(N)];
y_real=[real(y); imag(y)];   % Complex y -> Real y
y=[y_real;zeros(N,1)];
[Q,R,P,p] = SQRD(H);            % sorted QR decomposition
[W,L,T] = original_LLL(Q,R,N,delta);   % W*L = Q*R*T
H_tilde = H*P*T;                % H*P = Q*R
X_temp = inv(H_tilde'*H_tilde)*H_tilde'*y;   % MMSE detection
X_temp = round(X_temp);      % Slicing
X_temp = P*T*X_temp;
for i=1:Nr   % real x -> complex x
    X_estimate(i) = X_temp(i)+j*X_temp(i+4);
end
```

Program 11.16 "original_LLL"

```
function [Q,R,T] = original_LLL(Q,R,m,delta)
% Input
%     Q : Orthogonal matrix,   NRxNT
%     R : R with a large condition number
%     m : Column size of H
%     delta : Scaling variable
% Output
%     Q : Orthogonal matrix,   NRxNT
%     R : R with a small condition number
%     T : Unimodular matrix
P=eye(m);   T=P;   k=2;
while (k<=m)
    for j=k-1:-1:1
        mu = round(R(j,k)/R(j,j));
        if mu~=0
            R(1:j,k)=R(1:j,k)-mu*R(1:j,j);   T(:,k)=T(:,k)-mu*T(:,j);
        end
    end
    if (delta*R(k-1,k-1)^2 > R(k,k)^2+R(k-1,k)^2)   % column change
        R(:,[k-1 k])=R(:,[k k-1]);   T(:,[k-1 k])=T(:,[k k-1]);
        %calculate Givens rotation matrix such that R(k,k-1) becomes zero
        alpha = R(k-1,k-1)/sqrt(R(k-1:k,k-1).'*R(k-1:k,k-1));
        beta = R(k,k-1)/sqrt(R(k-1:k,k-1).'*R(k-1:k,k-1));
        theta = [alpha   beta; -beta   alpha];
        R(k-1:k,k-1:m)=theta*R(k-1:k,k-1:m);
        Q(:,k-1:k)=Q(:,k-1:k)*theta.';
```

```
         k=max([k-1 2]);
    else        k=k+1;
    end
end
```

Program 11.17 "SQRD"

```
function [Q,R,P,p] = SQRD(H)
% Sorted QR decomposition
% Input
%      H : complex channel matrix, NRxNT
% Output
%      Q : orthogonal matrix, NRxNT
%      P : permutation matrix
%      p : ordering information
Nt=size(H,2);   Nr=size(H,1)-Nt;   R=zeros(Nt);
Q=H;    p=1:Nt;
for i=1:Nt normes(i)=Q(:,i)'*Q(:,i); end
for i=1:Nt
    [mini,k_i]=min(normes(i:Nt)); k_i=k_i+i-1;
    R(:,[i k_i])=R(:,[k_i i]);
    p(:,[i k_i])=p(:,[k_i i]);
    normes(:,[i k_i])=normes(:,[k_i i]);
    Q(1:Nr+i-1,[i k_i])=Q(1:Nr+i-1,[k_i i]);
    % Wubben's algorithm: does not lead to
    % a true QR decomposition of the extended MMSE channel matrix
    % Q(Nr+1:Nr+Nt,:) is not triangular but permuted triangular
    R(i,i)=sqrt(normes(i));
    Q(:,i)=Q(:,i)/R(i,i);
    for k=i+1:Nt
        R(i,k)=Q(:,i)'*Q(:,k);
        Q(:,k)=Q(:,k)-R(i,k)*Q(:,i);
        normes(k)=normes(k)-R(i,k)*R(i,k)';
    end
end
P=zeros(Nt);   for i=1:Nt,   P(p(i),i)=1;   end
```

11.7 Soft Decision for MIMO Systems

In the previous sections, we discussed only hard-decision detection techniques. However, its performance can be further improved by using soft-decision values. If soft-input soft-output channel decoders are used, the output of signal detector must be given by soft-decision values. In this section, we study how the soft-output values can be produced by the signal detectors for the previous sections. Let us first briefly review how to generate soft-output values in single-input single-output (SISO) systems.

11.7.1 Log-Likelihood-Ratio (LLR) for SISO Systems

The received signal for the SISO system, y, can be represented by

$$y = hx + z \tag{11.77}$$

where x is the transmitted signal, h is the complex flat fading channel gain, and z is the additive white Gaussian noise. In this SISO system, the signal can be detected by the simple inverse processing on the received signal, that is,

$$\tilde{x} = \frac{y}{h} = x + \frac{z}{h} = x + \tilde{z} \tag{11.78}$$

where $\tilde{z} = z/h$. If z is Gaussian-distributed with zero mean and variance of σ_z^2, $\tilde{z}$ is also a Gaussian random variable $\tilde{z} \sim N(0, \sigma^2 \triangleq \sigma_z^2/|h|^2)$. Then the conditional PDF that $\tilde{x}$ is received given that a symbol x was transmitted is

$$f_{\tilde{X}}(\tilde{x}|x) = \frac{1}{\sqrt{2\pi\sigma^2}} \exp\left(-\frac{|\tilde{x}-x|^2}{2\sigma^2}\right) \tag{11.79}$$

In the following discussion, we consider the soft-decision detection for the gray-encoded 16-QAM symbols as an example. Figure 11.10 illustrates two different gray-encoded 16-QAM constellations.

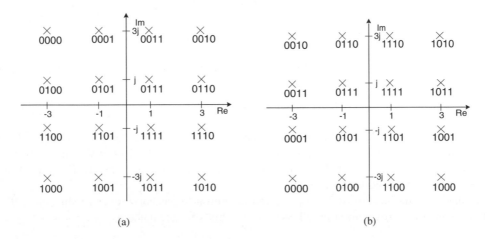

(a)   (b)

Figure 11.10   Gray-encoded 16-QAM constellations: examples.

Let S_l^+ and S_l^- denote a set of the symbols whose lth bit is 1, and a set of symbols whose lth bit is 0, respectively. Referring to 16-QAM constellations in Figure 11.10(a), for example, S_l^+ and S_l^- are given as

$$\begin{aligned}
S_l^+ &= \{-3+3j, -1+3j, 1+3j, 3+3j, -3+j, -1+j, 1+j, 3+j\} \\
S_l^- &= \{-3-3j, -1-3j, 1-3j, 3-3j, -3-j, -1-j, 1-j, 3-j\}
\end{aligned} \tag{11.80}$$

The log likelihood ratio (LLR) value or soft output for the lth bit is defined as

$$
LLR(b_l) = \ln \frac{\displaystyle\sum_{x \in S_l^+} f_X(x|\tilde{x})}{\displaystyle\sum_{x \in S_l^-} f_X(x|\tilde{x})} = \ln \frac{\displaystyle\sum_{x \in S_l^+} f_{\tilde{X}}(\tilde{x}|x)p(x)/f_{\tilde{X}}(\tilde{x})}{\displaystyle\sum_{x \in S_l^-} f_{\tilde{X}}(\tilde{x}|x)p(x)/f_{\tilde{X}}(\tilde{x})} \tag{11.81}
$$

If all symbols are equally likely (i.e., $p(x)$ is a constant), Equation (11.81) can be approximated as

$$
\begin{aligned}
LLR(b_l) &= \ln \frac{\displaystyle\sum_{x \in S_l^+} f_{\tilde{X}}(\tilde{x}|x)}{\displaystyle\sum_{x \in S_l^-} f_{\tilde{X}}(\tilde{x}|x)} \\[2ex]
&\approx \ln \frac{\displaystyle\max_{x \in S_l^+} f_{\tilde{X}}(\tilde{x}|x)}{\displaystyle\max_{x \in S_l^-} f_{\tilde{X}}(\tilde{x}|x)} \\[2ex]
&= \frac{1}{2\sigma^2} \left(\left| \tilde{x} - x_{l,opt}^- \right|^2 - \left| \tilde{x} - x_{l,opt}^+ \right|^2 \right)
\end{aligned} \tag{11.82}
$$

where $x_{l,opt}^+$ and $x_{l,opt}^-$ are defined as

$$
\begin{aligned}
x_{l,opt}^+ &= \arg\min_{x \in S_l^+} |\tilde{x} - x|^2 \\
x_{l,opt}^- &= \arg\min_{x \in S_l^-} |\tilde{x} - x|^2
\end{aligned} \tag{11.83}
$$

If σ^2 is constant over a coding block, the LLR value in Equation (11.82) can be further simplified as

$$
LLR(b_l) \approx \left| \tilde{x} - x_{l,opt}^- \right|^2 - \left| \tilde{x} - x_{l,opt}^+ \right|^2 \tag{11.84}
$$

The constant σ^2 indicates that the complex channel gain h is constant during a coding block. In the sequel, let us provide an example of LLR calculation for the 16-QAM constellation in Figure 11.10(a). Figure 11.11 shows that for a given value of $\tilde{x} = \tilde{x}_R + j\tilde{x}_I$, LLR computation for each bit is involved with two distances.

Considering the location of $\tilde{x} = \tilde{x}_R + j\tilde{x}_I$ and assuming a constant σ^2 over a coding block, the LLR values in approximation in (11.84) for all 4 bits are calculated as

$$
LLR(b_1) = \begin{cases}
(\tilde{x}_I - (3))^2 - (\tilde{x}_I - (-1))^2 = -8\tilde{x}_I + 8, & 2 \leq \tilde{x}_I \\
(\tilde{x}_I - (1))^2 - (\tilde{x}_I - (-1))^2 = -4\tilde{x}_I, & 0 \leq \tilde{x}_I < 2 \\
(\tilde{x}_I - 1)^2 - (\tilde{x}_I - (-1))^2 = -4\tilde{x}_I, & -2 \leq \tilde{x}_I < 0 \\
(\tilde{x}_I - 1)^2 - (\tilde{x}_I - (-3))^2 = -8\tilde{x}_I - 8, & \tilde{x}_I < -2
\end{cases} \tag{11.85}
$$

$$
LLR(b_2) = \begin{cases}
(\tilde{x}_I - 3)^2 - (\tilde{x}_I - 1)^2 = -4\tilde{x}_I + 8, & 2 \leq \tilde{x}_I \\
(\tilde{x}_I - 3)^2 - (\tilde{x}_I - 1)^2 = -4\tilde{x}_I + 8, & 0 \leq \tilde{x}_I < 2 \\
(\tilde{x}_I - (-3))^2 - (\tilde{x}_I - (-1))^2 = 4\tilde{x}_I + 8, & -2 \leq \tilde{x}_I < 0 \\
(\tilde{x}_I - (-3))^2 - (\tilde{x}_I - (-1))^2 = 4\tilde{x}_I + 8, & \tilde{x}_I < -2
\end{cases} \tag{11.86}
$$

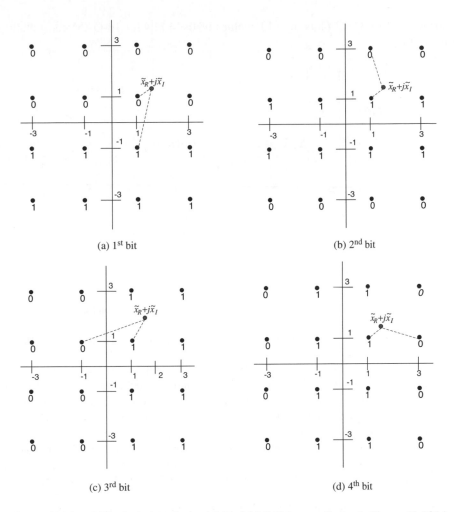

(a) 1st bit

(b) 2nd bit

(c) 3rd bit

(d) 4th bit

Figure 11.11 LLR calculations for each bit of 16-QAM constellation in Figure 11.10(a).

$$LLR(b_3) = \begin{cases} (\tilde{x}_R-(-1))^2-(\tilde{x}_R-(3))^2 = 8\tilde{x}_R-8, & 2 \le \tilde{x}_R \\ (\tilde{x}_R-(-1))^2-(\tilde{x}_R-(1))^2 = 4\tilde{x}_R, & 0 \le \tilde{x}_R < 2 \\ (\tilde{x}_R-(-1))^2-(\tilde{x}_R-1)^2 = 4\tilde{x}_R, & -2 \le \tilde{x}_R < 0 \\ (\tilde{x}_R-(-3))^2-(\tilde{x}_R-1)^2 = 8\tilde{x}_R+8, & \tilde{x}_R < -2 \end{cases} \quad (11.87)$$

$$LLR(b_4) = \begin{cases} (\tilde{x}_R-3)^2-(\tilde{x}_R-1)^2 = -4\tilde{x}_R+8, & 2 \le \tilde{x}_R \\ (\tilde{x}_R-3)^2-(\tilde{x}_R-1)^2 = -4\tilde{x}_R+8, & 0 \le \tilde{x}_R < 2 \\ (\tilde{x}_R-(-3))^2-(\tilde{x}_R-(-1))^2 = 4\tilde{x}_R+8, & -2 \le \tilde{x}_R < 0 \\ (\tilde{x}_R-(-3))^2-(\tilde{x}_R-(-1))^2 = 4\tilde{x}_R+8, & \tilde{x}_R < -2 \end{cases} \quad (11.88)$$

From Equations (11.85)–(11.88), it can be seen that the LLR values of the first and second bits depend on the imaginary part $\tilde{x}_I$, while the values of the third and fourth bits depend on the real part $\tilde{x}_R$. Figure 11.12 plots the LLR values of the 4 bits for 16-QAM constellations in Figure 11.10a.

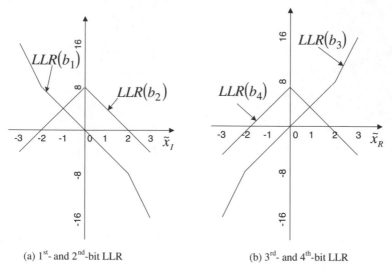

(a) 1st- and 2nd-bit LLR   (b) 3rd- and 4th-bit LLR

Figure 11.12   LLR as functions of $\tilde{x}_I$ or $\tilde{x}_R$ for 16-QAM constellation in Figure 11.10(a).

Program 11.18 ("soft_hard_SISO.m") can be used to compare the performances of hard-decision and soft-decision detection where it is assumed that a packet has a length of 1200 bits, forward-error-corrected by convolutional coder with a coding rate of 1/2 and constraint length of 7, implemented by the generator polynomials [1001111] and [1101101]. Therefore, the encoded data sequence is mapped into 600 16-QAM symbols. Furthermore, each symbol is subject to independent Rayleigh fading. The performance difference between hard-decision and soft-decision detection is shown in Figure 11.13. It is clear that soft-decision detection provides a significant performance gain in the packet error rate (PER) over the hard-decision one.

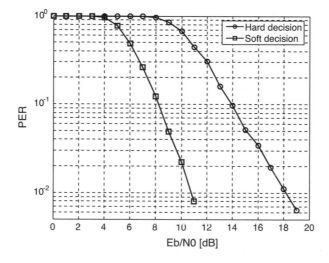

Figure 11.13   Packet error performance for SISO: hard decision vs. soft decision.

MATLAB® Programs: Performance of Soft/Hard-Decision Detection for SISO System

Program 11.18 "soft_hard_SISO.m": Hard/soft-decision detection for SISO system

```
% soft_hard_SISO.m
clear all, close all
decision = 0;   % Set to 0/1 for hard/soft decision
G = [1 0 1 1 0 1 1; 1 1 1 1 0 0 1]; K=1; N=2; Rc=K/N;
L_packet=1200; b=4; N_symbol=(L_packet*N+12)/b;
EbN0dBs = 0:19; sq05 = sqrt(1/2);
for i=1:length(EbN0dBs)
    EbN0dB=EbN0dBs(i); nope = 0;
    for N_packet = 1:1e10
        bit_strm = randint(1,L_packet);
        coded_bits = convolution_encoder(bit_strm); %2*(7-1)=12 tail bits
        symbol_strm = QAM16(coded_bits); % 16 QAM mapper
        h = sq05*(randn(1,N_symbol) + j*randn(1,N_symbol));
        faded_symbol = symbol_strm.*h; % Channel
        P_b = mean(abs(faded_symbol).^2)/b;
        noise_amp = sqrt(P_b/2*10^(-EbN0dB/10));
        faded_noisy_symbol = faded_symbol + noise_amp* ...
              (randn(1,N_symbol) + j*randn(1,N_symbol)); % Noise
        channel_compensated = faded_noisy_symbol./h;
        if decision==0
          sliced_symbol = QAM16_slicer(channel_compensated);
          hard_bits = QAM16_demapper(sliced_symbol);
          Viterbi_init;   bit_strm_hat = Viterbi_decode(hard_bits);
        else
          soft_bits = soft_decision_sigma(channel_compensated,h);
          Viterbi_init;   bit_strm_hat = Viterbi_decode_soft(soft_bits);
        end
        bit_strm_hat = bit_strm_hat(1:L_packet);
        nope = nope+(sum(bit_strm~=bit_strm_hat)>0); % # of packet errors
        if nope>50,   break;   end
    end
    PER(i) = nope/i_packet;   if PER(i)<1e-2,   break;   end
end
semilogy(EbN0dBs,PER,'k-o'); xlabel('Eb/N0[dB]'); ylabel('PER'); grid on
set(gca,'xlim',[0 EbN0dBs(end)],'ylim',[1e-3 1])
```

Program 11.19 "soft_decision_sigma": soft Viterbi decoding for SISO system

```
function [x4_soft]=soft_decision_sigma(x,h)
x=x(:).'; xr=real(x); xi=imag(x);
X=[xr; 2-abs(xr);   xi; 2-abs(xi)];   H=repmat(abs(h(:)).',4,1);
XH = X.*H; x4_soft = XH(:).';
```

11.7.2 LLR for Linear Detector-Based MIMO System

Linear signal detection in Section 11.1 intends to separate each of $\{x_i\}_{i=1}^{N_T}$ from the other symbols. Once they are separated, the bit-level LLR calculation for each symbol becomes similar to that of SISO systems. For example, let us consider the following 2×2 MIMO systems:

$$\mathbf{y} = \mathbf{h}_1 x_1 + \mathbf{h}_2 x_2 + \mathbf{z} \tag{11.89}$$

Then a linear detector such as linear MMSE or ZF detector can be applied to the received signal vector. Let $\{\mathbf{w}_{i,MMSE}\}_{i=1}^2$ denote the ith row vector of the MMSE weight in Equation (11.7). Then, the output of the MMSE detector for $\{x_i\}_{i=1}^2$ is given as

$$
\begin{aligned}
\tilde{x}_{i,MMSE} &= \mathbf{w}_{i,MMSE}\mathbf{y} \\
&= \mathbf{w}_{i,MMSE}\mathbf{h}_1 x_1 + \mathbf{w}_{i,MMSE}\mathbf{h}_2 x_2 + \mathbf{w}_{i,MMSE}\mathbf{z} , \qquad j \neq i \\
&= \rho x_i + I_j + \tilde{z}
\end{aligned}
\tag{11.90}
$$

where $\rho_i = \mathbf{w}_{i,MMSE}\mathbf{h}_i x_i$, $I_j = \mathbf{w}_{i,MMSE}\mathbf{h}_j x_j$, and $\tilde{z} = \mathbf{w}_{i,MMSE}\mathbf{z}$. From the perspective of $\tilde{x}_{i,MMSE}$, ρx_i corresponds to the signal component, $I_j = \mathbf{w}_{1,MMSE}\mathbf{h}_2 x_j$ is the interference component, and $\tilde{z} = \mathbf{w}_{i,MMSE}\mathbf{z}$ is the noise component. Assuming statistical independence among these three components, the post-detection SINR of x_i is expressed as

$$SINR_i = \frac{E\left\{|\rho x_i|^2\right\}}{E\left\{|I_j|^2\right\} + E\left\{|\tilde{z}|^2\right\}} = \frac{|\rho|^2 \mathsf{E}_x}{\left|\mathbf{w}_{i,MMSE}\mathbf{h}_j\right|^2 \mathsf{E}_x + \left\|\mathbf{w}_{i,MMSE}\right\|^2 \sigma_z^2} , \qquad j \neq i \tag{11.91}$$

If the interference and noise components in Equation (11.90) are assumed independent and Gaussian-distributed, $I_1 + \tilde{z}$ can be approximated by a Gaussian random variable with zero mean and variance of $\sigma_i^2 = |\mathbf{w}_{i,MMSE}\mathbf{h}_j|^2 \mathsf{E}_x + ||\mathbf{w}_{i,MMSE}||^2 \sigma_z^2$. Using this approximation, we have the following conditional PDF:

$$f_{\tilde{X}}\left(\tilde{x}_{i,MMSE}|x_i\right) = \frac{1}{\sqrt{2\pi\sigma_i^2}} \exp\left(-\frac{\left|\tilde{x}_{i,MMSE} - \rho x_i\right|^2}{2\sigma_i^2}\right) \tag{11.92}$$

As SNR becomes high, the value ρ approaches unity. Let $\mathsf{S}_{l,i}^+$ and $\mathsf{S}_{l,i}^-$ denote the set of vectors with their lth bit value of the ith symbols being either 1 or 0, respectively. Using Equation (11.92) for high SNR, the lth bit LLR of x_i is expressed as

$$
\begin{aligned}
LLR\left(b_{l,i}\right) &= \ln \frac{\displaystyle\sum_{x \in \mathsf{S}_{l,i}^+} f_{\tilde{X}}\left(\tilde{x}_{i,MMSE}|x\right)}{\displaystyle\sum_{x \in \mathsf{S}_{l,i}^-} f_{\tilde{X}}\left(\tilde{x}_{i,MMSE}|x\right)} \\
&\approx \ln \frac{\displaystyle\max_{x \in \mathsf{S}_{l,i}^+} f_{\tilde{X}}\left(\tilde{x}_{i,MMSE}|x\right)}{\displaystyle\max_{x \in \mathsf{S}_{l,i}^-} f_{\tilde{X}}\left(\tilde{x}_{i,MMSE}|x\right)} \\
&= \frac{1}{2\sigma_i^2}\left(\left|\tilde{x}_{i,MMSE} - x_{i,l,opt}^-\right|^2 - \left|\tilde{x}_{i,MMSE} - x_{i,l,opt}^+\right|^2\right), \qquad i = 1, 2
\end{aligned}
\tag{11.93}
$$

where $x_{i,l,opt}^+$ and $x_{i,l,opt}^-$ are defined as

$$
\begin{aligned}
x_{i,l,opt}^+ &= \underset{x \in S_l^+}{\arg\min} \left| \tilde{x}_{i,MMSE} - x \right|^2 \\
x_{i,l,opt}^- &= \underset{x \in S_l^-}{\arg\min} \left| \tilde{x}_{i,MMSE} - x \right|^2
\end{aligned}
\tag{11.94}
$$

We see a clear similarity between the bit-level LLR calculation for MIMO systems in Equation (11.89) and that for SISO systems in Equation (11.77). In Equation (11.93), however, the bit-level LLR values for x_i is a function of σ_i^2. As opposed to the case in Equation (11.82) where σ^2 is constant, $\sigma_1^2 \neq \sigma_2^2$ in general, the simplification in (11.84) for the SISO systems cannot be applied to the MIMO system in Equation (11.89) (i.e., see the simplification in Equation (11.93)).

Program 11.20 ("MMSE_detection_2x2.m") can be run to evaluate the performance of hard-decision and soft-decision linear MMSE detector for 2×2 MIMO systems. It simulates a block Rayleigh fading channel with each block of 81 symbols. Figure 11.14 shows the performance of the linear MMSE detector-based 2×2 MIMO systems with 16-QAM symbols that can be obtained by running Program 11.8. Block Rayleigh fading channels were simulated, where a block is composed of 81 symbol periods. A length of the coding block, excluding the 6 tail bits, is (2592–6) bits.

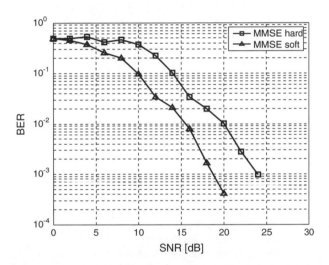

Figure 11.14 BER performance for MIMO with the MMSE linear detection: hard decision vs. soft decision.

MATLAB® Programs: Hard/Soft-Decision MMSE Detector for 2×2 MIMO System

Program 11.20 "MMSE_detection_2x2.m": MMSE detection for 2×2 MIMO system

```
% MMSE_detection_2x2.m
clear all; close all;
%%%%%% option %%%%%%
```

```
bits_option    = 2;    %% 0 : all zeros, 1 : all ones, 2: random binary
noise_option   = 1;    %% 0 : no noise addition, 1 : noise added
decision_scheme = 1;%% 0 : Hard decision, 1 : soft decision
b = 4; NT = 2;
SNRdBs =[0:2:25];    sq05=sqrt(0.5);
nobe_target = 500; BER_target = 1e-3;
raw_bit_len = 2592-6;
interleaving_num = 72; deinterleaving_num = 72;
N_frame = 1e8; %% maximum   generated bits #
for i_SNR=1:length(SNRdBs)
    SNRdB=SNRdBs(i_SNR); sig_power=NT;
    sigma2=sig_power*10^(-SNRdB/10)*noise_option;
    sigma1=sqrt(sigma2/2);
    nobe = 0; % Number of bit errors
    rand('seed',1); randn('seed',1); Viterbi_init;
    for i_frame=1:N_frame
        %%%%%%%%%%%%%%   Random data generation %%%%%%%%%%%%%%
        switch (bits_option)
            case {0} bits=zeros(1,raw_bit_len);
            case {1} bits=ones(1,raw_bit_len);
            case {2} bits=randint(1,raw_bit_len);
        end
        %%% Convolutional encoding %%%%%
        encoding_bits = convolution_encoder(bits);
        % Interleaving %%
        interleaved=[];
        for i=1:interleaving_num
            interleaved=[interleaved
                         encoding_bits([i:interleaving_num:end])];
        end
        temp_bit =[];
        for tx_time=1:648
            tx_bits=interleaved(1:8);
            interleaved(1:8)=[];
            %%%%%%%%%%%%%%   QAM16 modulation %%%%%%%%%%%%%%%%%
            QAM16_symbol = QAM16_mod(tx_bits, 2);
            %%%%%%%%%%%%%%   S/P   %%%%%%%%%%%%%%%%
            x(1,1) = QAM16_symbol(1); x(2,1) = QAM16_symbol(2);
            %%%%% Channel H and received y   %%%%%%%%%%%%
            if rem(tx_time-1,81)==0,
              H = sq05*(randn(2,2)+j*randn(2,2));
            end
            y = H*x;
            %%%   AWGN addition %%%
            noise=sigma1*(randn(2,1)+j*randn(2,1));
            if noise_option==1,   y = y + noise;   end
            %%%%%%%%%%%%%% MMSE Detector %%%%%%%%%%%%%%%%%%%
            W = inv(H'*H+sigma2*diag(ones(1,2)))*H';
            X_tilde = W*y;
            if decision_scheme==0 % Hard decision
```

```
            X_hat = QAM16_slicer(X_tilde, 2);
            temp_bit = [temp_bit QAM16_demapper(X_hat, 2)];
        else          % Soft decision
            soft_bits = soft_output2x2(X_tilde); Ps=1;
            SINR1=(Ps*(abs((W(1,:)*H(:,1)))^2)) / (Ps*(
                abs((W(1,:)*H(:,2)))^2 + W(1,:)*W(1,:)'*sigma2));
            SINR2=(Ps*(abs((W(2,:)*H(:,2)))^2)) / (Ps*(
                abs((W(2,:)*H(:,1)))^2 + W(2,:)*W(2,:)'*sigma2));
            soft_bits(1:4)=soft_bits(1:4)*SINR1;
            soft_bits(5:8)=soft_bits(5:8)*SINR2;
            temp_bit=[temp_bit soft_bits];
        end
    end
    %% Deinterleaving
    deinterleaved=[];
    for i=1:deinterleaving_num
        deinterleaved=[deinterleaved
            temp_bit([i:deinterleaving_num:end])];
    end
    %% Viterbi
    received_bit=Viterbi_decode(deinterleaved);
    %%%%%   Error check %%%%%
    for EC_dummy=1:1:raw_bit_len,
        if bits(EC_dummy)~=received_bit(EC_dummy), nobe=nobe+1;   end
        if nobe>=nobe_target,   break;   end
    end
    if nobe>=nobe_target,   break;   end
  end
  %%%%%%%%%%%%% save BER data & Display %%%%%%%%%%%%%%
  BER(i_SNR) = nobe/((i_frame-1)*raw_bit_len+EC_dummy);
  fprintf('\t%d\t\t%1.4f\n',SNR,BER(i_SNR));
  if BER(i_SNR)<BER_target, break; end
end
```

Program 11.21 "soft_output2x2": Soft output of MMSE detector for 2×2 MIMO system

```
function [x_soft] = soft_output2x2(x)
sq10=sqrt(10); sq10_2=2/sq10;
x=x(:).'; xr=real(x); xi=imag(x);
X=sq10*[-xi; sq10_2-abs(xi); xr; sq10_2-abs(xr)];   x4_soft = X(:).';
```

11.7.3 LLR for MIMO System with a Candidate Vector Set

Consider the $N_R \times N_T$ MIMO system subject to the AWGN noise:

$$\mathbf{y} = \mathbf{Hx} + \mathbf{z} \tag{11.95}$$

from which the noise can be expressed as

$$\mathbf{z} = \mathbf{y} - \mathbf{Hx}. \tag{11.96}$$

Note that the PDF of Gaussian noise vector $\mathbf{z}$ is given as

$$f_{\mathbf{Z}}(\mathbf{z}) = \frac{1}{2\pi\Delta^{1/2}} \exp\left(-\frac{1}{2} (\mathbf{z} - \boldsymbol{\mu})^T \Sigma^{-1} (\mathbf{z} - \boldsymbol{\mu}) \right) \tag{11.97}$$

where $\boldsymbol{\mu}$ is the mean vector, Σ is the covariance matrix of $\mathbf{z}$, and Δ is the determinant of the covariance matrix Σ. Assuming that the noise vector is a zero-mean circularly symmetric white Gaussian random vector, the PDF in Equation (11.97) can be re-expressed as

$$f_{\mathbf{Z}}(\mathbf{z}) = \frac{1}{2\pi\Delta^{1/2}} \exp\left(-\frac{1}{2\sigma_z^2} \|\mathbf{y} - \mathbf{Hx}\|^2 \right) = f_{\mathbf{Y}}(\mathbf{y}|\mathbf{x}). \tag{11.98}$$

Starting from the above PDF, we derive an expression for soft output. We first investigate the soft output at the symbol vector level, considering a pair of symbol vectors, $\mathbf{x}_i$ and $\mathbf{x}_j$. By Bayes' theorem, the following relation holds for log-likelihood ratio (LLR):

$$\ln\frac{p(\mathbf{x}_i|\mathbf{y})}{p(\mathbf{x}_j|\mathbf{y})} = \ln\frac{f_{\mathbf{Y}}(\mathbf{y}|\mathbf{x}_i)p(\mathbf{x}_i)/f_{\mathbf{Y}}(\mathbf{y})}{f_{\mathbf{Y}}(\mathbf{y}|\mathbf{x}_j)p(\mathbf{x}_j)/f_{\mathbf{Y}}(\mathbf{y})} = \ln\frac{f_{\mathbf{Y}}(\mathbf{y}|\mathbf{x}_i)p(\mathbf{x}_i)}{f_{\mathbf{Y}}(\mathbf{y}|\mathbf{x}_j)p(\mathbf{x}_j)} \tag{11.99}$$

If all the transmitted symbol vectors are equally likely (i.e., $p(\mathbf{x}_i) = 1/|\mathsf{C}|^{N_T}$, $\forall i$), Equation (11.99) is reduced to

$$\ln\frac{p(\mathbf{x}_i|\mathbf{y})}{p(\mathbf{x}_j|\mathbf{y})} = \ln\frac{f_{\mathbf{Y}}(\mathbf{y}|\mathbf{x}_i)}{f_{\mathbf{Y}}(\mathbf{y}|\mathbf{x}_j)}. \tag{11.100}$$

Using the PDF in Equation (11.98), the above log-likelihood ratio (LLR) for two symbol vectors is given as

$$\ln\frac{p(\mathbf{x}_i|\mathbf{y})}{p(\mathbf{x}_j|\mathbf{y})} = \ln\frac{\exp\left(-\frac{1}{2\sigma_z^2} \|\mathbf{y} - \mathbf{Hx}_i\|^2 \right)}{\exp\left(-\frac{1}{2\sigma_z^2} \|\mathbf{y} - \mathbf{Hx}_j\|^2 \right)} \tag{11.101}$$

Note that Equation (11.101) can be simplified as

$$\ln\frac{p(\mathbf{x}_i|\mathbf{y})}{p(\mathbf{x}_j|\mathbf{y})} = \frac{1}{2\sigma_z^2} \|\mathbf{y} - \mathbf{Hx}_j\|^2 - \frac{1}{2\sigma_z^2} \|\mathbf{y} - \mathbf{Hx}_i\|^2 \tag{11.102}$$

Positive value of Equation (11.102) indicates that transmission of $\mathbf{x}_i$ is more probable than that of $\mathbf{x}_k$. In fact, the larger the positive LLR value is, the more probable $\mathbf{x}_i$ is than $\mathbf{x}_k$.

Now let us discuss the calculation of a bit-level LLR. Let $b_{l,i}$ denote the lth bit of the symbol transmitted from the ith transmit antenna. If all bits are equally likely (i.e., $p(b_{l,i} = 1) = p(b_{l,i} = 0) = 1/2$), the bit-level LLR is given as

$$\ln \frac{p(b_{l,i} = 1|\mathbf{y})}{p(b_{l,i} = 0|\mathbf{y})} = \ln \frac{f_{\mathbf{Y}}(\mathbf{y}|b_{l,i} = 1)}{f_{\mathbf{Y}}(\mathbf{y}|b_{l,i} = 0)}$$

$$= \ln \frac{\displaystyle\sum_{\mathbf{x} \in S_{l,i}^+} f_{\mathbf{Y}}(\mathbf{y}|\mathbf{x})}{\displaystyle\sum_{\mathbf{x} \in S_{l,i}^-} f_{\mathbf{Y}}(\mathbf{y}|\mathbf{x})}$$

(11.103)

$$= \ln \frac{\displaystyle\sum_{\mathbf{x} \in S_{l,i}^+} \exp\left(-\frac{1}{2\sigma_z^2}\|\mathbf{y} - \mathbf{Hx}\|^2\right)}{\displaystyle\sum_{\mathbf{x} \in S_{l,i}^-} \exp\left(-\frac{1}{2\sigma_z^2}\|\mathbf{y} - \mathbf{Hx}\|^2\right)}$$

where $S_{l,i}^+$ and $S_{l,i}^-$ denote the set of vectors with their lth bit value of the ith symbol being either 1 or 0, respectively. Using the following max-log approximation,

$$\log\left(e^{X_1} + e^{X_2} + \ldots + e^{X_n}\right) \approx \max_i X_i$$

(11.104)

the LLR in Equation (11.103) can be approximated as

$$\ln \frac{p(\mathbf{y}|b_{l,i} = 1)}{p(\mathbf{y}|b_{l,i} = 0)} = \ln\left\{\sum_{\mathbf{x} \in S_{l,i}^+} \exp\left(-\frac{1}{2\sigma_z^2}\|\mathbf{y} - \mathbf{Hx}\|^2\right)\right\} - \ln\left\{\sum_{\mathbf{x} \in S_{l,i}^-} \exp\left(-\frac{1}{2\sigma_z^2}\|\mathbf{y} - \mathbf{Hx}\|^2\right)\right\}$$

$$\approx \ln\left\{\max_{\mathbf{x} \in S_{l,i}^+} \exp\left(-\frac{1}{2\sigma_z^2}\|\mathbf{y} - \mathbf{Hx}\|^2\right)\right\} - \ln\left\{\max_{\mathbf{x} \in S_{l,i}^-} \exp\left(-\frac{1}{2\sigma_z^2}\|\mathbf{y} - \mathbf{Hx}\|^2\right)\right\}$$

(11.105)

Using the fact that $e^{-g(x)}$ is a monotone decreasing function as long as $g(x) > 0$, Equation (11.103) can be approximated as

$$LLR(b_{l,i}|\mathbf{y}) \triangleq \ln \frac{p(b_{l,i} = 1|\mathbf{y})}{p(b_{l,i} = 0|\mathbf{y})}$$

$$\approx \frac{1}{2\sigma_z^2} \min_{\mathbf{x} \in S_{l,i}^-} \|\mathbf{y} - \mathbf{Hx}\|^2 - \frac{1}{2\sigma_z^2} \min_{\mathbf{x} \in S_{l,i}^+} \|\mathbf{y} - \mathbf{Hx}\|^2$$

(11.106)

$$= \min_{\mathbf{x} \in S_{l,i}^-} D(\mathbf{x}) - \min_{\mathbf{x} \in S_{l,i}^+} D(\mathbf{x})$$

where $D(\mathbf{x}) = \|\mathbf{y} - \mathbf{Hx}\|^2$.

11.7.4 LLR for MIMO System Using a Limited Candidate Vector Set

For the MIMO system, the number of elements in the candidate vector set, $|S_{l,i}^+| = |S_{l,i}^-|$, depends on the number of transmit antennas as well as the constellation size $|C|$. Note that the complexity of computing the LLR can be prohibitively high as the number of transmit antennas and/or constellation size increases. For example, if we assume 16-QAM constellation and $N_T = N_R = 3$, $|S_{l,i}^+| = |S_{l,i}^-| = |C|^{N_T}/2 = 16^3/2$. It implies that too much complexity is required for computing the LLR of Equation (11.106) in practice. We now discuss how the bit-level LLR can be computed for the complexity-reduced ML detection methods such as SD and QRM-MLD.

In the complexity-reduced ML detection methods such as SD and QRM-MLD methods, ML metric values for all the possible transmitted vectors are not available. Since the ML vector can be found after calculating ML metric values for a small set of vectors in SD, complexity can be reduced without hard-decision performance degradation. When soft output or LLR values are required, however, the SD performance is worse than that of ML detection, because only a small set of vectors is considered. Let $\mathbf{B}$ denote a set of candidate vectors obtained from the complexity-reduced ML detection methods. Recall that $S_{l,i}^+$ and $S_{l,i}^-$ denote the set of vectors that the lth bit value of the ith symbol is 1 or 0, respectively. If the ML metric values are available only for a subset of the transmitted vectors, the bit-level LLR values in Equation (11.106) must be approximated as

$$LLR(b_{l,i}|\mathbf{y}) \approx \min_{\mathbf{x} \in S_{l,i,\mathbf{B}}^-} D(\mathbf{x}) - \min_{\mathbf{x} \in S_{l,i,\mathbf{B}}^+} D(\mathbf{x}) \tag{11.107}$$

where $S_{l,i,\mathbf{B}}^- = S_{l,i}^- \cap \mathbf{B}$ and $S_{l,i,\mathbf{B}}^+ = S_{l,i}^+ \cap \mathbf{B}$. The LLR value approximated by Equation (11.107) may face the following two problems [241]:

Problem 1: There is a case that a candidate vector set can be empty, that is,

$$S_{l,i,\mathbf{B}}^- = S_{l,i}^- \cap \mathbf{B} = \phi \quad \text{or} \quad S_{l,i,\mathbf{B}}^+ = S_{l,i}^+ \cap \mathbf{B} = \phi \tag{11.108}$$

In this case, either one of the two terms in approximation in (11.107) cannot be computed. Meanwhile, note that at least one of the two ML metric values in (11.107) always exists, that is, $S_{l,i,\mathbf{B}}^-$ and $S_{l,i,\mathbf{B}}^+$ cannot be the empty sets at the same time, for a given l and i. Let $\mathbf{x}_{ML,\mathbf{B}}$ be defined as

$$\mathbf{x}_{ML,\mathbf{B}} = \arg\min_{\mathbf{x} \in \mathbf{B}} \|\mathbf{y} - \mathbf{Hx}\|^2 = \arg\min_{\mathbf{x} \in S_{l,i,\mathbf{B}}^- \cup S_{l,i,\mathbf{B}}^+} \|\mathbf{y} - \mathbf{Hx}\|^2 \tag{11.109}$$

Let us represent all the bit values of $\mathbf{x}_{ML,\mathbf{B}}$ by

$$[\underbrace{b_{1,1,ML,\mathbf{B}} \ b_{2,1,ML,\mathbf{B}} \cdots b_{k,1,ML,\mathbf{B}}} \ \underbrace{b_{1,2,ML,\mathbf{B}} \cdots b_{k,2,ML,\mathbf{B}}} \ \underbrace{b_{1,3,ML,\mathbf{B}} \cdots \cdots \cdots b_{k,N_T,ML,\mathbf{B}}}]$$

$$\tag{11.110}$$

where $b_{l,i,ML,\mathbf{B}}$ denotes the lth bit value of the ith symbol, $l = 1, 2, \ldots, k$, $i = 1, 2, \ldots, N_T$. In the course of calculating the LLR in Equation (11.107), the following relation holds:

$$\min_{\mathbf{x} \in S_{l,i,\mathbf{B}}^-} D(\mathbf{x}) = \|\mathbf{y}\text{-}\mathbf{Hx}_{ML,\mathbf{B}}\|^2 \quad \text{or} \quad \min_{\mathbf{x} \in S_{l,i,\mathbf{B}}^+} D(\mathbf{x}) = \|\mathbf{y}\text{-}\mathbf{Hx}_{ML,\mathbf{B}}\|^2 \tag{11.111}$$

To handle **Problem 1**, the ML metric value not available for a specific bit can be replaced with an arbitrarily large number Γ.

Problem 2: Even if $S^-_{l,i,B} \neq \phi$ and $S^+_{l,i,B} \neq \phi$, the corresponding ML metric values may not be correct due to a reduced set of candidate vectors. Let $\mathbf{x}_{ML}$ denote the ML solution vector among all possible transmit vectors and $\mathbf{x}_{ML,B}$ an ML solution vector among candidate vectors in $\mathbf{B}$. Similarly, let $b_{l,i,ML}$ and $b_{l,i,ML,B}$ denote the bit values of $\mathbf{x}_{ML}$ and $\mathbf{x}_{ML,B}$, respectively, for $l = 1, 2, \ldots, k, i = 1, 2, \ldots, N_T$. We first consider the bit-level LLR value when $\mathbf{x}_{ML} = \mathbf{x}_{ML,B}$, and thus $b_{l,i,ML} = b_{l,i,ML,B}$. As $D(\mathbf{x}_{ML}) = \min_{\mathbf{x} \in S^-_{l,i,B}} D(\mathbf{x})$ for $b_{l,i,ML} = b_{l,i,ML,B} = 0$ when $\mathbf{x}_{ML} = \mathbf{x}_{ML,B}$, the bit-level LLR is given as

$$LLR(b_{i,j}|\mathbf{y}) = \min_{\mathbf{x} \in S^-_{l,i,B}} D(\mathbf{x}) - \min_{\mathbf{x} \in S^+_{l,i,B}} D(\mathbf{x})$$
$$= D(\mathbf{x}_{ML}) - \min_{\mathbf{x} \in S^+_{l,i,B}} D(\mathbf{x}) < 0. \tag{11.112}$$

when $b_{l,i,ML} = b_{l,i,ML,B} = 0$ is assumed. Since it is probable that $\min_{\mathbf{x} \in S^+_{l,i,B}} D(\mathbf{x}) \geq \min_{\mathbf{x} \in S^+_{l,i}} D(\mathbf{x})$, the LLR value in Equation (11.112) might not be reliable. Now, we consider the bit-level LLR when $\mathbf{x}_{ML} \neq \mathbf{x}_{ML,B}$. Again assuming that $b_{i,j,ML,B} = 0$, then the bit-level LLR is given as

$$LLR(b_{i,j}|\mathbf{y}) = D(\mathbf{x}_{ML,B}) - \min_{\mathbf{x} \in S^+_{l,i,B}} D(\mathbf{x}) < 0 \tag{11.113}$$

The negativity in Equation (11.113) is attributed to the fact that for any l and i, the following inequality holds by the definition of $\mathbf{x}_{ML,B}$ in Equation (11.108):

$$\|\mathbf{y}\text{-}\mathbf{H}\mathbf{x}_{ML,B}\|^2 \leq \min_{\mathbf{x} \in S^-_{l,i,B}} \|\mathbf{y}\text{-}\mathbf{H}x\|^2 \text{ and } \|\mathbf{y}\text{-}\mathbf{H}\mathbf{x}_{ML,B}\|^2 \leq \min_{\mathbf{x} \in S^+_{l,i,B}} \|\mathbf{y}\text{-}\mathbf{H}x\|^2 \tag{11.114}$$

Since $D(\mathbf{x}_{ML,B}) \geq D(\mathbf{x}_{ML})$ and $\min_{\mathbf{x} \in S^+_{l,i,B}} D(\mathbf{x}) \geq \min_{\mathbf{x} \in S^+_{l,i}} D(\mathbf{x})$, both terms in (11.113) are subject to the positive errors and thus, the LLR value might not be reliable. The underlying unreliability problem can be handled by replacing each term (ML metric value) in Equation (11.113) with the predetermined value whenever it exceeds the given threshold. In other words, each term in Equation (11.113) can be truncated by an arbitrarily large number Γ. By limiting the ML metric value within the threshold, a critical performance degradation can be avoided.

MATLAB® Programs: Soft-Decision QRM-MLD Detector for 4×4 MIMO System

Program 11.22 ("QRM_MLD_simulation.m") can be used to evaluate the performance of hard/soft-decision with QRM-MLD detector for 4×4 MIMO system using 16-QAM ($M = 16$). It implements the multi-path channel with the power delay profile (PDP) as given in Figure 11.15. Figure 11.16 shows the performance of QRM-MLD with hard decision and soft decision. Table 11.2 summarizes the simulation parameters used for the results in Figure 11.16. In this example, there exists $|\mathbf{C}| \times M = 256$ candidate vectors in $\mathbf{B}$, and the LLR values are calculated using Equation (11.107). When Problem 1 occurs, $\Gamma = 2$ has been used as the non-existing ML metric values in the simulation.

Table 11.2 Simulation parameters.

	Multiplications
FFT size	64
CP size	16
Antenna configuration	4×4
Packet length	10 OFDM symbols
Symbol mapping	16-QAM
Channel coding	Convolutional coding
	- Rate: 1/2
	- Constraint length: $K = 7$
	- Generating polynomials: [1001111] and [1101101].

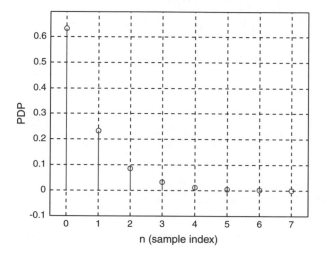

Figure 11.15 Power delay profile (PDP) for simulation.

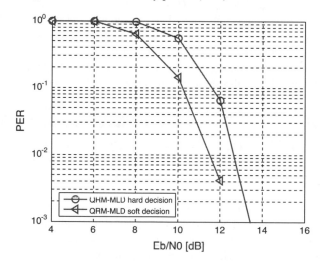

Figure 11.16 Performance of hard/soft-decision detection with QRM-MLD detector for 4×4 MIMO system using 16-QAM.

Program 11.22 "QRM_MLD_simulation.m"

```matlab
% QRM_MLD_simulation.m
clear all;
Rc=0.5; % Code rate
N_frame=100000; Nfft=64;
NT=4; NR=NT; b=4; N_block=10;
L_frame=NT*N_block*Nfft*b/2-6;
PDP=[6.3233e-001  2.3262e-001  8.5577e-002  3.1482e-002  1.1582e-002
     4.2606e-003  1.5674e-003  5.7661e-004];
N_candidate=16;  sq05=sqrt(0.5);   sq05PDP=sq05*PDP.^(1/2);
LPDP=length(PDP);
SNRdBs=[12:2:20];
for i_SNR=1:length(SNRdBs)
    SNRdB=SNRdBs(i_SNR); nofe=0; % Number of frame errors
    sig_power=NT; sigma2=sig_power*10^(-SNRdB/10); sigma1=sqrt(sigma2/2);
    rand('seed',1); randn('seed',1); Viterbi_init;
    for i_frame=1:N_frame
        LLR_estimate = zeros(N_block, NT, 4, Nfft);
        X_estimate_sym = zeros(N_block, NT, Nfft);
        s = randint(1,L_frame);
        coded_bits = transpose(convolution_encoder(s));
        interleaved = [];
        for i=1:128, interleaved=[interleaved coded_bits([i:128:end])]; end
        %ss = zeros(N_block,NT,Nfft,b);
        for i_bk=1:N_block
            for i_str=1:NT
                for i_sc=1:Nfft
                    ss(i_bk,i_str,i_sc,:)=interleaved(1:b);interleaved(1:b)=[];
                end
            end
        end
        for i_bk=1:N_block
            for i_str=1:NT
                for i_sc=1:Nfft
                    X(i_bk,i_str,i_sc)=QAM16_mod(ss(i_bk,i_str,i_sc,:),1);
                end
            end
        end
        for p=1:NR
            for q=1:NT
                tmp = sq05PDP.*(randn(1,LPDP)+j*randn(1,LPDP));
                Frame_H(p,q,:)=fft(tmp,Nfft);
            end
        end
        for i_bk=1:N_block
            for i_Rx=1:NR
```

```
            temp=0;
            for i_str=1:NT
                temp = temp + Frame_H(i_Rx,i_str,:).*X(i_bk,i_str,:);
            end
            Y(i_Rx,:)=reshape(temp,1,Nfft)+ ...
                sigma1*(randn(1,Nfft)+j*randn(1,Nfft));
        end
        for i_sc=1:Nfft
            H=Frame_H(:,:,i_sc);   y=Y(:,i_sc);
            x_test=X(i_bk,:,i_sc);
            LLR_estimate(i_bk,:,:,i_sc)=QRM_MLD_soft(y,H,N_candidate);
        end
    end
    soft_bits=[];   hard_bits=[];   s_hat=[];
    for i_bk=1:N_block
        for i_str=1:NT
            for i_sc=1:Nfft
                soft_bits=[soft_bits LLR_estimate(i_bk,i_str,1,i_sc),...
                           LLR_estimate(i_bk,i_str,2,i_sc),...
                           LLR_estimate(i_bk,i_str,3,i_sc),...
                           LLR_estimate(i_bk,i_str,4,i_sc)];
            end
        end
    end
    deinterleaved=[];
    for i=1:80
        deinterleaved=[deinterleaved soft_bits([i:80:end])];
    end
    s_hat=Viterbi_decode(deinterleaved);
    temp=find(xor(s,s_hat([1:L_frame]))==1);
    if length(temp)~=0,   nofe=nofe+1;   end
    if (nofe>200)&(i_frame>200),   break;   end
  end % End of frame index
  FER(i_SNR)=nofe/i_frame;
  if FER(i_SNR)<1e-3, break;   end
end % End of for loop with i_SNR
```

Program 11.23 "QRM_MLD_soft": Soft decision performance of QRM-MLD

```
function [LLR]=QRM_MLD_soft(y,H,M)
QAM_table=[-3-3j, -3-j, -3+3j, -3+j, -1-3j, -1-j, -1+3j, -1+j, 3-3j, ...
           3-j, 3+3j, 3+j, 1-3j, 1-j, 1+3j, 1+j]/sqrt(10);
norm_array=[norm(H(:,1))   norm(H(:,2))   norm(H(:,3))   norm(H(:,4))];
[X,I]=sort(norm_array); Reversed_order=wrev(I);
H_original=H;   H=H(:,Reversed_order);
X_hat=zeros(4,1);   X_hat_tmp=zeros(4,1);
```

```
X_LLR=zeros(4,4);   LLR=zeros(4,4);   X_LLR2=zeros(4,4);
% QR decomposition
[Q,R]=qr(H); y_tilde=Q'*y;
% 1st stage
for i=1:16,   norm_array(i)=abs(y_tilde(4)-R(4,4)*QAM_table(i))^2;   end
[T,sorted_index]=sort(norm_array); M_best_index_1=sorted_index(1:M);
% 2nd stage
M16_index=zeros(M*16,3); norm_array=zeros(M*16,1);
y_tmp=[y_tilde(3); y_tilde(4)];
R_tmp=[R(3,3) R(3,4);   0   R(4,4)];
count=1;
for i=1:M
    x4_tmp=QAM_table(M_best_index_1(i));
    for k=1:16
        x3_tmp=QAM_table(k);
        norm_array(count)=norm(y_tmp-R_tmp*[x3_tmp; x4_tmp])^2;
        M16_index(count,2:3) = [k   M_best_index_1(i)];
        count=count+1;
    end
end
clear sorted_index;
[T,sorted_index]=sort(norm_array);
M_best_index_2=M16_index(sorted_index(1:M),:);
% 3rd stage
norm_array=zeros(M*16,1);
y_tmp=[y_tilde(2); y_tilde(3); y_tilde(4)];
R_tmp=[R(2,2) R(2,3) R(2,4); 0   R(3,3) R(3,4); 0   0 R(4,4)];
count=1;
for i=1:M
    x4_tmp=QAM_table(M_best_index_2(i,3));
    x3_tmp=QAM_table(M_best_index_2(i,2));
    for k=1:16
        x2_tmp=QAM_table(k);
        norm_array(count)=norm(y_tmp-R_tmp*[x2_tmp;x3_tmp;x4_tmp])^2;
        M16_index(count,1:3) = [k M_best_index_2(i,2:3)];
        count=count+1;
    end
end
clear sorted_index;
[T,sorted_index]=sort(norm_array);
M_best_index_3=M16_index(sorted_index(1:M),:);
% 4th stage
y_tmp=y_tilde; R_tmp=R;
cost0=ones(16,1)*100; cost1=ones(16,1)*100;
LLR=zeros(4,4); X_bit=zeros(16,1);
LLR_0=zeros(16,1); LLR_1=zeros(16,1);
for i=1:M
    x4_tmp = QAM_table(M_best_index_3(i,3));
    x3_tmp = QAM_table(M_best_index_3(i,2));
```

```
    x2_tmp = QAM_table(M_best_index_3(i,1));
    X_bit(5:8) = QAM16_slicer_soft(x2_tmp);
    X_bit(9:12) = QAM16_slicer_soft(x3_tmp);
    X_bit(13:16) = QAM16_slicer_soft(x4_tmp);
    for k=1:16
        x1_tmp=QAM_table(k);
        X_bit(1:4)=QAM16_slicer_soft(x1_tmp);
        distance=norm(y_tmp-R_tmp*[x1_tmp;x2_tmp;x3_tmp;x4_tmp])^2;
        for kk=1:length(X_bit)
            if X_bit(kk)==0
                if distance<cost0(kk)
                    LLR_0(kk)=distance;   cost0(kk)=distance;
                end
            elseif X_bit(kk)==1
                if distance<cost1(kk)
                    LLR_1(kk)=distance;   cost1(kk)=distance;
                end
            end
        end
    end
end
LLR_0(find(LLR_0==0))=2; %2 is used for non-existing bit values
LLR_1(find(LLR_1==0))=2; %2 is used for non-existing bit values
LLR(Reversed_order(1),:)=(LLR_0(1:4)-LLR_1(1:4))';
LLR(Reversed_order(2),:)=(LLR_0(5:8)-LLR_1(5:8))';
LLR(Reversed_order(3),:)=(LLR_0(9:12)-LLR_1(9:12))';
LLR(Reversed_order(4),:)=(LLR_0(13:16)-LLR_1(13:16))';
```

Program 11.24 "QAM16_slicer_soft"

```
function [X_bits]=QAM16_slicer_soft(X);
QAM_table=[-3-3j, -3-j, -3+3j, -3+j, -1-3j, -1-j, -1+3j, -1+j, 3-3j,
...
          3-j, 3+3j, 3+j, 1-3j, 1-j, 1+3j, 1+j]/sqrt(10);
X_temp=dec2bin(find(QAM_table==X)-1,4);
for i=1:length(X_temp)
   X_bits(i) = bin2dec(X_temp(i));
end
```

Appendix 11.A Derivation of Equation (11.23)

We prove

$$\arg\min_{\mathbf{x}} \|\mathbf{y}-\mathbf{Hx}\|^2 = \arg\min_{\mathbf{x}} (\mathbf{x}-\hat{\mathbf{x}})^T \mathbf{H}^T \mathbf{H}(\mathbf{x}-\hat{\mathbf{x}}) \qquad (11.A.1)$$

where $\hat{\mathbf{x}}$ is the unconstrained LS solution (i.e., $\hat{\mathbf{x}} = (\mathbf{H}^H\mathbf{H})^{-1}\mathbf{H}^H\mathbf{y}$). Since this relationship holds for both complex and real systems, $\mathbf{H}$, $\mathbf{y}$, and $\mathbf{x}$ are used in place of $\bar{\mathbf{H}}$, $\bar{\mathbf{y}}$, and $\bar{\mathbf{x}}$, respectively. Consider the following expansion:

$$
\begin{aligned}
\|\mathbf{y} - \mathbf{Hx}\|^2 &= \|\mathbf{y} - \mathbf{Hx} - \mathbf{H\hat{x}} + \mathbf{H\hat{x}}\|^2 \\
&= (\mathbf{y} - \mathbf{Hx} - \mathbf{H\hat{x}} + \mathbf{H\hat{x}})^T (\mathbf{y} - \mathbf{Hx} - \mathbf{H\hat{x}} + \mathbf{H\hat{x}}) \\
&= \{(\mathbf{y} - \mathbf{H\hat{x}})^T + (\mathbf{H\hat{x}} - \mathbf{Hx})^T\}\{(\mathbf{y} - \mathbf{H\hat{x}}) + (\mathbf{H\hat{x}} - \mathbf{Hx})\} \quad (11.A.2) \\
&= (\mathbf{y} - \mathbf{H\hat{x}})^T (\mathbf{y} - \mathbf{H\hat{x}}) + (\mathbf{H\hat{x}} - \mathbf{Hx})^T (\mathbf{H\hat{x}} - \mathbf{Hx}) \\
&\quad + (\mathbf{H\hat{x}} - \mathbf{Hx})^T (\mathbf{y} - \mathbf{H\hat{x}}) + (\mathbf{y} - \mathbf{H\hat{x}})^T (\mathbf{H\hat{x}} - \mathbf{Hx})
\end{aligned}
$$

Since $\hat{\mathbf{x}}$ is the LS solution, $(\mathbf{H\hat{x}} - \mathbf{Hx})^T (\mathbf{y} - \mathbf{H\hat{x}}) = (\mathbf{y} - \mathbf{H\hat{x}})^T (\mathbf{H\hat{x}} - \mathbf{Hx}) = 0$ and thus, Equation (11.A.2) reduces to

$$
\|\mathbf{y} - \mathbf{Hx}\|^2 = (\mathbf{y} - \mathbf{H\hat{x}})^T (\mathbf{y} - \mathbf{H\hat{x}}) + (\mathbf{H\hat{x}} - \mathbf{Hx})^T (\mathbf{H\hat{x}} - \mathbf{Hx}) \quad (11.A.3)
$$

Substituting $\hat{\mathbf{x}}$ with $\left(\mathbf{H}^H \mathbf{H}\right)^{-1} \mathbf{H}^H \mathbf{y}$, Equation (11.A.3) is expressed as

$$
\|\mathbf{y} - \mathbf{Hx}\|^2 = \left\{\mathbf{y} - \mathbf{H}\left(\mathbf{H}^T \mathbf{H}\right)^{-1} \mathbf{H}^T \mathbf{y}\right\}^T \left\{\mathbf{y} - \mathbf{H}\left(\mathbf{H}^T \mathbf{H}\right)^{-1} \mathbf{H}^T \mathbf{y}\right\} + (\hat{\mathbf{x}} - \mathbf{x})^T \mathbf{H}^T \mathbf{H}(\hat{\mathbf{x}} - \mathbf{x}).
$$
$$(11.A.4)$$

Since $\mathbf{y} - \mathbf{H}\left(\mathbf{H}^T \mathbf{H}\right)^{-1} \mathbf{H}^T \mathbf{y} = \left\{\mathbf{I} - \mathbf{H}\left(\mathbf{H}^T \mathbf{H}\right)^{-1}\right\} \mathbf{y}$, the first term in Equation (11.A.4) becomes

$$
\begin{aligned}
&\mathbf{y}^T \left\{\mathbf{I} - \mathbf{H}\left(\mathbf{H}^T \mathbf{H}\right)^{-1} \mathbf{H}^T\right\}^T \left\{\mathbf{I} - \mathbf{H}\left(\mathbf{H}^T \mathbf{H}\right)^{-1} \mathbf{H}^T\right\} \mathbf{y} \\
&= \mathbf{y}^T \left\{\mathbf{I} - \mathbf{H}\left(\mathbf{H}^T \mathbf{H}\right)^{-T} \mathbf{H}^T\right\} \left\{\mathbf{I} - \mathbf{H}\left(\mathbf{H}^T \mathbf{H}\right)^{-1} \mathbf{H}^T\right\} \mathbf{y} \\
&= \mathbf{y}^T \left\{\mathbf{I} - \mathbf{H}\left(\mathbf{H}^T \mathbf{H}\right)^{-1} \mathbf{H}^T - \mathbf{H}\left(\mathbf{H}^T \mathbf{H}\right)^{-T} \mathbf{H}^T + \mathbf{H}\left(\mathbf{H}^T \mathbf{H}\right)^{-T} \mathbf{H}^T \mathbf{H}\left(\mathbf{H}^T \mathbf{H}\right)^{-1} \mathbf{H}^T\right\} \mathbf{y} \\
&= \mathbf{y}^T \left\{\mathbf{I} - \mathbf{H}\left(\mathbf{H}^T \mathbf{H}\right)^{-1} \mathbf{H}^T\right\} \mathbf{y}.
\end{aligned}
$$
$$(11.A.5)$$

which turns out to be constant with respect to $\mathbf{x}$. From Equations (11.A.4) and (11.A.5), our relationship in Equation (11.A.1) immediately follows:

$$
\arg\min_{\mathbf{x}} \|\mathbf{y} - \mathbf{Hx}\|^2 = \arg\min_{\mathbf{x}} (\mathbf{x} - \hat{\mathbf{x}}) \mathbf{H}^T \mathbf{H}(\mathbf{x} - \hat{\mathbf{x}}). \quad (11.A.6)
$$

12

Exploiting Channel State Information at the Transmitter Side

In Chapters 10 and 11, we assumed that only the receiver can track the channel. In this chapter, we will address transmission techniques that exploit the channel state information (CSI) on the transmitter side. The CSI can be completely or partially known on the transmitter side. Sometimes, only statistical information on the channel state may be available. Exploitation of such channel information allows for increasing the channel capacity, improving error performance, while reducing hardware complexity [242]. In the 4×2 MIMO system, for example, exploitation of the complete CSI may improve the system capacity by as much as 1.5 bps/Hz. In practice, however, full CSI may not be directly available due to feedback overhead and feedback delay. In particular, CSI for the time-varying channel cannot be tracked completely by the transmitter and thus, only partial information (e.g., the statistical information) can be exploited. In this chapter, we will first discuss how to obtain such channel information. Furthermore, we will mainly consider the precoding techniques and antenna selection techniques as the typical approaches that exploit CSI on the transmitter side.

12.1 Channel Estimation on the Transmitter Side

In general, a transmitter does not have direct access to its own channel state information. Therefore, some indirect means are required for the transmitter. In time division duplexing (TDD) system, we can exploit the channel reciprocity between opposite links (downlink and uplink). Based on the signal received from the opposite direction, it allows for indirect channel estimation. In frequency division duplexing (FDD) system, which usually does not have reciprocity between opposite directions, the transmitter relies on the channel feedback information from the receiver. In other words, CSI must be estimated at the receiver side and then, fed back to the transmitter side.

MIMO-OFDM Wireless Communications with MATLAB®   Yong Soo Cho, Jaekwon Kim, Won Young Yang and Chung G. Kang
© 2010 John Wiley & Sons (Asia) Pte Ltd

12.1.1 Using Channel Reciprocity

As long as the channel gains in both directions are highly correlated (i.e., *reciprocal* as shown in Figure 12.1), channel condition in one direction can be implicitly known from the other direction. In TDD systems, forward and backward channels tend to be reciprocal. There exists a non-negligible difference in their transmission time. However, if the difference is small relative to the coherence time, the reciprocity can be a useful property to exploit. In FDD systems, however, the two channels use different radio frequencies. Thus channel reciprocity does not hold.

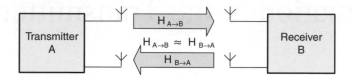

Figure 12.1   Reciprocity of wireless channel.

The actual effect of the wireless channels must include the characteristics of RF elements. In general, the RF characteristics in forward channels is different from those in backward channels. Such a difference must be compensated somehow in the course of taking advantage of the channel reciprocity.

12.1.2 CSI Feedback

One other possible approach in obtaining the channel condition in the transmit side is to use the explicit feedback from the receiver side, as illustrated in Figure 12.2. As opposed to exploiting the reciprocity, compensation for the RF difference is not necessary in this method. In order to warrant timely channel information, however, the feedback delay Δ_t must be less than the coherence time T_c, that is,

$$\Delta_t = T_c \tag{12.1}$$

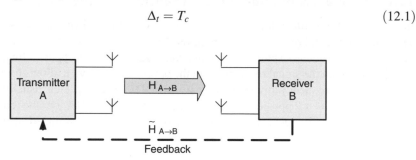

Figure 12.2   Feedback of channel state information.

Its main drawback is that additional resource is required for transmitting the feedback information. The amount of feedback information increases with the number of antennas. Therefore, the overhead problem can become critical when it comes to multiple antenna systems. When channels are subject to fast fading, the coherence time is small, which requires more frequent feedback to meet the constraint in Equation (12.1). The estimated CSI at the receiver can be compressed to reduce the feedback overhead. One particular approach is to quantize the channel gains. Let $Q_{quan}(\mathbf{H})$ represent the quantization function of the channel gain $\mathbf{H}$. Then, the channel gain can be quantized so as to minimize the mean square error $E\{||\mathbf{H}-Q_{quan}(\mathbf{H})||^2\}$. Another approach is to use the codebook that is shared by the transmitter and receiver. The codebook is a set of codewords, which are the quantized vectors to represent the states of channel condition. In this approach, channel gains are estimated at the receiver side. Then, the index of the appropriate codeword is selected to represent a state of estimated channel gain. Rather than the full CSI, only the corresponding index is fed back to the transmitter side. Each index can be represented with F_B bits, which allows for a total number of $L = 2^{F_B}$ codewords in the codebook. Note that L is referred to as a codebook size. Let $\mathbf{W}_i$ denote the ith codeword, $i = 1, 2, \cdots, L$. For a given codebook $\mathsf{F} = \{\mathbf{W}_1, \mathbf{W}_2, \mathbf{W}_3, \cdots, \mathbf{W}_L\}$, the codeword is selected by a mapping function $f(\cdot)$. For a given channel condition $\mathbf{H}$, the codebook method can be represented as

$$\mathbf{W}_{opt} = f(\mathbf{H}) \in \mathsf{F} = \{\mathbf{W}_1, \mathbf{W}_2, \mathbf{W}_3, \cdots, \mathbf{W}_L\} \tag{12.2}$$

where $\mathbf{W}_{opt}$ is the codeword that best represents $\mathbf{H}$ for a given mapping function $f(\cdot)$. However, the issue of designing a codebook remains. We are supposed to determine the codewords that quantize the channel space with the least distortion. We will discuss the codebook design methods in the following sections.

12.2 Precoded OSTBC

Consider the MISO system with N_T antennas, that is, $\mathbf{h} \in \mathbb{C}^{1 \times N_T}$. Let $\mathbf{C} \in \mathbb{C}^{M \times T}$ denote a space-time codeword with a length of M, which is represented as

$$\mathbf{C} = [\mathbf{c}_1\ \mathbf{c}_2 \cdots \mathbf{c}_T]$$

where $\mathbf{c}_k = [c_{k,1}\ c_{k,2} \cdots c_{k,M}]^T$, $k = 1, 2, \cdots, T$, and $M \leq N_T$. In the precoded OSTBC systems, the space-time codeword $\mathbf{C}$ is multiplied by a precoding matrix $\mathbf{W} \in \mathbb{C}^{N_T \times M}$, which is chosen from the codebook $\mathsf{F} = \{\mathbf{W}_1, \mathbf{W}_2, \mathbf{W}_3, \cdots, \mathbf{W}_L\}$. The objective is to choose an appropriate codeword that improves the overall system performance such as channel capacity or error performance. Assuming that N_T channels remain static over T, the received signal $\mathbf{y} \in \mathbb{C}^{1 \times T}$ can be expressed as

$$\mathbf{y} = \sqrt{\frac{E_x}{N_T}}\mathbf{hWC} + \mathbf{z} \tag{12.3}$$

Comparing the above equation with Equation (10.17), we can see that $\mathbf{X} \in \mathbb{C}^{N_T \times T}$ in Equation (10.17) is replaced by the product of $\mathbf{W} \in \mathbb{C}^{N_T \times M}$ and $\mathbf{C} \in \mathbb{C}^{M \times T}$ in Equation (12.3). We also note that in Equation (10.17), space-time codeword is composed of T column vectors, and the length of each vector is N_T. In Equation (12.3), however, the length of each vector is $M \leq N_T$ while the space-time codeword is still composed of T column vectors. The probability of codeword error for Equation (12.3) can be derived in a similar manner as Equation (10.17). For a given channel $\mathbf{h}$

and precoding matrix $\mathbf{W}$, we consider the pairwise codeword error probability $\Pr(\mathbf{C}_i \rightarrow \mathbf{C}_j|\mathbf{H})$. This is the probability that the space-time codeword $\mathbf{C}_i$ is transmitted whereas $\mathbf{C}_j$ with $j \neq i$ is decoded. Following the derivation in Section 10.2.2, the upper bound of the pairwise error probability is given as

$$\Pr(\mathbf{C}_i \rightarrow \mathbf{C}_j|\mathbf{H}) = Q\left(\sqrt{\frac{\rho\|\mathbf{HWE}_{i,j}\|_F^2}{2N_T}}\right) \leq \exp\left(-\frac{\rho\|\mathbf{HWE}_{i,j}\|_F^2}{4N_T}\right) \tag{12.4}$$

where ρ is the signal-to-noise ratio (SNR), given as $\rho = E_x/N_0$, and $\mathbf{E}_{i,j}$ is the error matrix between the codewords $\mathbf{C}_i$ and $\mathbf{C}_j$, which is defined as $\mathbf{E}_{i,j} = \mathbf{C}_i - \mathbf{C}_j$ for a given STBC scheme. From Equation (12.4), we see that $\|\mathbf{HWE}_{i,j}\|_F^2$ needs to be maximized in order to minimize the pairwise error probability [243, 244]. This leads us to the following codeword selection criterion:

$$\begin{aligned}
\mathbf{W}_{opt} &= \arg\max_{\mathbf{W}\in\mathsf{F}, i\neq j} \|\mathbf{HWE}_{i,j}\|_F^2 \\
&= \arg\max_{\mathbf{W}\in\mathsf{F}, i\neq j} \mathrm{Tr}\left(\mathbf{HWE}_{i,j}\mathbf{E}_{i,j}^H\mathbf{W}^H\mathbf{H}^H\right) \\
&= \arg\max_{\mathbf{W}\in\mathsf{F}} \mathrm{Tr}\left(\mathbf{HWW}^H\mathbf{H}^H\right) \\
&= \arg\max_{\mathbf{W}\in\mathsf{F}} \|\mathbf{HW}\|_F^2
\end{aligned} \tag{12.5}$$

In the course of deriving Equation (12.5), we have used the fact that the error matrix of OSTBC has the property of $\mathbf{E}_{i,j}\mathbf{E}_{i,j}^H = a\mathbf{I}$ with constant a. When the constraint $\mathbf{W} \in \mathsf{F}$ is not imposed, the above optimum solution $\mathbf{W}_{opt}$ is not unique, because $\|\mathbf{HW}_{opt}\|_F^2 = \|\mathbf{HW}_{opt}\mathbf{Z}\|_F^2$ where $\mathbf{Z}$ is a unitary matrix. The unconstrained optimum solution of Equation (12.5) can be obtained by a singular value decomposition (SVD) of channel $\mathbf{H} = \mathbf{U\Sigma V}^H$, where the diagonal entry of Σ is in descending order. It has been shown that the optimum solution of Equation (12.5) is given by the leftmost M columns of $\mathbf{V}$ [245], that is,

$$\mathbf{W}_{opt} = [\mathbf{v}_1 \, \mathbf{v}_2 \, \cdots \, \mathbf{v}_M] \triangleq \bar{\mathbf{V}} \tag{12.6}$$

Since $\bar{\mathbf{V}}$ is unitary, $\lambda_i(\mathbf{W}_{opt}) = 1$, $i = 1, 2, \cdots, M$, where $\lambda_i(\mathbf{A})$ denotes the ith largest eigenvalue of the matrix $\mathbf{A}$. In case that a channel is not deterministic, the following criterion is used for the codebook design:

$$E\left\{\min_{\mathbf{W}\in\mathsf{F}} \left(\|\mathbf{HW}_{opt}\|_F^2 - \|\mathbf{HW}\|_F^2\right)\right\}. \tag{12.7}$$

where the expectation is with regards to the random channel $\mathbf{H}$ [245]. $\mathbf{W}_{opt}$ in Equation (12.7) follows from Equation (12.6) for the given channel $\mathbf{H}$. The above expected value in Equation (12.7) is upper-bounded as

$$E\left\{\min_{\mathbf{W}\in\mathsf{F}} \left(\|\mathbf{HW}_{opt}\|_F^2 - \|\mathbf{HW}\|_F^2\right)\right\} \leq E\{\lambda_1^2\{\mathbf{H}\}\}E\left\{\min_{\mathbf{W}\in\mathsf{F}} \frac{1}{2}\left\|\mathbf{VV}^H - \mathbf{WW}^H\right\|_F^2\right\} \tag{12.8}$$

Since $\lambda_1^2\{\mathbf{H}\}$ is given, the codebook must be designed so as to minimize $E\left\{\min_{\mathbf{W}\in\mathsf{F}}\dfrac{1}{2}\left\|\bar{\mathbf{V}}\bar{\mathbf{V}}^H-\mathbf{W}\mathbf{W}^H\right\|_F^2\right\}$ in Equation (12.8). The corresponding minimization problem can be formulated into the Grassmannian subspace packing problem [245–247]. The performance measure in Grassmannian subspace packing is the chordal distance, which is defined as

$$d(\mathbf{W}_k,\mathbf{W}_l) = \frac{1}{\sqrt{2}}\left\|\mathbf{W}_k\mathbf{W}_k^H-\mathbf{W}_l\mathbf{W}_l^H\right\|_F \tag{12.9}$$

For random channels, the optimum codebook is designed to maximize the minimum chordal distance $\delta_{\min} = \min_{k\neq l,1\leq k,l\leq L}d(\mathbf{W}_k,\mathbf{W}_l)$ [245]. This particular situation is illustrated in Figure 12.3.

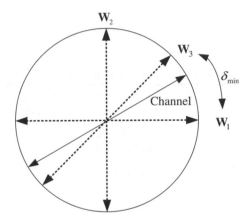

Figure 12.3   Precoding matrix and chordal distance.

Solving the Grassmannian packing problem for arbitrary N_T, codeword length M, and codebook size L is quite time-consuming and not straightforward [246, 248]. Instead, we will consider a suboptimal yet practical design method. One particular design method is to use DFT matrices given as [249]

$$\mathsf{F} = \{\mathbf{W}_{\mathrm{DFT}},\boldsymbol{\theta}\mathbf{W}_{\mathrm{DFT}},\cdots,\boldsymbol{\theta}^{L-1}\mathbf{W}_{\mathrm{DFT}}\} \tag{12.10}$$

The first codeword $\mathbf{W}_{\mathrm{DFT}}$ is obtained by selecting M columns of $N_T\times N_T$ DFT matrix, of which the (k,l)th entry is given as $e^{j2\pi(k-1)(l-1)/N_T}/\sqrt{N_T}$, $k,l = 1,2,\cdots,N_T$. Furthermore, $\boldsymbol{\theta}$ is the diagonal matrix given as

$$\boldsymbol{\theta} = \mathrm{diag}\left(\left[e^{j2\pi u_1/N_T}\quad e^{j2\pi u_2/N_T}\quad\cdots\quad e^{j2\pi u_{N_T}/N_T}\right]\right) \tag{12.11}$$

with free variables $\{u_l\}_{i=1}^{N_T}$ to be determined. Given the first codeword $\mathbf{W}_{\mathrm{DFT}}$, the remaining $(L-1)$ codewords are obtained by multiplying $\mathbf{W}_{\mathrm{DFT}}$ by $\boldsymbol{\theta}^i$, $i = 1,2,\cdots,L-1$. The free variables $\{u_i\}_{i=1}^{N_T}$ in Equation (12.11) is determined such that the minimum chordal distance is maximized, that is,

$$\mathbf{u} = \underset{\{u_1,u_2\cdots u_{N_T}\}}{\arg\max} \; \underset{l=1,2,\cdots,N-1}{\min} \; d(\mathbf{W}_{\mathrm{DFT}}, \boldsymbol{\theta}^l \mathbf{W}_{\mathrm{DFT}}) \qquad (12.12)$$

Note that IEEE 802.16e specification for the Mobile WiMAX system employs this particular design method. Table 12.1 shows the values of $\mathbf{u} = [u_1, u_2 \cdots u_{N_T}]^1$ that are adopted in IEEE 802.16e for various values of $N_T, M,$ and L. For example, when $N_T = 4, M = 3,$ and $L = 64, \mathbf{W}_1$ is given as

$$\mathbf{W}_1 = \frac{1}{\sqrt{4}} \begin{bmatrix} 1 & 1 & 1 \\ 1 & e^{j2\pi \cdot 1 \cdot 2/4} & e^{j2\pi \cdot 1 \cdot 3/4} \\ 1 & e^{j2\pi \cdot 2 \cdot 2/4} & e^{j2\pi \cdot 2 \cdot 3/4} \\ 1 & e^{j2\pi \cdot 3 \cdot 2/4} & e^{j2\pi \cdot 3 \cdot 3/4} \end{bmatrix} \qquad (12.13)$$

Table 12.1 Codebook design parameters for OSTBC in IEEE 802.16e specification.

N_T number of Tx antennas	M number of data streams	L/F_B codebook size (feedback bits)	$\mathbf{c}$ column indices	$\mathbf{u}$ rotation vector
2	1	8/(3)	[1]	[1,0]
3	1	32/(5)	[1]	[1,26,28]
4	2	32/(5)	[1,2]	[1,26,28]
4	1	64/(6)	[1]	[1,8,61,45]
4	2	64/(6)	[0,1]	[1,7,52,56]
4	3	64/(6)	[0,2,3]	[1,8,61,45]

The remaining precoding matrices $\mathbf{W}_i$ are obtained as

$$\mathbf{W}_i = \mathrm{diag}\left(\left[e^{j2\pi \cdot 1/4} \; e^{j2\pi \cdot 8/4} \; e^{j2\pi \cdot 61/4} \; e^{j2\pi \cdot 45/4}\right]\right)^{i-1} \mathbf{W}_1, \quad i = 2,3,\cdots,64 \qquad (12.14)$$

Program 12.1 ("codebook_generator") generates the codebook using the design method in Equation (12.10) with $N_T = 4, M = 2,$ and $L = 64$ as shown in Table 12.1. To simulate the BER performance of the precoded OSTBC using the Alamouti coding scheme, Program 12.2 ("Alamouti_2x1_precoding.m") has been run to yield Figure 12.4, which compares the performance of STBC with and without precoding for $N_T = 2$ and $N_R = 1$ in a block flat Rayleigh fading channel. It demonstrates that the precoded STBC scheme outperforms the traditional STBC scheme without increasing transmit power or increasing spectral bandwidth.

---

[1]The free variable vector $\mathbf{u} = [u_1, u_2 \cdots u_{N_T}]$ in Equation (12.11) is referred to as the rotation vector in IEEE 802.16e specification.

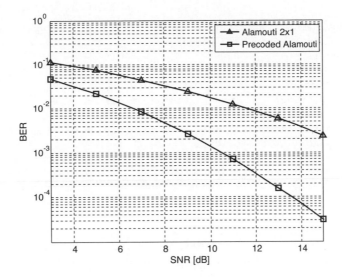

Figure 12.4 BER performance of OSTBC with and without precoding in Rayleigh fading channel.

MATLAB® Programs: Codebook Generation and Precoding

Program 12.1 "codebook_generator" for codebook generation

```
function [code_book]=codebook_generator
% N_Nt: number of Tx antennas
% N_M : codeword length
% N_L : codebook size
N_Nt=4; N_M=2; N_L=64;
cloumn_index=[1 2]; rotation_vector=[1 7 52 56];
kk=0:N_Nt-1; ll=0:N_Nt-1;
w = exp(j*2*pi/N_Nt*kk.'*ll)/sqrt(N_Nt);
w_1 = w(:,cloumn_index([1 2]));
theta = diag(exp(j*2*pi/N_L*rotation_vector));
code_book(:,:,1) = w_1 ;
for i=1:N_L-1,   code_book(:,:,i+1) = theta*code_book(:,:,i);   end
```

Program 12.2 "Alamouti_2x1_precoding.m" for Alamouti coding with precoded OSTBC

```
% Alamouti_2x1_precoding.m
clear all; clf
%%%%% Parameter Setting %%%%%%%%%
N_frame=1000; N_packet=100; %Number of frames/packet and Number of packets
b=2; M=2^b; % Number of bits per symbol and Modulation order
```

```
mod_obj=modem.qammod('M',M,'SymbolOrder','Gray','InputType','bit');
demod_obj = modem.qamdemod(mod_obj);
% MIMO Parameters
T_TX=4; code_length=64;
NT=2; NR=1; % Numbers of transmit/receive antennas
N_pbits=NT*b*N_frame; N_tbits=N_pbits*N_packet;
code_book = codebook_generator;
fprintf('==================================================\n');
fprintf(' Precoding transmission');
fprintf('\n %d x %d MIMO\n %d QAM', NT,NR,M);
fprintf('\n Simulation bits : %d', N_tbits);
fprintf('\n==================================================\n');
SNRdBs = [0:2:10]; sq2=sqrt(2);
for i_SNR=1:length(SNRdBs)
  SNRdB = SNRdBs(i_SNR);
  noise_var = NT*0.5*10^(-SNRdB/10); sigma = sqrt(noise_var);
  rand('seed',1); randn('seed',1); N_ebits=0;
  for i_packet=1:N_packet
    msg_bit = randint(N_pbits,1); % Bit generation
    %%%%%%%%%%%%%% Transmitter %%%%%%%%%%%%%%%%%%%%
    s = modulate(mod_obj, msg_bit );
    Scale = modnorm(s,'avpow',1); % Normalization
    S = reshape(Scale*s,NT,1,N_frame); % Transmit symbol
    Tx_symbol = [S(1,1,:) -conj(S(2,1,:)); S(2,1,:) conj(S(1,1,:))];
    %%%%%%%%%%%%%% Channel and Noise %%%%%%%%%%%%%%
    H = (randn(NR,T_TX)+j*randn(NR,T_TX))/sq2;
    for i=1:code_length
      cal(i) = norm(H*code_book(:,:,i),'fro');
    end
    [val,Index] = max(cal);
    He = H*code_book(:,:,Index);
    norm_H2 = norm(He)^2; % H selected and its norm2
    for i=1:N_frame
      Rx(:,:,i)=He*Tx_symbol(:,:,i)+sigma*(randn(NR,2)+j*randn(NR,2));
    end
    %%%%%%%%%%%%%% Receiver %%%%%%%%%%%%%%%%%%%%%%%
    for i=1:N_frame
      y(1,i) = (He(1)'*Rx(:,1,i)+He(2)*Rx(:,2,i)')/norm_H2;
      y(2,i) = (He(2)'*Rx(:,1,i)-He(1)*Rx(:,2,i)')/norm_H2;
    end
    S_hat = reshape(y/Scale,NT*N_frame,1);
    msg_hat = demodulate(demod_obj,S_hat);
    N_ebits = N_ebits + sum(msg_hat~=msg_bit);
  end
  BER(i_SNR) = N_ebits/N_tbits;
end
semilogy(SNRdBs,BER,'-k^', 'LineWidth',2); hold on; grid on;
xlabel('SNR[dB]'), ylabel('BER'); legend('Precoded Alamouti');
```

12.3 Precoded Spatial-Multiplexing System

As in Section 12.2, CSI can be exploited at the transmitter not only for OSTBC systems, but also for spatial-multiplexing MIMO systems with the channel gain of $\mathbf{H} \in \mathbb{C}^{N_R \times N_T}$ with $N_R \geq N_T$. An obvious method to use CSI on the transmitter side is the modal decomposition in Section 9.2.1. For the modal decomposition, the matrix $\mathbf{V} \in \mathbb{C}^{N_T \times N_T}$ in Equation 9.16 must be used as a precoding matrix on the transmitter side. Then, the interference-free modes are obtained as described in Equation 9.17.

Among various possible methods that use CSI for the spatial-multiplexing system, we will focus on the linear pre-equalization method. As illustrated in Figure 12.5, it employs *pre-equalization* on the transmitter side, which is equivalent to precoding in the previous section.

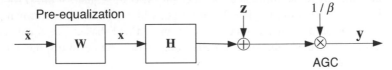

Figure 12.5 Linear pre-equalization.

The pre-equalization can be represented by a pre-equalizer weight matrix $\mathbf{W} \in \mathbb{C}^{N_T \times N_T}$ and thus, the precoded symbol vector $\mathbf{x} \in \mathbb{C}^{N_T \times 1}$ can be expressed as

$$\mathbf{x} = \mathbf{W}\tilde{\mathbf{x}} \tag{12.15}$$

where $\tilde{\mathbf{x}}$ is the original symbol vector for transmission. In case where the zero-forcing (ZF) equalization is employed, the corresponding weight matrix (assuming that the channel matrix is square) is given as

$$\mathbf{W}_{ZF} = \beta \mathbf{H}^{-1} \tag{12.16}$$

where β is a constant to meet the total transmitted power constraint after pre-equalization and it is given as

$$\beta = \sqrt{\frac{N_T}{\text{Tr}\left(\mathbf{H}^{-1}\left(\mathbf{H}^{-1}\right)^H\right)}}. \tag{12.17}$$

To compensate for the effect of amplification by a factor of β at the transmitter, the received signal must be divided by β via automatic gain control (AGC) at the receiver, as illustrated in Figure 12.5. The received signal $\mathbf{y}$ is given by

$$\begin{aligned}
\mathbf{y} &= \frac{1}{\beta}\left(\mathbf{H}\mathbf{W}_{ZF}\tilde{\mathbf{x}} + \mathbf{z}\right) \\
&= \frac{1}{\beta}\left(\mathbf{H}\beta\mathbf{H}^{-1}\tilde{\mathbf{x}} + \mathbf{z}\right) \\
&= \tilde{\mathbf{x}} + \frac{1}{\beta}\mathbf{z} \\
&= \tilde{\mathbf{x}} + \tilde{\mathbf{z}}.
\end{aligned} \tag{12.18}$$

Other than ZF pre-equalization, MMSE pre-equalization can also be used. In this case, the weight matrix is given as

$$\mathbf{W}_{MMSE} = \beta \times \underset{\mathbf{W}}{\arg\min} \ \mathrm{E}\left\{\left\|\beta^{-1}(\mathbf{HW\tilde{x}}+\mathbf{z})-\tilde{x}\right\|^2\right\}$$

$$= \beta \times \mathbf{H}^H \left(\mathbf{HH}^H + \frac{\sigma_z^2}{\sigma_x^2}\mathbf{I}\right)^{-1} \qquad (12.19)$$

where the constant β is used again to meet the total transmitted power constraint. It is calculated by Equation (12.17), but we now replace $\mathbf{H}^{-1}$ with $\mathbf{H}^H\left(\mathbf{HH}^H + \frac{\sigma_z^2}{\sigma_x^2}\mathbf{I}\right)^{-1}$ [250]. We note that the pre-equalization scheme on the transmitter side outperforms the receiver-side equalization. It is attributed to the fact that the receiver-side equalization suffers from noise enhancement in the course of equalization.

To simulate the BER performance of the pre-MMSE equalization, Program 12.3 ("pre_MMSE.m") has been run to yield Figure 12.6, which shows the performances of ZF/MMSE-based equalizations on the receiver side and MMSE-based pre-equalization with Equation (12.19). It is clear that pre-MMSE equalization outperforms the receiver-side equalization schemes.

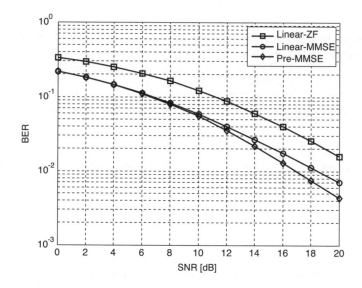

Figure 12.6 Performance comparison: receiver-side ZF/MMSE equalization vs. pre-MMSE equalization.

MATLAB® Program: Pre-MMSE Equalization

Program 12.3 "pre_MMSE.m" for Pre-MMSE equalization

```
% pre_MMSE.m
clear all; clf
%%%%%% Parameter Setting %%%%%%%%%%
```

```
N_frame=100; N_packet=1000; %Number of frames/packet & Number of packets
b=2; M=2^b; % Number of bits per symbol and Modulation order
mod_obj=modem.qammod('M',M,'SymbolOrder','Gray','InputType','bit');
demod_obj = modem.qamdemod(mod_obj);
NT=4; NR=4; sq2=sqrt(2); I=eye(NR,NR);
N_pbits = N_frame*NT*b;
N_tbits = N_pbits*N_packet;
fprintf('================================================\n');
fprintf(' Pre-MMSE transmission');
fprintf('\n %d x %d MIMO\n %d QAM', NT,NR,M);
fprintf('\n Simulation bits : %d',N_tbits);
fprintf('\n================================================\n');
SNRdBs = [0:2:20];
for i_SNR=1:length(SNRdBs)
  SNRdB = SNRdBs(i_SNR);
  noise_var = NT*0.5*10^(-SNRdB/10);
  sigma = sqrt(noise_var);
  rand('seed',1); randn('seed',1); N_ebits = 0;
  %%%%%%%%%%%%% Transmitter %%%%%%%%%%%%%%%%%%
  for i_packet=1:N_packet
    msg_bit = randint(N_pbits,1); % bit generation
    symbol = modulate(mod_obj,msg_bit).';
    Scale = modnorm(symbol,'avpow',1); % normalization
    Symbol_nomalized = reshape(Scale*symbol,NT,N_frame);
    H = (randn(NR,NT)+j*randn(NR,NT))/sq2;
    temp_W = H'*inv(H*H'+noise_var*I);
    beta = sqrt(NT/trace(temp_W*temp_W')); % Eq.(12.17)
    W = beta*temp_W; % Eq.(12.19)
    Tx_signal = W*Symbol_nomalized;
    %%%%%%%%%%%%% Channel and Noise %%%%%%%%%%%%%
    Rx_signal = H*Tx_signal+sigma*(randn(NR,N_frame)+j*randn(NR,N_frame));
    %%%%%%%%%%%%% Receiver %%%%%%%%%%%%%%%%%%%%%
    y = Rx_signal/beta; % Eq.(12.18)
    Symbol_hat = reshape(y/Scale,NT*N_frame,1);
    msg_hat = demodulate(demod_obj,Symbol_hat);
    N_ebits = N_ebits + sum(msg_hat~=msg_bit);
  end
  BER(i_SNR) = N_ebits/N_tbits;
end
semilogy(SNRdBs,BER,'-k^','LineWidth',2); hold on; grid on;
xlabel('SNR[dB]'), ylabel('BER');
legend('Pre-MMSE transmission')
```

12.4 Antenna Selection Techniques

The advantage of MIMO systems is that better performance can be achieved without using additional transmit power or bandwidth extension. However, its main drawback is that additional high-cost RF modules are required as multiple antennas are employed. In general, RF modules include low noise amplifier (LNA), frequency down-converter, and analog-to-digital converter (ADC). In an effort to reduce the cost associated with the multiple RF modules, antenna selection techniques can be used to employ a smaller number of RF modules than the number of transmit antennas. Figure 12.7 illustrates the end-to-end configuration of the antenna selection in which only Q RF modules are used to support N_T transmit antennas ($Q < N_T$). Note that Q RF modules are selectively mapped to Q of N_T transmit antennas.

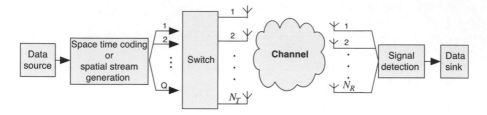

Figure 12.7 Antenna selections with Q RF modules and N_T transmit antennas ($Q < N_T$).

Since Q antennas are used among N_T transmit antennas, the effective channel can now be represented by Q columns of $\mathbf{H} \in \mathbb{C}^{N_R \times N_T}$. Let p_i denote the index of the ith selected column, $i = 1, 2, \cdots, Q$. Then, the corresponding effective channel will be modeled by $N_R \times Q$ matrix, which is denoted by $\mathbf{H}_{\{p_1, p_2, \cdots, p_Q\}} \in \mathbb{C}^{N_R \times Q}$. Let $\mathbf{x} \in \mathbb{C}^{Q \times 1}$ denote the space-time-coded or spatially-multiplexed stream that is mapped into Q selected antennas. Then, the received signal $\mathbf{y}$ is represented as

$$\mathbf{y} = \sqrt{\frac{E_x}{Q}} \mathbf{H}_{\{p_1, p_2, \cdots, p_Q\}} \mathbf{x} + \mathbf{z} \tag{12.20}$$

where $\mathbf{z} \in \mathbb{C}^{N_R \times 1}$ is the additive noise vector. The channel capacity of the system in Equation (12.20) will depend on which transmit antennas are chosen as well as the number of transmit antennas that are chosen. In the following subsections, we will discuss how the channel capacity can be improved by the antenna selection technique.

12.4.1 Optimum Antenna Selection Technique

A set of Q transmit antennas must be selected out of N_T transmit antennas so as to maximize the channel capacity. When the total transmitted power is limited by P, the channel capacity of the system using Q selected transmit antennas is given by

$$C = \max_{\mathbf{R}_{xx}, \{p_1, p_2, \cdots, p_Q\}} \log_2 \det \left(\mathbf{I}_{N_R} + \frac{E_x}{Q N_0} \mathbf{H}_{\{p_1, p_2, \cdots, p_Q\}} \mathbf{R}_{xx} \mathbf{H}^H_{\{p_1, p_2, \cdots, p_Q\}} \right) \text{ bps/Hz} \tag{12.21}$$

where $\mathbf{R}_{xx}$ is $Q \times Q$ covariance matrix. If equal power is allocated to all selected transmit antennas, $\mathbf{R}_{xx} = \mathbf{I}_Q$, which yields the channel capacity for the given $\{p_i\}_{i=1}^{Q}$ as

$$C_{\{p_1,p_2,\cdots,p_Q\}} \triangleq \log_2 \det\left(\mathbf{I}_{N_R} + \frac{\mathbf{E}_x}{Q\mathbf{N}_0}\mathbf{H}_{\{p_1,p_2,\cdots,p_Q\}}\mathbf{H}^H{}_{\{p_1,p_2,\cdots,p_Q\}}\right) \text{ bps/Hz} \qquad (12.22)$$

The optimal selection of P antennas corresponds to computing Equation (12.22) for all possible antenna combinations. In order to maximize the system capacity, one must choose the antenna with the highest capacity, that is,

$$\{p_1^{opt}, p_2^{opt}, \cdots, p_Q^{opt}\} = \underset{\{p_1,p_2,\cdots,p_Q\}\in\mathsf{A}_Q}{\arg\max} C_{\{p_1,p_2,\cdots,p_Q\}} \qquad (12.23)$$

where A_Q represents a set of all possible antenna combinations with Q selected antennas. Note that $|\mathsf{A}_Q| = \binom{N_T}{Q}$, that is, considering all possible antenna combinations in Equation (12.23) may involve the enormous complexity, especially when N_T is very large. Therefore, some methods of reducing the complexity need to be developed. In the following subsection, we will consider this particular issue. Figure 12.8 shows the channel capacity with antenna selection for $N_T = 4$ and $N_R = 4$ as the number of the selected antennas varies by $Q = 1, 2, 3, 4$. It is clear that the channel capacity increases in proportion to the number of the selected antennas. When the SNR is less than 10dB, the selection of three antennas is enough to warrant the channel capacity as much as the use of all four antennas. Program 12.4 ("MIMO_channel_cap_ant_sel_optimal.m") can be used to compute the MIMO channel capacity using optimal antenna selection scheme. The number of selected antennas can be set by varying the parameter `sel_ant=1,2,3,4`.

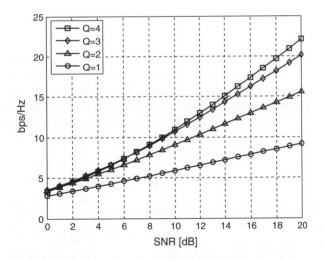

Figure 12.8 Channel capacity with optimal antenna selection: $N_T = N_R = 4$, and $Q = 1, 2, 3, 4$.

Program 12.4 "MIMO_channel_cap_ant_sel_optimal.m" for channel capacity with the optimal antenna selection method

```
% MIMO_channel_cap_ant_sel_optimal.m
clear all; clf
NT=4; NR=4; MaxIter=1000;
sel_ant=1; I=eye(NR,NR); sq2=sqrt(2);
SNRdBs=[0:2:20];
for i_SNR=1:length(SNRdBs)
    SNRdB = SNRdBs(i_SNR);
    SNR_sel_ant = 10^(SNRdB/10)/sel_ant;
    rand('seed',1); randn('seed',1); cum = 0;
    for i=1:MaxIter
      H = (randn(NR,NT)+j*randn(NR,NT))/sq2;
      if sel_ant>NT|sel_ant<1
       error('sel_ant must be between 1 and NT!');
       else indices = NCHOOSEK([1:NT],sel_ant);
      end
      for n=1:size(indices,1)
          Hn = H(:,indices(n,:));
          log_SH(n)=log2(real(det(I+SNR_sel_ant*Hn*Hn'))); % Eq.(12.22)
      end
      cum = cum + max(log_SH);
    end
    sel_capacity(i_SNR) = cum/MaxIter;
end
plot(SNRdBs,sel_capacity,'-ko','LineWidth',2); hold on;
xlabel('SNR[dB]'), ylabel('bps/Hz'), grid on;
```

12.4.2 Complexity-Reduced Antenna Selection

As mentioned in the previous subsection, optimal antenna selection in Equation (12.23) may involve too much complexity depending on the total number of available transmit antennas. In order to reduce its complexity, we may need to resort to the sub-optimal method. For example, additional antenna can be selected in ascending order of increasing the channel capacity. More specifically, one antenna with the highest capacity is first selected as

$$
\begin{aligned}
p_1^{subopt} &= \arg \max_{p_1} \; C_{\{p_1\}} \\
&= \arg \max_{p_1} \log_2 \det\left(\mathbf{I}_{N_R} + \frac{\mathsf{E}_\mathsf{x}}{Q\mathsf{N}_0} \mathbf{H}_{\{p_1\}} \mathbf{H}_{\{p_1\}}^H \right)
\end{aligned}
\tag{12.24}
$$

Given the first selected antenna, the second antenna is selected such that the channel capacity is maximized, that is,

$$
\begin{aligned}
p_2^{subopt} &= \arg \max_{p_2 \neq p_1^{subopt}} \; C_{\{p_1^{subopt},p_2\}} \\
&= \arg \max_{p_2 \neq p_1^{subopt}} \log_2 \det\left(\mathbf{I}_{N_R} + \frac{\mathsf{E}_\mathsf{x}}{Q\mathsf{N}_0} \mathbf{H}_{\{p_1^{subopt},p_2\}} \mathbf{H}_{\{p_1^{subopt},p_2\}}^H \right)
\end{aligned}
\tag{12.25}
$$

After the nth iteration which provides $\{p_1^{subopt}, p_2^{subopt}, \ldots, p_n^{subopt}\}$, the capacity with an additional antenna, say antenna l, can be updated as

$$C_l = \log_2 \det \left\{ \mathbf{I}_{N_R} + \frac{E_x}{QN_0} \left(\mathbf{H}_{\{p_1^{subopt}, \ldots, p_n^{subopt}\}} \mathbf{H}^H_{\{p_1^{subopt}, \ldots, p_n^{subopt}\}} + \mathbf{H}_{\{l\}} \mathbf{H}^H_{\{l\}} \right) \right\}$$

$$= \log_2 \det \left\{ \mathbf{I}_{N_R} + \frac{E_x}{QN_0} \mathbf{H}_{\{p_1^{subopt}, \ldots, p_n^{subopt}\}} \mathbf{H}^H_{\{p_1^{subopt}, \ldots, p_n^{subopt}\}} \right\} \qquad (12.26)$$

$$+ \log_2 \left\{ 1 + \frac{E_x}{QN_0} \mathbf{H}_{\{l\}} \left(\mathbf{I}_{N_R} + \frac{E_x}{QN_0} \mathbf{H}_{\{p_1^{subopt}, \ldots, p_n^{subopt}\}} \mathbf{H}^H_{\{p_1^{subopt}, \ldots, p_n^{subopt}\}} \right)^{-1} \mathbf{H}^H_{\{l\}} \right\}$$

It can be derived using the following identities:

$$\det(\mathbf{A} + \mathbf{u}\mathbf{v}^H) = (1 + \mathbf{v}^H \mathbf{A}^{-1} \mathbf{u}) \det(\mathbf{A})$$
$$\log_2 \det(\mathbf{A} + \mathbf{u}\mathbf{v}^H) = \log_2(1 + \mathbf{v}^H \mathbf{A}^{-1} \mathbf{u}) \det(\mathbf{A}) = \log_2 \det(\mathbf{A}) + \log_2(1 + \mathbf{v}^H \mathbf{A}^{-1} \mathbf{u}) \qquad (12.27)$$

where

$$\mathbf{A} = \mathbf{I}_{N_R} + \frac{E_x}{QN_0} \mathbf{H}_{\{p_1^{subopt}, \ldots, p_n^{subopt}\}} \mathbf{H}^H_{\{p_1^{subopt}, \ldots, p_n^{subopt}\}}$$

$$\text{and} \quad \mathbf{u} = \mathbf{v} = \sqrt{\frac{E_x}{QN_0}} \mathbf{H}_{\{l\}}.$$

The additional $(n+1)$th antenna is the one that maximizes the channel capacity in Equation (12.26), that is,

$$p_{n+1}^{subopt} = \arg\max_{l \notin \{p_1^{subopt}, \ldots, p_n^{subopt}\}} C_l$$

$$= \arg\max_{l \notin \{p_1^{subopt}, \ldots, p_n^{subopt}\}} \mathbf{H}_{\{l\}} \left(\frac{QN_0}{E_x} \mathbf{I}_{N_R} + \mathbf{H}_{\{p_1^{subopt}, \ldots, p_n^{subopt}\}} \mathbf{H}^H_{\{p_1^{subopt}, \ldots, p_n^{subopt}\}} \right)^{-1} \mathbf{H}^H_{\{l\}} \qquad (12.28)$$

This process continues until all Q antennas are selected (i.e., continue the iteration Equation (12.28) until $n+1 = Q$). Note that only one matrix inversion is required for all $l \in \{1, 2, \cdots, N_T\} - \{p_1^{subopt}, p_2^{subopt}, \ldots, p_n^{subopt}\}$ in the course of the selection process.

Meanwhile, the same process can be implemented by deleting the antenna in descending order of decreasing channel capacity. Let $\mathbf{S}_n$ denote a set of antenna indices in the nth iteration. In the initial step, we consider all antennas, $\mathbf{S}_1 = \{1, 2, \cdots, N_T\}$, and select the antenna that contributes least to the capacity, that is,

$$p_1^{deleted} = \arg\max_{p_1 \in \mathbf{S}_1} \log_2 \det \left(\mathbf{I}_{N_R} + \frac{E_x}{QN_0} \mathbf{H}_{\mathbf{S}_1 - \{p_1\}} \mathbf{H}^H_{\mathbf{S}_1 - \{p_1\}} \right) \qquad (12.29)$$

The antenna selected from Equation (12.29) will be deleted from the antenna index set, and the remaining antenna set is updated to $\mathbf{S}_2 = \mathbf{S}_1 - \{p_1^{deleted}\}$. If $|\mathbf{S}_2| = N_T - 1 > Q$, we choose another antenna to delete. This will be the one that contributes least to the capacity now for the current antenna index set S_2, that is,

$$p_2^{deleted} = \arg \max_{p_2 \in S_2} \log_2 \det\left(I_{N_R} + \frac{E_x}{QN_0} H_{S_2-\{p_2\}} H^H_{S_2-\{p_2\}}\right). \qquad (12.30)$$

Again, the remaining antenna index set is updated to $S_3 = S_2 - \{p_2^{deleted}\}$. This process will continue until all Q antennas are selected, that is, $|S_n| = Q$. Note that the complexity of selection method in descending order is higher than that in ascending order. From the performance perspective, however, the selection method in descending order outperforms that in ascending order when $1 < Q < N_T$. This is due to the fact that the selection method in descending order considers all correlations between the column vectors of the original channel gain before choosing the first antenna to delete. When $Q = N_T - 1$, the selection method in descending order produces the same antenna index set as the optimal antenna selection method produces Equation (12.23). When $Q = 1$, however, the selection method in ascending order produces the same antenna index as the optimal antenna selection method in Equation (12.23) and achieves better performance than any other selection methods. In general, however, all these methods are just suboptimal, except for the above two special cases.

The channel capacities with two suboptimal selection methods can be computed by using Program 12.5 ("MIMO_channel_cap_subopt_ant_sel.m"), where `sel_ant=1,2,...,` N_T-1 is used for setting the number of selected antennas, and variable `sel_method = 0` or 1 indicates whether selection is done in descending or ascending order. Figure 12.9 shows the channel capacity with the selection method in descending order for various numbers of selected antennas with $N_T = 4$ and $N_R = 4$. Comparing the curves in Figure 12.9 with those in Figure 12.8, we can see that the suboptimal antenna selection method in Equation (12.28) achieves almost the same channel capacity as the optimal antenna selection method in Equation (12.23).

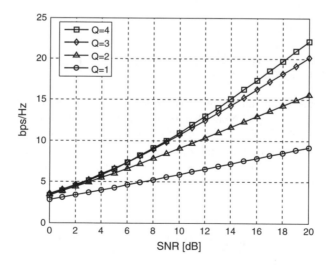

Figure 12.9 Channel capacities for antenna selection method in descending order.

MATLAB® Program: Channel Capacity with Suboptimal Antenna Selection

Program 12.5 "MIMO_channel_cap_ant_sel_subopt.m" for channel capacity with the suboptimal antenna selection method

```
% MIMO_channel_cap_ant_sel_subopt.m
clear all; clf
sel_ant=2; % Number of antennas to select
sel_method=0; % 0/1 for increasingly/decreasingly ordered selection
NT=4; NR=4; % Number of transmit/receive antennas
I=eye(NR,NR); sq2=sqrt(2);
SNRdBs = [0:10]; MaxIter=1000;
for i_SNR=1:length(SNRdBs)
  SNRdB = SNRdBs(i_SNR);
  SNR_sel_ant = 10^(SNRdB/10)/sel_ant;
  rand('seed',1); randn('seed',1);
  cum = 0;
  for i=1:MaxIter
      if sel_method==0
         sel_ant_indices=[]; rem_ant_indices=[1:NT];
      else
         sel_ant_indices=[1:NT]; del_ant_indices=[];
      end
      H = (randn(NR,NT)+j*randn(NR,NT))/sq2;
      if sel_method==0 % increasingly ordered selection method
        for current_sel_ant_number=1:sel_ant
          clear log_SH;
          for n=1:length(rem_ant_indices)
            Hn = H(:,[sel_ant_indices rem_ant_indices(n)]);
            log_SH(n)=log2(real(det(I+SNR_sel_ant*Hn*Hn')));
          end
          maximum_capacity = max(log_SH);
          selected = find(log_SH==maximum_capacity);
          sel_ant_index = rem_ant_indices(selected);
          rem_ant_indices = [rem_ant_indices(1:selected-1) ...
                          rem_ant_indices(selected+1:end)];
          sel_ant_indices = [sel_ant_indices sel_ant_index];
        end
      else % decreasingly ordered selection method
        for current_del_ant_number=1:NT-sel_ant
          clear log_SH;
          for n=1:length(sel_ant_indices)
            Hn = H(:,[sel_ant_indices(1:n-1) sel_ant_indices(n+1:end)]);
            log_SH(n)=log2(real(det(I+SNR_sel_ant*Hn*Hn')));
          end
          maximum_capacity = max(log_SH);
          selected = find(log_SH==maximum_capacity);
          sel_ant_indices = [sel_ant_indices(1:selected-1) ...
                          sel_ant_indices(selected+1:end)];
```

```
      end
    end
    cum = cum + maximum_capacity;
  end
  sel_capacity(i_SNR) = cum/MaxIter;
end
plot(SNRdBs,sel_capacity,'-ko','LineWidth',2); hold on;
xlabel('SNR[dB]'), ylabel('bps/Hz'), grid on;
title('Capacity of suboptimally selected antennas')
```

12.4.3 Antenna Selection for OSTBC

In the previous subsection, channel capacity has been used as a design criterion for antenna selection. Error performance can also be used as another design criterion. In other words, transmit antennas can be selected so as to minimize the error probability. Let $\Pr(\mathbf{C}_i \rightarrow \mathbf{C}_j | \mathbf{H}_{\{p_1,p_2,\cdots,p_Q\}})$ denote the pairwise error probability when a space-time codeword $\mathbf{C}_i$ is transmitted but $\mathbf{C}_j$ is decoded for the given channel $\mathbf{H}_{\{p_1,p_2,\cdots,p_Q\}}, j \neq i$. For an effective channel $\mathbf{H}_{\{p_1,p_2,\cdots,p_Q\}}$ with Q columns of $\mathbf{H}$ chosen, an upper bound for the pairwise error probability for orthogonal STBC (OSTBC) is given as

$$\Pr(\mathbf{C}_i \rightarrow \mathbf{C}_j | \mathbf{H}_{\{p_1,p_2,\cdots,p_Q\}}) = Q\left(\sqrt{\frac{\rho \left\| \mathbf{H}_{\{p_1,p_2,\cdots,p_Q\}}\mathbf{E}_{i,j} \right\|_F^2}{2N_T}} \right) \leq \exp\left(-\frac{\rho \left\| \mathbf{H}_{\{p_1,p_2,\cdots,p_Q\}}\mathbf{E}_{i,j} \right\|_F^2}{4N_T} \right)$$

$$(12.31)$$

The above upper bound follows from a similar way as in Section 10.2.2. The Q transmit antennas can be selected to minimize the upper bound in Equation (12.31), or equivalently

$$
\begin{aligned}
\left\{ p_1^{opt}, p_2^{opt}, \cdots, p_Q^{opt} \right\} &= \underset{p_1,p_2,\cdots,p_Q \in \mathsf{A}_Q}{\arg \max} \left\| \mathbf{H}_{\{p_1,p_2,\cdots,p_Q\}}\mathbf{E}_{i,j} \right\|_F^2 \\
&= \underset{p_1,p_2,\cdots,p_Q \in \mathsf{A}_Q}{\arg \max} \operatorname{tr}\left[\mathbf{H}_{\{p_1,p_2,\cdots,p_Q\}}\mathbf{E}_{i,j}\mathbf{E}_{i,j}^H\mathbf{H}_{\{p_1,p_2,\cdots,p_Q\}}^H \right] \\
&= \underset{p_1,p_2,\cdots,p_Q \in \mathsf{A}_Q}{\arg \max} \operatorname{tr}\left[\mathbf{H}_{\{p_1,p_2,\cdots,p_Q\}}\mathbf{H}_{\{p_1,p_2,\cdots,p_Q\}}^H \right] \\
&= \underset{p_1,p_2,\cdots,p_Q \in \mathsf{A}_Q}{\arg \max} \left\| \mathbf{H}_{\{p_1,p_2,\cdots,p_Q\}} \right\|_F^2
\end{aligned}
$$

$$(12.32)$$

In deriving Equation (12.32), we have used the fact that the error matrix $\mathbf{E}_{i,j}$ has the property $\mathbf{E}_{i,j}\mathbf{E}_{i,j}^H = a\mathbf{I}$ with constant a. From Equation (12.32), we can see that the antennas corresponding to high column norms are selected for minimizing the error rate. The average SNR on the receiver side with Q selected antennas of $\{p_i\}_{i=1}^Q$ is given as

$$\eta_{\{p_1,p_2,\cdots,p_Q\}} = \frac{\rho}{Q} \left\| \mathbf{H}_{\{p_1,p_2,\cdots,p_Q\}} \right\|_F^2$$

$$(12.33)$$

Equations (12.32) and (12.33) imply that the antennas with the highest SNR on the receiver side must be chosen. Denoting the indices with the highest Q column norms of $\mathbf{H}$ by $\left\{ p_1^{opt}, p_2^{opt}, \cdots, p_Q^{opt} \right\}$, we have the following inequality:

$$\frac{\left\| \mathbf{H}_{\{p_1^{opt}, p_2^{opt}, \cdots, p_Q^{opt}\}} \right\|_F^2}{Q} \geq \frac{\|\mathbf{H}\|_F^2}{N_T} \tag{12.34}$$

Since $Q \leq N_T$, we also have the following inequality:

$$\begin{aligned}
\left\| \mathbf{H}_{\{p_1^{opt}, p_2^{opt}, \cdots, p_Q^{opt}\}} \right\|_F^2 &= \left\| \mathbf{H}_{\{p_1^{opt}\}} \right\|^2 + \left\| \mathbf{H}_{\{p_2^{opt}\}} \right\|^2 + \cdots + \left\| \mathbf{H}_{\{p_Q^{opt}\}} \right\|^2 \\
&\leq \left\| \mathbf{H}_{\{1\}} \right\|^2 + \left\| \mathbf{H}_{\{2\}} \right\|^2 + \cdots + \left\| \mathbf{H}_{\{N_T\}} \right\|^2 \\
&= \|\mathbf{H}\|_F^2
\end{aligned} \tag{12.35}$$

where $\mathbf{H}_{\{k\}}$ represents the kth column of $\mathbf{H}$. From Equations (12.34) and (12.35), the average SNR on the receiver side with the optimally selected antennas is ranged by

$$\frac{\rho}{Q} \|\mathbf{H}\|_F^2 \geq \eta_{\{p_1^{opt}, p_2^{opt}, \cdots, p_Q^{opt}\}} \geq \frac{\rho}{N_T} \|\mathbf{H}\|_F^2 \tag{12.36}$$

From the inequality in (12.36), we can see that the upper and lower bounds of the average received SNR are functions of $\|\mathbf{H}\|_F^2$. This implies that a diversity order of $N_T N_R$ is achieved with optimal antenna selection in Equation (12.23) when the entries of $\mathbf{H}$ are i.i.d. Gaussian-distributed.

To simulate the antenna selection method in Equation (12.32) for the Alamouti STBC scheme, Program 12.6 ("Alamouti_2x1_ant_selection.m") has been run to yield Figure 12.10,

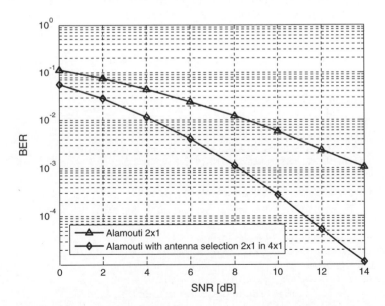

Figure 12.10 BER performance of Alamouti STBC scheme with antenna selection: $Q = 2$ and $N_T = 4$.

which shows its BER performance with $Q = 2$ and $N_T = 4$. Note that the further diversity gain has been achieved without using additional RF modules on the transmitter side. It is also interesting to compare the results in Figure 12.4 with those in Figure 12.10, which demonstrate that the antenna selection method provides more gain over the precoding method.

MATLAB® Program: Antenna Selection for OSTBC

Program 12.6 "Alamouti_2x1_ant_selection.m": Antenna selection for OSTBC

```
% Alamouti_2x1_ant_selection.m
clear all; clf
%%%%%% Parameter Setting %%%%%%%%%
N_frame=100; N_packet=100; %Number of frames/packet & Number of packets
b=2; M=2^b;
mod_obj=modem.qammod('M',M,'SymbolOrder','Gray','InputType','bit');
demod_obj = modem.qamdemod(mod_obj);
% MIMO Parameters
T_TX=4; NT=2; NR=1;
N_pbits=NT*b*N_frame; N_tbits=N_pbits*N_packet;
fprintf('====================================================\n');
fprintf(' Ant_selection transmission');
fprintf('\n %d x %d MIMO\n %d QAM', NT,NR,M);
fprintf('\n Simulation bits : %d',N_tbits);
fprintf('\n====================================================\n');
SNRdBs = [0:2:20]; sq2=sqrt(2);
for i_SNR=1:length(SNRdBs)
    SNRdB= SNRdBs(i_SNR);
    noise_var = NT*0.5*10^(-SNRdB/10); sigma = sqrt(noise_var);
    rand('seed',1); randn('seed',1); N_ebits = 0;
    %%%%%%%%%%%%% Transmitter %%%%%%%%%%%%%%%%%%%%
    for i_packet=1:N_packet
      msg_bit = randint(N_pbits,1); % Bit generation
      s = modulate(mod_obj,msg_bit);
      Scale = modnorm(s,'avpow',1); % Normalization factor
      S=reshape(Scale*s,NT,1,N_frame);
      Tx_symbol=[S(1,1,:) -conj(S(2,1,:));
                 S(2,1,:) conj(S(1,1,:))];
      %%%%%%%%%%%%% Channel and Noise %%%%%%%%%%%%%
      H = (randn(NR,T_TX)+j*randn(NR,T_TX))/sq2;
      for TX_index=1:T_TX
        ch(TX_index)=norm(H(:,TX_index),'fro');
      end
      [val,Index] = sort(ch,'descend');
      Hs = H(:,Index([1 2]));
      norm_H2=norm(Hs,'fro')^2; % H selected and its norm2
      for i=1:N_frame
        Rx(:,:,i) = Hs*Tx_symbol(:,:,i) + ...
                    sigma*(randn(NR,2)+j*randn(NR,2));
      end
```

```
        %%%%%%%%%%%% Receiver %%%%%%%%%%%%%%%%%%%%
        for i=1:N_frame
            y(1,i) = (Hs(1)'*Rx(:,1,i)+Hs(2)*Rx(:,2,i)')/norm_H2;
            y(2,i) = (Hs(2)'*Rx(:,1,i)-Hs(1)*Rx(:,2,i)')/norm_H2;
        end
        S_hat = reshape(y/Scale,NT*N_frame,1);
        msg_hat = demodulate(demod_obj,S_hat);
        N_ebits = N_ebits + sum(msg_hat~=msg_bit);
    end
    BER(i_SNR) = N_ebits/N_tbits;
end
semilogy(SNRdBs,BER,'-k^', 'LineWidth',2); hold on; grid on;
xlabel('SNR[dB]'), ylabel('BER');
legend('Ant-selection transmission');
```

13

Multi-User MIMO

In Chapter 9, we have shown that the channel capacity of the single-user $N_R \times N_T$ MIMO systems is proportional to $N_{\min} = \min(N_T, N_R)$ [211, 251–253]. In fact, MIMO technique is an essential means of increasing capacity in the high SNR regime, providing at most $N_{\min}$ spatial degrees of freedom. In the single-user MIMO system, a point-to-point high data rate transmission can be supported by spatial multiplexing while providing spatial diversity gain. However, most communication systems deal with multiple users who are sharing the same radio resources. Figure 13.1 illustrates a typical multi-user communication environment in which the multiple mobile stations are served by a single base station in the cellular system. In Figure 13.1, three out of four users are selected and allocated communication resource such as time, frequency, and spatial stream. Suppose that the base station and each mobile station are equipped with N_B and N_M antennas, respectively. As K independent users form a virtual set of $K \cdot N_M$ antennas which communicate with a single BS with N_B antennas, the end-to-end configuration can be considered as a $(K \cdot N_M) \times N_B$ MIMO system for downlink, or $N_B \times (K \cdot N_M)$ MIMO system for uplink. In this multi-user communication system, multiple antennas allow the independent users to transmit their own data stream in the uplink (many-to-one) at the same time or the base station to transmit the multiple user data streams to be decoded by each user in the downlink (one-to-many). This is attributed to the increase in degrees of freedom with multiple antennas as in the single-user MIMO system.

In the multi-user MIMO system, downlink and uplink channels are referred to as broadcast channel (BC) and multiple access channel (MAC), respectively. Since all data streams of K independent users are available for a single receiver of the base station in the multiple access channel, the multi-user MIMO system is equivalent to a single user $(K \cdot N_M) \times N_B$ MIMO system in the uplink. Similar to the single-user MIMO system, therefore, it can be shown that the uplink capacity of multi-user MIMO system is proportional to $\min(N_B, K \cdot N_M)$.

In this chapter, we first discuss a mathematical model of the multi-user MIMO system and its capacity. Then, we present the precoded transmission schemes, which is a specific means of implementing the multi-user MIMO system for the downlink.

MIMO-OFDM Wireless Communications with MATLAB® Yong Soo Cho, Jaekwon Kim, Won Young Yang and Chung G. Kang
© 2010 John Wiley & Sons (Asia) Pte Ltd

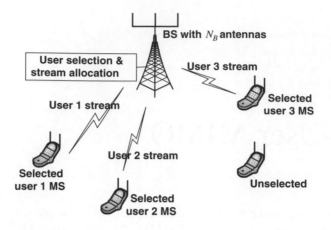

Figure 13.1 Multi-user MIMO communication systems: $K = 4$.

13.1 Mathematical Model for Multi-User MIMO System

Consider K independent users in the multi-user MIMO system. We assume that the BS and each MS are equipped with N_B and N_M antennas, respectively. Figure 13.2 shows the uplink channel, known as a multiple access channel (MAC) for K independent users. Let $\mathbf{x}_u \in \mathbb{C}^{N_M \times 1}$ and $\mathbf{y}_{MAC} \in \mathbb{C}^{N_B \times 1}$ denote the transmit signal from the u th user, $u = 1, 2, \cdots, K$, and the received signal at the BS, respectively. The channel gain between the u th user MS and BS is represented by $\mathbf{H}_u^{UL} \in \mathbb{C}^{N_B \times N_M}$, $u = 1, 2, \cdots, K$. The received signal is expressed as

$$
\begin{aligned}
\mathbf{y}_{MAC} &= \mathbf{H}_1^{UL}\mathbf{x}_1 + \mathbf{H}_2^{UL}\mathbf{x}_2 + \cdots + \mathbf{H}_K^{UL}\mathbf{x}_K + \mathbf{z} \\
&= \underbrace{\left[\mathbf{H}_1^{UL}\, \mathbf{H}_2^{UL} \cdots \mathbf{H}_K^{UL}\right]}_{=\mathbf{H}^{UL}} \begin{bmatrix} \mathbf{x}_1 \\ \vdots \\ \mathbf{x}_K \end{bmatrix} + \mathbf{z} = \mathbf{H}^{UL} \begin{bmatrix} \mathbf{x}_1 \\ \vdots \\ \mathbf{x}_K \end{bmatrix} + \mathbf{z}
\end{aligned} \tag{13.1}
$$

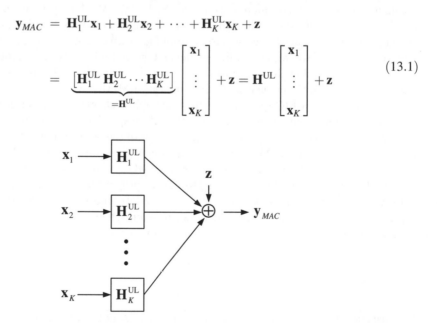

Figure 13.2 Uplink channel model for multi-user MIMO system: multiple access channel (MAC).

where $\mathbf{z} \in \mathbb{C}^{N_B \times 1}$ is the additive noise in the receiver and it is modeled as a zero-mean circular symmetric complex Gaussian (ZMCSCG) random vector.

On the other hand, Figure 13.3 shows the downlink channel, known as a broadcast channel (BC) in which $\mathbf{x} \in \mathbb{C}^{N_B \times 1}$ is the transmit signal from the BS and $\mathbf{y}_u \in \mathbb{C}^{N_M \times 1}$ is the received signal at the u th user, $u = 1, 2, \cdots, K$. Let $\mathbf{H}_u^{\mathrm{DL}} \in \mathbb{C}^{N_M \times N_B}$ represent the channel gain between BS and the u th user. In MAC, the received signal at the u th user is expressed as

$$\mathbf{y}_u = \mathbf{H}_u^{\mathrm{DL}} \mathbf{x} + \mathbf{z}_u, \quad u = 1, 2, \cdots, K \tag{13.2}$$

where $\mathbf{z}_u \in \mathbb{C}^{N_M \times 1}$ is the additive ZMCSCG noise at the u th user. Representing all user signals by a single vector, the overall system can be represented as

$$\underbrace{\begin{bmatrix} \mathbf{y}_1 \\ \mathbf{y}_2 \\ \vdots \\ \mathbf{y}_K \end{bmatrix}}_{\mathbf{y}_{BC}} = \underbrace{\begin{bmatrix} \mathbf{H}_1^{\mathrm{DL}} \\ \mathbf{H}_2^{\mathrm{DL}} \\ \vdots \\ \mathbf{H}_K^{\mathrm{DL}} \end{bmatrix}}_{\mathbf{H}_{\mathrm{DL}}} \mathbf{x} + \underbrace{\begin{bmatrix} \mathbf{z}_1 \\ \mathbf{z}_2 \\ \vdots \\ \mathbf{z}_K \end{bmatrix}}_{\mathbf{z}} \tag{13.3}$$

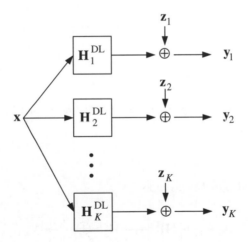

Figure 13.3 Downlink channel model for multi-user MIMO system: broadcast channel (BC).

13.2 Channel Capacity of Multi-User MIMO System

Based on the mathematical models in the previous section, we discuss some examples for the channel capacity of MAC and BC in the AWGN channel.

13.2.1 Capacity of MAC

Capacity region of MAC was first introduced in [280]. Let P_u and R_u denote the power and data rate of the u th user in the K-user MIMO system, $u = 1, 2, \cdots, K$. Referring to Figure 13.2, the MAC capacity region for $K = 2$ and $N_M = 1$ is given as

$$R_1 \leq \log_2\left(1 + \left\|\mathbf{H}_1^{\text{UL}}\right\|^2 P_1\right)$$

$$R_2 \leq \log_2\left(1 + \left\|\mathbf{H}_2^{\text{UL}}\right\|^2 P_2\right) \tag{13.4}$$

$$R_1 + R_2 \leq \log_2\left(1 + \left\|\mathbf{H}_1^{\text{UL}}\right\|^2 P_1 + \left\|\mathbf{H}_2^{\text{UL}}\right\|^2 P_2\right)$$

and it is also illustrated in Figure 13.4 [254].

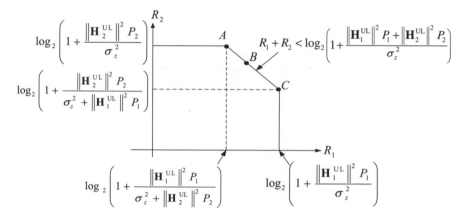

Figure 13.4   Capacity region of MAC: $K = 2$ and $N_M = 1$.

In this case, the received signal is given as

$$
\begin{aligned}
\mathbf{y}_{MAC} &= \mathbf{H}_1^{\text{UL}} x_1 + \mathbf{H}_2^{\text{UL}} x_2 + \begin{bmatrix} z_1 \\ z_2 \end{bmatrix} \\
&= \begin{bmatrix} \mathbf{H}_1^{\text{UL}} & \mathbf{H}_2^{\text{UL}} \end{bmatrix} \begin{bmatrix} x_1 \\ x_2 \end{bmatrix} + \begin{bmatrix} z_1 \\ z_2 \end{bmatrix}
\end{aligned} \tag{13.5}
$$

where x_u is the transmitted signal from the u th user, $u = 1, 2$. In order to achieve the point A in Figure 13.4, signal x_1 is detected by assuming that it is interfered with the signal from the user 2. Once x_1 is detected correctly, which is possible as long as the transmission rate is less than the corresponding channel capacity $R_1 = \log_2\{1 + \|\mathbf{H}_1^{\text{UL}}\|^2 P_1/(1 + \|\mathbf{H}_2^{\text{UL}}\|^2 P_2)\}$, it can be canceled from the received signal as follows:

$$\tilde{\mathbf{y}}_{MAC} = \mathbf{y}_{MAC} - \mathbf{H}_1^{UL} x_1 = \mathbf{H}_2^{UL} x_2 + \begin{bmatrix} z_1 \\ z_2 \end{bmatrix} \tag{13.6}$$

From the above interference-free signal $\tilde{\mathbf{y}}_{MAC}$, x_2 is detected to achieve $R_2 = \log_2(1 + \|\mathbf{H}_2^{UL}\|^2 P_2)$. The point C can be achieved in the other way around. All the other rate points on the segment AC, for example, point B in Figure 13.4, can be achieved by time-sharing or rate-splitting between the multiple access schemes in point A and point C. In [255], it was shown that the sum rate capacity of MAC channel is proportional to $\min(N_B, K \cdot N_M)$.

13.2.2 Capacity of BC

The capacity region of Gaussian broadcast channel remains as an unsolved problem. In this subsection, we discuss an achievable downlink channel capacity for a special case of $N_B = 2$, $N_M = 1$ and $K = 2$, using dirty paper coding (DPC) [256–258]. In this case, the received signal in Equation (13.3) is expressed as

$$\underbrace{\begin{bmatrix} y_1 \\ y_2 \end{bmatrix}}_{\mathbf{y}_{BC}} = \underbrace{\begin{bmatrix} \mathbf{H}_1^{DL} \\ \mathbf{H}_2^{DL} \end{bmatrix}}_{\mathbf{H}^{DL}} \begin{bmatrix} x_1 \\ x_2 \end{bmatrix} + \begin{bmatrix} z_1 \\ z_2 \end{bmatrix} \tag{13.7}$$

where $\mathbf{H}_u^{DL} \in \mathbb{C}^{1 \times 2}$ denotes the channel matrix between the BS and the u th user, $u = 1, 2$, $\tilde{x}_u$ represents the u th user signal, while x_i is the signal transmitted by the i th transmit antenna, $i = 1, 2$. If the channel information is completely available at BS, the overall channel can be LQ-decomposed as

$$\mathbf{H}^{DL} = \underbrace{\begin{bmatrix} l_{11} & 0 \\ l_{21} & l_{22} \end{bmatrix}}_{\mathbf{L}} \underbrace{\begin{bmatrix} \mathbf{q}_1 \\ \mathbf{q}_2 \end{bmatrix}}_{\mathbf{Q}} \tag{13.8}$$

where

$$l_{11} = \|\mathbf{H}_1^{DL}\|, \mathbf{q}_1 = \frac{1}{l_{11}} \mathbf{H}_1^{DL},$$

$$l_{21} = \mathbf{q}_1 \cdot (\mathbf{H}_2^{DL})^H,$$

$$l_{22} = \|\mathbf{H}_2^{DL} - l_{21}\mathbf{q}_1\|,$$

$$\text{and} \quad \mathbf{q}_2 = \frac{1}{l_{22}} (\mathbf{H}_2^{DL} - l_{21}\mathbf{q}_1).$$

We note that $\{\mathbf{q}_i\}_{i=1}^2$ in Equation (13.8) are the orthonormal row vectors. Given $\{\mathbf{q}_i\}_{i=1}^2$ from the channel information, we can precode the transmitted signal as

$$\underbrace{\begin{bmatrix} x_1 \\ x_2 \end{bmatrix}}_{\mathbf{x}} = \mathbf{Q}^H \begin{bmatrix} \tilde{x}_1 \\ \tilde{x}_2 - \dfrac{1}{l_{22}} l_{21} \tilde{x}_1 \end{bmatrix} \tag{13.9}$$

Transmitting the above precoded signal, the received signal is given as

$$
\begin{aligned}
\mathbf{y}_{\text{BC}} &= \mathbf{H}^{\text{DL}} \mathbf{x} + \mathbf{z} \\[2mm]
&= \begin{bmatrix} l_{11} & 0 \\ l_{21} & l_{22} \end{bmatrix} \begin{bmatrix} \mathbf{q}_1 \\ \mathbf{q}_2 \end{bmatrix} \begin{bmatrix} \mathbf{q}_1^H & \mathbf{q}_2^H \end{bmatrix} \begin{bmatrix} \tilde{x}_1 \\ \tilde{x}_2 - \dfrac{1}{l_{22}} l_{21} \tilde{x}_1 \end{bmatrix} + \mathbf{z} \\[2mm]
&= \begin{bmatrix} l_{11} & 0 \\ 0 & l_{22} \end{bmatrix} \begin{bmatrix} \tilde{x}_1 \\ \tilde{x}_2 \end{bmatrix} + \mathbf{z} \\[2mm]
&= \begin{bmatrix} \left\| \mathbf{H}_1^{\text{DL}} \right\| & 0 \\ 0 & \left\| \mathbf{H}_2^{\text{DL}} - l_{12} \mathbf{q}_1 \right\| \end{bmatrix} \begin{bmatrix} \tilde{x}_1 \\ \tilde{x}_2 \end{bmatrix} + \mathbf{z}
\end{aligned} \tag{13.10}
$$

From Equation (13.10), we can see that two virtual interference-free channels have been created. Assume that the total power P is divided into αP and $(1-\alpha)P$ for the first and second users, respectively, that is,

$$E\left\{ |x_1|^2 \right\} = E\left\{ |\tilde{x}_1|^2 \right\} = \alpha P$$

and $\quad E\left\{ |x_2|^2 \right\} = E\left\{ \left| \tilde{x}_2 - \dfrac{l_{21}}{l_{22}} \tilde{x}_1 \right|^2 \right\} = (1-\alpha)P, \quad \alpha \in [0,1].$

Then, the capacities for the first and second user are respectively given as

$$R_1 = \log\left(1 + \left\| \mathbf{H}_1^{\text{DL}} \right\|^2 \frac{\alpha P}{\sigma_z^2} \right), \tag{13.11}$$

and

$$R_2 = \log_2\left(1 + \left\| \mathbf{H}_2^{\text{DL}} - l_{21} \mathbf{q}_1 \right\|^2 \frac{(1-\alpha)P}{\sigma_z^2} \right). \tag{13.12}$$

If the second user is selected such that $l_{21} = 0$, then its capacity becomes as

$$R_2 = \log_2\left(1 + \left\| \mathbf{H}_2^{\text{DL}} \right\|^2 \frac{(1-\alpha)P}{\sigma_z^2} \right). \tag{13.13}$$

In [258–263], the duality of the uplink and downlink channel capacities was used to derive the capacity of broadcast channel. It was also shown that the downlink channel capacity using

DPC is the same as that of using the multiple access channel. In [264, 265], it was shown that the sum rate capacity is proportional to $\min(N_B, K \cdot N_M)$ where N_B is the number of antennas at BS, N_M is the number of antennas at MS, and K is the number of users.

13.3 Transmission Methods for Broadcast Channel

The main difficulty in data transmission in BC is that the coordinated signal detection on the receiver side is not straightforward, and thus, interference cancellation at BS is required. In this section, we consider four different transmission methods: channel inversion, block diagonalization, dirty paper coding (DPC), and Tomlinson-Harashima precoding (THP).

13.3.1 Channel Inversion

In this section, we assume $N_M = 1$ for all users and $K = N_B$. Let $\tilde{x}_u$ denote the u th user signal while $\mathbf{H}_u^{DL} \in \mathbb{C}^{1 \times K}$ denotes the channel matrix between BS and the u th user, $u = 1, 2, \ldots, K$. The received signal of the u th user can be expressed as

$$
y_u = \mathbf{H}_u^{DL} \begin{bmatrix} \tilde{x}_1 \\ \tilde{x}_2 \\ \vdots \\ \tilde{x}_K \end{bmatrix} + z_u, \quad u = 1, 2, \cdots, K. \tag{13.14}
$$

The received signals of all users can be represented as

$$
\underbrace{\begin{bmatrix} y_1 \\ y_2 \\ \vdots \\ y_K \end{bmatrix}}_{\mathbf{y}_{BC}} = \underbrace{\begin{bmatrix} \mathbf{H}_1^{DL} \\ \mathbf{H}_2^{DL} \\ \vdots \\ \mathbf{H}_K^{DL} \end{bmatrix}}_{\mathbf{H}^{DL}} \underbrace{\begin{bmatrix} \tilde{x}_1 \\ \tilde{x}_2 \\ \vdots \\ \tilde{x}_K \end{bmatrix}}_{\mathbf{x}} + \underbrace{\begin{bmatrix} z_1 \\ z_2 \\ \vdots \\ z_K \end{bmatrix}}_{\mathbf{z}} \tag{13.15}
$$

The received signal at each user terminal in Equation (13.15) is a scalar while each user's received signal in Equation (13.2) is a vector. Since each user is equipped with a single antenna, interferences due to other signals cannot be canceled. Instead, precoding techniques such as channel inversion and regularized channel inversion can be considered [266–273]. In the multi-user MIMO scenario, channel inversion is the same processing as the ZF pre-equalization discussed in Chapter 12. The only difference is that $\mathbf{H}$ in Equation (12.16) is replaced with $\mathbf{H}^{DL}$ from Equation (13.15) [266–270]. As in the single user MIMO case, noise enhancement can be mitigated using the MMSE criterion. Again $\mathbf{H}^{DL}$ is used instead of $\mathbf{H}$ in Equation (12.19), which is referred to as a regularized channel inversion in the context of multi-user MIMO.

Program 13.1 ("multi_user_MIMO.m") can be used to get the BER performance of a multi-user MIMO system where either channel inversion or regularized channel inversion can be selected by setting mode = 0 or mode = 1, respectively. In Figure 13.5, the BER performance of

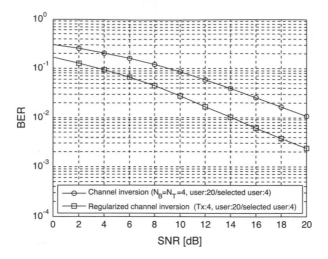

Figure 13.5   BER performance of two channel inversion methods.

channel inversion is compared to that of regularized channel inversion for $N_B = 4$ and $N_M = 1$, in which four users with the highest channel norm values are selected out of $K = 20$. As can be seen in Figure 13.5, regularized channel inversion achieves better performance than channel inversion method, thanks to mitigating noise enhancement.

MATLAB® Programs: Channel Inversion Methods for Multi-User MIMO

Program 13.1 "multi_user_MIMO.m" for a multi-user MIMO system with channel inversion

```
% multi_user_MIMO.m
clear all; clf
mode=1; % Set 0/1 for channel inversion or regularized channel inversion
N_frame=10; N_packet=200; % Number of frames/packet and Number of packets
b=2; NT=4; N_user=20; N_act_user=4; I=eye(N_act_user,NT);
N_pbits = N_frame*NT*b; % Number of bits in a packet
N_tbits = N_pbits*N_packet; % Number of total bits
SNRdBs = [0:2:20]; sq2=sqrt(2);
for i_SNR=1:length(SNRdBs)
    SNRdB=SNRdBs(i_SNR); N_ebits = 0; rand('seed',1); randn('seed',1);
    sigma2 = NT*0.5*10^(-SNRdB/10); sigma = sqrt(sigma2);
    for i_packet=1:N_packet
      msg_bit = randint(1,N_pbits); % Bit generation
      symbol = QPSK_mapper(msg_bit).'; x = reshape(symbol,NT,N_frame);
      for i_user=1:N_user
          H(i_user,:) = (randn(1,NT)+j*randn(1,NT))/sq2;
          Channel_norm(i_user)=norm(H(i_user,:));
      end
      [Ch_norm,Index]=sort(Channel_norm,'descend');
      H_used = H(Index(1:N_act_user),:);
```

```
    temp_W = H_used'*inv(H_used*H_used'+(mode==1)*sigma2*I);
    beta = sqrt(NT/trace(temp_W*temp_W')); % Eq.(12.17)
    W = beta*temp_W; % Eq.(12.19)
    Tx_signal = W*x; % Pre-equalized signal at Tx
    %%%%%%%%%%%%%% Channel and Noise %%%%%%%%%%%%%%
    Rx_signal = H_used*Tx_signal + ...
        sigma*(randn(N_act_user,N_frame)+j*randn(N_act_user,N_frame));
    %%%%%%%%%%%%%%% Receiver %%%%%%%%%%%%%%%%%%%%%%%
    x_hat = Rx_signal/beta; % Eq.(12.18)
    symbol_hat = reshape(x_hat,NT*N_frame,1);
    symbol_sliced = QPSK_slicer(symbol_hat);
    demapped=QPSK_demapper(symbol_sliced);
    N_ebits = N_ebits + sum(msg_bit~=demapped);
  end
  BER(i_SNR) = N_ebits/N_tbits;
end
semilogy(SNRdBs,BER,'-o'), grid on
```

Program 13.2 "QPSK_mapper"

```
function [QPSK_symbols] = QPSK_mapper(bitseq);
QPSK_table = [1 j -j -1]/sqrt(2);
for i=1:length(bitseq)/2
    temp = bitseq(2*(i-1)+1)*2 +bitseq(2*(i-1)+2);
    QPSK_symbols(i) =QPSK_table(temp+1);
end
```

Program 13.3 "QPSK_slicer"

```
function [x_sliced] = QPSK_slicer(x)
sq05=1/sqrt(2); jsq05=j*sq05;
for i=1:length(x)
    if imag(x(i))>real(x(i))
      if imag(x(i))>-real(x(i)),   x_sliced(i)=jsq05;
      else     x_sliced(i)=-sq05;
      end
    else
      if imag(x(i))>-real(x(i)),   x_sliced(i)=sq05;
      else     x_sliced(i)=-jsq05;
      end
    end
end
```

Program 13.4 "QPSK_demapper"

```
function [bit_seq] = QPSK_demapper(x)
QPSK_table = [1 j -j -1]/sqrt(2);
Nx=length(x);
for i=1:Nx
    x_temp(2*(i-1)+1:2*i)=dec2bin(find(QPSK_table==x(i))-1,2);
end
for i=1:Nx*2, bit_seq(i)=bin2dec(x_temp(i)); end
```

13.3.2 Block Diagonalization

In the previous subsection, we have seen channel inversion methods that deal with the multiple users, each with a single antenna. In the channel inversion methods, all signals other than target signal x_{u_T} (i.e., x_u, $u \neq u_T$) is considered as interference, and are canceled from y_{u_T} via precoding. A similar method can be applicable to multiple users, each with multiple antennas. Since the inter-antenna interference in its own signal as well as other user interference are canceled or mitigated in the channel inversion processes, noise enhancement becomes more severe from the perspective of the target user. In this situation, a block diagonalization (BD) method is more suitable [271–273]. In the BD method, unlike in the channel inversion methods, only the interference from other user signals is canceled in the process of precoding. Then, the inter-antenna interference for each user can be canceled by various signal detection methods in Chapter 11.

Let $N_{M,u}$ denote the number of antennas for the u th user, $u = 1, 2, \cdots, K$. For the u th user signal $\tilde{\mathbf{x}}_u \in \mathbb{C}^{N_{M,u} \times 1}$, the received signal $\mathbf{y}_u \in \mathbb{C}^{N_{M,u} \times 1}$ is given as

$$
\begin{aligned}
\mathbf{y}_u &= \mathbf{H}_u^{\mathrm{DL}} \sum_{k=1}^{K} \mathbf{W}_k \tilde{\mathbf{x}}_k + \mathbf{z}_u \\
&= \mathbf{H}_u^{\mathrm{DL}} \mathbf{W}_u \tilde{\mathbf{x}}_u + \sum_{k=1, \, k \neq u}^{K} \mathbf{H}_u^{\mathrm{DL}} \mathbf{W}_k \tilde{\mathbf{x}}_k + \mathbf{z}_u
\end{aligned}
\tag{13.16}
$$

where $\mathbf{H}_u^{\mathrm{DL}} \in \mathbb{C}^{N_{M,u} \times N_B}$ is the channel matrix between BS and the u th user, $\mathbf{W}_u \in \mathbb{C}^{N_B \times N_{M,u}}$ is the precoding matrix for the u th user, and $\mathbf{z}_u$ is the noise vector. Consider the received signals for the three-user case (i.e., $K = 3$),

$$
\begin{aligned}
\begin{bmatrix} \mathbf{y}_1 \\ \mathbf{y}_2 \\ \mathbf{y}_3 \end{bmatrix} &= \underbrace{\begin{bmatrix} \mathbf{H}_1^{\mathrm{DL}} & \mathbf{H}_1^{\mathrm{DL}} & \mathbf{H}_1^{\mathrm{DL}} \\ \mathbf{H}_2^{\mathrm{DL}} & \mathbf{H}_2^{\mathrm{DL}} & \mathbf{H}_2^{\mathrm{DL}} \\ \mathbf{H}_3^{\mathrm{DL}} & \mathbf{H}_3^{\mathrm{DL}} & \mathbf{H}_3^{\mathrm{DL}} \end{bmatrix}}_{\mathbf{H}_{\mathrm{DL}}} \underbrace{\begin{bmatrix} \mathbf{W}_1 \tilde{\mathbf{x}}_1 \\ \mathbf{W}_2 \tilde{\mathbf{x}}_2 \\ \mathbf{W}_3 \tilde{\mathbf{x}}_3 \end{bmatrix}}_{\mathbf{x}} + \begin{bmatrix} \mathbf{z}_1 \\ \mathbf{z}_2 \\ \mathbf{z}_3 \end{bmatrix} \\
&= \begin{bmatrix} \mathbf{H}_1^{\mathrm{DL}} \mathbf{W}_1 & \mathbf{H}_1^{\mathrm{DL}} \mathbf{W}_2 & \mathbf{H}_1^{\mathrm{DL}} \mathbf{W}_3 \\ \mathbf{H}_2^{\mathrm{DL}} \mathbf{W}_1 & \mathbf{H}_2^{\mathrm{DL}} \mathbf{W}_2 & \mathbf{H}_2^{\mathrm{DL}} \mathbf{W}_3 \\ \mathbf{H}_3^{\mathrm{DL}} \mathbf{W}_1 & \mathbf{H}_3^{\mathrm{DL}} \mathbf{W}_2 & \mathbf{H}_3^{\mathrm{DL}} \mathbf{W}_3 \end{bmatrix} \begin{bmatrix} \tilde{\mathbf{x}}_1 \\ \tilde{\mathbf{x}}_2 \\ \tilde{\mathbf{x}}_3 \end{bmatrix} + \begin{bmatrix} \mathbf{z}_1 \\ \mathbf{z}_2 \\ \mathbf{z}_3 \end{bmatrix}
\end{aligned}
\tag{13.17}
$$

where $\{\mathbf{H}_u^{\text{DL}}\mathbf{W}_k\}$ form an effective channel matrix for the u th-user receiver and the kth-user transmit signal $(u, k = 1, 2, \cdots, K)$. Note that $\{\mathbf{H}_u^{\text{DL}}\mathbf{W}_k\}_{u \neq k}$ incurs interference to the u th user unless $\mathbf{H}_u^{\text{DL}}\mathbf{W}_k = \mathbf{0}_{N_{M,u} \times N_{M,u}}, \forall u \neq k$ in Equation (13.16), where $\mathbf{0}_{N_{M,u} \times N_{M,u}}$ is a zero matrix. In other words, the interference-free transmission will be warranted as long as the effective channel matrix in Equation (13.17) can be block-diagonalized, that is,

$$\mathbf{H}_u^{\text{DL}}\mathbf{W}_k = \mathbf{0}_{N_{M,u} \times N_{M,u}}, \forall u \neq k \qquad (13.18)$$

In order to meet the total transmit power constraint, the precoders $\mathbf{W}_u \in \mathbb{C}^{N_B \times N_{M,u}}$ must be unitary, $u = 1, 2, \cdots, K$. Under the condition of Equation (13.18), the received signal in Equation (13.16) is now interference-free, that is,

$$\mathbf{y}_u = \mathbf{H}_u^{\text{DL}}\mathbf{W}_u\tilde{\mathbf{x}}_u + \mathbf{z}_u, \quad u = 1, 2, \cdots, K \qquad (13.19)$$

Once we construct the interference-free signals in Equation (13.19), various signal detection methods in Chapter 11 can be used to estimate $\tilde{\mathbf{x}}_u$.

We now discuss how to obtain $\{\mathbf{W}_k\}_{k=1}^{K}$ that satisfy the condition in Equation (13.18). Let us construct the following channel matrix that contains the channel gains of all users except for the u th user:

$$\tilde{\mathbf{H}}_u^{\text{DL}} = \left[\left(\mathbf{H}_1^{\text{DL}}\right)^H \cdots \left(\mathbf{H}_{u-1}^{\text{DL}}\right)^H \left(\mathbf{H}_{u+1}^{\text{DL}}\right)^H \cdots \left(\mathbf{H}_K^{\text{DL}}\right)^H \right]^H \qquad (13.20)$$

When $N_{M,total} = \sum_{u=1}^{K} N_{M,u} = N_B$, that is, the total number of antennas used by all active users are the same as the number of BS antennas, Equation (13.18) is equivalent to

$$\tilde{\mathbf{H}}_u^{\text{DL}}\mathbf{W}_u = \mathbf{0}_{(N_{M,total}-N_{M,u}) \times N_{M,u}}, \quad u = 1, 2, \cdots, K \qquad (13.21)$$

This implies that the precoder matrix $\mathbf{W}_u \in \mathbb{C}^{N_B \times N_{M,u}}$ must be designed to lie in the null space[1] of $\tilde{\mathbf{H}}_u^{\text{DL}}$. If Equation (13.21) is satisfied for the case of $K = 3$, the received signal in Equation (13.17) is expressed as

$$\begin{bmatrix} \mathbf{y}_1 \\ \mathbf{y}_2 \\ \mathbf{y}_3 \end{bmatrix} = \begin{bmatrix} \mathbf{H}_1^{\text{DL}}\mathbf{W}_1 & \mathbf{0} & \mathbf{0} \\ \mathbf{0} & \mathbf{H}_2^{\text{DL}}\mathbf{W}_2 & \mathbf{0} \\ \mathbf{0} & \mathbf{0} & \mathbf{H}_3^{\text{DL}}\mathbf{W}_3 \end{bmatrix} \begin{bmatrix} \tilde{\mathbf{x}}_1 \\ \tilde{\mathbf{x}}_2 \\ \tilde{\mathbf{x}}_3 \end{bmatrix} + \begin{bmatrix} \mathbf{z}_1 \\ \mathbf{z}_2 \\ \mathbf{z}_3 \end{bmatrix} \qquad (13.22)$$

where the zero matrices are of appropriate dimensions.

We now discuss how to design the precoders that satisfy Equation (13.21). We note that the dimension of the matrix $\tilde{\mathbf{H}}_u^{\text{DL}} \in \mathbb{C}^{(N_{M,total}-N_{M,u}) \times N_B}$ is less than $\min(N_{M,total}-N_{M,u}, N_B)$. If we assume $N_{M,total} = N_B$, $\min(N_{M,total}-N_{M,u}, N_B) = N_B - N_{M,u}$. Then the singular value decomposition (SVD) of $\tilde{\mathbf{H}}_u^{\text{DL}}$ can be expressed as

---

[1] The null space of a matrix $A \in \mathbb{C}m \times n$ is defined as $\text{Null}(A) = (x \in \mathbb{C}1 \times n)Ax = 0m$. If a matrix lies in the null space of $A \in \mathbb{C}m \times n$, then all column vectors of the matrix lie in $\text{Null}(A)$.

$$\tilde{\mathbf{H}}_u^{DL} = \tilde{\mathbf{U}}_u \tilde{\Lambda}_u \left[\tilde{\mathbf{V}}_u^{\text{non-zero}} \ \tilde{\mathbf{V}}_u^{\text{zero}} \right]^H \tag{13.23}$$

where $\tilde{\mathbf{V}}_u^{\text{non-zero}} \in \mathbb{C}^{(N_{M,total} - N_{M,u}) \times N_B}$ and $\tilde{\mathbf{V}}_u^{\text{zero}} \in \mathbb{C}^{N_{M,u} \times N_B}$ are composed of right singular vectors that correspond to non-zero singular values and zero singular values, respectively. Multiplying $\tilde{\mathbf{H}}_u^{DL}$ with $\tilde{\mathbf{V}}_u^{\text{zero}}$, we have the following relationship:

$$
\begin{aligned}
\tilde{\mathbf{H}}_u^{DL} \tilde{\mathbf{V}}_u^{\text{zero}} &= \tilde{\mathbf{U}}_u \left[\tilde{\Lambda}_u^{\text{non-zero}} \ \mathbf{0} \right] \left[\begin{array}{c} \left(\tilde{\mathbf{V}}_u^{\text{non-zero}} \right)^H \\ \left(\tilde{\mathbf{V}}_u^{\text{zero}} \right)^H \end{array} \right] \tilde{\mathbf{V}}_u^{\text{zero}} \\
&= \tilde{\mathbf{U}}_u \tilde{\Lambda}_u^{\text{non-zero}} \left(\tilde{\mathbf{V}}_u^{\text{non-zero}} \right)^H \tilde{\mathbf{V}}_u^{\text{zero}} \\
&= \tilde{\mathbf{U}}_u \tilde{\Lambda}_u^{\text{non-zero}} \mathbf{0} \\
&= \mathbf{0}
\end{aligned} \tag{13.24}
$$

From Equation (13.24), it can be seen that $\tilde{\mathbf{V}}_u^{\text{zero}}$ is in the null space of $\tilde{\mathbf{H}}_u^{DL}$, that is, when a signal is transmitted in the direction of $\tilde{\mathbf{V}}_u^{\text{zero}}$, all but the u th user receives no signal at all. Thus, $\mathbf{W}_u = \tilde{\mathbf{V}}_u$ can be used for precoding the u th user signal.

Let us take an example of $N_B = 4$, $K = 2$, and $N_{M,1} = N_{M,2} = 2$:

$$
\begin{aligned}
\tilde{\mathbf{H}}_1^{DL} &= \tilde{\mathbf{U}}_1 \tilde{\Lambda}_1 \left[\tilde{\mathbf{V}}_1^{\text{non-zero}} \ \tilde{\mathbf{V}}_1^{\text{zero}} \right]^H \\
&= \left[\tilde{\mathbf{u}}_{11} \ \tilde{\mathbf{u}}_{12} \right] \begin{bmatrix} \tilde{\lambda}_{11} & 0 & 0 & 0 \\ 0 & \tilde{\lambda}_{12} & 0 & 0 \end{bmatrix} \left[\tilde{\mathbf{v}}_{11} \ \tilde{\mathbf{v}}_{12} \ \tilde{\mathbf{v}}_{13} \ \tilde{\mathbf{v}}_{14} \right]^H
\end{aligned} \tag{13.25}
$$

$$
\begin{aligned}
\tilde{\mathbf{H}}_2^{DL} &= \tilde{\mathbf{U}}_2 \tilde{\Lambda}_2 \left[\tilde{\mathbf{V}}_2^{\text{non-zero}} \ \tilde{\mathbf{V}}_2^{\text{zero}} \right]^H \\
&= \left[\tilde{\mathbf{u}}_{21} \ \tilde{\mathbf{u}}_{22} \right] \begin{bmatrix} \tilde{\lambda}_{21} & 0 & 0 & 0 \\ 0 & \tilde{\lambda}_{22} & 0 & 0 \end{bmatrix} \left[\tilde{\mathbf{v}}_{21} \ \tilde{\mathbf{v}}_{22} \ \tilde{\mathbf{v}}_{23} \ \tilde{\mathbf{v}}_{24} \right]^H
\end{aligned} \tag{13.26}
$$

From Equations (13.25) and (13.26), we have the following precoding matrices $\mathbf{W}_u \in \mathbb{C}^{4 \times 2}$, $u = 1, 2$:

$$
\begin{aligned}
\mathbf{W}_1 &= \tilde{\mathbf{V}}_1^{\text{zero}} = \left[\tilde{\mathbf{v}}_{13} \ \tilde{\mathbf{v}}_{14} \right] \\
\mathbf{W}_2 &= \tilde{\mathbf{V}}_2^{\text{zero}} = \left[\tilde{\mathbf{v}}_{23} \ \tilde{\mathbf{v}}_{24} \right]
\end{aligned} \tag{13.27}
$$

which are used to construct the following transmitted signal $\mathbf{s} \in \mathbb{C}^{4 \times 1}$:

$$\mathbf{x} = \mathbf{W}_1 \tilde{\mathbf{x}}_1 + \mathbf{W}_2 \tilde{\mathbf{x}}_2 \qquad (13.28)$$

where $\tilde{\mathbf{x}}_u \in \mathbb{C}^{2 \times 1}$ is the u th user signal, $u = 1, 2$. Then the received signal of the first user is given as

$$\begin{aligned}
\mathbf{y}_1 &= \mathbf{H}_1^{DL} \mathbf{x} + \mathbf{z}_1 \\
&= \mathbf{H}_1^{DL} (\mathbf{W}_1 \tilde{\mathbf{x}}_1 + \mathbf{W}_2 \tilde{\mathbf{x}}_2) + \mathbf{z}_1 \\
&= \tilde{\mathbf{H}}_2^{DL} \left(\tilde{\mathbf{V}}_1^{zero} \tilde{\mathbf{x}}_1 + \tilde{\mathbf{V}}_2^{zero} \tilde{\mathbf{x}}_2 \right) + \mathbf{z}_1 \qquad (13.29) \\
&= \tilde{\mathbf{H}}_2^{DL} \tilde{\mathbf{V}}_1^{zero} \tilde{\mathbf{x}}_1 + \mathbf{z}_1 \\
&= \mathbf{H}_1^{DL} \tilde{\mathbf{V}}_1^{zero} \tilde{\mathbf{x}}_1 + \mathbf{z}_1
\end{aligned}$$

In deriving Equation (13.29), we have used the fact that $\mathbf{H}_1^{DL} = \tilde{\mathbf{H}}_2^{DL}$ and $\mathbf{H}_2^{DL} = \tilde{\mathbf{H}}_1^{DL}$. From Equation (13.29), we can see that the received signal is composed of the desired signal only. The received signal of the second user is found in a similar way.

Running Program 13.5 ("Block_diagonalization.m") yields Figure 13.6, which shows the BER performance of block diagonalization method for the example of $N_B = 4$, $K = 2$, and $N_{M,1} = N_{M,2} = 2$ where the average BER is taken for both users while employing a zero-forcing detection at the receiver. Note that more advanced signal detection methods in Chapter 11 can also be used to improve the BER performance.

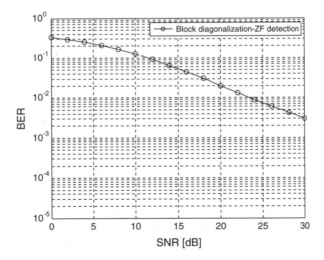

Figure 13.6 BER performance of block diagonalization method using zero-forcing detection at the receiver: $N_B = 4$, $K = 2$, and $N_{M,1} = N_{M,2} = 2$.

MATLAB® Program: Block Diagonalization Method for a Multi-User MIMO System

Program 13.5 "Block_diagonalization.m" for BD method using zero-forcing detection

```
% Block_diagonalization.m
clear all; clf
N_frame=10; N_packet=100; % Number of frames/packet and Number of packets
b=2; % Number of bits per QPSK symbol
NT=4; NR=2; N_user=2;
N_pbits = N_frame*NT*b; % Number of bits per packet
N_tbits = N_pbits*N_packet; % Number of total bits
SNRdBs = [0:2:30]; sq2=sqrt(2);
for i_SNR=1:length(SNRdBs)
   SNRdB=SNRdBs(i_SNR); N_ebits=0; rand('seed',1); randn('seed',1);
   sigma2 = NT*0.5*10^(-SNRdB/10); sigma = sqrt(sigma2);
   for i_packet=1:N_packet
       msg_bit = randint(1,N_pbits); % bit generation
       symbol = QPSK_mapper(msg_bit).';
       x = reshape(symbol,NT,N_frame);
       H1 = (randn(NR,NT)+j*randn(NR,NT))/sq2;
       H2 = (randn(NR,NT)+j*randn(NR,NT))/sq2;
       [U1,S1,V1] = svd(H1); W2 = V1(:,3:4);
       [U2,S2,V2] = svd(H2); W1 = V2(:,3:4);
       Tx_Data = W1*x(1:2,:) + W2*x(3:4,:);
       Rx1 = H1*Tx_Data + sigma*(randn(2,N_frame)+j*randn(2,N_frame));
       Rx2 = H2*Tx_Data + sigma*(randn(2,N_frame)+j*randn(2,N_frame));
       W1_H1=H1*W1;
       EQ1 = W1_H1'*inv(W1_H1*W1_H1'); % Equalizer for the 1st user
       W2_H2=H2*W2;
       EQ2 = W2_H2'*inv(W2_H2*W2_H2'); % Equalizer for the 2nd user
       y = [EQ1*Rx1; EQ2*Rx2];
       symbol_hat = reshape(y,NT*N_frame,1);
       symbol_sliced = QPSK_slicer(symbol_hat);
       demapped = QPSK_demapper(symbol_sliced);
       N_ebits = N_ebits + sum(msg_bit~=demapped);
   end
   BER(i_SNR) = N_ebits/N_tbits;
end
semilogy(SNRdBs,BER,'-o'), grid on
```

13.3.3 Dirty Paper Coding (DPC)

We already have illustrated an idea of dirty paper coding (DPC) in the course of deriving the channel capacity of broadcast channel (BC) in Section 13.2.2, showing that an interference-free transmission can be realized by subtracting the potential interferences before transmission. In theory, DPC would be implemented when channel gains are completely known on the transmitter side. Dirty paper coding (DPC) is a method of precoding the data such that the effect of the interference can be canceled subject to some interference that is known to the transmitter. More specifically, the interferences due to the first up to $(k-1)$ th user signals are canceled in the course of precoding the kth user signal. To simplify the exposition, we just consider the case of $N_B = 3, K = 3$, and $N_{M,u} = 1, u = 1, 2, 3$. If the u th user signal is given by $\tilde{x}_u \in \mathbb{C}$, then the received signal is given as

$$
\begin{bmatrix} y_1 \\ y_2 \\ y_3 \end{bmatrix} = \underbrace{\begin{bmatrix} \mathbf{H}_1^{DL} \\ \mathbf{H}_2^{DL} \\ \mathbf{H}_3^{DL} \end{bmatrix}}_{\mathbf{H}^{DL}} \begin{bmatrix} \tilde{x}_1 \\ \tilde{x}_2 \\ \tilde{x}_3 \end{bmatrix} + \begin{bmatrix} z_1 \\ z_2 \\ z_3 \end{bmatrix} \tag{13.30}
$$

where $\mathbf{H}_u^{DL} \in \mathbb{C}^{1 \times 3}$ is the channel gain between BS and the u th user. The channel matrix $\mathbf{H}^{DL}$ can be LQ-decomposed as

$$
\mathbf{H}^{DL} = \underbrace{\begin{bmatrix} l_{11} & 0 & 0 \\ l_{21} & l_{22} & 0 \\ l_{31} & l_{32} & l_{33} \end{bmatrix}}_{\mathbf{L}} \underbrace{\begin{bmatrix} \mathbf{q}_1 \\ \mathbf{q}_2 \\ \mathbf{q}_3 \end{bmatrix}}_{\mathbf{Q}} \tag{13.31}
$$

where $\{\mathbf{q}_i\}_{i=1}^{3} \in \mathbb{C}^{1 \times 3}$ are orthonormal row vectors. Let $\mathbf{x} = [x_1 \ x_2 \ x_3]^T$ denote a precoded signal for $\tilde{\mathbf{x}} = [\tilde{x}_1 \ \tilde{x}_2 \ \tilde{x}_3]^T$. By transmitting $\mathbf{Q}^H \mathbf{x}$, the effect of $\mathbf{Q}$ in Equation (13.31) is eliminated through the channel. Leaving the lower-triangular matrix after transmission, the received signal is given as

$$
\begin{bmatrix} y_1 \\ y_2 \\ y_3 \end{bmatrix} = \underbrace{\begin{bmatrix} \mathbf{H}_1^{DL} \\ \mathbf{H}_2^{DL} \\ \mathbf{H}_3^{DL} \end{bmatrix}}_{\mathbf{H}^{DL}} \mathbf{Q}^H \mathbf{x} + \begin{bmatrix} z_1 \\ z_2 \\ z_3 \end{bmatrix} = \begin{bmatrix} l_{11} & 0 & 0 \\ l_{21} & l_{22} & 0 \\ l_{31} & l_{32} & l_{33} \end{bmatrix} \begin{bmatrix} x_1 \\ x_2 \\ x_3 \end{bmatrix} + \begin{bmatrix} z_1 \\ z_2 \\ z_3 \end{bmatrix}. \tag{13.32}
$$

From Equation (13.32), the received signal of the first user is given as

$$
y_1 = l_{11} x_1 + z_1. \tag{13.33}
$$

From the first-user perspective, therefore, the following condition needs to be met for the interference-free data transmission:

$$
x_1 = \tilde{x}_1 \tag{13.34}
$$

From Equation (13.34), it can be seen that the precoded signal x_1 is solely composed of the first-user signal $\tilde{x}_1$. From Equations (13.32) and (13.34), the received signal of the second user is given as

$$
y_2 = l_{21} x_1 + l_{22} x_2 + z_2 = l_{21} \tilde{x}_1 + l_{22} x_2 + z_2. \tag{13.35}
$$

From Equation (13.35), it can be seen that the following precoding cancels the interference component, $l_{21}x_1$ or $l_{21}\tilde{x}_1$, on the transmitter side:

$$x_2 = \tilde{x}_2 - \frac{l_{21}}{l_{22}}x_1 = \tilde{x}_2 - \frac{l_{21}}{l_{22}}\tilde{x}_1 \tag{13.36}$$

From Equation (13.36), it can be seen that the precoded signal x_2 is now composed of the user signals, $\tilde{x}_1$ and $\tilde{x}_2$. Finally, the received signal of the third user is given as

$$y_3 = l_{31}x_1 + l_{32}x_2 + l_{33}x_3 + z_3. \tag{13.37}$$

where the precoded signals, x_1 and x_2, are composed of the known user signals, $\tilde{x}_1$ and $\tilde{x}_2$, given in Equations (13.34) and (13.36). From the perspective of the third user, the precoded signals, x_1 and x_2, are interference components in Equation (13.37), which can be canceled by the following precoding on the transmitter side:

$$x_3 = \tilde{x}_3 - \frac{l_{31}}{l_{33}}x_1 - \frac{l_{32}}{l_{33}}x_2 \tag{13.38}$$

The precoded signals in Equations (13.34), (13.36), and (13.38) can be expressed in a matrix as

$$\begin{bmatrix} x_1 \\ \tilde{x}_2 \\ \tilde{x}_3 \end{bmatrix} = \begin{bmatrix} 1 & 0 & 0 \\ 0 & 1 & 0 \\ 0 & 0 & 1 \end{bmatrix} \begin{bmatrix} \tilde{x}_1 \\ \tilde{x}_2 \\ \tilde{x}_3 \end{bmatrix}, \tag{13.39}$$

$$\begin{bmatrix} x_1 \\ x_2 \\ \tilde{x}_3 \end{bmatrix} = \begin{bmatrix} 1 & 0 & 0 \\ -\dfrac{l_{21}}{l_{22}} & 1 & 0 \\ 0 & 0 & 1 \end{bmatrix} \begin{bmatrix} x_1 \\ \tilde{x}_2 \\ \tilde{x}_3 \end{bmatrix}, \tag{13.40}$$

and

$$\begin{bmatrix} x_1 \\ x_2 \\ x_3 \end{bmatrix} = \begin{bmatrix} 1 & 0 & 0 \\ 0 & 1 & 0 \\ \dfrac{l_{31}}{l_{33}} & \dfrac{l_{32}}{l_{33}} & 1 \end{bmatrix} \begin{bmatrix} x_1 \\ x_2 \\ \tilde{x}_3 \end{bmatrix} \tag{13.41}$$

Combining the above three precoding matrices, we can express the DPC in the following matrix form:

$$
\begin{bmatrix} x_1 \\ x_2 \\ x_3 \end{bmatrix} = \begin{bmatrix} 1 & 0 & 0 \\ 0 & 1 & 0 \\ -\dfrac{l_{31}}{l_{33}} & -\dfrac{l_{32}}{l_{33}} & 1 \end{bmatrix} \begin{bmatrix} 1 & 0 & 0 \\ -\dfrac{l_{21}}{l_{22}} & 1 & 0 \\ 0 & 0 & 1 \end{bmatrix} \begin{bmatrix} 1 & 0 & 0 \\ 0 & 1 & 0 \\ 0 & 0 & 1 \end{bmatrix} \begin{bmatrix} \tilde{x}_1 \\ \tilde{x}_2 \\ \tilde{x}_3 \end{bmatrix}
$$

$$
= \begin{bmatrix} 1 & 0 & 0 \\ -\dfrac{l_{21}}{l_{22}} & 1 & 0 \\ -\dfrac{l_{31}}{l_{33}} + \dfrac{l_{32}}{l_{33}}\dfrac{l_{21}}{l_{22}} & -\dfrac{l_{32}}{l_{33}} & 1 \end{bmatrix} \begin{bmatrix} \tilde{x}_1 \\ \tilde{x}_2 \\ \tilde{x}_3 \end{bmatrix}. \tag{13.42}
$$

Using the above precoding matrix, Equation (13.32) can be re-written as

$$
\begin{bmatrix} y_1 \\ y_2 \\ y_3 \end{bmatrix} = \begin{bmatrix} l_{11} & 0 & 0 \\ l_{21} & l_{22} & 0 \\ l_{31} & l_{32} & l_{33} \end{bmatrix} \begin{bmatrix} x_1 \\ x_2 \\ x_3 \end{bmatrix} + \begin{bmatrix} z_1 \\ z_2 \\ z_3 \end{bmatrix}
$$

$$
= \begin{bmatrix} l_{11} & 0 & 0 \\ l_{21} & l_{22} & 0 \\ l_{31} & l_{32} & l_{33} \end{bmatrix} \begin{bmatrix} 1 & 0 & 0 \\ -\dfrac{l_{21}}{l_{22}} & 1 & 0 \\ -\dfrac{l_{31}}{l_{33}} + \dfrac{l_{32}}{l_{33}}\dfrac{l_{21}}{l_{22}} & -\dfrac{l_{32}}{l_{33}} & 1 \end{bmatrix} \begin{bmatrix} \tilde{x}_1 \\ \tilde{x}_2 \\ \tilde{x}_3 \end{bmatrix} + \begin{bmatrix} z_1 \\ z_2 \\ z_3 \end{bmatrix}
$$

$$
= \begin{bmatrix} l_{11} & 0 & 0 \\ 0 & l_{22} & 0 \\ 0 & 0 & l_{33} \end{bmatrix} \begin{bmatrix} \tilde{x}_1 \\ x_2 \\ \tilde{x}_3 \end{bmatrix} + \begin{bmatrix} z_1 \\ z_2 \\ z_3 \end{bmatrix} \tag{13.43}
$$

From Equation (13.43), it is obvious that the interference-free detection can be made for each user. We can see from Equation (13.43) that the precoding matrix in DPC is a scaled inverse matrix of the lower triangular matrix which is obtained from the channel gain matrix, that is,

$$
\begin{bmatrix} 1 & 0 & 0 \\ -\dfrac{l_{21}}{l_{22}} & 1 & 0 \\ -\dfrac{l_{31}}{l_{33}} + \dfrac{l_{32}}{l_{33}}\dfrac{l_{21}}{l_{22}} & -\dfrac{l_{32}}{l_{33}} & 1 \end{bmatrix} = \begin{bmatrix} l_{11} & 0 & 0 \\ l_{21} & l_{22} & 0 \\ l_{31} & l_{32} & l_{33} \end{bmatrix}^{-1} \begin{bmatrix} l_{11} & 0 & 0 \\ 0 & l_{22} & 0 \\ 0 & 0 & l_{33} \end{bmatrix} \tag{13.44}
$$

Program 13.6 ("Dirty_or_TH_precoding.m") with mode set to 0 can be used to simulate the DPC for $N_B = 4$ and $K = 10$. In this example, four best users, whose indices are denoted by $(u_1^*, u_2^*, u_3^*, u_4^*)$, are selected out of ten users by using the following selection criterion: Select $(u_1^*, u_2^*, u_3^*, u_4^*) \in \{1, 2, \ldots, 10\}$ such that

$$l_{u_1^* u_1^*} \geq l_{u_2^* u_2^*} \geq l_{u_3^* u_3^*} \geq l_{u_4^* u_4^*} \geq l_{uu} \tag{13.45}$$

where $u \notin (u_1^*, u_2^*, u_3^*, u_4^*) \in \{1, 2, \ldots, 10\}$.

Note that selection of the users with the largest l_{ii} ensures to minimize the noise enhancement on the receiver side (e.g., as can be justified by Equations (13.33), (13.35) and (13.37)). The BER performance will be provided in the next section for comparison with Tomlinson-Harashima precoding.

13.3.4 Tomlinson-Harashima Precoding

DPC on the transmitter side is very similar to decision feedback equalization (DFE) on the receiver side. In fact, combination of DPC with symmetric modulo operation turns out to be equivalent to Tomlinson-Harashima (TH) precoding [256, 274–276]. TH precoding was originally invented for reducing the peak or average power in the decision feedback equalizer (DFE), which suffers from error propagation. The original idea of TH precoding in DFE is to cancel the post-cursor ISI in the transmitter, where the past transmit symbols are known without possibility of errors. In fact, it requires a complete knowledge of the channel impulse response, which is only available by a feedback from the receiver for time-invariant or slowly time-varying channel. To facilitate exposition of the idea, consider the precoding in the one-dimensional case, in which the data symbol x is drawn from the M-ary PAM constellation, $\{-(A-1), -(A-3), \cdots, -3, -1, 1, 3, \cdots, (A-3), (A-1)\}$, where A is an even integer given by $A = \sqrt{M}$. By adding $2A \cdot m$ to the data symbol x, where m is an integer, an expanded symbol c can be defined as

$$c = x + 2A \cdot m \tag{13.46}$$

In order to reduce the peak or average power, m in Equation (13.46) must be chosen to minimize the magnitude of the expanded symbol c in the transmitter. Note that the original data symbol x can be recovered from the expanded symbol c by the symmetric modulo operation defined as

$$x = \text{mod}_A(c) \triangleq c - 2A\lfloor (c+A)/2A \rfloor \tag{13.47}$$

In order to address TH precoding for the multi-user MIMO system, we discuss the symmetric modulo operation for M-ary QAM modulated symbols, which is an extension of Equation (13.47) to the two-dimensional case. In M-ary QAM with a square constellation, the real and imaginary parts of a symbol are bounded by $[-A, A)$, with $A = \sqrt{M}$. Figure 13.7 illustrates a 16-QAM constellation with $A = \sqrt{16} = 4$. As illustrated in Figure 13.8, the symmetric modulo operation is defined as

$$\text{mod}_A(x) = x - 2A\lfloor (x+A+jA)/2A \rfloor \tag{13.48}$$

The above modulo operation can be interpreted as a method to find integer values, m and n, such that the following inequalities are satisfied:

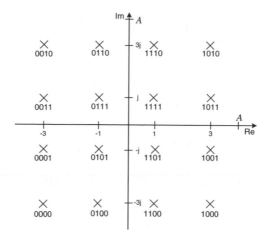

Figure 13.7   16-QAM constellation.

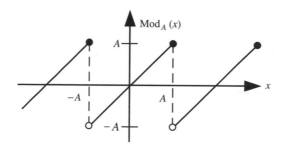

Figure 13.8   Illustration of symmetric modulo operation (only for the real part of x).

$$-A-jA \leq \mathrm{mod}_A(x) = x+2A \cdot m+j\,2A \cdot m < A+jA \qquad (13.49)$$

Note that inequality of complex numbers x_1 and x_2 in Equation (13.49) is defined as

$$x_1 < x_2 \Leftrightarrow \mathrm{Re}\{x_1\} < \mathrm{Re}\{x_2\} \text{ and } \mathrm{Im}\{x_1\} < \mathrm{Im}\{x_2\}. \qquad (13.50)$$

Then, the modulo operation in Equation (13.48) can be expressed as

$$\mathrm{mod}_A(x) = x+2A \cdot m+j\,2A \cdot n \qquad (13.51)$$

Let us take an example of TH precoding for $K=3$. Let $\{x_u^{\mathrm{TH}}\}_{u=1}^{3}$ denote the TH precoded signal for the u th user. Referring to Equations (13.34), (13.36), and (13.38), by the above modulo operation, TH-precoded data symbols are represented as

$$x_1^{\text{TH}} = \text{mod}_A(\tilde{x}_1) = \tilde{x}_1 \tag{13.52}$$

$$x_2^{\text{TH}} = \text{mod}_A\left(\tilde{x}_2 - \frac{l_{21}}{l_{22}} x_1^{\text{TH}}\right) \tag{13.53}$$

$$x_3^{\text{TH}} = \text{mod}_A\left(\tilde{x}_3 - \frac{l_{31}}{l_{33}} x_1^{\text{TH}} - \frac{l_{32}}{l_{33}} x_2^{\text{TH}}\right) \tag{13.54}$$

Furthermore, the interpretation in Equation (13.51) gives the following expressions for the TH-precoded signals:

$$x_1^{\text{TH}} = \tilde{x}_1 \tag{13.55}$$

$$x_2^{\text{TH}} = \tilde{x}_2 - \frac{l_{21}}{l_{22}} \tilde{x}_1 + 2A \cdot m_2 + j\, 2A \cdot n_2 \tag{13.56}$$

$$x_3^{\text{TH}} = \tilde{x}_3 - \frac{l_{31}}{l_{33}} x_1^{\text{TH}} - \frac{l_{32}}{l_{33}} x_2^{\text{TH}} + 2A \cdot m_3 + j\, 2A \cdot n_3 \tag{13.57}$$

For the transmitted signal $\mathbf{Q}^H \mathbf{x}^{\text{TH}} = \mathbf{Q}^H [x_1^{\text{TH}}\ x_2^{\text{TH}}\ x_3^{\text{TH}}]^T$, the received signal is given as

$$\begin{bmatrix} y_1 \\ y_2 \\ y_3 \end{bmatrix} = \underbrace{\begin{bmatrix} \mathbf{H}_1^{\text{DL}} \\ \mathbf{H}_2^{\text{DL}} \\ \mathbf{H}_3^{\text{DL}} \end{bmatrix}}_{\mathbf{H}^{\text{DL}}} \mathbf{Q}^H \mathbf{x}^{\text{TH}} + \begin{bmatrix} z_1 \\ z_2 \\ z_3 \end{bmatrix} = \begin{bmatrix} l_{11} & 0 & 0 \\ l_{21} & l_{22} & 0 \\ l_{31} & l_{32} & l_{33} \end{bmatrix} \begin{bmatrix} x_1^{\text{TH}} \\ x_2^{\text{TH}} \\ x_3^{\text{TH}} \end{bmatrix} + \begin{bmatrix} z_1 \\ z_2 \\ z_3 \end{bmatrix} \tag{13.58}$$

Since $x_1^{\text{TH}} = \tilde{x}_1$, the signal detection for the first user is obvious. The received signal of the second user is given as

$$y_2 = l_{21} x_1^{\text{TH}} + l_{22} x_2^{\text{TH}} + z_2 = l_{21}\tilde{x}_1 + l_{22} x_2^{\text{TH}} + z_2. \tag{13.59}$$

Using Equation (13.56), Equation (13.59) can be expressed as

$$\begin{aligned} y_2 &= l_{21}\tilde{x}_1 + l_{22}\left(\tilde{x}_2 - \frac{l_{21}}{l_{22}}\tilde{x}_1 + 2A \cdot m_2 + j\, 2A \cdot n_2\right) + z_2 \\ &= l_{22}(\tilde{x}_2 + 2A \cdot m_2 + j\, 2A \cdot n_2) + z_2 \end{aligned} \tag{13.60}$$

Defining $\tilde{y}_2$ as a scaled version of y_2, that is,

$$\tilde{y}_2 = \frac{y_2}{l_{22}} = \tilde{x}_2 + 2A \cdot m_2 + j\,2A \cdot n_2 + \frac{z_2}{l_{22}} \tag{13.61}$$

the second-user signal $\tilde{x}_2$ can be detected with the modular operation

$$\hat{\tilde{x}}_2 = \mathrm{mod}_A(\tilde{y}_2). \tag{13.62}$$

If the noise component in Equation (13.62) is small enough to meet the following condition:

$$-A \le \tilde{x}_2 + \frac{z_2}{l_{22}} < A, \tag{13.63}$$

then Equation (13.62) turns out to be

$$\mathrm{mod}_A(\tilde{y}_2) = \tilde{y}_2 - 2A \left\lfloor \frac{(\tilde{y}_2 + A + jA)}{2A} \right\rfloor = \tilde{y}_2 - 2A(m_2 + jn_2) = \tilde{x}_2 + \frac{z_2}{l_{22}}. \tag{13.64}$$

From Equation (13.58), the received signal of the third user is given as

$$y_3 = l_{31}s_1^{\mathrm{TH}} + l_{32}s_2^{\mathrm{TH}} + l_{33}s_3^{\mathrm{TH}} + z_3. \tag{13.65}$$

Using Equation (13.57), the above received signal is expressed as

$$
\begin{aligned}
y_3 &= l_{31}x_1^{\mathrm{TH}} + l_{32}x_2^{\mathrm{TH}} + l_{33}x_3^{\mathrm{TH}} + z_3 \\
&= l_{31}x_1^{\mathrm{TH}} + l_{32}x_2^{\mathrm{TH}} + l_{33}\left(\tilde{x}_3 - \frac{l_{31}}{l_{33}}x_1^{\mathrm{TH}} - \frac{l_{32}}{l_{33}}x_2^{\mathrm{TH}} + 2A \cdot m_3 + j\,2A \cdot n_3 \right) + z_3 \quad (13.66) \\
&= l_{33}(\tilde{x}_3 + 2A \cdot m_3 + j\,2A \cdot n_3) + z_3.
\end{aligned}
$$

As in detection of $\tilde{x}_2$, the third-user signal $\tilde{x}_3$ can be detected as

$$\hat{\tilde{x}}_3 = \mathrm{mod}_A(\tilde{y}_3) \tag{13.67}$$

where

$$\tilde{y}_3 = \frac{y_3}{l_{33}} = x_3 + 2A \cdot m_3 + j\,2A \cdot n_3 + \frac{z_3}{l_{33}}.$$

Running Program 13.6 ("Dirty_or_TH_precoding.m") with mode $= 0$ or 1 yields the BER curve with DPC or TH precoding for $N_B = 4$ and $K = 10$ as depicted in Figure 13.9 where four users are selected out of the ten users by using the criterion in Equation (13.45). As can be seen in Figure 13.9, the DPC outperforms the THP. In this comparison, however, transmitted power of DPC is higher than that of THP. Note that the reduced transmit power of THP is attributed to modulo operations in the precoding process.

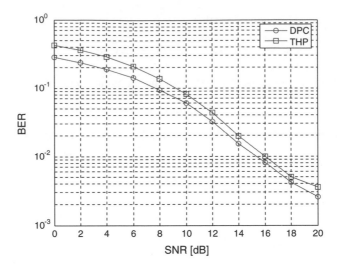

Figure 13.9   BER performance: DPC vs. THP.

See [277–279] for additional information about multi-user MIMO capacity. For further information on user selection and resource allocation for multi-user MIMO systems, the reader may consult the references [280, 281]. See [282–284] for additional information about more advanced downlink transmission schemes.

MATLAB® Programs: Dirty Paper Coding and Tomlinson-Harashima Precoding

Program 13.6 "Dirty_or_TH_precoding.m": DPC or THP for a multi-user MIMO system

```
% Dirty_or_TH_precoding.m
clear all; clf
mode=1; % Set 0/1 for Dirty or TH precoding
N_frame=10; N_packet=200; % Number of frames/packet and Number of packets
b=2; % Number of bits per QPSK symbol
NT=4; N_user=10; N_act_user=4; I=eye(N_act_user,NT);
N_pbits = N_frame*NT*b; % Number of bits in a packet
N_tbits = N_pbits*N_packet; % Number of total bits
SNRdBs=[0:2:20]; sq2=sqrt(2);
for i_SNR=1:length(SNRdBs)
   SNRdB = SNRdBs(i_SNR); N_ebits = 0;
   rand('seed',1); randn('seed',1);
   sigma2 = NT*0.5*10^(-SNRdB/10); sigma = sqrt(sigma2);
   for i_packet=1:N_packet
       msg_bit = randint(1,N_pbits); % Bit generation
       %——————— Transmitter ———————
       symbol = QPSK_mapper(msg_bit).';
       x = reshape(symbol,NT,N_frame);
       H = (randn(N_user,NT)+j*randn(N_user,NT))/sq2;
       Combinations = nchoosek([1:N_user],N_act_user)'; % User selection
```

```
        for i=1:size(Combinations,2)
            H_used = H(Combinations(:,i),:);
            [Q_temp,R_temp] = qr(H_used);
            % Diagonal entries of R_temp are real
            minimum_l(i) = min(diag(R_temp));
        end
        [max_min_l,Index] = max(minimum_l);
        H_used = H(Combinations(:,Index),:);
        [Q_temp,R_temp] = qr(H_used');
        L=R_temp'; Q=Q_temp';
        %————————— TH precoding —————————
        xp = x;
        if mode==0 % Dirty precoding
          for m=2:4 % Eqs.(13.39)~(13.41)
              xp(m,:) = xp(m,:) - L(m,1:m-1)/L(m,m)*xp(1:m-1,:);
          end
        else % TH precoding
          for m=2:4 % Eqs.(13.52)~(13.54)
              xp(m,:) = modulo(xp(m,:)-L(m,1:m-1)/L(m,m)*xp(1:m-1,:),sq2);
          end
        end
        Tx_signal = Q'*xp; % DPC/TH encoder
        %————————— Channel and Noise —————————
        Rx_signal = H_used*Tx_signal + ...
        sigma*(randn(N_act_user,N_frame)+j*randn(N_act_user,N_frame));
        %————————— Receiver —————————
        y = inv(diag(diag(L)))*Rx_signal;
        symbol_hat = reshape(y,NT*N_frame,1);
        if mode==1 % in the case of TH precoding
            symbol_hat = modulo(symbol_hat,sq2);
        end
        symbol_sliced = QPSK_slicer(symbol_hat);
        demapped = QPSK_demapper(symbol_sliced);
        N_ebits = N_ebits + sum(msg_bit~=demapped);
    end
    BER(i_SNR) = N_ebits/N_tbits;
end
semilogy(SNRdBs,BER,'-o'), grid on
```

Program 13.7 "modulo": Modulo function for Tomlinson-Harashima Precoding (THP)

```
function [y]=modulo(x,A)
temp_real = floor((real(x)+A)/(2*A));
temp_imag = floor((imag(x)+A)/(2*A));
y = x -temp_real*(2*A) -j*temp_imag*(2*A);
```

The readers who are interested in the commercial wireless communication systems based on MIMO-OFDM may consult the references [285, 286] for IEEE 802.11a/n, [287–296] for DAB, DMB, DVB, DVB-H, [297] for MB-OFDM, [298–303] for Mobile WiMAX, and [304–309] for 3GPP-LTE.

References

1. Sklar, B. (2002) *Digital Communications: Fundamentals and Applications 2/E*, Prentice Hall.
2. Rappaport, T.S. (2001) *Wireless Communications: Principles and Practice 2/E*, Prentice Hall.
3. Greenwood, D. and Hanzo, L. (1994) Characterization of mobile radio channels. Chapter 2, *Mobile Radio Communications* (ed. R. Steele), Pentech Press-IEEE Press, London.
4. Friis, H.T. (1946) A note on a simple transmission formula. *Proc. IRE*, **34**(5), 254–256.
5. Lee, W.C.Y. (1985) *Mobile Communications Engineering*, McGraw Hill, New York.
6. Okumura, Y., Ohmori, E., Kawano, T., and Fukuda, K. (1968) Field strength and its variability in VHF and UHF land mobile radio service. *Rev. Elec. Commun. Lab.*, **16**, 825–873.
7. Hata, M. (1980) Empirical formula for propagation loss in land mobile radio services. *IEEE Trans. Veh. Technol.*, **29**(3), 317–325.
8. Erceg, V., Greenstein, L.J., Tjandra, S.Y. *et al.* (1999) An empirically based path loss model for wireless channels in suburban environments. *IEEE J. Select. Areas Commun.*, **17**(7), 1205–1211.
9. IEEE (2007) 802.16j-06/013r3. *Multi-Hop Relay System Evaluation Methodology (Channel Model and Performance Metric)*.
10. IEEE (2001) 802.16.3c-01/29r4. *Channel Models for Fixed Wireless Applications*.
11. IST (2004) 4-027756. *WINNER II, D1.1.1 WINNER II Interim Channel Models*.
12. Recommendation (1997) ITU-R M.1225. *Guidelines for Evaluation of Radio Transmission Technologies for IMT-2000*.
13. Clarke, R.H. (1968) A statistical theory of mobile radio reception. *Bell System Tech. J.*, **47**, 987–1000.
14. Capoglu, I.R., Li, Y., and Swami, A. (2005) Effect of doppler spread in OFDM based UWB systems. *IEEE Trans. Wireless Commun.*, **4**(5), 2559–2567.
15. Stuber, G.L. (1996) *Principles of Mobile Communication*, Kluwer Academic Publishers.
16. Tepedelenliglu, C. and Giannakis, G.B. (2001) On velocity estimation and correlation properties of narrow-band mobile communication channels. *IEEE Trans. Veh. Technol.*, **50**(4), 1039–1052.
17. Andersen, J.B., Rappaport, T.S., and Yoshida, S. (1995) Propagation measurements and models for wireless communications channels. *IEEE Commun. Mag.*, **33**(1), 42–49.
18. Bajwa, A.S. and Parsons, J.D. (1982) Small-area characterisation of UHF urban and suburban mobile radio propagation. *Inst. Elec. Eng. Proc.*, **129**(2), 102–109.
19. Bello, P.A. (1963) Characterization of randomly time-variant linear channels. *IEEE Trans. Commun.*, **11**(4), 360–393.
20. Black, D.M. and Reudink, D.O. (1972) Some characteristics of mobile radio propagation at 836 MHz in the Philadelphia area. *IEEE Trans. Veh. Technol.*, **21**(2), 45–51.
21. Corazza, G.E. and Vatalaro, F. (1994) A statistical model for land mobile satellite channels and its application to nongeostationary orbit systems systems. *IEEE Trans. Veh. Technol.*, **43**(3), 738–742.
22. Akki, A.S. and Haber, F. (1986) A statistical model of mobile-to-mobile land communication channel. *IEEE Trans. Veh. Technol.*, **35**(1), 2–7.
23. IEEE (1996) P802.11-97/96. *Tentative Criteria for Comparison of Modulation Methods*.

24. Saleh, A.M. and Valenzuela, R.A. (1987) A statistical model for indoor multipath propagation. *IEEE J. Select. Areas Commun.*, **5**(2), 128–137.

25. IEEE (2003) 802.15-02/490R-L. *Channel Modeling sub-committee. Report finals.*

26. Smith, J.I. (1975) A computer generated multipath fading simulation for mobile radio. *IEEE Trans. Veh. Technol.*, **24**(3), 39–40.

27. Jakes, W.C. (1974) *Microwave Mobile Communications*, John Wiley & Sons, Inc., New York.

28. 3GPP (2007) TR 25.996, v7.0.0. *3rd Generation Partnership Project; Technical Specification Group Radio Access Network; Spatial Channel Model For Multiple Input Multiple Mutput Simulations (Release 7).*

29. SCM (2002) 065v2. *SCM Model Correlations.*

30. SCM (2002) 033-R1. *Spatial Channel Model Issues.*

31. COST 207 (1989) Digital land mobile radio communications, ©ECSC-EEC-EAEC, Brussels-Luxembourg, 1989.

32. Fleury, B.H. and Leuthold, P.E. (1996) Radiowave propagation in mobile communications: An overview of European research. *IEEE Commun. Mag.*, **34**(2), 70–81.

33. Greenstein, L.J. (1978) A multipath fading channel model for terrestrial digital radio systems. *IEEE Trans. Commun.*, **26**(8), 1247–1250.

34. Loo, C. (1985) A statistical model for a land mobile satellite link. *IEEE Trans. Veh. Technol.*, **34**(3), 122–127.

35. Lutz, E. and Plochinger, E. (1985) Generating Rice processes with given spectral properties. *IEEE Trans. Veh. Technol.*, **34**(4), 178–181.

36. Schilling, D.L. *et al.* (1991) Broadband CDMA for personal communications systems. *IEEE Commun. Mag.*, **29**(11), 86–93.

37. Seidel, S.Y. *et al.* (1991) Path loss, scattering and multipath delay statistics in four european cities for digital celluarl and microcellular radiotelephone. *IEEE Trans. Veh. Technol.*, **40**(4), 721–730.

38. Pedersen, K.I., Mogensen, P.E., and Fleury, B.H. (2000) A stochastic model of the temporal and azimuthal dispersion seen at the base station in outdoor propagation environments. *IEEE Trans. Veh. Technol.*, **49**(2), 437–447.

39. Schumacher, L., Pedersen, K.I., and Mogensen, P.E. (2002) From antenna spacings to theoretical capacities-guidelines for simulating MIMO systems. PIMRC'02, vol. 2, pp. 587–592.

40. I-METRA, D2 (Feb. 1999) IST-1999-11729, MIMO channel characterisation.

41. Lee, W. (1973) Effect on correlation between two mobile radio base-station antennas. *IEEE Trans. Commun.*, **21**(11), 1214–1224.

42. IST-METRA project, [Online] Available: http://www.ist-imetra.org.

43. I-METRA, D2 v1.2 (Oct. 2002) IST-2000-30148, Channel characterisation.

44. I-METRA (Oct. 2003) IST-2000-30148, Final Report.

45. Laurent Schumacher (March 2002) Description of the MATLAB implementation of a MIMO channel model suited for link-level simulations.

46. Pedersen, K.I., Andersen, J.B., Kermoal, J.P., and Mogensen, P. (Sept. 2000) A stochastic multiple-input-multiple-output radio channel model for evaluation of space-time coding algorithms. IEEE VTC'00, vol. 2, pp. 893–897.

47. 3GPP (2002) TR 25.876 V1.1.0. *3rd Generation Partnership Project; Technical Specification Group Radio Access Network; Multiple Input Multiple Output Antenna Processing for HSDPA.*

48. 3GPP (2006) TR 25.814 V1.2.2. *3rd Generation Partnership Project; Technical Specification Group Radio Access Network; Physical Layer Aspects for Evolved UTRA (Release 7).*

49. 3GPP (March 2003) Correlation properties of SCM. SCM-127, SCM Conference Call.

50. Available: http://legacy.tkk.fi/Units/Radio/scm/.

51. Wang, J.G., Mohan, A.S., and Aubrey, T.A. (1996) Angles-of-Arrival of multipath signals in indoor environments. IEEE VTC'96, pp. 155–159.

52. Pedersen, K., Mogensen, P., and Fleury, B. (May 1998) Spatial channel characteristics in outdoor environments and their impact on BS antenna system performance. IEEE VTC'98, Ottawa, Canada, pp. 719–723.

53. Nilsson, M., Lindmark, B., Ahlberg, M. *et al.* (Sep. 1998) Measurements of the spatio-temporal polarization characteristics of a radio channel at 1800 MHz. PIMRC'99, pp. 1278–1283.

54. Eggers, P.C.F. (Sep. 1995) Angular dispersive mobile radio environments sensed by highly directive base station antennas. PIMRC'95, pp. 522 526.

55. Martin, U. (1998) Spatio-temporal radio channel characteristics in urban macrocells. *IEE P-Radar. Son. Nav.*, **145**(1), 42–49.

56. Pettersen, M., Lehne, P.H., Noll, J. *et al.* (Sep. 1999) Characterisation of the directional wideband radio channel in urban and suburban areas. IEEE VTC'99, Amsterdam, Netherlands, pp. 1454–1459.

57. Kalliola, K. and Vainikainen, P. (May 1998) Dynamic wideband measurement of mobile radio channel with adaptive antennas. IEEE VTC'98, Ottawa, Canada, pp. 21–25.

58. Pajusco, P. (May 1998) Experimental characterization of DoA at the base station in rural and urban area. IEEE VTC'98, Ottawa, Canada, pp. 993–997.

59. Spencer, Q., Jeffs, B., Jensen, M., and Swindlehurst, A.L. (2000) Modeling the statistical time and angle of arrival characteristics of an indoor multipath channel. *IEEE J. Select. Areas Commun.*, **18**(3), 347–360.

60. Baum, D.S. and El-Sallabi, H. *et al.* (Oct. 2005) Final Report on Link Level and System Level Channel Models, WINNER Deliverable D5.4.

61. Baum, D.S., Salo, J., Del Galdo, G. *et al.* (May 2005) An interim channel model for beyond-3G systems. IEEE VTC'05, Stochholm, vol. 5, pp. 3132–3136.

62. COST 231 (Sep. 1991) Urban transmission loss models for mobile radio in the 900- and 1800 MHz bands, TD(90) 119 Rev. 2.

63. Erceg, V., Schumacher, L., Kyritsi, P. *et al.* (Jan. 2004) TGn Channel Models, IEEE, 802.11-03/940r2.

64. Chong, C.C., Tan, C.M., Laurenson, D.I. *et al.* (2003) A new statistical wideband spatio-temporal channel model for 5 GHz band WLAN systems. *IEEE J. Select. Areas Commun.*, **21**(2), 139–150.

65. Kyosti, P., Meinila, J. *et al.* (Nov. 2006) WINNER II Interim Channel Models, WINNER Deliverable D1.1.1.

66. Zhao, X., Kivinen, J., Vainikainen, P., and Skog, K. (2002) Propagation characteristics for wideband outdoor mobile communications at 5.3 GHz. *IEEE J. Select. Areas Commun.*, **20**(3), 507–514.

67. Zhao, X., Rautiainen, T., Kalliola, K., and Vainikainen, P. (Sep. 2004) Path Loss Models for Urban Microcells at 5.3 GHz, COST 273 TD(04)207, Duisburg, Germany.

68. Haykin, S. (2001) *Communication Systems 4/E*, John Wiley & Sons, Inc., New York.

69. Nyquist, H. (1928) Certain topics in telegraph transmission theory. *Trans. AIEE*, **477**, 617–644.

70. Turin, G.L. (1960) An introduction to matched filters. *IEEE Trans. Info. Theor*, **6**(3), 311–329.

71. Goldsmith, A. (2005) *Wireless Communications*, Cambridge University Press.

72. Bingham, J.A.C. (1990) Multi-carrier modulation for data transmission: an idea whose time has come. *IEEE Commun. Mag.*, **28**(5), 17–25.

73. Chang, R.W. (1966) Synthesis of band-limited orthogonal signals for multi-channel data transmission. *Bell System Tech. J.*, **46**, 1775–1796.

74. Vaidyanathan, P.P. (1993) *Multi-rate System and Filter Banks*, Prentice Hall.

75. Saltzberg, B.R. (1967) Performance of an efficient parallel data transmission system. *IEEE Trans. Commun., Technol.*, **COM-15**, 805–811.

76. Weinstein, S.B. (1971) Data transmission by frequency division multiplexing using the discrete Fourier transform. *IEEE Trans. Commun. Technol.*, **COM-19**(5), 628–634.

77. Bingham, J.A.C. (2000) *ADSL, VDSL, and Multicarrier Modulation*, John Wiley & Sons, Inc., New York.

78. Peled, A. (1980) Frequency domain data transmission using reduced computational complexity algorithms. *IEEE ICASSP*, **5**, 964–967.

79. Cherubini, G. (2000) Filter bank modulation techniques for very high-speed digital subscriber lines. *IEEE Commun. Mag.*, **38**(5), 98–104.

80. Lottici, V. (2005) Blind carrier frequency tracking for filter bank multicarrier wireless communications. *IEEE Trans. Commun.*, **53**(10), 1762–1772.

81. Sandberg, S.D. (1995) Overlapped discrete multi-tone modulation for high speed copper wire communications. *IEEE J. Select. Areas Commun.*, **13**(9), 1571–1585.

82. 3GPP (Feb. 2003) TSG-RAN-1 Meeting #31, Technical description of the OFDM/IOTA modulation.

83. Alard, M. and Lassalle, R. (1987) Principles of modulation and channel coding for digital broadcasting for mobile receivers. *EBU Tech. Review* (224), 47–69.

84. Sari, H., Karam, G., and Jeanclaude, I. (1995) Transmission techniques for digital terrestrial TV broadcasting. *IEEE Commun. Mag.*, **33**(2), 100–109.

85. Mestdagh, D.G. *et al.* (2000) Zipper VDSL: a solution for robust duplex communication over telephone lines. *IEEE Commun. Mag.*, **38**(5), 90–96.

86. Alliance, W. (Feb. 2007) Multiband OFDM Physical Layer Specification, Release 1.2.

87. Proakis, J.G. (2008) *Digital Communications 5/E*, McGraw-Hill, New York.

88. Chow, P.S., Cioffi, J.M., and Bingham, J.A.C. (1995) A practical discrete multitone transceiver loading algorithm for data transmission over spectrally shaped channels. *IEEE Trans. Commun.*, **43**(234), 773–775.

89. Krongold, B.S., Ramchandran, K., and Jones, D.L. (June 1998) Computationally efficient optimal power allocation algorithm for multicarrier communication systems. IEEE ICC'98, vol. 2, pp. 1018–1022.

90. Kalet, I. (1989) The multitone channel. *IEEE Trans. Commun.*, **37**(2), 119–124.

91. Willink, T.J. and Witteke, P.H. (1997) Optimization and performance evaluation of multicarrier transmission. *IEEE Trans. Commun.*, **43**(2), 426–440.

92. Czylwik, A. (Nov. 1996) Adaptive OFDM for wideband radio channels. IEEE GLOBECOM'96, vol. 1, pp. 713–718.

93. Lai, S.K., Cheng, R.S., Letaief, K.B., and Tsui, C.Y. (1999) Adaptive tracking of optimal bit and power allocation for OFDM systems in time-varying channels. *IEEE Wireless Commun. and Networking Conf.*, **2**, 776–780.

94. Pollet, T., van Bladel, M., and Moeneclaey, M. (1995) BER sensitivity of OFDM systems to carrier frequency offset and wiener phase noise. *IEEE Trans. on Commun.*, **43**(2/3/4), 191–193.

95. Pauli, M. and Kuchenbecker, H.P. (1997) Minimization of the intermodulation distortion of a nonlinearly amplified OFDM signal. *Wireless Personal Commun.*, **4**(1), 93–101.

96. Rapp, C. (Oct. 1991) Effects of HPA-nonlinearity on a 4-DPSK/OFDM signal for a digital sound broadcasting signal. Proc. of the Second European Conf. on Satellite Commun, N92-15210 06-32, pp. 179–184.

97. Rohling, H. and Grunheid, R. (May 1997) Performance comparison of different multiple access scheme for the downlink of an OFDM communication system. IEEE VTC'97, vol. 3, pp. 1365–1369.

98. Hara, S. and Prasad, R. (1997) Overview of multicarrier CDMA. *IEEE Commun. Mag.*, **35**(12), 126–133.

99. IEEE (2005) 802.16e-2005. *Part 16: Air Interface for Fixed and Mobile Broadband Wireless Access Systems.*

100. Muschallik, C. (1995) Influence of RF oscillator on an OFDM signal. *IEEE Trans. Consumer Elect.*, **41**(3) 592–603.

101. Pollet, T. and Moeneclaey, M. (Nov. 1995) Synchronizability of OFDM signals. IEEE GLOBECOM'95, pp. 2054–2058.

102. Tomba, L. (1998) On the effect of Wiener phase noise in OFDM systems. *IEEE Trans. Commun.*, **46**(5), 580–583.

103. Minn, H., Zeng, M., and Bhargava, V.K. (2000) On timing offset estimation for OFDM systems. *IEEE Trans. Commun.*, **4**(5), 242–244.

104. Tourtier, P.J., Monnier, R., and Lopez, P. (1993) Multicarrier modem for digital HDTV terrestrial broadcasting. *Signal Process.*, **5**(5), 379–403.

105. Speth, M., Classen, F., and Meyr, H. (May 1997) Frame synchronization of OFDM systems in frequency selective fading channels. IEEE VTC'97, pp. 1807–1811.

106. Van de Beek, J.J., Sandell, M., Isaksson, M., and Börjesson, P.O. (Nov. 1995) Low-complex frame synchronization in OFDM systems. IEEE ICUPC, pp. 982–986.

107. Van de Beek, J.J., Sandell, M., and Börjesson, P.O. (1997) ML estimation of time and frequency offset in OFDM systems. *IEEE Trans. Commun.*, **45**(7), 1800–1805.

108. Schmidl, T.M. and Cox, D.C. (June 1996) Low-overhead, low-complexity burst synchronization for OFDM. IEEE ICC'96, pp. 1301–1306.

109. Schmidl, T.M. and Cox, D.C. (1997) Robust frequency and timing synchronization for OFDM. *IEEE Trans. Commun.*, **45**(12), 1613–1621.

110. Taura, K., Tsujishta, M., Takeda, M. *et al.* (1996) A digital audio broadcasting(DAB) receiver. *IEEE Trans. Consumer Electronics*, **42**(3), 322–326.

111. Daffara, F. and Adami, O. (July 1995) A new frequency detector for orthogonal multi-carrier transmission techniques. IEEE VTC'95, pp. 804–809.

112. Moose, P.H. (1994) A technique for orthogonal frequency division multiplexing frequency offset correction. *IEEE Trans. Commun.*, **42**, 2908–2914.

113. Classen, F. and Myer, H. (June 1994) Frequency synchronization algorithm for OFDM systems suitable for communication over frequency selective fading channels. IEEE VTC'94, pp. 1655–1659.

114. Yang, B., Letaief, K.B., Cheng, R.S., and Cao, Z. (2000) Timing recovery for OFDM transmission. *IEEE Journal on Selected Areas in Commun.*, **18**(11), 2278–2291.

115. Speth, M., Daecke, D., and Meyr, H. (1998) Minimum overhead burst synchronization for OFDM based broadcasting transmission. IEEE GLOBECOM'98, pp. 3227–3232.

116. Pollet, T., Spruyt, P., and Moeneclaey, M. (Nov. 1994) The BER performance of OFDM systems using non-synchronized sampling. IEEE GLOBECOM'94, pp. 253–257.

117. Zepernick, H.J. and Finger, A. (2005) *Pseudo Random Signal Processing*, John Wiley & Sons, Ltd., Chichester, UK.

118. Kim, K.S., Kim, S.W., Cho, Y.S., and Ahn, J.Y. (2007) Synchronization and cell search technique using preamble for OFDM cellular systems. *IEEE Trans. Veh. Tech.*, **56**(6), 3469–3485.

119. IEEE (2006) Std 802.16e. *IEEE Standard for Local and Metropolitan Area Networks Part 16.*

120. Dahlman, E., Parkvall, S., Skold, J., and Beming, P. (2007) *3G Evolution: HSPA and LTE for Mobile Broadband*, Academic Press.

121. Tsai, Y., Zhang, G., Grieco, D. *et al.* (2007) Cell search in 3GPP long term evolution systems. *IEEE Vehicular Technology Magazine*, **2**(2), 23–29.

122. Nogami, H. and Nagashima, T. (Sep. 1995) A frequency and timing period acquisition technique for OFDM systems. IEEE PIMRC, pp. 1010–1015.

123. Schilpp, M., S-Greff, W., Rupprecht, W., and Bogenfeld, E. (June 1995) Influence of oscillator phase noise and clipping on OFDM for terrestrial broadcasting of digital HDTV. IEEE ICC'95, pp. 1678–1682.

124. Song, H.K., You, Y.H., Paik, J.H., and Cho, Y.S. (2000) Frequency-offset synchronization and channel estimation for OFDM-based transmission. *IEEE Trans. Commun.*, **4**(3), 95–97.

125. Warner, W.D. and Leung, C. (1993) OFDM/FM Frame synchronization for mobile radio data communication. *IEEE Trans. Veh. Technol.*, **42**(3), 302–313.

126. Cimini, L.J. (1985) Analysis and simulation of a digital mobile channel using orthogonal frequency-division multiplexing. *IEEE Trans. Commun.*, **33**(7), 665–675.

127. Tufvesson, F. and Maseng, T. (May 1997) Pilot assisted channel estimation for OFDM in mobile cellular systems. IEEE VTC'97, vol. 3, pp. 1639–1643.

128. van de Beek, J.J., Edfors, O., Sandell, M. *et al.* (July 1995) On channel estimation in OFDM systems. IEEE VTC'95, vol. 2, pp. 815–819.

129. Coleri, S., Ergen, M., Puri, A., and Bahai, A. (2002) Channel estimation techniques based on pilot arrangement in OFDM systems. *IEEE Trans. on Broadcasting*, **48**(3), 223–229.

130. Heiskala, J. and Terry, J. (2002) *OFDM Wireless LANs: A Theoretical and Practical Guide*, SAMS.

131. Hsieh, M. and Wei, C. (1998) Channel estimation for OFDM systems based on comb-type pilot arrangement in frequency selective fading channels. *IEEE Trans. Consumer Electron.*, **44**(1), 217–228.

132. van Nee, R. and Prasad, R. (2000) *OFDM for Wireless Multimedia Communications*, Artech House Publishers.

133. Lau, H.K. and Cheung, S.W. (May 1994) A pilot symbol-aided technique used for digital signals in multipath environments. IEEE ICC'94, vol. 2, pp. 1126–1130.

134. Yang, W.Y., Cao, W., Chung, T.S., and Morris, J. (2005) *Applied Numerical Methods Using MATLAB*, John Wiley & Sons, Inc., New York.

135. Minn, H. and Bhargava, V.K. (1999) An investigation into time-domain approach for OFDM channel estimation. *IEEE Trans. on Broadcasting*, **45**(4), 400–409.

136. Fernandez-Getino Garcia, M.J., Paez-Borrallo, J.M., and Zazo, S. (May 2001) DFT-based channel estimation in 2D-pilot-symbol-aided OFDM wireless systems. IEEE VTC'01, vol. 2, pp. 810–814.

137. van de Beek, J.J., Edfors, O., Sandell, M. *et al.* (2000) Analysis of DFT-based channel estimators for OFDM. *Personal Wireless Commun.*, **12**(1), 55–70.

138. Zhao, Y. and Huang, A. (May 1998) A novel channel estimation method for OFDM mobile communication systems based on pilot signals and transform-domain processing. IEEE VTC'98, vol. 46, pp. 931–939.

139. Lyman, R.J. and Edmonson, W.W. (1999) Decision-directed tracking of fading channels using linear prediction of the fading envelope. 33rd Asilomar Conference on Signal Processing, Systems, and Computers, vol. 2, pp. 1154–1158.

140. Ran, J.J., Grunbeid, R., Rohling, H. *et al.* (Apr. 2003) Decision-directed channel estimation method for OFDM systems with high velocities. IEEE VTC'03, vol. 4, pp. 2358–2361.

141. Xinmin, D., Haimovich, A.M., and Garcia-Frias, J. (May 2003) Decision-directed iterative channel estimation for MIMO systems. IEEE ICC'03, vol. 4, pp. 2326–2329.

142. Daofeng, X. and Luxi, Y. (June 2004) Channel estimation for OFDM systems using superimposed training. International Conference in Central Asia on Internet, pp. 26–29.

143. Hoeher, P. and Tufvesson, F. (1999) Channel estimation with superimposed pilot sequence. IEEE GLOBECOM'99, pp. 2162–2166.

144. Hu, D. and Yang, L. (Dec. 2003) Time-varying channel estimation based on pilot tones in OFDM systems. IEEE Int. Conf. Neural Networks & Signal Processing, vol. 1, pp. 700–703.

145. Jeon, W.G., Chang, K.H., and Cho, Y.S. (1999) An equalization technique for orthogonal frequency-division multiplexing system in time-variant multipath channels. *IEEE Trans. Commun.*, **47**(1), 27–32.

146. Li, Y., Cimini, L. Jr, and Sollenberger, N.R. (1998) Robust channel estimation for OFDM systems with rapid dispersive fading channels. *IEEE Trans. on Commun.*, **46**(7), 902–915.

147. Tang, Z., Cannizzaro, R.C., Leus, G., and Banelli, P. (2007) Pilot-assisted time-varying channel estimation for OFDM systems. *IEEE Tran. Signal Processing*, **55**(5), 2226–2238.

148. Park, K.W. and Cho, Y.S. (2005) An MIMO-OFDM technique for high-speed mobile channels. *IEEE Commun. Letters*, **9**(7), 604–606.

149. Zhao, Y. and Haggman, S.G. (2001) Intercarrier interference self-cancellation scheme for OFDM mobile communication systems. *IEEE Tran. Commun.*, **49**(7), 1158–1191.

150. Aldana, C.H., Carvalho, E., and Cioffi, J.M. (2003) Channel estimation for multicarrier multiple input single output systems using the EM algorithm. *IEEE Trans. on Signal Processing*, **51**(12), 3280–3292.

151. Dempster, A.P., Laird, N.M., and Rubin, D.B. (1977) Maximum likelihood from incomplete data via the EM algorithm. *J. Roy. Soc.*, **39**(1), 1–38.

152. Lee, K.I., Woo, K.S., Kim, J.K. *et al.* (Sep. 2007) Channel estimation for OFDM based cellular systems using a DEM algorithm. PIMRC'07, Athens.

153. Ma, X., Kobayashi, H., and Schwartz, S.C. (2004) EM-based channel estimation algorithms for OFDM. *EURASIP J. Appl. Signal Process.*, **10**, 1460–1477.

154. Moon, T.K. (1996) The expectation-maximization algorithm. *IEEE Signal Proc. Mag.*, **13**(6), 47–60.

155. Yongzhe, X. and Georghiades, C.N. (2003) Two EM-type channel estimation algorithms for OFDM with transmitter diversity. *IEEE Trans. on Commun*, **51**(1), 106–115.

156. Ding, Z. and Li, Y. (2001) *Blind Equalization and Identification*, Marcel Dekker.

157. Godard, D.N. (1980) Self-recovering equalization and carrier tracking in two-dimensional data communication systems. *IEEE Trans. Commun.*, **28**(11), 1867–1875.

158. Haykin, S. (2001) *Adaptive Filter Theory*, 4th edn, PHIPE.

159. Sato, Y. (1975) A method of self-recovering equalization for multilevel amplitude-modulation systems. *IEEE Trans. Commun.*, **23**(6), 679–682.

160. Shin, C. and Power, E.J. (2004) Blind channel estimation for MIMO-OFDM systems using virtual carriers. IEEE GLOBECOM'04, vol. 4, pp. 2465–2469.

161. Tong, L., Xu, G., and Kailath, T. (Nov. 1991) A new approach to blind identification and equalization of multipath channel. Proc. 25th Asilomar Conf. Signals, Systems, and Computers, CA, vol. 2, pp. 856–860.

162. Zhang, R. (Nov. 2002) Blind OFDM channel estimation through linear precoding: a subspace approach. IEEE ACSSC, vol. 2, pp. 631–633.

163. Litsyn, S. (2007) *Peak Power Control in Multicarrier Communications*, Cambridge University Press.

164. Han, S.H. and Lee, J.H. (2005) An overview of peak-to-average power ratio reduction techniques for multicarrier transmission. *IEEE Wireless Commun.*, **12**(2), 56–65.

165. Palicot, J. and Louët, Y. (Sep. 2005) Power ratio definitions and analysis in single carrier modulations. EUSIPCO, Antalya, Turkey.

166. Ochiai, H. and Imai, H. (1997) Block coding scheme based on complementary sequences for multicarrier signals. *IEICE Trans. Fundamentals*, **E80-A**(11), 2136–2143.

167. Ochiai, H. and Imai, K. (2001) On the distribution of the peak-to-average power ratio in OFDM signals. *IEEE Trans. Commun.*, **49**(2), 282–289.

168. van Nee, R. and de Wild, A. (May 1998) Reducing the peak-to-average power ratio of OFDM. IEEE VTC'98, vol. 3, pp. 18–21.

169. Ochiai, H. and Imai, K. (Dec. 2000) On clipping for peak power reduction of OFDM signals. IEEE GTC, vol. 2, pp. 731–735.

170. Ifeachor, E.C. and Jervis, B.W. (2002) *Digital Signal Processing-a Practical Approach 2/E*, Prentice Hall.

171. Chu, D.C. (1972) Polyphase codes with periodic correlation properties. *IEEE Trans. Info. Theory*, **18**(4), 531–532.

172. Lüke, H.D., Schotten, H.D., and Mahram, H.H. (2003) Binary and quadriphase sequences with optimal autocorrelation properties. *IEEE Trans. Info. Theory*, **49**(12), 3271–3282.

173. Li, X. and Cimini, L.J. (1998) Effects of clipping and filtering on the performance of OFDM. *IEEE Commun. Letter*, **2**(20), 131–133.

174. Slimane, S.B. (Dec. 2000) Peak-to-average power ratio reduction of OFDM signals using pulse shaping. IEEE GTC, vol. 3, pp. 1412–1416.

175. Wilkinson, T.A. and Jones, A.E. (July 1995) Minimization of the peak-to-mean envelope power ratio of multicarrier transmission scheme by block coding. IEEE VTC'95, Chicago, vol. 2, pp. 825–829.
176. Golay, M.J.E. (1961) Complementary series. *IEEE Trans. Inf. Theory*, **7**(2), 82–87.
177. Popovic, B.M. (1991) Synthesis of power efficient multitone signals with flat amplitude spectrum. *IEEE Trans. Commun.*, **39**(7), 1031–1033.
178. van Nee, R.D.J. (Nov. 1996) OFDM codes for peak-to-average power reduction and error correction. IEEE GTC, London, vol. 2, pp. 740–744.
179. Davis, J.A. and Jedwab, J. (1999) Peak-to-mean power control in OFDM, Golay complementary sequences, and Reed-Muller codes. *IEEE Trans. Info. Theory*, **45**(7), 2397–2417.
180. Davis, J.A. and Jedwab, J. (1997) Peak-to-mean power control and error correction for OFDM transmission using Golay sequences and Reed-Muller codes. *Electron. Lett.*, **33**(4), 267–268.
181. Urbanke, R. and Krishnakumar, A.S. (Oct. 1996) Compact description of Golay sequences and their extensions. Proc. of the Thirty-Fourth Annual Allerton Conference on Commun., Control and Computing Pagination, Urbana, IL, pp. 693–701.
182. Li, X. and Ritcey, J.A. (1997) M-sequences for OFDM peak-to-average power ratio reduction and error correction. *Electron. Lett.*, **33**(7), 554–555.
183. Tellambura, C. (1997) Use of m-sequences for OFDM peak-to-average power ratio reduction. *Electron. Lett.*, **33**(15), 1300–1301.
184. Park, M.H. *et al.* (2000) PAPR reduction in OFDM transmission using Hadamard transform. IEEE ICC'00, vol. 1, pp. 430–433.
185. Bauml, R.W., Fischer, R.F.H., and Huber, J.B. (1996) Reducing the peak-to-average power ratio of multicarrier modulation by selective mapping. *Electron. Lett.*, **32**(22), 2056–2057.
186. Ohkubo, N. and Ohtsuki, T. (Apr. 2003) Design criteria for phase sequences in selected mapping. IEEE VTC'03, vol. 1, pp. 373–377.
187. Muller, S.H. *et al.* (1997) OFDM with reduced peak-to-average power ratio by multiple signal representation. *In Annals of Telecommun.*, **52**(1–2), 58–67.
188. Muller, S.H. and Huber, J.B. (1996) OFDM with reduced peak-to-average power ratio by optimum combination of partial transmit sequences. *Electron. Lett.*, **32**(22), 2056–2057.
189. Muller, S.H. and Huber, J.B. (Sep. 1997) A novel peak power reduction scheme for OFDM. PIMRC'97, vol. 3, pp. 1090–1094.
190. Cimini, L.J. Jr (2000) Peak-to-average power ratio reduction of an OFDM signal using partial transmit sequences. *IEEE Commun. Letters*, **4**(3), 86–88.
191. Tellambura, C. (1998) A coding technique for reducing peak-to-average power ratio in OFDM. IEEE GLOBECOM'98, vol. 5, pp. 2783–2787.
192. Tellado-Mourelo, J. (Sep. 1999) Peak to average power reduction for multicarrier modulation, Ph.D. Dissertation, Stanford Univ.
193. Jeon, W.G., Chang, K.H., and Cho, Y.S. (1997) An adaptive data predistorter for compensation of nonlinear distortion in OFDM systems. *IEEE Trans. on Commun.*, **45**(10), 1167–1171.
194. Bruninghaus, K. and Rohling, H. (May 1998) Multi-carrier spread spectrum and its relationship to single-carrier transmission. IEEE VTC'98, vol. 3, pp. 2329–2332.
195. Galda, D. and Rohling, H. (2002) A low complexity transmitter structure for OFDM-FDMA uplink systems. IEEE VTC'02, vol. 4, pp. 1737–1741.
196. Myung, H.G., Lim, J., and Goodman, D.J. (2006) Single carrier FDMA for uplink wireless transmission. *IEEE Veh. Technol. Mag.*, **1**(3), 30–38.
197. Myung, H.G., Lim, J., and Goodman, D.J. (Sept. 2006) Peak-to-average power ratio of single carrier FDMA signals with pulse shaping. PIMRC'06, pp. 1–5.
198. Qualcomm (2005) R1-050896. *Description and Simulations of Interference Management Technique for OFDMA Based E-UTRA Downlink Evaluation*, 3GPP RAN WG1 #42, London, UK.
199. Nokia (2006) R1-060298. *Uplink Inter Cell Interference Mitigation and Text Proposal*, 3GPP RAN WG1 #44, Denver, USA.
200. Samsung (2005) R1-051341. *Flexible Fractional Frequency Reuse Approach*, 3GPP RAN WG1 #43, Seoul, Korea.
201. Cimini, L.J., Chuang, J.C., and Sollenberger, N.R. (1998) Advanced cellular internet service (ACIS). *IEEE Commun. Mag.*, **36**(10), 150–159.

202. Yun, S.B., Park, S.Y., Lee, Y.W. *et al.* (2007) Hybrid division duplex system for next-generation cellular services. *IEEE Trans. Veh. Technol.*, **56**(5), 3040–3059.
203. Jeong, D.G. and Jeon, W.S. (2000) Comparison of time slot allocation strategies for CDMA/TDD systems. *IEEE J. Select. Areas in Commun.*, **18**(7), 1271–1278.
204. Auer, G. (Oct. 2003) On modeling cellular interference for multi-carrier based communication systems including a synchronization offset. WPMC'03, pp. 290–294.
205. Ericsson (2005) R1-050764. *Inter-cell Interference Handling for E-UTRA*, 3GPP RAN WG1 #42, London, UK.
206. RITT (2005) R1-050608. *Inter-cell Interference Mitigation based on IDMA*, 3GPP TSG RAN WG1 Ad Hoc on LTE, Sophia Antipolis, France.
207. Ericsson (2006) R1-062851. *Frequency hopping for E-UTRA uplink*, 3GPP TSG RAN WG1 #46, Seoul, Korea.
208. IEEE (2005) 802.16e-2005. *IEEE Std 802.16e-2005 and IEEE Std 802.16-2004/Cor 1-2005*, Mobile WiMAX forum, New York, USA.
209. Bottomley, G.E. (July 1995) Adaptive arrays and MLSE equalization. IEEE VTC'95, vol. 1, pp. 50–54.
210. Paulraj, A., Nabar, R., and Gore, D. (2003) *Introduction to Space-Time Wireless Communications*, Cambridge University Press.
211. Foschini, G.J. (1996) Layered space-time architecture for wireless communication in a fading environment when using multi-element antennas. *Bell Labs Tech. J.*, **1**(2), 41–59.
212. IEEE (2006) Std 802.16e™-2005. EEE Std 802.16™-2004/Cor1-2005. *Part 16: Air Interface for Fixed and Mobile Broadband Wireless Access Systems*.
213. Golden, G.D., Foschini, C.J., Valenzuela, R.A., and Wolniansky, P.W. (1999) Detection algorithm and initial laboratory result using V-BLAST space-time communication architecture. *Electron. Lett.*, **35**(1), 14–15.
214. Alamounti, S.M. (1998) A simple transmit diversity scheme for wireless communications. *IEEE J. Select. Areas Commun.*, **16**(8), 1451–1458.
215. Tarokh, V., Jafrakhani, H., and Calderbank, A.R. (1999) Space-time block codes from orthogonal designs. *IEEE Trans. Inform. Theory*, **45**(5), 1456–1467.
216. Sanhdu, S. and Paulraj, A. (2000) Space-time block codes: a capacity perspective. *IEEE Commun. Letters*, **4**(12), 384–386.
217. Telatar, I. (1999) Capacity of multi-antenna Gaussian channels. *European Trans. Tel.*, **10**(6), 585–595.
218. Traveset, J.V., Caire, G., Biglieri, E., and Taricco, G. (1997) Impact of diversity reception on fading channels with coded modulation–Part I: coherent detection. *IEEE Trans. Commun.*, **45**(5), 563–572.
219. Tarokh, V., Seshadri, N., and Calderbank, A.R. (1998) Space-time codes for high data rate wireless communication: performance criterion and code construction. *IEEE Trans. Inform. Theory*, **44**(2), 744–765.
220. Hughes, B.L. (2000) Differential space-time modulation. *IEEE Trans. Info. Theory*, **46**(7), 2567–2578.
221. Hochwald, B.M. and Sweldens, W. (2000) Differential unitary space-time modulation. *IEEE Trans. Commun.*, **48**(12), 2041–2052.
222. Taricco, G. and Biglieri, E. (2002) Exact pairwise error probability of space-time codes. *IEEE Trans. Info. Theory*, **48**(2), 510–513.
223. Uysal, M. and Georghiades, C.N. (2000) Error performance analysis of space-time codes over Rayleigh fading channels. *Journal of Commun. and Networks*, **2**(4), 351–355.
224. Tarokh, V., Naguib, A., Seshadri, N., and Calderbank, A.R. (1999) Combined array processing and space-time coding. *IEEE Trans. Info. Theory*, **45**(4), 1121–1128.
225. Grimm, J., Fitz, M.P., and Krogmeier, J.V. (Sep. 1998) Further results on space-time coding for Rayleigh fading. 36th Allerton Conf., pp. 391–400.
226. Baro, S., Bauch, G., and Hansmann, A. (2000) Improved codes for space-time trellis coded modulation. *IEEE Commun. Letters*, **4**(1), 20–22.
227. Simon, M.K. (2001) Evaluation of average bit error probability for space-time coding based on a simpler exact evaluation of pairwise error probability. *Journal of Commun. and Networks*, **3**(3), 257–264.
228. Tarokh, V., Jafarkhani, H., and Calderbank, A.R. (1999) Space-time bock coding for wireless communications: performance results. *IEEE J. Sel. Areas Commun.*, **17**(3), 451–460.
229. Chen, Z., Yuan, J., and Vucetic, B. (2001) Improved space-time trellis coded modulation scheme on slow Rayleigh fading channels. *IEE Electron. Lett.*, **37**(7), 440–442.
230. Firmanto, W., Vucetic, B., and Yuan, J. (2001) Space-time TCM with improved performance on fast fading channels. *IEEE Commun. Letters*, **5**(4), 154–156.
231. Vucetic, B. and Yuan, J. (2003) *Space-Time Coding*, John Wiley & Sons, Ltd., Chichester, UK.

232. Wolniansky, P., Foschini, G., Golden, G., and Valenzuela, R. (Sep. 1998) V-BLAST: an architecture for realizing very high data rates over the rich-scattering wireless channel. Proc. ISSSE'98, Pisa, Italy, pp. 295–300.

233. Kim, S. and Kim, K. (2006) Log-likelihood ratio based detection ordering in V-BLAST. *IEEE Trans. Commun.*, **54**(2), 302–307.

234. Kim, J., Kim, Y., and Kim, K. (2007) Computationally efficient signal detection method for next generation mobile communications using multiple antennas. *SK Telecommun. Review*, **17**(1C), 183–191.

235. Viterbo, E. and Boutros, J. (1999) A universal lattice code decoder for fading channels. *IEEE Trans. Info. Theory*, **45**(5), 1639–1642.

236. Hochwald, B.M. and Brink, S. (2003) Achieving near-capacity on a multiple-antennas channel. *IEEE Trans. Commun.*, **51**(3), 389–399.

237. Kim, K.J. and Yue, J. (2002) Joint channel estimation and data detection algorithms for MIMO-OFDM systems. Proc. 36th Asilomar Conf. Signals, System Comput, vol. 2, pp. 1857–1861.

238. Kawai, H., Higuichi, K., Maeda, N. *et al.* (2005) Likelihood function for QRM-MLD suitable for soft-decision turbo decoding and its performance for OFCDM MIMO multiplexing in multipath fading channel. *IEICE Trans. Commun.*, **E88-B**(1), 47–57.

239. Yao, H. and Wornell, G.W. (Nov. 2002) Lattice-reduction-aided detectors for MIMO communication systems. IEEE GLOBECOM'02, vol. 1, pp. 424–428.

240. Windpassinger, C., Lampe, L., and Fischer, R.F.H. (July 2003) From lattice-reduction-aided detection towards maximum-likelihood detection in MIMO systems. WOC'03, pp. 144–148.

241. Im, T.H., Kim, J.K., Yi, J.H. *et al.* (May 2008) MMSE-OSIC^2 signal detection method for spatially multiplexed MIMO systems. IEEE VTC'08, pp. 1468–1472.

242. Vu, M. (July 2006) Exploiting Transmit Channel Side Information in MIMO Wireless Systems, PhD Thesis, Stanford University.

243. Larsson, E.G. and Stoica, P. (2003) *Space-Time Block Coding for Wireless Communications*, Cambridge University., New York Press.

244. Larsson, E.G., Ganesan, G., Stoica, P., and Wong, W.H. (2002) On the performance of orthogonal space-time block coding with quantized feedback. *IEEE Commun. Letters*, **12**(6), 487–489.

245. Love, D.J. and Heath, R.W. Jr (2005) Limited feedback unitary precoding for orthogonal space-time block codes. *IEEE Trans. Signal. Proc.*, **53**(1), 64–73.

246. Conway, J.H., Hardin, R.H., and Sloane, N.J.A. (1996) Packing lines, planes, etc.: packings in Grassmannian spaces. *Experimental Math.*, **5**, 139–159.

247. Barg, A. and Nogin, D.Y. (2002) Bounds on packings of spheres in the Grassmann manifold. *IEEE Trans. Info. Theory*, **48**(9), 2450–2454.

248. Strohmer, T. and Heath, R.W. Jr (2003) Grassmannian frames with applications to coding and communications. *Appl. Comput. Harmon. Anal.*, **14**, 257–275.

249. Hochwald, B.M., Marzetta, T.L., Richardson, T.J. *et al.* (2000) Systematic design of unitary space-time constellations. *IEEE Trans. Info. Theory*, **46**, 1962–1973.

250. Joham, M., Utschick, W., and Nossek, J.A. (2005) Linear transmit processing in MIMO communications systems. *IEEE Trans. Signal Processing*, **53**(8), 2700–2712.

251. Caire, G. and Shamai, S. (2003) On the capacity of some channels with channel state information. *IEEE Trans. Info. Theory*, **49**(7), 1691–1706.

252. Chuah, C., Tse, D., Kahn, J., and Valenzuela, R. (2002) Capacity scaling in MIMO wireless systems under correlated fading. *IEEE Trans. Info. Theory*, **48**(3), 637–650.

253. Goldsmith, A., Jafar, S., Jindal, N., and Vishwanath, S. (2003) Capacity limits of MIMO channels. *IEEE J. Select Areas Commun.*, **21**(3), 684–702.

254. Cover, T.M. and Thomas, J.A. (1991) *Elements of Information Theory, 2/E*, John Wiley & Sons, Inc., New York.

255. Jindal, N. and Goldsmith, A. (2005) Dirty paper coding vs. TDMA for MIMO broadcast channel. *IEEE Trans. Info. Theory*, **51**(5), 1783–1794.

256. Costa, M.H.M. (1983) Writing on dirty paper. *IEEE Trans. Info. Theory*, **29**(3), 439–441.

257. Schubert, M. and Boche, H. (Sept. 2002) Joint 'dirty-paper' pre-coding and downlink beamforming. IEEE Int. Sym. Spread Spectrum Tech. and App., Prague, Czech Republic.

258. Caire, G. and Shamai, S. (2003) On the achievable throughput of a multi-antenna Gaussian broadcast channel. *IEEE Trans. Info. Theory*, **43**(7), 1691–1706.

259. Love, D.J., Heath, R.W. Jr, Santipach, W., and Honig, M.L. (2004) What is the value of limited feedback for MIMO channels? *IEEE Commun. Mag.*, **42**(10), 54–59.

260. Vishwanath, P. and Tse, D. (2003) Sum capacity of the vector Gaussian broadcast channel and uplink-downlink duality. *IEEE Trans. Info. Theory*, **49**(8), 1912–1921.

261. Yu, W. and Cioffi, J. (2002) Sum capacity of a Gaussian vector broadcast channel. *IEEE Trans. Info. Theory*, **50**(9), 1875–1892.

262. Vishwanath, S., Jindal, N., and Goldsmith, A. (2003) Duality, achievable rates and sum capacity of Gaussian MIMO broadcast channels. *IEEE Trans. Info. Theory*, **49**(10), 2658–2668.

263. Weingarten, H., Steinberg, Y., and Shamai, S. (2006) The capacity region of the Gaussian MIMO broadcast channel. *IEEE Trans. Info. Theory*, **52**(9), 3936–3964.

264. Aftas, D., Bacha, M., Evans, J., and Hanly, S. (Oct. 2004) On the sum capacity of multiuser MIMO channels. Intl. Symp. on Inform. Theory and its Applications, pp. 1013–1018.

265. El Gama, A. and Cover, T. (1980) Multiple user information theory. *IEEE*, **68**(12), 1466–1483.

266. Peel, C.B., Hochwald, B.M., and Swindlehurst, A.L. (2005) A vector-perturbation technique for near-capacity multiantenna multiuser communication-part I: channel inversion and regularization. *IEEE Tran. Commun.*, **53**(1), 195–202.

267. Hochwald, B.M., Peel, C.B., and Swindelhurst, A.L. (2005) A vector-perturbation technique for near-capacity multiple-antenna multi-user communication-part II: perturbation. *IEEE Tran. Commun.*, **53**(3), 537–544.

268. Haustein, T., Helmolt, C.V., Jorwieck, E. *et al.* (May 2002) Performace of MIMO systems with channel inversion. IEEE VTC'02, vol. 1, pp. 35–39.

269. Stojnic, M., Vikalo, H., and Hassibi, B. (Nov. 2004) Rate maximization in multi-antenna broadcast channels with linear preprocessing. IEEE GLOBECOM'04, vol. 4, pp. 3957–3961.

270. Spencer, Q.H., Peel, C.B., Swindlehurst, A.L., and Haardt, M. (2004) An introduction to the multi-user MIMO downlink. *IEEE Commun. Mag.*, **42**(10), 60–67.

271. Spencer, Q.H., Swindlehurst, A.L., and Haardt, M. (2004) Zero-forcing methods for downlink spatial multiplexing in multi-user MIMO channels. *IEEE Trans. Signal Processing*, **52**(2) 461–471.

272. Pan, Z., Wong, K.K., and Ng, T. (May 2003) MIMO antenna system for multi-user multi-stream orthogonal space division multiplexing. IEEE ICC'03, vol. 5, pp. 3220–3224.

273. Choi, R. and Murch, R. (2003) A transmit preprocessing technique for multiuser MIMO systems using a decomposition approach. *IEEE Trans. Wireless Commun.*, **2**(2), 20–24.

274. Tomlinson, M. (1971) New automatic equalizer employing modulo arithmetic. *Electron. Lett.*, **7**, 138–139.

275. Harashima, H. and Miyakawa, H. (1972) Matched-transmission technique for channels with intersymbol interference. *IEEE Trans. Commun.*, **20**(4), 774–780.

276. Fischer, R.F.H., Windpassinger, C., Lampe, A., and Huber, J.B. (Jan. 2002) Space time transmission using Tomlinson-Harashima precoding. ITG Conference on Source and Channel Coding, pp. 139–147.

277. Mun, C., Lee, M.W., Yook, J.G., and Park, H.K. (2004) Exact capacity analysis of multiuser diversity combined with transmit diversity. *Electron. Lett.*, **40**(22), 1423–1424.

278. Chen, C. and Wang, L. (June 2004) A unified capacity analysis for wireless systems with joint antenna and multiuser diversity in Nakagami fading channels. IEEE ICC'04, vol. 6, pp. 3523–3527.

279. Viswanath, P., Tse, N.C., and Rajiv, R. (2002) Opportunistic beamforming using dumb antennas. *IEEE Trans. on Info. Theory*, **48**(6), 1277–1294.

280. Shin, O.S. and Lee, K.B. (2003) Antenna-assisted round robin scheduling for MIMO cellular systems. *IEEE Commun. Letters*, **7**(3), 109–111.

281. Chaskar, H.M. and Madhow, U. (2003) Fair scheduling with tunable latency: a round-robin approach, Networking. *IEEE/ACM Trans.*, **11**(4), 592–601.

282. Bourdoux, A. and Khaled, N. (Sep. 2002) Joint Tx-Rx optimization for MIMO-SDMA based on a null-space constraint. IEEE VTC'02, vol. 1, pp. 171–174.

283. Rim, M. (2002) Multi-user downlink beamforming with multiple transmit and receive antennas. *Electron. Lett.*, **38**(25), 1725–1726.

284. Stankovic, V., Haardt, M., and Fuchs, M. (2004) Combination of block diagonalization and THP transmit filtering for downlink beamforming in multi-user MIMO systems. Proceedings of the 7th European Conference on Wireless Technology, 145–148.

285. IEEE (1999) 802.11a, *Part 11: Wireless LAN Medium Access Control(MAC) and Physical Layer(PHY) Specifications: High-Speed Physical Layer in the 5 GHz Band, Supplement to IEEE 802.11 Standard.*

286. IEEE (2007) 802.11nTM/D2.00. *Part 11: Wireless LAN Medium Access Control(MAC) and Physical Layer (PHY) specifications.*

287. ETSI (2004) EN 300 401. *Radio Broadcasting Systems; Digital Audio Broadcasting (DAB) to Mobile, Portable and Fixed Receivers.*

288. Sieber, A. and Weck, F.C. (2004) What's the difference between DVB-H and DAB - in the mobile environment? *EBU Tech. Rev.*, 299.

289. Woo, K.S., Lee, K.I., Paik, J.H. *et al.* (2007) A DSFBC-OFDM for a next-generation broadcasting system with multiple antennas. *IEEE Trans. Broadcasting*, **53**(2), 539–546.

290. Kozamernik, F. (2004) DAB-from digital radio towards mobile multimedia. *EBU Tech. Rev.*, 297.

291. ETSI (2005) EN 300 744. *Digital Video Broadcasting (DVB); Framing Structure, Channel Coding and Modulation for Digital Terrestrial Television.*

292. ETSI (2005) EN 301 192. *Digital Video Broadcasting (DVB); DVB Specification for Data Broadcasting.*

293. ETSI (2005) EN 302 304. *Digital Video Broadcasting (DVB); Transmission System for Handheld Terminals.*

294. ETSI (2004) EN 101 191. *Digital Video Broadcasting (DVB); DVB Mega-Frame for Single Frequency Network (SFN) Synchronization.*

295. ETSI (2005) EN 300 468. *Digital Video Broadcasting (DVB); Specification for Service Information (SI) in DVB Systems.*

296. Kornfeld, M. and Reimers, U. (2005) DVB-H – the emerging standard for mobile data communication. *EBU Tech. Rev.* 301.

297. (2007) ECMA-368. *High Rate Ultra Wideband PHY and MAC Standard, 2/E.*

298. TTAS (2005) KO-06.0082/R1. *2.3 GHz Portable Internet Specification – PHY and MAC Layers.*

299. WiMAX Forum™ (2007) Mobile System Profile Release 1.0 Approved Specification, Revision 1.4.0.

300. TTA (2004) 2.3 GHz Portable Internet (WiBro) Overview.

301. Andrews, J.G., Ghosh, A., and Muhamed, R. (2007) *Fundamentals of WiMAX: Understanding Broadband Wireless Networking*, Prentice Hall.

302. IEEE (2005) C802.16maint-05/083. *Hit Ratio Problems with PUSC Permutation.*

303. Segal, Y. (2005) *Tutorial on Multi Access OFDM (OFDMA) Technology*, Runcom.

304. 3GPP (2007) TS 36.201. *E-UTRA; LTE Physical Layer-General Description, R8.*

305. 3GPP (2008) TS 36.211. *E-UTRA; Physical Channels and Modulation, R8.*

306. 3GPP (2008) TS 36.212. *E-UTRA; Multiplexing and Channel Coding, R8.*

307. 3GPP (2008) TS 36.213. *E-UTRA; Physical Layer Procedures, R8.*

308. 3GPP (2008) TS 36.213. *E-UTRA; Physical Layer-Measurements, R8.*

309. 3GPP (2008) TS 36.300. *E-UTRA and E-UTRAN; Overall Description, R8.*

Index

MIMO-OFDM Wireless Communications with MATLAB® Yong Soo Cho, Jaekwon Kim, Won Young Yang and Chung G. Kang
© 2010 John Wiley & Sons (Asia) Pte Ltd